W. C. Trogler

Organometallic Radical Processes

Organometallic Radical Processes

by

Prof. Dr. William C. Trogler
(Editor)
San Diego (USA)

With 80 Figures and 58 Tables

AKADEMIE-VERLAG BERLIN
1990

Die Originalausgabe erscheint im Verlag
Elsevier Science Publishers B. V.
innerhalb der Serie
"Journal of Organometallic Chemistry Library"

Vertrieb ausschließlich für alle Staaten
mit Ausnahme der sozialistischen Länder:
Elsevier Science Publishers B. V., Amsterdam

Vertrieb für die sozialistischen Länder:
Akademie-Verlag Berlin

ISBN 3-05-500770-0

Erschienen im Akademie-Verlag Berlin, Leipziger Straße 3–4, Berlin, DDR-1086

Printed in the German Democratic Republic
Gesamtherstellung: VEB Druckhaus Köthen, Köthen, DDR-4370
LSV: 1255
Bestellnummer: 764 084 9 (9281)
12500

Journal of Organometallic Chemistry Library

1 D. SEYFERTH (Editor)
NEW APPLICATIONS OF ORGANOMETALLIC REAGENTS IN ORGANIC SYNTHESIS
Proceedings of a symposium at the American Chemical Society National Meeting held in New York City, April 6–9th, 1976

2 D. SEYFERTH (coordinating editor), A. G. DAVIES, E. O. FISCHER, J. F. NORMANT and O. A. REUTOV (editors)
ORGANOMETALLIC CHEMISTRY REVIEWS: ORGANOSILICON REVIEWS

3 D. SEYFERTH (coordinating editor), A. G. DAVIES, E. O. FISCHER, J. F. NORMANT and O. A. REUTOV (editors)
ORGANOMETALLIC CHEMISTRY REVIEWS

4 D. SEYFERTH and R. B. KING (editors)
ORGANOMETALLIC CHEMISTRY REVIEWS; ANNUAL SURVEYS: SILICON - TIN - LEAD

5 D. SEYFERTH (coordinating editor), A. G. DAVIES, E. O. FISCHER, J. F. NORMANT and O. A. REUTOV (editors)
ORGANOMETALLIC CHEMISTRY REVIEWS

6 D. SEYFERTH and R. B. KING (editors)
ORGANOMETALLIC CHEMISTRY REVIEWS; ANNUAL SURVEYS: SILICON - GERMANIUM - TIN - LEAD

7 D. SEYFERTH (coordinating editor), A. G. DAVIES, E. O. FISCHER, J. F. NORMANT and O. A. REUTOV (editors)
ORGANOMETALLIC CHEMISTRY REVIEWS

8 D. SEYFERTH and R. B. KING (editors)
ORGANOMETALLIC CHEMISTRY REVIEWS; ANNUAL SURVEYS: SILICON - GERMANIUM - TIN - LEAD

9 D. SEYFERTH (coordinating editor), A. G. DAVIES, E. O. FISCHER, J. F. NORMANT and O. A. REUTOV (editors)
ORGANOMETALLIC CHEMISTRY REVIEWS; including Selected Plenary Lectures from the fifth International Symposium on Organosilicon Chemistry held in Karlsruhe, August 14–18, 1978

10 D. SEYFERTH and R. B. KING (editors)
ORGANOMETALLIC CHEMISTRY REVIEWS; ANNUAL SURVEYS. SILICON - GERMANIUM - TIN - LEAD

11 D. SEYFERTH and R. B. KING (editors)
ORGANOMETALLIC CHEMISTRY REVIEWS; ANNUAL SURVEYS: SILICON - TIN - LEAD

12 D. SEYFERTH (coordinating editor), A. G. DAVIES, E. O. FISCHER, J. F. NORMANT and O. A. REUTOV (editors)
ORGANOMETALLIC CHEMISTRY REVIEWS; including Plenary Lectures from the 3rd International Conference on the Organometallic and Coordination Chemistry of Germanium, Tin and Lead (Dortmund, July 21st–25th, 1980)

13 D. SEYFERTH and R. B. KING (editors)
ORGANOMETALLIC CHEMISTRY REVIEWS; ANNUAL SURVEYS: SILICON - GERMANIUM - TIN - LEAD

14 R. B. KING and J. P. OLIVER (editors)
ORGANOMETALLIC CHEMISTRY REVIEWS; ANNUAL SURVEYS: SILICON - LEAD

15 O. ŠTROUF, B. ČÁSENSKÝ and V. KUBÁNEK
SODIUM DIHYDRIDO-BIS(2-METHOXYETHOXO)ALUMINATE (SDMA). A Versatile Organometallic Hydride

16 C. J. EVANS and S. KARPEL
ORGANOTIN COMPOUNDS IN MODERN TECHNOLOGY

17 R. B. KING and J. P. OLIVER (editors)
ORGANOMETALLIC CHEMISTRY REVIEWS; ANNUAL SURVEYS: Si, Sb, Bi, Mn, Tc, Re, Heteronuclear complexes
18 I. OMAE
ORGANOMETALLIC INTRAMOLECULAR-COORDINATION COMPOUNDS
19 R. B. KING and J. P. OLIVER (editors)
ORGANOMETALLIC CHEMISTRY REVIEWS; ANNUAL SURVEYS: TRANSITION METALS IN ORGANIC SYNTHESIS; ORGANIC REACTIONS OF SELECTED π-COMPLEXES
20 A. G. DAVIES, E. O. FISCHER and O. A. REUTOV (editors)
ORGANOMETALLIC CHEMISTRY REVIEWS
21 I. OMAE
ORGANOTIN CHEMISTRY
22 W. C. TROGLER (editor)
ORGANOMETALLIC RADICAL PROCESSES

Foreword

Odd electron organotransition metal complexes, once viewed as isolated exceptions to the 18-electron rule, are now recognized as a subset of compounds with unique reactivities. The identification of organometallic radicals as intermediates in the reaction mechanisms of even electron systems makes them increasingly important to the nonradical chemist as well. This volume focuses on the generation, spectroscopic properties, structures, and reaction mechanisms exhibited by 17- and 19-electron organometallic complexes. The topics discussed center on transition metal complexes, although analogies to the chemistry of main group radicals are illustrated where appropriate. It is hoped that this volume will provide an overview of progress in the field for the research specialist, as well as for the advanced graduate student. Some measure of the rapid growth of interest in the reactions and properties of organometallic radicals can be judged from the references cited. These predominantly cover the literature since 1978, the year that the influential book "*Organometallic Mechanisms and Catalysis*" by J. K. Kochi appeared and emphasized the importance of radical processes to the solution chemistry of organometallic complexes.

Contents

CHAPTER 1

Electrochemical Approaches for the Determination of Organometallic Reactivity

C. Amatore

Ecole Normale Supérieure, Laboratoire de Chimie, UA CNRS 1110
"Activation Moléculaire", 24, rue Lhomond, 75231 Paris Cedex 05, France

Table of Contents

1. Introduction

Although the electrochemical behavior of a large number of organometallic compounds has been examined [1], it is an indisputable fact that the field of organometallic electrochemistry is considerably less extended than the organic one [2]. This is not due to a more recent introduction of electrochemistry in the organometallic field, as attested by the impressive body of work by Vlček [3] or Dessy [4] and their co-workers in the sixties. These groups established that a broad variety of organometallic or inorganic complexes were electroactive, and that their reactivity in electron transfer initiated reactions could be advantageously studied by electrochemical techniques. Such studies are of comparable quality and interest to those which have been simultaneously produced in organic electrochemistry and demonstrate the equivalent abilities of electrochemical techniques in both domains.

However the importance of transient paramagnetic species in reaction mechanisms has been recognized earlier in the organic field [5]. This stimulated the interest of the electroanalytical chemists – who are mainly responsible for the development of the method [6, 7, and refs. therein] – in mechanistic problems relevant to organic reactions. This interest was even reinforced by the success of a few electroorganic industrial processes. This explains the rapid growth of organic electrochemistry, and the high level of sophistication reached in the mechanistic studies of organic reactions by electrochemical techniques [8].

On the other hand the important role of transient paramagnetic species in the inorganic or organometallic field has been evidenced only in the very recent years [9]. It is thus not a surprise that the field of organometallic electrochemistry did not develop extensively, and was mainly restricted to the study of reversible electron transfer processes, with thermodynamic goals. Indeed, electrochemical techniques, especially cyclic voltammetry, are particularly adequate for determination of standard potentials, when the electrode processes are uncomplicated by fast follow-up chemical reactions or electron transfer kinetics [1–8]. The existence of many stable redox states for a large number of inorganic or organometallic molecules [1] certainly explains why e.g. cyclic voltammetry has been adopted in several research groups in the area. By comparison organic electrochemistry did not obtain a similar success in penetrating organic laboratories, the usual extremely short lifetime of the intermediates of interest for synthesis precluding the use of "standard" cyclic voltammetry as a spectroscopic method for the determination of redox properties. A rapid survey of the main chemical journals will demonstrate unambiguously that inorganic or organometallic chemists resort more frequently to electrochemical data than organic chemists although, up to the recent years, most electrochemists have addressed mainly organic problems.

Yet this success is only apparent since, as explained above, electrochemical methods are mainly used as "redox spectroscopic" methods, in a similar way as e.g. n.m.r. or e.s.r. are used for gaining structural information. Restricting electrochemical techniques to these applications unfortunately discards their largest advantages, i.e., their unique potentialities in solving mechanistic problems. These capabilities stem from the two following properties of electrodes. First, they can be viewed as extremely large molecules whose reducing or oxidizing properties can be tuned continuously and with precision to adjust at will the driving force of a reaction. Second, the resulting current flow is a direct measure of the kinetics undergoing at the electrode. It is noteworthy that the same device (the electrode) is used to generate intermediates (potential) and to analyze simultaneously their chemical behavior (current). This associated with the fact that the accessible time scale extends [8, 10] from a few seconds to a few nanoseconds for *direct* electrochemical methods, or from a few milliseconds to a few picoseconds for *indirect* methods, clearly evidences the unique suitability of electrochemical methods for the investigation of mechanisms of chemical reactions.

It is the aim of the present chapter to present and discuss some aspects of the hidden part of the electrochemical "iceberg". In a first section we will discuss the relevance of kinetic or mechanistic data determined by electrochemistry to homogeneous chemistry. The second consists in a presentation of the most usual electrochemical methods, whereas the third is an illustration, based on some typical examples selected from our research, of their use for solving mechanistic problems in organometallic chemistry.

In the following we chose to present most electrochemical concepts and methods in terms of words and references to analogous concepts well known by non-electrochemists, rather than in terms of equations. Thus we hope that this chapter may constitute a guide and an incentive to further more specialized readings [6–8, 11].

2. Electrochemical vs. Homogeneous Chemical Reactivities

2.1 Introduction

As mentioned above, electrochemistry affords a facile and precise way to generate highly energetic intermediates by continuous tuning of the electrode potential, without resorting to a variety of reductants or oxidants of various strength. A second advantage is related to the fact that the current flowing through the electrode is a direct measure of the production/consumption rate of the generated intermediates. It ensues that this parameter, particularly critical in catalysis or in product selectivity, is easily controlled and adjusted at will under electrochemical conditions. On the other hand, and from a kinetic point of view, the current-potential-time relationship contains all the information on the fate of the electrogenerated intermediates, yet under a form which may appear cryptic to non-specialists. It must be emphasized that this is not particular to electrochemical techniques, the same being true for most of the usual spectroscopic methods. Thus one may learn to decipher cyclic voltammograms to obtain kinetic information, in the same way as he deciphers n.m.r. or i.r. spectra to obtain structural information.

In this perspective it is worth mentioning that most of the electrochemical "elementary" mechanisms have been described in the electrochemical literature and compiled in several books or reviews on the topic [6, 8, 12]. The fact that these mechanisms, which involve various combinations of single electron transfer and chemical steps, have been generally presented in the context of organic electrochemistry should not be perceived by organometallic or inorganic chemists as an obstacle. Indeed as for any kinetic method, the behavior observed does not depend on a particular chemical system but on the mechanism followed, i.e. on the relationships existing between the various intermediates. For example a simple ECE [13] (*E*lectrochemical-*C*hemical-*E*lectrochemical) sequence such as:

$$A + ne \rightleftharpoons B \quad (E) \tag{1}$$

$$B \rightarrow C \quad (C) \tag{2}$$

$$C + ne \rightleftharpoons D \quad (E) \tag{3}$$

leads to the same electrochemical behavior independent of the organic or organometallic nature of A, B, etc, or of the oxidative (n = −1) or reductive (n = +1) nature of electron transfer. Thus the reduction of aromatic halides [14]:

$$ArX + e \rightleftharpoons ArX^{\cdot -} \quad (E) \tag{1a}$$

$$ArX^{\cdot -} \rightarrow Ar^{\cdot} + X^{-} \quad (C) \tag{2a}$$

$$Ar^{\cdot} + e \rightleftharpoons Ar^{-} \quad (E) \tag{3a}$$

or that of carbonylmanganese(I) cations [15]:

$$(CO)_5Mn(MeCN)^{+} + e \rightleftharpoons (CO)_5Mn(MeCN)^{\cdot} \quad (E) \tag{1b}$$

$$(CO)_5Mn(MeCN)^{\cdot} \rightarrow (CO)_5Mn^{\cdot} + MeCN \quad (C) \tag{2b}$$

$$(CO)_5Mn^{\cdot} + e \rightleftharpoons (CO)_5Mn^{-} \quad (E) \tag{3b}$$

do not differ in terms of their general electrochemical behaviors, although the chemistry involved is different.

Most of the kinetic analysis for the main electrochemical mechanisms have been produced under the form of general laws or working curves [8]. However a variety of numerical approaches to the solution of electrochemical kinetics have been made available. Moreover, computer programs have been designed which enable the simulation of almost any mechanism, provided that the rate constants involved are not too large [16]. Most of these versatile programs, executable on Personal Computers, are easily adapted to a particular kinetic situation, without particular skill in programing or electrochemistry, to afford e.g. simulated cyclic voltammograms to be compared with the experimental ones [17]. Such a situation is reminiscent of that existing e.g. in e.s.r. which enables experimental chemists to characterize intermediates and derive structural informations by comparison between experimental and simulated spectra, without the need of being an expert in computer sciences.

2.2 Homogeneous vs. Electrochemical Reactivities

From the above discussion it should be understood that a large variety of tools exist today, which allow non-electrochemists to extract the rich kinetic information contained in transient electrochemical data. Yet at this point the following questions must be addressed: (i) are electrochemical intermediates identical to those generated under homogeneous conditions? (ii) how do reactivities observed under electrochemical conditions compare to those found in homogeneous chemistry? In other words do electrochemical techniques produce mechanistic information which is valid under non-electrochemical conditions?

The answer to the first question is obviously positive. Outer-sphere electron transfer to (or from) a molecule results in the creation of the same intermediate, independent of the origin – an electrode or another molecule – of the electron (or hole) transferred [18]. The primary intermediates thus formed may undergo selected cascades of reactions to afford most of the intermediates observed or postulated in homogeneous chemistry. For example the reduction of the stable 18e carbonylmanganese(I) cation in eqns. 1–3, yields sequentially a 19e and then a 17e intermediate before reaching the thermodynamically stable 18e $Mn(CO)_5^-$ anion. Virtually any intermediate of interest may thus be generated under electrochemical conditions provided an electroactive precursor exists.

The answer to the second question is more difficult. One is tempted to give a positive answer when considering that the electrical perturbation resulting from the existence of a large potential drop at the electrode surface extends only over a few Å from the electrode (see section 2.3.1). This conclusion must however be tempered since the composition of usual electrochemical media may differ from those found in homogeneous chemistry. Indeed because one wants to minimize the problems related to conduction of currents through the solution, rather conducting media and concentrations in the millimolar range are needed. Thus the electrochemists prefer solvents with rather large dielectric constants and large ionic strength [19]. However the recent development of "ultramicroelectrodes", i.e. of electrodes whose dimensions do not exceed a few micrometers [20], has allowed one to overcome most of these conduction problems [20, 21]. Thus concentrated solutions may be investigated at ultramicroelectrodes [22]. Similarly solvents such as ethers, arenes or even alkanes, [23] or media with low ionic strength [24] can now be used. Thus although most of the reported electrochemical data were obtained in a limited class of media, conditions akin to those of organometallic chemistry [23f, 24d, 25] are now accessible to electrochemical investigations (see e.g. sections 3.2.2., 4.1. and 4.2.).

To conclude this section we discuss more deeply the problems related to the structure of solutions and the electrical field near an electrode surface.

2.3 Are Electrochemical Reactions more Heterogeneous than "Homogeneous" Ones?

Under electrochemical conditions, electrons are exchanged at a bidimensional surface – the electrode surface – whereas electroactive materials are contained in a volumic phase – the bulk solution –. Therefore the electrochemical act of electron transfer occurs in essentially heterogeneous conditions. On the other hand the electrochemical driving force originates in the electrical potential difference imposed between the electrode and the solution. Both factors may lead to the implicit notion that electrochemical mechanisms take place under conditions that are quite different from those encountered in other fields of chemistry.

2.3.1 The Double-Layer

For the sake of simplicity we will restrict the presentation to the case of a cathode; obviously it transposes immediately to the case of an anode provided the appropriate changes of polarities and charges are made. The negatively charged surface of a cathode results in an important electrical field which tends to "capture" positive ions from the solution. The positive layer thus formed on the electrode surface exerts an electrical field of opposite direction, which therefore attracts negative ions. Repetition of this phenomenon should result in the formation of alternating positive and negative layers. Yet the absolute charge density in the successive layers rapidly decays with the distance from the electrode, because of partial compensation between the resulting opposite electrical fields. The large decrease of the overall electrical field results also in the layers being less and less bound. Thermal agitation is then sufficient to make the layers interpenetrate each other, except for the first one, which remains tightly bound to the electrode surface. This is indicated by the usual names of "compact layer" and "diffuse layer" for the two successive structures; both of them constitute the "double layer". Their dimensions (a few Å for the compact layer; a few tens of Å for the diffuse layer) can be evaluated via the Gouy-Chapman-Stern theory, which is developed in most basic electrochemical textbooks [26]. The main conclusions of this model are the following. (i) Because of the existence of the compact layer, most of the potential difference between the electrode and the solution takes place over only a few Å from the electrode surface. (ii) Beyond a few tens of Å of the electrode surface the electrical potential is identical to that of the bulk solution. As a result, as soon as their lifetime exceeds a few nanoseconds (see section 2.3.2.), the electrogenerated intermediates react in regions of the solution where essentially no electrical potential applies.

2.3.2 The Diffusion Layer

As mentioned above, in order to react molecules must be carried from the solution to the electrode surface. This transport can be achieved through three different modes. "Convection", the most ubiquituous mode of transport in chemistry, consists in movements of microscopic volumes of solution, arising because of thermal agitation or stirring. Thus the molecules are transported according to the exact hydrodynamics of the solution. Under usual stirring conditions, hydrodynamics result in the formation of a stagnant liquid film at the surface of any wall or particle of sufficient size [27]. In this stagnant layer, whose thickness ranges from a hundred micrometers for unstirred solutions (thermal convection) to a few tens of micrometers for magnetic stirring (forced convection), no convection can occur. As a result the solution near an electrode can be viewed as being formed of a stagnant liquid layer (10 μm to 100 μm) adjacent to the electrode, and of the remaining part of the solution – the bulk solution – where concentrations are macroscopically homogeneous (see however section 2.3.3). Within the stagnant layer transport is achieved by "diffusion" and "migration".

Migration results from electrical forces applied to charged molecules [24, 28]. As soon as the ionic strength is slightly larger than the concentration of electroactive molecules, no significant electrical field exists in electrochemical solutions [24a, b] outside of the double layer (see section 2.3.2). Thus, almost everywhere in the stagnant layer, molecules are transported by diffusion [29].

Since diffusion of molecules is not instantaneous the concentration perturbations imposed at the electrode cannot be transmitted instantaneously to the solution, i.e. to the end of the stagnant layer. Thus the perturbation propagates into the solution with a time delay depending on the distance from the electrode. This results in the creation of concentration profiles, i.e. of variations of concentration with the distance from the electrode surface, similar to those represented in figure 1. The fraction of the stagnant layer which is modified, i.e. that where local concentrations differ from those in the bulk solution, is referred to as the "diffusion layer". Its thickness depends on the time elapsed since the beginning of the perturbation. A rough estimate of the diffusion layer thickness can be obtained

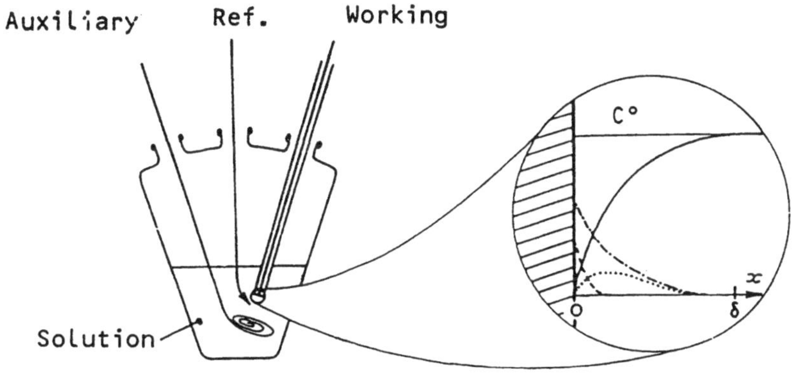

Fig. 1. Schematic representation of an electrochemical cell with a three electrode configuration (Ref.: Reference electrode), indicating the concentration profiles of the reactant A (—), product D (–·–·–) and of the two intermediates B (- - -) and C (....) at the working electrode surface for an ECE sequence (Eqns. 1–3). C^0: initial concentration profile of the reactant before electrolysis.

by considering that a molecule, whose diffusion coefficient is D, reaches a distance $\delta = 2 \cdot (Dt)^{1/2}$ after a time t has elapsed since its generation at the electrode surface [29]. For most molecules and solvents considered in usual organic or organometallic electrochemistry, diffusion coefficients are on the order of 10^{-6} to 10^{-5} cm^2 s^{-1}, and diffusion times range between a millisecond and a few seconds. Thus diffusion layers generally extend over distances ranging from a few tenths to a few tens of a micrometer from the electrode.

From what precedes it should be understood that electrochemically initiated reactions take place in a region where large concentration gradients exist. This feature is included in the treatment of electrochemical data, the rate laws being then time and space dependent [6, 8, 12] rather than simply time dependent as in usual kinetics.

When there is only one chemical route for a given intermediate, the existence of these concentration profiles has obviously no role in the product distribution. However it may greatly influence the selectivity when several reaction paths are possible for an intermediate. This is a situation akin to that observed in homogeneous chemistry when different experimental conditions are modified. Moreover, the fact that these spatial structures are perfectly described and may be controlled gives additional means to "drive" a reaction to the target product [30]. As an illustration of this point consider the following example related to the reactivity of 17e manganese radicals $Mn(CO)_3P_2^{\cdot}$, where P is a phosphine ligand, in the presence of tributyltinhydride as an H-atom donor [15, 31].

Such radicals may be generated by a photochemical route from the parent dimer [31]. Their reaction with $HSnBu_3$ under these conditions is reported to afford the manganese hydride and hexabutylditin, according to the following reaction scheme (**Mn** = $Mn(CO)_3P_2$, and **Sn** = $SnBu_3$):

$$\mathbf{Mn}-\mathbf{Mn} \xrightarrow{h\nu} 2\,\mathbf{Mn}^{\cdot} \quad (4)$$

$$\mathbf{Mn}^{\cdot} + \mathrm{H}-\mathbf{Sn} \rightleftharpoons \mathbf{Mn}-\mathrm{H} + \mathbf{Sn}^{\cdot} \quad (5)$$

$$2\,\mathbf{Sn}^{\cdot} \longrightarrow \mathbf{Sn}-\mathbf{Sn} \quad (6)$$

The same radicals may be generated by electrochemical oxidation of the stable $Mn(CO)_3P_2^-$, 18e anion [15]. Under these conditions transient electrochemistry shows that a similar mechanism is followed except that step 6 is replaced by the cross coupling reaction 7 [15, 32]:

$$\mathbf{Mn^{\cdot} + Sn^{\cdot} \rightarrow Mn{-}Sn} \qquad (7)$$

The two studies afford essentially identical results for the rate constants of the H-atom transfer reactions. The discrepancy observed (eqn. 6 vs eqn. 7) then does not arise from a different intrinsic reactivity of the 17e manganese radicals under electrochemical conditions. However, under electrochemical conditions large amounts of these radicals are produced at the electrode. Thus the tin radicals are formed (step 5) in a region of solution, where their own concentration is extremely low and controlled by steady state conditions, but where that of the manganese radicals is large. The resulting concentration effect is then sufficient to favor the cross-coupling step 7 vis a vis the homo-coupling in eqn. 6, although the former is intrinsically slower in terms of relative rate constants. Under photochemical conditions, the manganese radical concentration is lower so that the rapid dimerization of tin radicals (eqn. 6) overwhelms reaction 7.

2.3.3 Are Homogeneous Reactions Really Homogeneous?

At first glance the phenomena described above appear to be specific to electrochemistry, because of its essentially interfacial nature. However most of them exist also in "homogeneous chemistry" although not identified or taken into consideration by most chemists.

2.3.3.a Diffusion Layers. Do They Exist in Homogeneous Chemistry

The similarity of electrode reactions to homogeneous chemistry is particularly obvious when reactions involve a solid phase as e.g. when one of the reactants is a dispersed powder (reduction by metal particles, insoluble reactant, ...) or an amalgam. Thus the particle or solid surface evidently plays an identical role to that of an electrode, the reaction occuring or being initiated at the solid / liquid interface. They are less obvious but still exist when two non miscible liquid phases are involved as e.g. in phase transfer catalysis. In all these cases as soon as the particles or liquid droplets sizes exceed a few molecular dimensions, the transport of molecules to or from their surface is achieved by the same physical means of convection, migration and diffusion. Thus the concentrations near the interphase vary in an identical fashion to that described above for electrodes.

Although the reactions pertaining to the above class represent a significant proportion of "homogeneous" chemistry, the larger class consists of "one-phase" reactions. However, most of these reactions involve the initial mixing of at least two solutions of different compositions, e.g. one containing the starting material and the other being or containing a reagent. In such cases homogeneity is valid only at the macroscale level. At the microscale level the solution, depending on its rheological properties, consists of a collection of segregated lumps or of imbricated layers of each initial solution [33], as schematized in figure 2b, for the latter situation. In order to react the two reactants must coexist in a region of molecular dimension. A simplistic view of a chemical reaction at the microscopic level is then given by the superimposition of the three following processes [33].

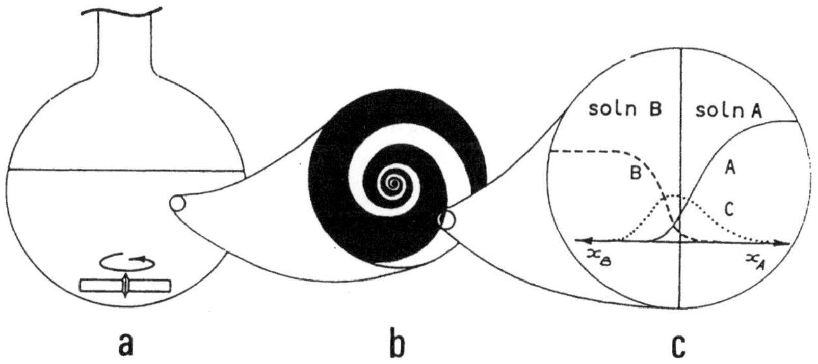

Fig. 2. Schematic representation of microscopic heterogeneity in an "homogeneous" solution made from two stock solutions of A and B. (a): macroscopic homogeneity in a reaction vessel. (b): microscopic heterogeneity in a local vortex indicating the imbrication of each solution layer (solution A: white area; solution B: black area). (c): concentration profiles of the two reactants A, B and of the product C, at the interface between two layers, for a bimolecular $A + B \rightarrow C$ reaction.

(i) The chemicals must cross the interfaces existing between the microscopic structures; this involves their diffusion from the core of the microscopic structures to the external boundaries. (ii) Chemical reaction occurs at the interface for reactions whose rates are large vis a vis the diffusion, or interferes with diffusion when both rates are comparable (compare figure 2c). (iii) Macroscopic mixing imposed by the stirring conditions, is hastened by large scale motions of the above mentioned lumps or layers, which results in extending their interfaces (compare figure 2b).

These phenomena are responsible, for example, for the lower time limit in "stopped-flow" techniques. Although not usually rationalized under these terms, this is intuitively known by experimental chemists. Indeed this is the reason why stirring of homogeneous reactions is maintained up to completion, although macroscopic mixing occurs in the early stages. Among other reasons this is also one of the factors responsible for acceleration of homogeneous reactions by ultrasound, which results in smashing of the microscopic structures and thus in considerable extension of their reacting interfaces.

At this point of the discussion one may wonder whether microscopic homogeneity is ever achieved in "homogeneous" chemistry? It is easily understood from what precedes that a positive answer to the question requires molecular mixing be considerably faster than chemical reactions. This is e.g. true for extremely slow reactions. Another example is given by photochemically initiated reactions, since the reactive centers are then generated statistically within the microscopic lumps or layers of solution. Thus analogies exist between most "homogeneous" reactions and reactions at interfaces such as electrochemical ones, at least for the existence of microscopic spatial structures in which chemical reactions take place.

2.3.3.b Double Layers. Do They Exist in Homogeneous Chemistry?

As explained above the large electrical field existing at an electrode surface drops to negligible values at distances of a few molecular diameters from the electrode surface. Nevertheless it affects the very act of electron transfer, which occurs in this electrically perturbed region. It should also affect the reactivity of electrogenerated intermediates as soon as their lifetime is in the range of a few nanoseconds or less. Indeed a rough estimate of the time needed to "escape" from this region is given by that required for a molecule to move over ca. 50 Å. As deduced from $\delta = 2 \cdot (Dt)^{1/2}$, the diffusion length during a time interval t (see section 2.3.2), this requires ca. 10 to 100 nanoseconds, a time range now accessible by direct electrochemical techniques [34].

Do such molecular structures exist in homogeneous chemistry? Again a positive answer has to be given as soon as the reaction involves different macroscopic phases (solid reagents, amalgams, phase transfer catalysis, ...). It is also true for microscopic heterogeneities such as the lumps or layer described in the preceding section, for the case of electron transfer reactions as well as for those resulting in charge creation or anihilation. Indeed since these reactions involve charge modifications, the ionic content of each phase or of their interfaces differ either at the beginning or during the reaction. Thus, their electrical potentials, as given by Nernst law [35], also differ, which results in the creation of ionic structures [36] akin to the double layer described above for an electrode / solution interface.

The same phenomenon occurs also within each phase or microscopic structure, in the molecular vicinity of any charged molecule. The electrical field created by a molecular charge at its surface has a comparable magnitude to that created by an electrode. It results in a molecular ordering of the ionic content of the surrounding space, very reminiscent of that achieved by an electrode [37]. This is well recognized in most theories of ionic reactions, where this phenomenon is partly responsible for the so-called "work-terms", in e.g. Marcus theory for electron or proton transfer [38]. It is also at the basis of the dependency of diffusion limit rate constants on the charges of reactants or products [39], or of some of the "ionic strength effects" on rates of chemical reactions involving charged species.

3. Electrochemical Methods for the Investigation of Chemical Kinetics

From the above presentation one is led to the conclusion that with the exception of a few cases, electrochemical conditions do not differ considerably from those considered in "macroscopically homogeneous chemistry". However one of the advantages of electrochemistry is that the region of space where microscopic or molecular structures of space develop is physically accessible, because of the macroscopic nature of an electrode. For example monitoring the chemical exchanges or reactions at the surface of a reducing metal particle is not an easy task. Yet this is feasible at an electrode under conditions that do not significantly differ and are easily transposed to the particle case (see for example section 4.4.3).

3.1 Principle of Kinetic Measurements in Electrochemistry

The basis of any kinetic determination is to oppose the rate of a chemical process to a reference clock. This reference clock can be a physical external object as the chronometer used for slow reactions, or the time scale of an oscilloscope, or of a strip chart recorder. It can be a physical process as e.g. a flow rate as in stopped-flow experiments. It can be also chemical and internal to the experiment as e.g. when an unknown reaction is opposed to a known one in competitive experiments, as for the well-known "radical-clocks".

The principle of most electrochemical methods for the investigation of chemical kinetics consists in using the rate of mass transfer to or from the electrode as an internal reference clock. Its obliged involvement in any electrochemical experiment makes it a natural built-in clock. Indeed as explained above the very physical nature of mass transfer makes it easily quantifiable and adjustable continuously from equivalent times of a few seconds to a few tens of nanoseconds [34]. The basis of the method is easily understood when one thinks in terms of concentration profiles of intermediates in the diffusion layer (section 2.3.2). Let us consider for example that an intermediate is generated at an electrode surface. Diffusion, and possibly other means of mass transport, will remove it from the electrode surface, which results in the creation of a concentration gradient for this intermediate. When its lifetime is large by comparison to the mass transfer rate, the concentration profile, i.e. the concentration distribution of the intermediate, is only a function of the mass transfer rate. Conversely when the lifetime is considerably smaller than the mass transfer rate, its concentration profile extends only in a tiny fraction of the diffusion layer. In the intermediate range the concentration profile of the intermediate is controlled by both processes (see e.g. the concentration profiles of species B and C in Fig. 1, for an ECE sequence similar to that in Eqns. 1–3, in Section 2.1). Thus in each of the above three cases different relationships are observed between the concentration gradient (i.e. the current) of the intermediate and its concentration at the electrode surface, which is related to the electrode potential. The resulting current-time-potential relationships are therefore characteristic of the nature and of the level of interference of the chemical reaction with the process of mass transfer, which allows the pertinent kinetic information to be extracted [8, 40].

The different electrochemical methods based on direct competition between chemical kinetics and mass transfer rates, as explained above, are denoted "direct" electrochemical methods. They differ by the nature (current or potential) and the type or amplitude of the electrical perturbation applied to the electrode, as will be developed in a forthcoming section (see Sections 3.2–3.4) and summarized in Fig. 3. Although they have the great advantages of simplicity and of affording easily recognizable current / potential patterns, they suffer of being limited to the tens of nanosecond time scale.

Thus "indirect" methods have been devised which extend the accessible time window to the picosecond level, although based on conventional electrochemical equipment. The principle of these "indirect" methods is to oppose the rate of a chemical reaction to that of the diffusion limit rate constant, i.e. that of molecular transport over atomic distances [39].

Among indirect methods the most general is "redox-catalysis", which involves an "electron transfer mediator". The latter is any stable electrogenerated species, whose standard potential is selected so that electron transfer from the mediator to the investigated molecule is uphill by a few tenths of a volt. Thus in the absence of a follow-up reaction no electron transfer can take place, and the concentration profile of the mediator is controlled by diffusion only. Conversely, in the presence of a fast follow-up reaction the uphill electron transfer is continuously displaced, which consumes the mediator, and affects therefore its concentration profile. The degree to which the concentration profile is affected depends on the magnitude of the follow-up reaction, which affords an indirect way of measuring the corresponding rate constant. To illustrate this point let us consider a case in reduction, where $M/M^{\cdot-}$ is the mediator couple, and A the electroactive precursor to the

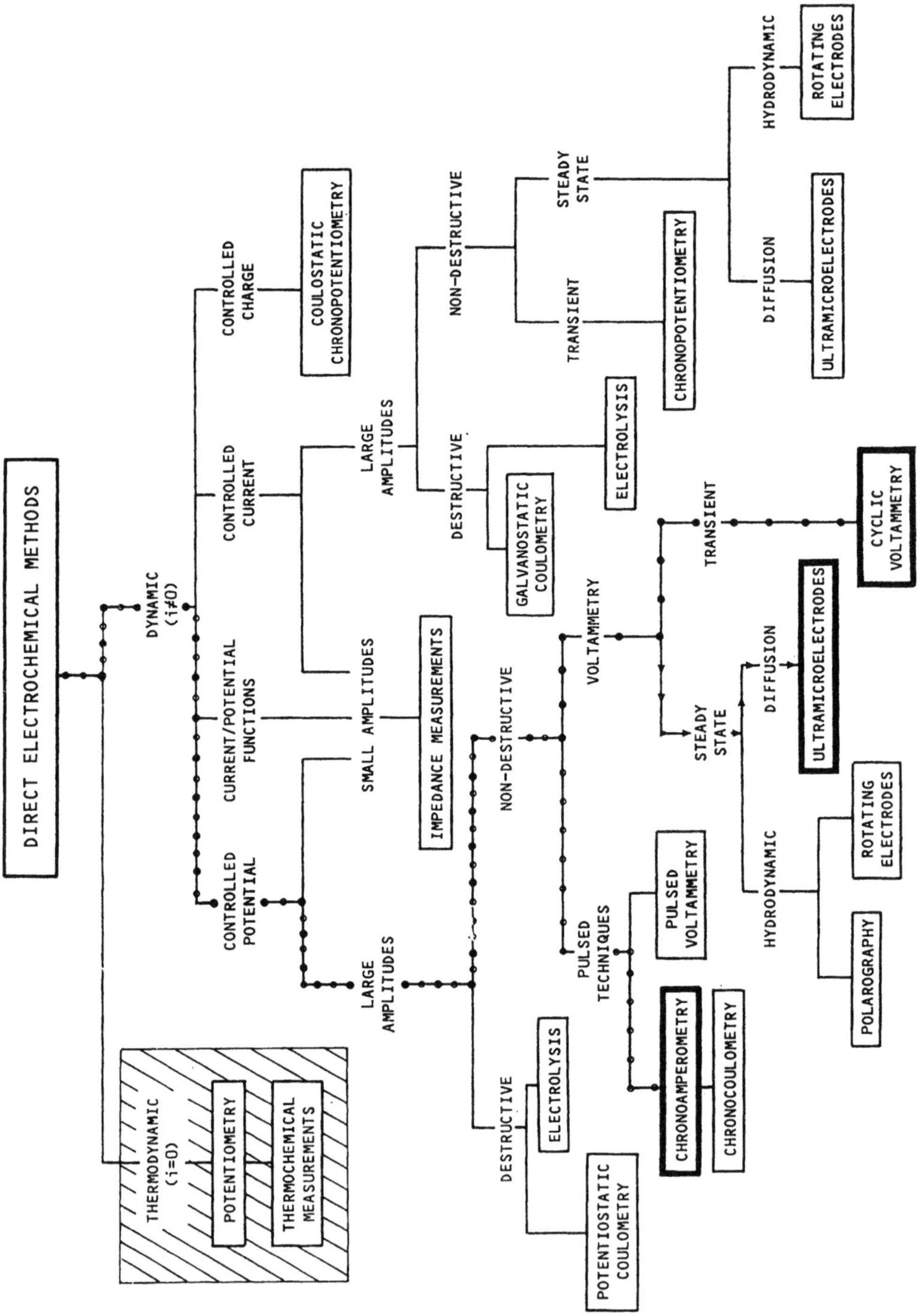
DIRECT ELECTROCHEMICAL METHODS
THERMODYNAMIC (i=0)
POTENTIOMETRY
THERMOCHEMICAL MEASUREMENTS
DYNAMIC (i≠0)
CONTROLLED POTENTIAL
CURRENT/POTENTIAL FUNCTIONS
CONTROLLED CURRENT
CONTROLLED CHARGE
COULOSTATIC CHRONOPOTENTIOMETRY
LARGE AMPLITUDES
SMALL AMPLITUDES
IMPEDANCE MEASUREMENTS
SMALL AMPLITUDES
LARGE AMPLITUDES
NON-DESTRUCTIVE
DESTRUCTIVE
ELECTROLYSIS
GALVANOSTATIC COULOMETRY
TRANSIENT
STEADY STATE
CHRONOPOTENTIOMETRY
DIFFUSION
ULTRAMICROELECTRODES
HYDRODYNAMIC
ROTATING ELECTRODES
NON-DESTRUCTIVE
DESTRUCTIVE
ELECTROLYSIS
POTENTIOSTATIC COULOMETRY
VOLTAMMETRY
PULSED TECHNIQUES
PULSED VOLTAMMETRY
CHRONOAMPEROMETRY
CHRONOCOULOMETRY
TRANSIENT
CYCLIC VOLTAMMETRY
STEADY STATE
DIFFUSION
ULTRAMICROELECTRODES
HYDRODYNAMIC
ROTATING ELECTRODES
POLAROGRAPHY

investigated intermediate $A^{\cdot-}$, which evolves e.g. via a first order reaction of rate constant k. In the absence of A the mediator exhibits a "reversible" behavior corresponding to the reversible electron transfer in Eqn. 8. In the presence of A the sequence in Eqns. 9 and 10 may take place:

$$M + e \rightleftarrows M^{\cdot-} \tag{8}$$

$$M^{\cdot-} + A \underset{k_b}{\overset{k_f}{\rightleftarrows}} A^{\cdot-} + M \tag{9}$$

$$A^{\cdot-} \xrightarrow{k} \text{etc} \tag{10}$$

Since application of the method corresponds to highly reactive intermediates, steady state conditions apply to $A^{\cdot-}$. Thus the rate law for the consumption of $M^{\cdot-}$ is given in Eqn. 11:

$$d[M^{\cdot-}]/dt = -k^{ap}[M^{\cdot-}] \tag{11}$$

where:

$$k^{ap} = k\{k_f[A]/(k + k_b[M])\} \tag{12}$$

For concentrations of A and M in the same range, the bracketed term in Eqn. 12 is considerably smaller than unity, owing to the large endergonicity of the homogeneous electron transfer in Eqn. 9. Thus k^{ap} may be determined by direct electrochemical methods, which allows an access to the value of k. However extraction of k from Eqn. 12 supposes that the latter is not considerably larger than $(k_b[M])$, otherwise Eqn. 12 simplifies to $k^{ap} \approx k_f[A]$. In practice k_b is close to the diffusion limit rate constant, since the backward electron transfer in Eqn. 9 is largely exergonic. Thus straightforward application of the method enables the determination of rate constants from ca 10^6 s^{-1} up to ca. 10^8 to 10^9 s^{-1} with usual electrochemical equipment. Other methods of similar principle but involving other specifically designed competitions, allow the picosecond time-scale to be reached [10, 41].

3.2 The Electrochemical Cell

3.2.1 Monitoring One Half-Reaction

For electron transfer reactions or electron transfer initiated reactions, electrodes afford an important advantage over homogeneous methods, besides the obvious one of continuous adjustment of the driving force. In homogeneous chemistry an electron transfer reaction involves two reactants, an acceptor, A, and a donor, D. It consists then of the interactive superimposition of two elementary acts: the acceptor reduction and the donor oxidation. Consequently the overall kinetics observed is not characteristic of one of the elementary half-reactions but of their interactive superimposition. Under electrochemical conditions reductions take place at cathodes and oxidations at anodes, i.e. at two different physical locations.

Fig. 3. Main electrochemical methods and their relationships presented under the form of a "family tree" (adapted from an original idea by P. T. Kissinger in Ref. 7, p. 6). The three methods presented in this chapter are indicated in the bold face boxes, as well as their "genealogical paths". The shaded box in the upper left corner corresponds to thermodynamical methods, i.e. to those presented in most chemistry textbooks.

This physical separation allows the monitoring of one of the two elementary acts without interference of the other.

This separation is achieved in practice via electronic control of the potential difference between one of the electrodes – the "working electrode" – and a "reference electrode" whose potential is identical within a constant to that of the bulk solution. This is achieved by a "potentiostat" [42–44] in the following way. A potential difference ($E_W - E_A$) is applied between the working electrode and the third electrode – the "auxiliary" or "counter" electrode –. The resulting potential difference ($E_W - E_{Ref}$), between the working and reference electrode, is measured and compared to the desired value, E, imposed by a signal generator. If ($E_W - E_{Ref}$) is too large ($E_W - E_A$) is decreased; conversely if ($E_W - E_{Ref}$) is too small ($E_W - E_A$) is increased. Repetition of this procedure allows one to impose the desired potential difference, E, within a microsecond for most potentiostats and within a few tens of nanoseconds for fast scan potentiostats [43, 44]. These times are even smaller when the potential is varied continuously as it is for cyclic voltammetry (see Section 3.3.1).

3.2.2 *Ultramicroelectrodes*

In the preceding section we explained how a given potential, E with respect to a given reference electrode, can be applied to a working electrode. However in practice there is a serious difficulty introduced by the fact that the working and the reference electrodes are electrically connected via a solution element whose resistance, R, is in the range of several $k\Omega$ for usual aprotic solvents and electrodes of millimetric dimensions. Thus the measured potential difference between the two electrodes involves an ohmic contribution, Ri, which depends on the current i flowing through the cell. For thermodynamic measurements (Fig. 3) this is not a problem since the currents are vanishingly small, the Ri term being then negligible as compared to E. However dynamic measurements at electrodes of millimetric or larger sizes involve currents of several μA to a few mA, which makes Ri a significant fraction of the measured potential difference, $(E_W - E_{Ref})_{mes}$, which is imposed to be equal to E. Thus the real potential difference usable for electrochemical purposes is $(E_W - E_{Ref})_{true} = E + Ri$, and may differ significantly from E, its desired value. [Note that with the electrochemical convention used in this chapter reduction currents are positive and oxidation currents are negative; thus in the expression $E + Ri$, the two terms have generally opposed signs, which corresponds then to a decrease in the absolute value of the potential driving force]. Although electronic feedback can be used to decrease the "ohmic drop contribution" [44], it can not be totally cancelled because of electronic instabilities. This factor is the main experimental reason for which electrochemists usually favored rather conducting media, i.e. solvents with large permitivities and large ionic strength, since this minimizes R.

However in the very recent years a new strategy has appeared for minimizing ohmic drop contribution. This approach, made possible by the concomitant development of conducting fibers, is based on the following observations. For a disc electrode of sufficiently small radius, r_0, vis a vis the reference electrode tip and the distance between the working and reference electrodes, R varies as $(1/r_0)$. On the other hand, for transient electrochemical methods, the current is proportional to the working electrode surface area, i.e. to r_0^2. Therefore the ohmic drop contribution, Ri, varies as r_0 and tends to zero when r_0 is decreased [20, 21]. In the experimental practice we feel that a good compromise between ease of fabrication, detection of currents and decrease in ohmic drop, is achieved for upper time scales at band electrodes [20, 24a–c] whose width is in the range of a few micrometers (whereas their length is of a few millimeters), and for lower time scales at disc electrodes of micrometric radii [20, 24d].

These electrodes, which are referred to as "ultramicroelectrodes", have opened new frontiers to electrochemistry. Indeed the considerable decrease in ohmic drop contributions allows electrochemistry to be performed in poorly conducting media, i.e. solvents such as alkanes or arenes or media of low ionic strength (see e.g. Sections 4.1 and 4.2). These electrodes are also responsible for the extension of direct electrochemical methods time-scale to the nanosecond regime or for allowing steady state measurements over a wide time range, without resorting to hydrodynamics (see Section 3.4).

3.3 Transient Electrochemical Methods

As explained above (Section 3.1) the principle of electrochemical methods for investigation of kinetics is based on the dependence of the current-time-potential relationships on the overall processes (diffusion / chemical reactions) at the electrode vicinity. Since these processes affect the current-time-potential relationships through modification of the concentration profiles of intermediates, all dynamic electrochemical methods presented in Fig. 3 should contain the same information. Although true in principle, this is not exact in practice. Indeed the experimental problem depends on the precision in extracting and visualizing these information, i.e. on the sensitivity of the method. Also important are its versatility and the accessible time range. For these reasons two complementary methods, at least in the opinion of this author, "cyclic voltammetry" and "chronoamperometry" have received considerable attention.

3.3.1 Cyclic Voltammetry

Among the various techniques presented in Fig. 3, cyclic voltammetry is certainly the most adequate for the investigation of chemical reactivity because it has the considerable advantage of affording easily recognizable patterns. Thus, with only a limited training, characteristic kinetic situations or trends can be identified by simple inspection of a cyclic voltammogram. Also, the involvement of given intermediates in a reaction mechanisms is easily detected by direct observation of their voltammetric "signatures" (see e.g. Section 4.2). The fact that these advantages are not earned at the expense of precision, explains the success of cyclic voltammetry.

These particularities stem from the intimate coupling of the potential and time variables. The coupling is achieved by imposing a triangular dependence with time to the potential of the working electrode, as sketched in Fig. 4a. Thus the potential is varied linearly with time from a value, E_i, where the investigated molecule is not electroactive, to a value, E_f, which exceeds the standard redox potential, E^0, of the molecule. The potential is then scanned back to its initial value. The slope of the linear variation of the potential with time, the scan rate v, in V s^{-1}, is generally identical for the forward and backward scans. The current response, which is evidently a function of time and potential, is usually plotted as a function of the potential, to afford a characteristic pattern which is referred to

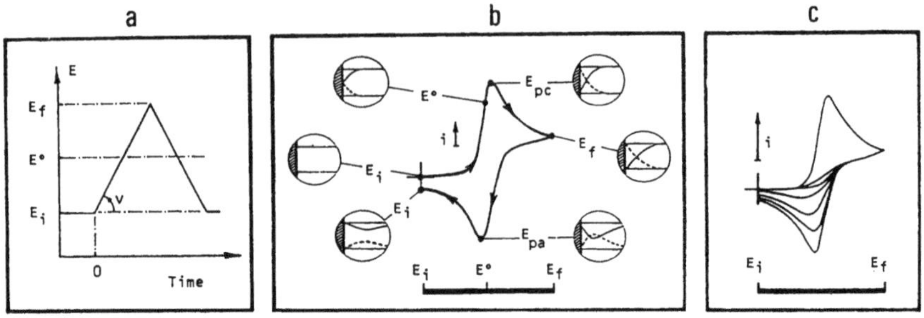

Fig. 4. Cyclic Voltammetry. (a): Triangular potential scan applied to the working electrode as a function of time; E_i: initial or rest potential, E_f: final potential, v: scan rate; E^0: standard potential of the investigated $0 + e \rightleftharpoons R$ electrochemical reaction. (b): Cyclic voltammogram indicating the concentration profiles of 0 (—) and R (----) at characteristic points of the voltammogram: E_{pc}: cathodic potential peak, E_{pa}: anodic potential peak. Note that according to the electrochemical convention used in this chapter negative potentials are plotted to the right abscissa axis and reduction currents are plotted to the upper ordinate axis. The direction of the potential scan is indicated by the arrows on the voltammogram. (c) Schematic deformation of a cyclic voltammogram when the chemical stability of the reduction product, R, is decreased.

as the "cyclic voltammogram" of the investigated compound at a given scan rate. Such a voltammogram is presented in Fig. 4b for the reduction of a reversible couple, $0 + e \rightleftharpoons R$, with a standard potential E^0; the upper trace corresponds to the forward or cathodic scan, whereas the lower trace corresponds to the backward or anodic scan. Simple inspection of the cyclic voltammogram in Fig. 4b shows the presence of two current peaks, a maximum, the "cathodic peak" and a minimum, the "anodic peak". These peaks which are characteristic of cyclic voltammetry result from the existence of two opposite effects which control successively the magnitude of the current flow at the electrode. Thus when the potential of the electrode is made more and more negative the concentration of 0 at the electrode surface decreases rapidly, particularly around E^0. This results in the concentration profile of 0 being steeper and steeper, i.e. in an increase of the current, since the latter is proportional to the gradient of the concentration profile at the electrode surface. However when the potential is more cathodic than E^0 by a few tenth of a volt, its influence is minimized because the concentration of 0 at the electrode surface is already negligible. Thus the propagation of the perturbation into the solution, which results in a decrease of the concentration gradient at the electrode surface, becomes the main phenomenon controlling the current, which therefore decreases. Convolution of the two phenomena results in the observation of a cathodic peak. During the anodic scan the same phenomenology applies to the product, R, of the reaction and results then in the observation of an inverted peak, the anodic peak. When R reacts chemically during the scan its concentration in the diffusion layer is evidently smaller, which results then in the observation of a smaller anodic peak, as sketched in Fig. 4c.

From this brief presentation of cyclic voltammetry it is understood that the overall shape of a cyclic voltammogram is indicative of the chemical processes occuring at the electrode and in the solution. Other information can be easily gathered from a cyclic voltammogram. For example, the difference betwen the cathodic and anodic peak potentials, $E_{pc} - E_{pa}$, is directly related to the intrinsic

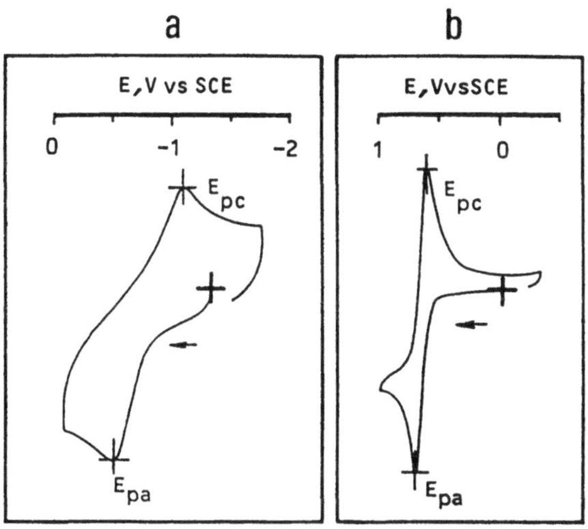

Fig. 5. Comparison of the cyclic voltammograms of the one-electron chemically reversible oxidation of (a) $Fe_2(CO)_8^{2-}$ and (b) ferrocene (see text). The initial direction of each scan, starting from the cross, is indicated by arrows.

activation barrier, $\Delta G_0^{\neq}$, of the electron transfer, i.e. of the molecular reorganization observed on electron transfer. Such an example is shown in Fig. 5, where the cyclic voltammograms corresponding to the oxidations of ferrocene and $Fe_2(CO)_8^{2-}$ are compared. Both cyclic voltammograms correspond to a one-electron exchange at the electrode:

$$Cp_2Fe - e \rightleftharpoons Cp_2Fe^{\cdot+} \tag{13}$$

$$Fe_2(CO)_8^{2-} - e \rightleftharpoons Fe_2(CO)_8^{\cdot-} \tag{14}$$

The molecular reorganization of ferrocene upon electron transfer is small, which results in a peak-to-peak separation close to the theoretical 60 mV [6]. In the $Fe_2(CO)_8^{2-}$ case the structure of the anion radical considerably differs from that of the dianion. Indeed IR and X-ray structural studies have shown that the dianion $Fe_2(CO)_8^{2-}$ has an unbridged metal-metal bond whereas a recent single crystal esr study has established that the anion radical $Fe_2(CO)_8^{\cdot-}$ has two bridging CO ligands [45]. This results in the large peak-to-peak separation ($\approx$600 mV) easily observed in the cyclic voltammogram in Fig. 5a [24d].

3.3.2 Double Step Chronoamperometry

In chronoamperometric methods the potential applied to the working electrode is a step function or combination of step functions of time. Thus at time zero the potential is stepped from a value, E_1, where the investigated molecule is not electroactive, to a value E_2, where its concentration at the electrode surface is zero (Fig. 6b). After a time θ, the potential is stepped back, for example to its initial value, as represented in Fig. 6a. In "double step chronoamperometry" a single pulse, i.e. two potential steps, is considered. Yet for some applications a succession of pulses may be used. The electrochemical information is contained in the current-time dependence of the kind shown in Fig. 6c. Variations of the method

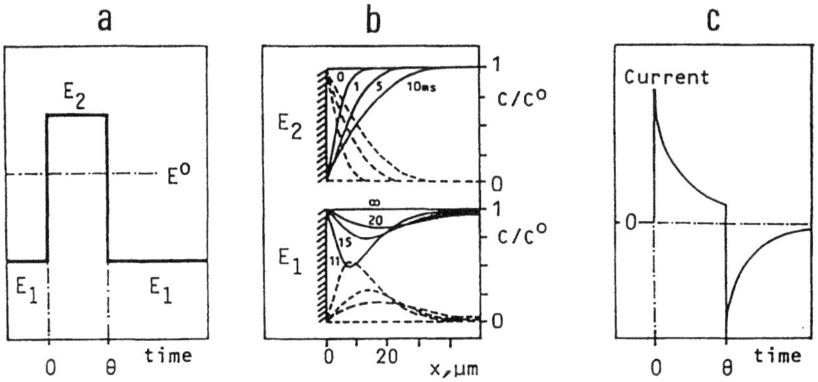

Fig. 6. Double Step Chronoamperometry. (a): potential perturbation applied to the working electrode (see text). E^0: standard potential of the investigated $0 + e \rightleftharpoons R$ electrochemical reaction. (b): resulting concentration profiles at the working electrode surface for 0 (—) and R (----). The numbers on each curve are the values of the time in ms for $\theta = 10$ ms; upper $0 < t < \theta$, $E = E_2$; lower $t > \theta$, $E = E_1$. (c): variations of the current as a function of time.

consist in recording the charge, i.e. the integral of the current with respect to time, as a function of time. The method is then referred to as "chronocoulometry".

Because of the instant concentration changes at the electrode surface introduced for each step, "chronoamperograms" are not peak-shaped and lose the rich "eye-accessible" informations contained in cyclic voltammograms. Thus, at least in the opinion of this author, the method is not well suited to the investigation of unknown kinetics. However, its combination with cyclic voltammetry is extremely useful.

Indeed in cyclic voltammetry the mixing of informations related to electron transfer kinetics and homogeneous chemical reactions makes difficult the extraction of the homogeneous kinetics data, without sophisticated treatments of the whole cyclic voltammogram [46]. In chronoamperometries electron transfer kinetics do not interfere provided that the potential steps are made on the electrochemical wave "plateau", i.e. the concentration of the electroactive material is zero at the electrode surface. As a result the sensivity to homogeneous kinetics is very often more important than for cyclic voltammetry [47]. This may appear as contradictory with our above statment on the suitability of both methods; yet the contradiction is only apparent. Indeed a caricatural description of the two methods would amount to say that cyclic voltammetry allows a perfect qualitative understanding of the processes undergoing at an electrode and in its vicinity, yet quantitative information is often difficult to extract except for simple mechanisms, i.e. for those involving a single rate determining step [48]. On the other hand double step chronoamperometry affords quantitative information with a high precision, but without indicating directly what is measured. We thus believe that one should more often take advantage of the potentialities of each technique by their combined use. Note that this is not a problem since both methods require an identical electrochemical set-up.

Combined use of both methods is extremely useful for e.g. quantifying product distributions observed in transient electrochemistry. For example the voltammo-

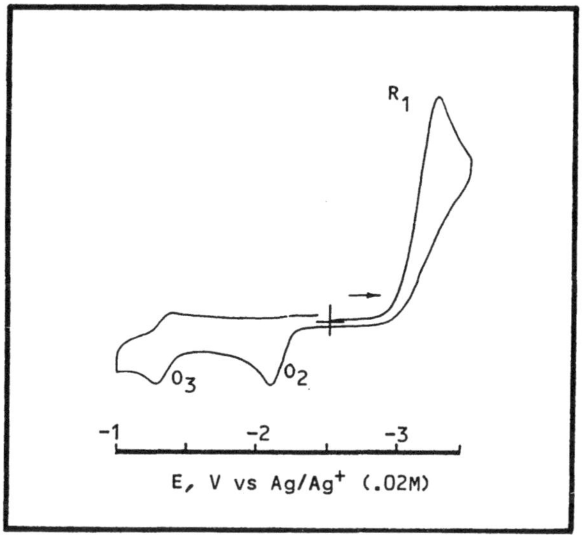

Fig. 7. Cyclic voltammogram of the reduction of a 2 mM solution of $Cr(CO)_6$ in THF (0.3 M NBu_4BF_4) at $v = 10\ V\ s^{-1}$. Platinum electrode, $\phi = 125\ \mu m$. Potentials are referred to the $Ag/AgBF_4$ (0.02 M in THF 0.3 M NBu_4BF_4) reference electrode. The direction of the initial scan, starting from the cross, is indicated by the arrow.

gram in Fig. 7 shows that reduction of $Cr(CO)_6$ at wave R_1 results in the appearance of two oxidation waves, O_2 and O_3, during the backward scan [49]. Comparison to authentic samples establishes that the two waves correspond to the oxidation of $Cr(CO)_5{}^{2-}$ (wave O_2) and of $Cr_2(CO)_{10}{}^{2-}$ (wave O_3). However independent oxidation of $Cr(CO)_5{}^{2-}$ results in the duplication of the anion radical which affords $Cr_2(CO)_{10}{}^{2-}$:

$$Cr(CO)_5{}^{2-} - e \rightleftharpoons Cr(CO)_5{}^{\cdot -} \quad \text{(wave } O_2\text{)} \tag{15}$$

$$2\,Cr(CO)_5{}^{\cdot -} \rightarrow Cr_2(CO)_{10}{}^{2-} \tag{16}$$

$$Cr_2(CO)_{10}{}^{2-} - e \rightleftharpoons Cr_2(CO)_{10}{}^{\cdot -} \quad \text{(wave } O_3\text{)} \tag{17}$$

It is thus extremely difficult to decide whether the O_3 wave corresponds to the $Cr_2(CO)_{10}{}^{2-}$ dianion created at wave O_2 or if some has been already electrogenerated at wave R_1, as e.g. observed in the reduction of the related $Fe(CO)_5$ [24d] (see Section 4.2). However the problem is quite simple to solve with double step chronoamperometry. A first potential step is made cathodic and located on the plateau of wave R_1, the second being located in between waves O_2 and O_3 in a first series of experiments; thus a series of anodic currents $i_a^1(\theta)$ is determined as a function of the pulse duration θ. In a second series of experiments the cathodic step is kept identical, but the anodic step is now located on the plateau of wave O_3; a second series of anodic currents $i_a^2(\theta)$ is then measured, allowing the ratio $R_e = i_a^2(\theta)/i_a^1(\theta)$ to be evaluated for each θ value. On the other hand since the two oxidations processes at wave O_2 and O_3 are monoelectronic, R_e is given by $R_e = 1.5 + 0.5[y/(1 - y)]$, where y is the yield in $Cr_2(CO)_{10}{}^{2-}$ during the cathodic process occurring at wave R_1, according to the following balanced equation:

$$Cr(CO)_6 + (2 - y)\,e \rightarrow (1 - y)\,Cr(CO)_5{}^{2-} + (y/2)\,Cr_2(CO)_{10}{}^{2-} + CO$$

The experimental value of R_e is determined to be 1.5, within the accuracy of measurements, for θ values ranging from ca 1 s to 1 ms. Thus the yield in $Cr_2(CO)_{10}^{2-}$ is negligible at wave R_1, which corresponds therefore to the overall two-electron process [49]:

$$Cr(CO)_6 + 2e \rightarrow Cr(CO)_5^{2-} + CO \quad (\text{wave } R_1) \tag{18}$$

The reader may find it surprising that this latter result had not been obtained by direct measurement of the number of electrons exchanged at wave R_1. However, as will be explained in a next section (3.5), determination of number of electrons exchanged is not easy for non-destructive electrochemical methods.

3.4 Steady State Electrochemical Methods

One of the greatest advantages of transient electrochemical methods is related to their high sensitivity to time-dependent phenomena. In this respect "steady state" electrochemical methods seem disadvantaged. However, when used in conjunction with transient methods, they may be of a considerable utility for the investigation of chemical reactivity, as will be shown in the following example, and at various instances in other sections of this chapter (see e.g. Sections 4.1 and 4.4.2). Their main advantage is that they give access to the composition of a solution at a given potential without depending on the "history" of the electrode. Thus they can be viewed as direct probes eliminating possible artefacts, which may arise from time dependence of transient methods. A simple but important application of this unique property is given in the following.

Let us assume that a solution contains ferrocene. Steady state voltammetry of the solution will then afford the sigmoidal voltammogram [50] in Fig. 8a. The latter displays two "horizontal" limits: one at zero current observed for potentials negative to the E^0 of ferrocene, the other, the wave "plateau", corresponding to an oxidation current is attained at potentials anodic to E^0. These limiting behaviors are separated by a rather sharp transition around E^0. Each of the three zones correspond to a given ferrocene concentration at the electrode surface. Thus for $E \ll E^0$, ferrocene is stable at the electrode surface and therefore no current flows.

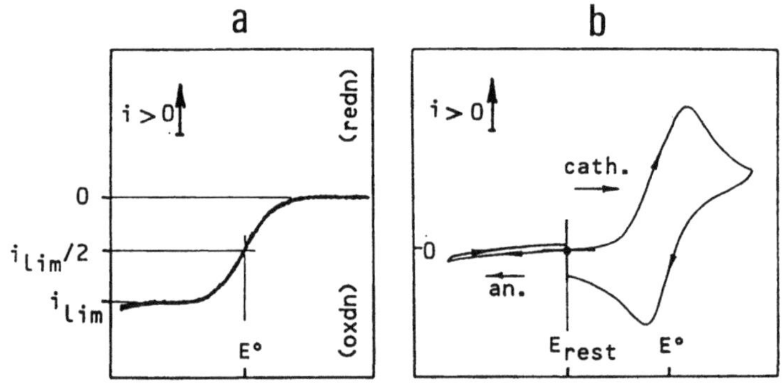

Fig. 8. (a) Steady state voltammogram of a 2 mM solution of ferrocene in THF. Platinum electrode, $\phi = 10$ µm; $v = 0.1$ V s^{-1}. (b) Cyclic voltammetry of the same solution, at a platinum electrode, $\phi = 125$ µm; $v = 10$ V s^{-1}, when the electrode rest potential is anodic to the ferrocene E^0 (see text); the two arrows indicate the direction of each individual scans; an.: anodic scan; cath.: cathodic scan.

For $E \gg E^0$, the ferrocene concentration at the electrode surface is negligible, when compared to that in the bulk solution; the difference between the two concentrations, i.e. that in the bulk and that at the electrode surface, no more depends significantly on the electrode potential (compare e.g. Fig. 6b), which results in nearly constant concentration profiles and therefore a constant current with E. When $E \approx E^0$, slight variations in potential result in drastic changes of the ferrocene concentration at the electrode surface and of its concentration profile; thus one observes rather sharp current variations with E.

Let us now examine what a transient electrochemical method, e.g. cyclic voltammetry, gives for the same system. When the rest potential of the electrode is cathodic to the E^0 of ferrocene, one obtains a classical chemically reversible cyclic voltammogram indicative of an oxidation process, as discussed above (Section 3.3.1, Fig. 5b). Yet when the rest potential of the electrode is anodic to E^0, and the initial scan made anodically no current is observed because when the electrode has been polarized for a sufficient time (i.e. larger than a few tens of seconds) at values anodic to E^0, all the ferrocene has been depleted in the diffusion layer and therefore none can be oxidized. If now, the first scan is made cathodically, still from the same rest potential anodic to E^0, a classical chemically reversible wave (Fig. 8b) is observed, which corresponds to reduction of the ferricinium ion accumulated in the diffusion layer. In the latter case a reduction voltammogram is thus observed although the bulk solution contains only ferrocene, an oxidizable species.

Obviously the simplicity of this example is such that nobody may ever believe that the reduction wave observed in the last situation pertains to the electrochemistry of ferrocene, because the simple thought that ferrocene may be reduced in this potential range is ridiculous. Yet consider now that the investigation concerned a compound that may undergo several reduction or oxidation processes. It would be almost impossible to decide whether the reduction wave observed is that of the starting material or of one of its oxidized forms! However, the answer is easily given by the steady state voltammogram of the same solution as explained above. Indeed the magnitude and sign of the current possibly observed at the selected rest potential of the electrode will easily indicate an electrochemical process and its nature (compare Fig. 8a). Thus a correct rest potential may be chosen by simply varying the potential so that the current is negligible. It is thus seen that steady state methods are considerably helpful in deciding a correct rest potential, i.e. a potential at which no electrochemical process takes place.

Other applications of steady state methods are given in other places of this chapter, therefore to conclude this section we want only to describe a few characteristics of the main "steady state" electrochemical methods. In order to obtain steady state current-potential curves it is necessary to impose a constant rate of mass transport to and from the electrode. Also mass transport must be fast with respect to the rate of perturbation occurring at the electrode. This is achieved by two different means: forced convection or forced diffusion.

Polarography and Rotating Disk Electrodes (RDE) voltammetries are based on forced convection [50]. This results in decreasing the thickness of the stagnant layer existing at the electrode surface to ca. 5 to 10 μm, and therefore in a considerable increase in the mass transfer rate to or from the electrode surface. Yet there are obvious difficulties in using these methods in controlled atmospheres, except may be in a glove box. The same difficulties exist also for polarography, yet with the additional one that the use of a mercury electrode may introduce severe perturbations in organometallic mechanisms. In our opinion the largest disadvantage of these methods is that they allow a rather small time scale to be investigated with the obligation of using specifically designed cells (for RDE) compatible with usual organometallic experimental conditions.

The recent development of ultramicroelectrodes [20] has allowed one to circumvent the above problems. At these electrodes steady state conditions are achieved by means of forced diffusion, i.e. cylindrical or spherical diffusion, without any contribution of convection. Thus, although their phenomenological behavior is very close to that observed at e.g. a RDE, it has different origins. Yet from a practical point of view it is sufficient to recall that the electrochemical significance of most of the characteristics (i.e. half-wave potentials, $E_{\frac{1}{2}}$, limiting current plateau, i_{lim}) of steady state voltammograms at ultramicroelectrodes is similar to those obtained e.g. at the RDE (see e.g. Eqn. 23 in Section 3.5).

3.5 Determination of Electron Stoichiometry in Non-Destructive Electrochemical Methods

As explained at various instances the current flowing through an electrode depends on the concentration gradients at the electrode surface (x being the distance from the electrode):

$$i = nFDA(\delta C/\delta x)_{x=0} \tag{19}$$

The proportionality constant between current and gradient involves the electrode surface area, A, the number, n, of electron(s) exchanged, the diffusion coefficient, D, of the investigated species, and the Faraday, F. In practice there is no problem in the determination of A within reasonable accuracy, yet n and D are a priori unknown and $(\delta C/\delta x)_{x=0}$ depends on the – a priori unknown – mechanism, and on the electrochemical method. It is thus easily understood that determination of the electron stoichiometry, i.e. of n, is not simple. However, variations of n are easily determined on relative scales since then the D coefficient in Eqn. 19 is eliminated. For example a variation by two is measured with large precision (compare e.g. Fig. 12e in Section 4.2) yet one can not determine from such variations whether n varies from 1 to 2 or e.g. from 2 to 4, without measuring separately at least one of the limiting values.

One could think of by-passing the problem by resorting to preparative scale electrolysis, since n is then directly proportional to the charge consumed, Q, which is easily determined. Indeed one has:

$$Q = nFC^0V \tag{20}$$

where C^0 is the concentration of the material electrolyzed and V the volume of the solution. However, this very frequent approach has no validity, except when one is sure that no mechanistic change occurs between the two different time scales [32, 51]. This is often not true in practice. For instance, the reduction of hexacarbonylchromium at short time scale, i.e. under a min, involves two electrons and affords the dianion $Cr(CO)_5^{2-}$, as explained above [49]. However, preparative electrolysis results in a one-electron [52] consumption and yields the dimer dianion $Cr_2(CO)_{10}^{2-}$, because of the simultaneous reaction [49]:

$$Cr(CO)_5^{2-} + Cr(CO)_6 \rightarrow Cr_2(CO)_{10}^{2-} + CO \tag{21}$$

taking place in the bulk solution during the half-an-hour duration of an electrolysis.

The above example emphazises the necessity of determining n on the same time scale where the mechanistic analysis is performed. This can be achieved by comparing the results obtained in a transient electrochemical method and in steady state methods as explained hereafter.

In chronoamperometry the current at time θ since the beginning of the pulse is given by [53]:

$$i_{\text{chrono.}} = nFAC^0\, D^{1/2}/(\pi\theta)^{1/2} \tag{22}$$

whereas the limiting plateau of a steady state voltammogram is [54]:

$$i_{\text{lim}} = nFAC^0\, D/\delta \tag{23}$$

where δ is the diffusion layer thickness. For a rotating disk electrode of angular velocity w (in rotations per second, i.e. in s^{-1}), $\delta = 1.61\, D^{1/3} W^{-1/2} \nu^{1/6}$, ν being the kinematic viscosity of the solution [54]; for a disc ultramicroelectrode of radius r_0, $\delta = 0.25\pi r_0$ [20].

Thus the ratio of Eqns. 22 and 23, eliminates the unknown n value allowing then the independent determination of D. Introduction of the latter in either equation yields the desired n value. Straightforward application of the method presents some difficulties; one prefers to use the relay of a "reference" compound. The latter is generally chosen to be a reversible compound whose n value is known, such as e.g. ferrocene ($|n| = 1$). Determination of the quantities in Eqns. 22 and 23, corrected for the concentration ratio, for the investigated compound, and e.g. ferrocene allows the following ratio to be determined:

$$R_{\text{chrono}} = (i_{\text{chrono}}/i_{\text{chrono}}^{\text{Ferrocene}})\,(C^0_{\text{Ferrocene}}/C^0) = n(D/D^{\text{Ferrocene}})^{1/2} \tag{24}$$

and for e.g. a disc ultramicroelectrode:

$$R_{\text{lim}} = (i_{\text{lim}}/i_{\text{lim}}^{\text{Ferrocene}})\,(C^0_{\text{Ferrocene}}/C^0) = n(D/D^{\text{Ferrocene}}) \tag{25}$$

from which n and D are easily determined:

$$n = R^2_{\text{chron}}/R_{\text{lim}} \tag{26}$$

$$D = D^{\text{Ferrocene}}(R_{\text{lim}}/R_{\text{chrono}})^2 \tag{27}$$

The currents used in the determination of the ratios must correspond to the same time-scale, i.e. as explained above to the same mass transport rate. This is achieved by using couples of data such as $\theta \approx (D/\delta^2)$. In this respect ultramicroelectrodes are particularly adequate since they allow the observation of steady state voltammograms even at rather large mass transport rates.

A similar approach can be derived for cyclic voltammetry instead of chronoamperometry as the transient electrochemical method. However, it suffers because the current peak value in cyclic voltammetry is given by [8]:

$$i_p = Bn^{\beta}FAC^0(DFv/RT)^{1/2} \tag{28}$$

where the two constants B and β depend on the – a priori unknown – mechanism followed. When the mechanism can be determined independently, e.g. from peak potential analysis, the method becomes equivalent to that described for chronoamperometry.

For example, application of the method to the reduction of the phenylnickel(II) complex in Eqn. 29 establishes that in transient methods, the latter is reduced by a one-electron process to afford a phenylnickel(I) species [25], in contradiction with previous reports of a two-electron consumption [55] based on preparative scale experiments (see e.g. Section 4.4.2 for the mechanistic importance of such data). [Note that "dppe" stands for 1,2-bisdiphenylphosphinoethane].

$$Ph{-}Ni(II)Br(dppe) + e \rightarrow Ph{-}Ni(I)(dppe) + Br^- \qquad (29)$$

4. Selected Examples of Electrochemical Approaches to Organometallic Reactivity

In the following section we would like to show how electrochemical approaches may be effective in solving a variety of problems in organometallic chemistry. Although several related studies can be found in the literature, we decided to restrict this presentation to some selected examples investigated in our research group in Paris. Each of these examples is illustrative of a specific approach to a particular problem. However, their succession shows how electrochemistry may afford quantitative descriptions of mechanisms of increasing complexity. Thus the first study, oxidative addition to zerovalent transition metals, deals with an "elementary" reaction, whereas the second one shows how the mechanism of an important reaction involving several steps can be established. The third is related to electron transfer catalysis of a non-redox chemical reaction. The last one describes the role of transient organometallic intermediates in the catalysis of an organic reaction.

4.1 Rates of Oxidative Addition to Zerovalent Nickel or Palladium Complexes

Oxidative addition to zerovalent transition metals such as nickel or palladium is thought to proceed via coordinatively unsaturated species, M^0L_2, although such intermediates are too short lived for their reaction with organic halides to be observed directly [56]. Thus all the compiled data on the reaction relate to the overall process in reaction 30, which involves two rapid pre-equilibria:

$$M^0L_4 \rightleftharpoons L + M^0L_3 \rightleftharpoons 2L + M^0L_2 \xrightarrow{RX} R\text{-}M^{II}L_2X \qquad (30)$$

For M = Ni or Pd the first equilibrium is totally displaced to the right in most solvents of interest. Although it has been established that the second dissociation is considerably endergonic, its equilibrium constant, K_2, has been impossible to measure. Thus kinetic investigations of the overall process in Eqn. 30 led to the determination of apparent rate constant, $k^{ap} = kK_2$, where k is the real date constant of the oxidative addition elementary step. This did not hamper the study of effects arising from the organic part [56], since K_2 is independent of variations in the latter. Studies of steric or electronic effects of ligands were precluded since a ligand change should affect both k and K_2 to different extents.

On the other hand electrochemistry affords an easy way to generate low ligated species such as M^0L_2, by reduction of the corresponding divalent salt $M^{II}L_2X_2$ [25, 57]. For example Fig. 9 shows that $Ni^{II}Cl_2$(dppe) is reduced in two successive irreversible one-electron steps to afford a zerovalent low-ligated nickel complex:

$$Ni^{II}Cl_2(dppe) + e \rightarrow Ni^{I}Cl(dppe) + Cl^- \quad (\text{Wave } R_1) \qquad (31)$$

$$Ni^{I}Cl(dppe) + e \rightarrow Ni^{0}(dppe) + Cl^- \quad (\text{Wave } R_2) \qquad (32)$$

In the presence of added bromobenzene a third wave (R_3) develops in a potential range negative to wave R_2 (Fig. 10a), which features the one-electron reduction of the in-situ generated phenylnickel(II) complex:

$$Ni^0(dppe) + PhBr \xrightarrow{k} Ph{-}Ni^{II}Br(dppe) \quad (33)$$

$$Ph{-}Ni^{II}Br(dppe) + e \longrightarrow Ph{-}Ni^{I}(dppe) + Br^- \quad (\text{Wave } R_3) \quad (34)$$

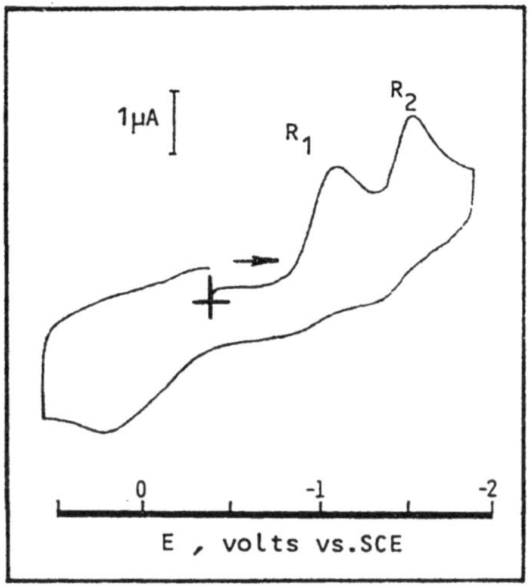

Fig. 9. Cyclic voltammetry of the reduction of a 2 mM solution of $NiCl_2$(dppe) in THF/HMPA (2/1 vols.) at 500 V s^{-1}. Gold electrode (ϕ: 125 µm). The cross indicates the rest potential of the electrode, and the arrow the direction of the initial scan.

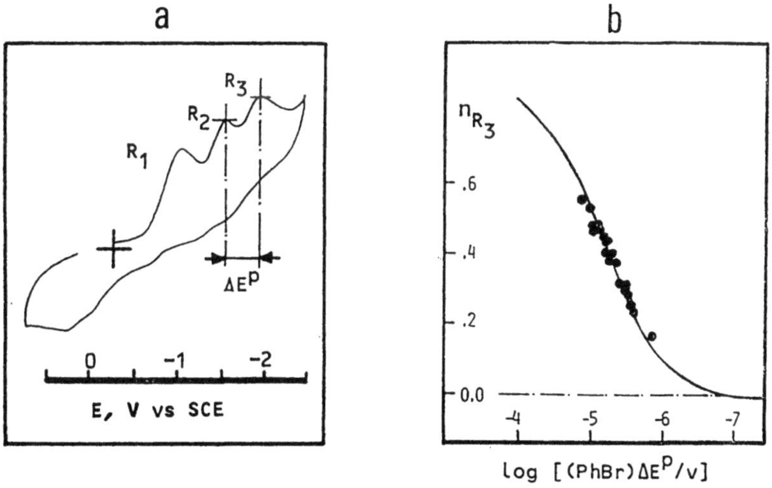

Fig. 10. (a) Cyclic voltammetry of a 2 mM $NiCl_2$(dppe), 10 mM PhBr solution in THF/HMPA (2/1 vols.) at 50 V s^{-1}, indicating the monoelectronic reduction wave (R_3) of the *insitu* electrogenerated phenylnickel(II) complex. (b) Variation of the apparent number of electron, n_{R_3}, involved at wave R_3, as a function of the scan rate, v, and the bromobenzene concentration. See (a) for definition of ΔE_p. Solid line: theoretical variation for the reaction sequence in Eqns. 31–34, with $k = 1.1\ 10^{5}\ M^{-1}\ s^{-1}$.

as evidenced by comparison to an authentic sample. The rate constant of the reaction of the electrogenerated zerovalent nickel with PhBr (Eqn. 33) can then be derived from the magnitude of wave R_3. Indeed for a given scan rate, v, and a given PhBr concentration the current peak of wave R_3, relative to that of e.g. the one-electron wave R_1, is a direct measure of the fraction of the electrogenerated Ni^0(dppe) which has reacted in Eqn. 33. Comparison to the working curve in Fig. 10b, allowed then the determination of a rate constant $k = 10^5\ M^{-1}\ s^{-1}$ for the oxidative addition [25].

Such a rate constant magnitude corresponds to a lifetime ca. 50,000 times smaller for the lowligated Ni^0(dppe) than for the saturated analog $Ni^0(PEt_3)_4$ [56a] and about 5,000,000 times smaller than for the more closely related $Ni^0(PPh_3)_4$ [58]; no reaction was observed on our time-scale with the saturated $Ni^0(dppe)_2$. These dramatic effects emphasize the considerable enhancement in the reactivity of lowligated zerovalent transition metals vis a vis their stable chemical analogs M^0L_4. Although this study was the first one to quantify this effect, the results obtained are in good agreement with previous estimations; for example Kochi et al. [56] evaluated K_2 to be smaller than 10^{-6} M for $Ni^0(PEt_3)_4$.

The above example is a very representative illustration of the great advantages offered by electrochemistry, which allows one to generate highly reactive intermediates (potential) and study simultaneously (current) their chemistry. The following shows how it allows the intimate nature of an "elementary" step to be discussed.

Oxidative addition of organic halides to zerovalent transition metals is usually considered as a true elementary, step, i.e. as a reaction with a single transition state. Several years ago it was proposed that it may proceed via the SET (Single Electron Transfer) mechanism outlined in the following sequence [56a]:

$$M^0L_2 + ArX \rightarrow Ar\text{-}M^{II}L_2X$$

$$\downarrow \qquad\qquad\qquad \uparrow$$

$$[M^IL_2, ArX^{\cdot -}] \rightarrow [M^IL_2, Ar^{\cdot}, X^-]$$

The difficulty in discussing the real nature of the reaction arises from the fact that the rate laws for both routes are identical. Moreover the electronic factors, which govern the reaction kinetics, are expected to result in the same trends for each pathway. For example, an increase of the electron affinity of the organic halide or of the electron density at the metal center should increase the overall rate of either route. Another approach to the problem is based on the consideration that the charge separation is totally different for the two pathways. The direct reaction should proceed through a transition state with minimal zwitterionic character, whereas the SET mechanism involves an important charge separation. As a result changes in the polar properties of the solvent should have almost no effect on the direct route but greatly influence the SET one.

We thus decided to investigate the kinetics of oxidative addition of a series of organic halides to $Pd^0(PPh_3)_4$, in toluene ($\varepsilon \approx 2.4$) and to compare it to that previously observed in THF ($\varepsilon \approx 7.6$). Taking advantage of the unique properties of ultramicroelectrodes, the kinetics of the reaction was followed by monitoring the current plateau of the oxidation wave of $Pd^0(PPh_3)_4$. The perfectly defined voltammograms, such as those presented in Fig. 11a, allowed the determination of the overall rate constant of the sequence in Eqns. 35, 36:

$$Pd^0(PPh_3)_3 \rightleftharpoons Pd^0(PPh_3)_2 + PPh_3 \tag{35}$$

$$Pd^0(PPh_3)_2 + ArX \rightarrow Ar\text{-}Pd^{II}(PPh_3)_2X \tag{36}$$

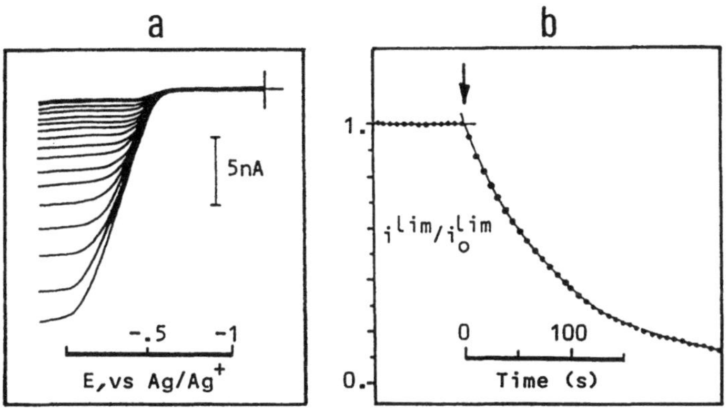

Fig. 11. (a) Steady state voltammograms, as a function of time, of a 1.5 mM solution of $Pd^0(PPh_3)_4$ in toluene (0.06 M $NHex_4BF_4$) in the presence of iodobenzene (10 mM) and PPh_3 (50 mM). Potentials are referred to the $Ag/AgBF_4$ 0.01 M in THF. Platinum disk electrode (ϕ = 10 μm); scan rate 0.1 V s^{-1}. (b) Variation of the limiting current, i^{lim}, on the voltammograms plateau as a function of time. i_0^{lim}: limiting current before introduction of iodobenzene. The arrow indicates the time (taken as the origin of time) at which iodobenzene is introduced in the $Pd^0(PPh_3)_4$ solution. The solid line in (b) corresponds to the theoretical kinetics for the reaction sequence in Eqns. 35–36.

Essentially identical rates were determined in both solvents; also identical Hammet's slopes ($\varrho_Y^{THF} = +2.0$; $\varrho_Y^{Toluene} = +2.3 \pm 0.2$) were obtained for a series of Y-PhX as the organic halides. The identity was furthermore confirmed by that of the activation parameters of the reaction with iodobenzene [23f]: $\Delta H^{\neq}_{toluene} =$ 18 kcal mol^{-1}; $\Delta S^{\neq}_{toluene} = 2$ cal $mol^{-1} \cdot K^{-1}$; $\Delta H^{\neq}_{THF} = 18.5$ kcal mol^{-1}; $\Delta S^{\neq}_{THF} = 3$ cal $mol^{-1} \cdot K^{-1}$. These rates and activation parameters incorporate thermodynamic effects arising from the rapid preequilibrium in Eqn. 35, yet their striking identity strongly favors the direct pathway. Indeed for the SET mechanism, the variation by a factor three of ε between the two solvents should result in a ca. threefold increase for the electrostatic contribution to $\Delta H^{\neq}$, when passing from THF to toluene if identical geometries are assumed for the ion-pairs in both solvents.

4.2 Mechanism of Collman's Reagent Formation from Ironpentacarbonyl Reduction

There is general agreement that reduction of ironpentacarbonyl is chemically irreversible and that it is followed by rapid chemical reactions [59]. The intermediacy of the 19-electron $Fe(CO)_5^{\cdot -}$ anion radical formed upon electron transfer has been demonstrated by its interception by H-atom transfer from $(n\text{-}Bu)_3SnH$ to afford the corresponding formyl anion $(CO)_4Fe{-}CHO^-$ [59f]:

$$Fe(CO)_5 + e \rightarrow Fe(CO)_5^{\cdot -} \xrightarrow{(n\text{-}Bu)_3SnH} (CO)_4Fe{-}CHO^- \quad (37)$$

In the absence of an H-atom donor different reaction products have been observed in transient electrochemical methods. In aprotic media preparative electrolysis was shown to involve an overall one-electron consumption and the major product was the $Fe_2(CO)_8^{2-}$ carbonylate dianion. It has been implicitly or explicitly [59] assumed that this product arises from dimerization of the transient 17-electron $Fe(CO)_4^{\cdot -}$ formed by CO loss from $Fe(CO)_5^{\cdot -}$:

$$Fe(CO)_5^{\cdot -} \xrightarrow{-CO} Fe(CO)_4^{\cdot -} \rightarrow \tfrac{1}{2} Fe_2(CO)_8^{2-} \quad (38)$$

In the presence of acetone [59c] or in the presence of a proton donor [59e], the reduction is shown to involve an overall two-electron process. In the latter case this was tentatively ascribed to fast protonation of the $Fe(CO)_4^{\cdot -}$ anion radical followed by a second transfer to afford finally the hydride, which was identified electrochemically and by IR spectroscopy [59e]:

$$Fe(CO)_4^{\cdot -} \xrightarrow{H^+} HFe(CO)_4^{\cdot} \xrightarrow{+e} HFe(CO)_4^{-} \quad (39)$$

Furthermore, in synthetic work using sodium naphthalenide as a one-electron reducing agent, it was found that the reduction of $Fe(CO)_5$ by one equivalent of this reagent yielded $Fe_2(CO)_8^{2-}$, whereas two equivalents afforded Collman's reagent $Fe(CO)_4^{2-}$ [60].

One could be tempted to rationalize most of these apparently conflicting results by considering that $Fe(CO)_4^{\cdot -}$ evolves either by a fast dimerization or by a fast protonation, according to the proticity of the reaction medium, as outlined in Scheme I [59e, f].

Scheme I

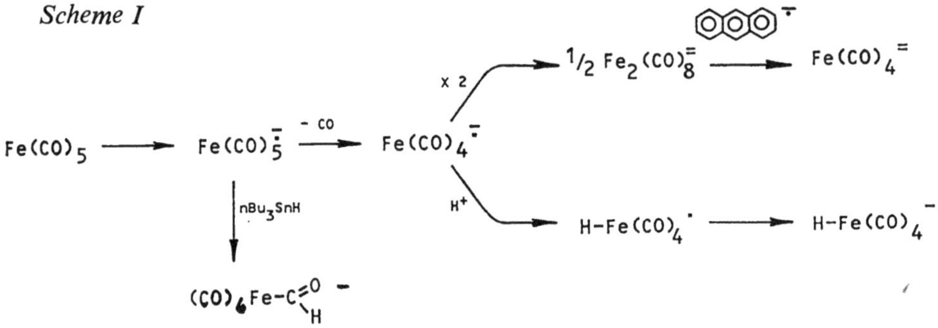

Scheme I considers that Collman's reagent is never formed at the reduction potential of $Fe(CO)_5$. Its formation in the sodium naphthalenide reduction would arise from further reduction of the $Fe_2(CO)_8^{2-}$ carbonylate dianion, made possible by the strong reducing character of this reagent. Besides purely mechanistic considerations we were interested in a complete understanding of the chemistry initiated by one-electron transfer to $Fe(CO)_5$, since Scheme I bears important synthetic consequences. For example, according to its formulation one is led to the conclusion that formation of Collman's reagent requires strong reducing agents, and as matter of fact it cannot be electrogenerated.

Two cyclic voltammograms for the reduction of $Fe(CO)_5$ in THF are presented in Figs. 12a, b [24d]. At a scan rate of 5 V s^{-1} (Fig. 12a) the results are essentially equivalent to those reported previously [59]: an irreversible reduction wave R_1, involving the overall exchange of 1.1 electron, is followed by three associated anodic peaks O_2, O_3, and O_4. Wave O_3 is associated with a reduction wave R_3 observed on scan reversal. At faster scan rates the O_2 wave increases at the expense of O_3 and O_4, and starts showing chemical reversibility at 20 V s^{-1} (Fig. 12b) Concomitantly the apparent number of electrons exchanged at wave R_1 (Fig. 12e) increases to approach the two-electron limit at scan rates above 1000 V s^{-1}. The

O_3/R_3 couple and the chemically irreversible wave O_4 correspond to the $Fe_2(CO)_8^{2-}$ electrochemical oxidation and feature the slow one-electron oxidation in Eqn. 40 (see Section 3.3.1) followed by the irreversible one-electron oxidation of the stable anion radical $Fe_2(CO)_8^{\cdot-}$, in Eqn. 41 (Fig. 12c):

$$Fe_2(CO)_8^{2-} - e \rightleftharpoons Fe_2(CO)_8^{\cdot-} \quad \text{(Wave } O_3\text{)} \qquad (40)$$

$$Fe_2(CO)_8^{\cdot-} - e \rightleftharpoons Fe_2(CO)_8 \rightarrow \text{etc.} \quad \text{(Wave } O_4\text{)} \qquad (41)$$

$$Fe(CO)_4^{2-} - e \rightleftharpoons Fe(CO)_4^{\cdot-} \quad \text{(Wave } O_2\text{)} \qquad (42)$$

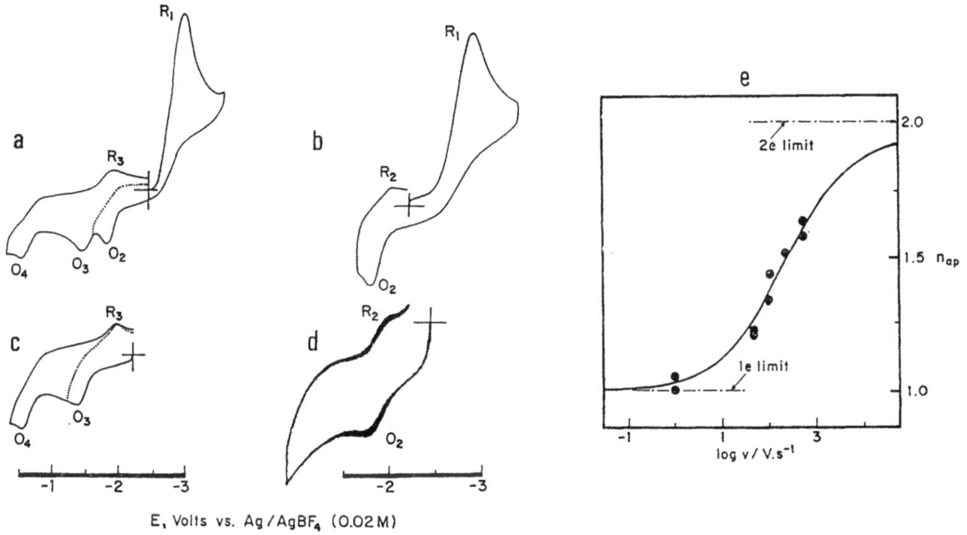

Fig. 12. Cyclic voltammetry of the reduction of a 1 mM solution of $Fe(CO)_5$ in THF (0.3 NBu_4BF_4) at $v = 5$ (a) or 20 (b) $V\,s^{-1}$, at a gold electrode ($\phi = 0.8$ mm). (c) Cyclic voltammetry of the oxidation of a 1 mM solution of $Na_2Fe_2(CO)_8$ in THF (0.3 M NBu_4BF_4) at $v = 5\ V\,s^{-1}$, at a gold electrode ($\phi = 0.8$ mm). (d) Cyclic voltammetry of the oxidation of a 1 mM solution of $Na_2Fe(CO)_4$ in THF/HMPA (95/5 vols) at 20 $V\,s^{-1}$, without supporting electrolyte; gold electrode ($\phi = 125$ μm), $v = 20\ V\,s^{-1}$. (e) Variations of the apparent number of electron, n_{ap}, exchanged at wave R_1 as a function of the scan rate v; the solid line corresponds to the theoretical variations predicted for the reaction sequence in Eqns. 44–47, with $k = 6.0 \times 10^6\ M^{-1}\ s^{-1}$. (Note that the current scales in Figs. a–d, are not respected in order to facilitate the comparison between the four voltammograms. Potentials are referred to the $Ag/AgBF_4$ (0.02 M) in THF).

The O_2/R_2 couple pertains to the one-electron oxidation of $Fe(CO)_4^{2-}$ to the transient radical anion $Fe(CO)_4^{\cdot-}$, in Eqn. 42, as is demonstrated by the independent cyclic voltammogram of authentic $Na_2Fe(CO)_4$ in Fig. 12d. Note that owing to the instability of the latter in usual electrochemical media, particularly toward tetraalkylammonium cations, its cyclic voltammetry was performed in the absence of supporting electrolyte in 95:5 v/v THF/HMPA, at a disc electrode of 62.5 μm radius. At scan rates under 20 $V\,s^{-1}$ wave O_2 becomes chemically irreversible but remains mono-electronic, which features the facile – although not extremely, fast, $k \approx 10^5\ M^{-1}\ s^{-1}$ – dimerization of $Fe(CO)_4^{\cdot-}$ in Eqn. 43:

$$2Fe(CO)_4^{\cdot-} \rightarrow Fe_2(CO)_8^{2-} \qquad (43)$$

No other mono or polynuclear iron carbonylates are formed on the time scale of our experiments as judged by a comparison of authentic samples of $HFe(CO)_4^-$, $Fe_3(CO)_{11}^{2-/\cdot-}$ and $Fe_4(CO)_{13}^{2-/\cdot-}$ [24d].

The above results indicate that at sufficiently fast scan rates, i.e. in the microsecond or less time-scale, the electrochemical reduction of $Fe(CO)_5$ in THF is an overall two-electron ECE process leading to the formation of Collman's reagent $Fe(CO)_4^{2-}$:

$$Fe(CO)_5 + e \rightleftharpoons Fe(CO)_5^{\cdot-} \quad (44)$$

$$Fe(CO)_5^{\cdot-} \rightarrow Fe(CO)_4^{\cdot-} + CO \quad (45)$$

$$Fe(CO)_4^{\cdot-} + e \rightleftharpoons Fe(CO)_4^{2-} \quad (46)$$

The second electron transfer in Eqn. 46 is observed because the reduction potential of $Fe(CO)_5$ is much more negative (≈ -3.0 V) than the standard potential (-1.95 V) for the couple $Fe(CO)_4^{2-}/Fe(CO)_4^{\cdot-}$ [61]. Thus $Fe(CO)_4^{\cdot-}$, formed at a potential much more negative than its E^0, must necessarily be reduced to $Fe(CO)_4^{2-}$. Under such conditions even if the dimerization rate constant (Eqn. 43) were diffusion controlled, $Fe(CO)_4^{\cdot-}$ anion radicals cannot dimerize [62]. The apparent one-electron consumption observed at slower scan rates, which results in formation of $Fe_2(CO)_8^{2-}$ as the main product, must therefore result from further chemical reaction of $Fe(CO)_4^{2-}$. To the best of our knowledge the only plausible explanation is that $Fe(CO)_4^{2-}$ reacts with unreduced $Fe(CO)_5$. This is indeed a convenient synthetic route to the $Fe_2(CO)_8^{2-}$ dianion [63]:

$$Fe(CO)_4^{2-} + Fe(CO)_5 \xrightarrow{k} Fe_2(CO)_8^{2-} + CO \quad (47)$$

In order to verify this hypothesis we compared in Fig. 12e the experimental variations with the scan rate of the apparent number of electrons, n_{ap}, involved in $Fe(CO)_5$ voltammograms with the theoretical predictions based on the sequence of reactions 44–47. The very good fit obtained further supports the involvement of reaction 47, as the source for $Fe_2(CO)_8^{2-}$, and allows the determination of its rate constant to $(6.0 \pm 2)\ 10^6\ M^{-1}\ s^{-1}$ [24d].

Beyond its quantitative agreement with our own data, the mechanism in reactions 44–47 reconciles all the apparently conflicting results previously reported. Thus the $Fe(CO)_5^{\cdot-}$ electrogenerated in Eqn. 44 may be trapped by an H-atom donor reagent to afford the formyl derivative (Eqn. 37), as soon as the effective rate of H-atom transfer exceeds that of CO loss. In the absence of the H-atom donor, the $Fe(CO)_4^{\cdot-}$ anion radical formed in Eqn. 45 is quantitatively reduced to $Fe(CO)_4^{2-}$ (Eqn. 46). In the presence of a proton donor the latter, a powerful base, is protonated (Eqn. 48) to yield the hydride $H\text{-}Fe(CO)_4^-$, thus leading to an overall two-electron consumption through an ECEC sequence.

$$Fe(CO)_4^{2-} + BH \rightarrow H{-}Fe(CO)_4^- + B^- \quad (48)$$

Besides mechanistic consequences the above study shows that electrochemistry is a convenient method for the in-situ generation of Collman's reagent and that it might be used for synthesis with reactions that can compete with reaction 47.

4.3 Electron Transfer Catalysis of Dithiocarbamate Chelation in Cyclopentadienylironcarbonyl Complexes

4.3.1 General Considerations on Electron Transfer Catalysis

The principle of electron transfer catalysis of a non-redox reaction is outlined in Scheme II, where $n = +1$ for reductive activations, and $n = -1$ for oxidative ones [9b, 17, 64]. Thus, an exergonic but slow chemical reaction may be considerably accelerated, when carried at the level of the reduced or oxidized materials.

Scheme II

$$\begin{array}{ccc} \text{Reactant} & \xrightarrow{\text{slow}} & \text{Product} \\ +\text{ne} \downarrow\uparrow & & \downarrow\uparrow -\text{ne} \\ \text{Reactant}^{n-} & \xrightarrow{\text{fast}} & \text{Product}^{n-} \end{array}$$

Most of the reported examples, starting from the original observations by Taube [65] in the inorganic field or Kornblum and Russel [66] in organic chemistry, involve single electron transfer activation, yet a few examples which involve overall two-electron activation steps have also been reported [67].

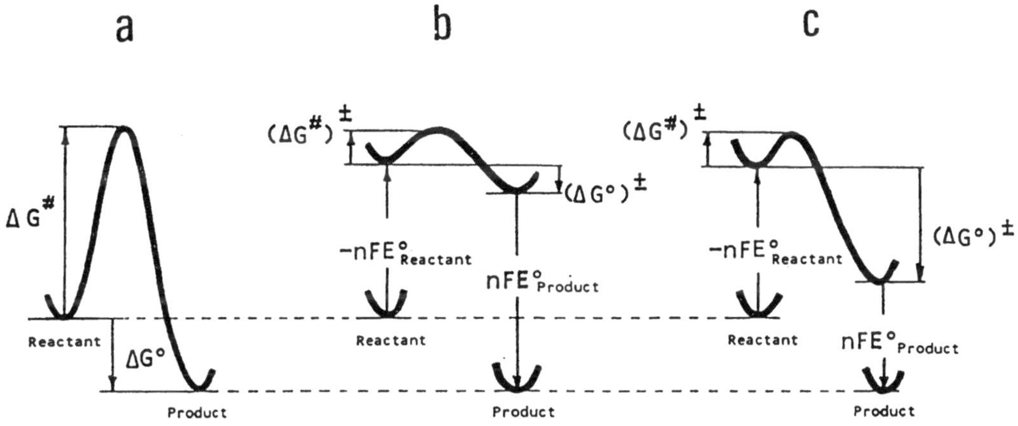

Fig. 13. Schematic representation of the activation barrier for a reaction with single transition state. (a) Reaction between diamagnetic reactant and product ($\Delta G^{\#}$: activation free enthalpy; ΔG^0: free energy of the reaction). (b, c) Electron transfer catalysis of the reaction in (a). $(\Delta G^{\#})^{\pm}$: activation free enthalpy between the paramagnetic intermediates; $(\Delta G^0)^{\pm}$: free energy of the reaction between the paramagnetic intermediates. $[nF(E^0_{\text{Reactant}} - E^0_{\text{Product}})] < 0$ (b), or >0 (c).

Although not established on solid grounds, the reason for the dramatic increase in rates is certainly related to lower activation barriers in the reduced or oxidized forms, as sketched in Fig. 13, and not to thermodynamic considerations. For example the oxidatively catalyzed ligand substitution in Eqn. 49 (where Py = pyridine) [17a, b]:

$$(\text{Me—Cp})\text{Mn(CO)}_2\text{NCMe} + \text{Py} \rightarrow (\text{Me—Cp})\text{Mn(CO)}_2\text{Py} + \text{MeCN} \quad (49)$$

involves a $\Delta G^0 = 2.6$ kcal mol^{-1} between the neutral complexes and $(\Delta G^0)^+ = 2.8$ kcal mol^{-1} between the 17-electron paramagnetic cations:

$$(Me{-}Cp)Mn(CO)_2NCMe^{\cdot +} + Py \rightarrow (Me{-}Cp)Mn(CO)_2Py^{\cdot +} + MeCN \qquad (50)$$

Thus the thermodynamical benefit, if any, is negative and yet the reaction rate is increased dramatically (compare Figs. 13b and c) and associated with large coulombic efficiencies [17].

Note in this respect that electrochemistry provides a straightforward method to evaluate these thermodynamical variations [17]. Let ΔG^0 and $(\Delta G^0)^{\pm}$ be the respective free enthalpies of the reaction between the stable forms or between the electron transfer activated ones. Thus from the thermodynamic cycle in Scheme II one obtains:

$$(\Delta G^0)^{\pm} = \Delta G^0 + nF(E^0_{Reactant} - E^0_{Product}) \qquad (51)$$

where again $n > 0$ for a reductive ($n < 0$ for an oxidative) activation. The two pertinent E^0 may be generally easily determined from e.g. cyclic voltammetry. Interestingly Eqn. 51 predicts that when the product is more difficulty electrolyzed than the reactant, i.e. when $[nF(E^0_{Reactant} - E^0_{Product})] > 0$, the reaction between the paramagnetic species is disadvantaged on thermodynamical grounds, while it is favored from an operative point of view. Indeed as examplified, e.g., for an oxida-

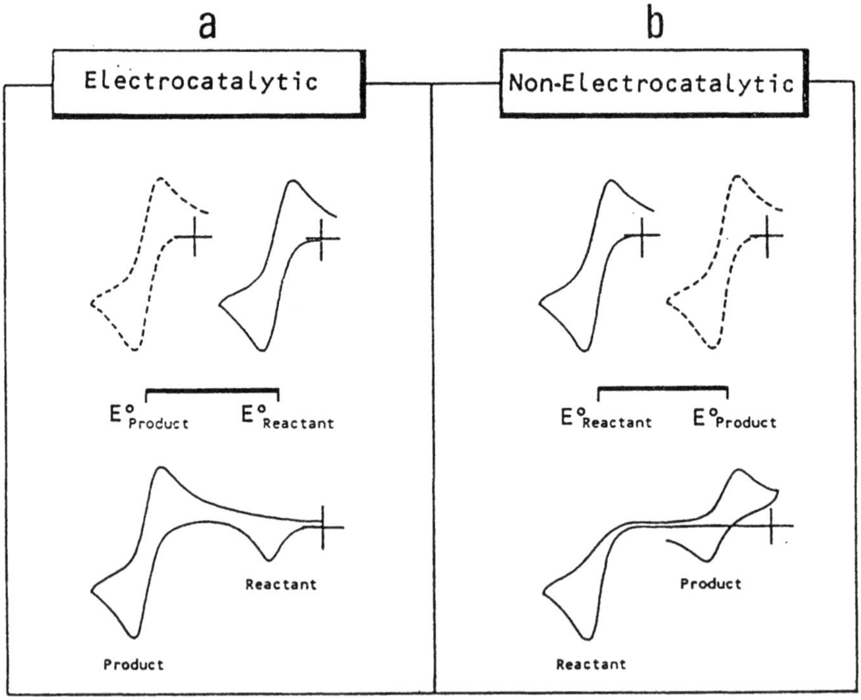

Fig. 14. Electron transfer catalysis initiated by oxidation ($n = -1$), for an electrocatalytic ($[nF(E^0_{Reactant} - E^0_{Product})] > 0$) or a non-electrocatalytic ($[nF(E^0_{Reactant} - E^0_{Product})] < 0$) system in cyclic voltammetry (lower part of each box). In the upper part of each box are featured the cyclic voltammograms of the reactant (—) and the product (----) in the absence of electron transfer catalysis.

tive ($n = -1$) activation in Fig. 14a, the condition $[nF(E^0_{\text{Reactant}} - E^0_{\text{Product}})] > 0$ results in the product being stable under its non-oxidized form, when the electrode is polarized at the level of the reactant oxidation wave. Thus, when the electrode potential is set on the reactant wave, the product oxidized form must necessarily reduce as soon as it is formed. This consumes an infinitely small charge, which is the reason why the process is termed "electrocatalytic". Conversely when $[nF(E^0_{\text{Reactant}} - E^0_{\text{Product}})] < 0$, the product is obtained in its oxidized form (Fig. 14b) and needs to be reduced independently; then the process is termed "non-electrocatalytic" and should involve a stoichiometric electron consumption. However, on thermodynamical grounds, it is favored by promotion to the oxidized state (Eqn. 51).

The above statement concerning the "electrocatalytic" or "non-electrocatalytic" nature of the systems must be taken with great caution. Indeed it is based on thermodynamic considerations, i.e. on the respective stabilities of the product or its reduced or oxidized form, Product$^{\pm}$, vis a vis the electrode potential. However the involvement of the homogeneous electron transfer step in Eqn. 52 may alter drastically these conclusions.

$$\text{Product}^{\pm} + \text{Reactant} \rightleftharpoons \text{Product} + \text{Reactant}^{\pm} \rightarrow \ldots \qquad (52)$$

Although this step is necessarily endergonic for "non-electrocatalytic" systems, its rate and equilibrium constants are not exceedingly small, except for extremely large differences between the product and reactant E^0's. Thus, the irreversible chemical reaction of the activated reactant may continuously displace Eqn. 52 to the right-hand-side. This effects the overall reaction "electrocatalytically" in the solution, although it is performed "non-electrocatalytically" at the electrode. Obviously the homogeneous process exists also for the "electrocatalytic" systems since then Eqn. 52 is thermodynamically favored. However, in such cases it will be hampered by the electrode, which now continuously destroys the chain carrier, i.e. Product$^{\pm}$.

The opposite trends of the homogeneous and heterogeneous processes in the above discussion may seem rather confusing. Yet, let us consider the electrode reactions as two coupled "activation / termination" processes, and those occurring in the solution as a "propagation" sequence. Thus, as for any catalytic chain reaction, the efficiency of the catalysis in terms of turn-over depends on the respective importance of the two processes in the conversion of the reactant [68].

When $[nF(E^0_{\text{Reactant}} - E^0_{\text{Product}})] > 0$, the homogeneous propagation is expected to be fast, but also the termination reaction. In practice this is not a serious limitation, except in the presence of competitive reactions, since both propagation and activation / termination are electrocatalytic.

When $[nF(E^0_{\text{Reactant}} - E^0_{\text{Product}})] < 0$, the homogeneous propagation is expected to be slower but may nevertheless convert the reactant in a reasonable time period. On the other hand, in the absence of other competitive routes, there is no termination step. Thus, the turn-over number in terms of coulombic efficiency depends on the rate of electrolysis (i.e. the activation step, which is electron consuming) vis a vis that of the homogeneous propagation, which involves no electron consumption. In other words for a given value of the latter, and a given set-up a decrease in the current density at the electrode results in a decrease of the electrolysis rate and therefore in an increase of the coulombic efficiency or catalytic turnover [17].

However, if additional termination steps, e.g. other chemical routes branched at different stages of the propagating chain, are involved these conclusions may be altered [68, 69]. Indeed such competitive reactions destroy part of the chain carriers; the delicate balance between activation and propagation steps should be reconsidered as a function of the additional termination steps. Let assume for example that the additional termination is first-order in chain carrier. Then an increase of the activation rate will compensate the loss in chain carrier and the homogeneous propagation will proceed as in the absence of the additional termination step. Conversely, when the latter is second-order in chain carrier, an increase of the activation rate tends to increase the chain carriers concentration, which favors the termination step at the expense of propagation [69].

From what preceeds it is understood that although electron transfer catalysis is extremely appealing, its optimization generally requires a complete understanding of all the processes involved, i.e. of the activation / propagation / termination sequence, and not only of the propagation one. In this respect electrochemical methods, which allow a perfect control of the activation rate (potential) and afford a direct measure of the overall conversion rate (current), seem particularly suited to the study of electron transfer catalytic processes. It is no wonder that most of the quantitative mechanistic work in the area has been produced on the basis of electrochemical approaches [17, 64, 70, 71].

4.3.2 Chelation of a Dithiocarbamate (dtc) Ligand

Most of the electrocatalytic processes reported up to now in the organometallic field involve ligand substitution in mono or polynuclear complexes [64e, 71]. Indeed thermal ligand exchange from 18-electron complexes are generally slow, which is attributed to the large activation energies associated with 16-electron diamagnetic intermediates involved in dissociative mechanisms, or with 20-electron configurations for transition states in associative mechanisms. On the other hand, when conducted at the level of the paramagnetic 17- or 19-electron analogs these reactions are expected to involve lower activation barriers. Associative mechanisms from 17-electron complexes should proceed via 19-electron transition states, whereas dissociative mechanisms from a 19-e species should involve 17-e intermediates. Thus one expects the intermediates or transition states to have energies similar to their paramagnetic precursors, which should lead to low activation barriers (compare Fig. 13). Although somewhat simplistic these considerations are certainly at the basis of the considerable accelerations observed in ligand exchange reactions when performed at the level of paramagnetic intermediates.

Bidentate ligands usually prefer the chelating mode of coordination owing to the large thermodynamic gain arising because of entropic factors [72]. Yet, this thermodynamic statement may often be invalid because of kinetic reasons. For example dithiocarbamate (dtc) ligands are generally chelated but several synthetic routes afford the monodentate complexes, which are then chelated with difficulty. For instance, Fp(η^1-dtc)CO, where Fp = CpFeCO or (Me_5Cp)FeCO, are easily obtained but their thermal conversion to the more stable chelates, Fp(η^2-dtc), proceeds with poor yields ($<5\%$ at 100 °C) and is generally accompanied by decomposition [73]. However, it was found that the addition of ca. 10% of ferricinium was sufficient to give 50% yields of chelated complexes within a few minutes at room temperature [74]. We also observed that exposure to daylight for a few hours resulted in nearly quantitative conversion to chelates. Both observations were strongly indicative of electron transfer catalysis.

$$\mathrm{Fp}(\eta^1\text{-dtc})\mathrm{CO} \xrightarrow{[-e] \text{ or } h\nu} \mathrm{Fp}(\eta^2\text{-dtc}) + \mathrm{CO} \qquad (53)$$

The chelated, Fp(η^2-dtc), product is more electron rich than the monodentate, Fp(η^1-dtc)CO, because a CO ligand has been replaced by an electron releasing sulfur ligand. Thus one expects the product to be more easily oxidized than the starting material, leading thus to a "non-electrocatalytic" situation as described above. Cyclic voltammetry of Fp(η^1-dtc)CO, in Fig. 15 (compare Fig. 14b), demonstrates the nevertheless facile conversion of the cation radical Fp(η^1-dtc)$CO^{\cdot+}$ formed at wave O_1, into the stable cation radical of the chelate, which is reduced on scan reversal at wave R_2. Thus, upon repetitive cycling, wave O_2/R_2 associated with the electrochemical couple Fp(η^2-dtc)$^{0/\cdot+}$, grows at the expense of the irreversible wave O_1, relative to the monodentate complex [74a].

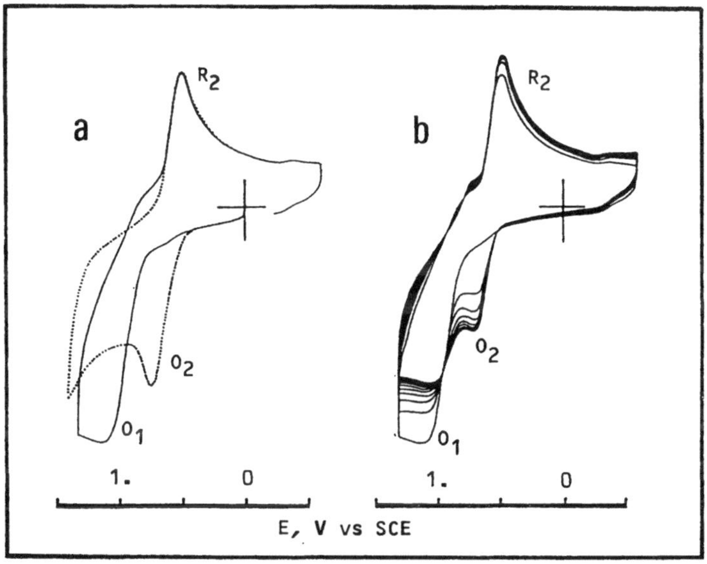

Fig. 15. (a) Cyclic voltammetry of a 2 mM solution of Fe(η^5-Me_5Cp)$(CO)_2$[σ-SC(S)NMe_2] (solid line) or of a 2 mM solution of Fe(η^5-Me_5Cp)(CO)[η^2-SC(S)NMe_2] (dotted line) in THF (0.3 M NBu_4BF_4) at $v = 20\ \mathrm{V\ s^{-1}}$. (b) Repetitive cyclic voltammetry at $v = 20\ \mathrm{V\ s^{-1}}$ (same conditions as in (a)) of a 2 mM solution of Fe(η^5-Me_5Cp)$(CO)_2$[σ-SC(S)NMe_2], indicating the progessive decay of wave O_1 and the concomitant growing of the O_2/R_2 wave system.

However, in order that electrocatalytic conversion to the chelate is effected efficiently at a scan rate of $20\ \mathrm{V\ s^{-1}}$, as in Fig. 16a, the electrode potential needs to be made sufficiently anodic. Thus when only the foot of wave O_1 is attained almost no conversion is observed, as indicated by the small magnitude of waves O_2/R_2, in Figs. 16a, A–C. Such a behavior establishes that for time scales in the millisecond range, which correspond to $v \approx 20\ \mathrm{V\ s^{-1}}$, no homogeneous propagation sequence can develop, the conversion being controlled entirely by the electrode [68, 74a, 75]:

$$Fp(\eta^1\text{-dtc})CO - e \rightarrow Fp(\eta^1\text{-dtc})CO^{\cdot +} \qquad \text{(Wave } O_1\text{)} \qquad (54)$$

$$Fp(\eta^1\text{-dtc})CO^{\cdot +} \rightarrow Fp(\eta^2\text{-dtc})^{\cdot +} + CO \qquad (55)$$

$$Fp(\eta^2\text{-dtc})^{\cdot +} + e \rightleftharpoons Fp(\eta^2\text{-dtc}) \qquad \text{(Wave } R_2\text{)} \qquad (56)$$

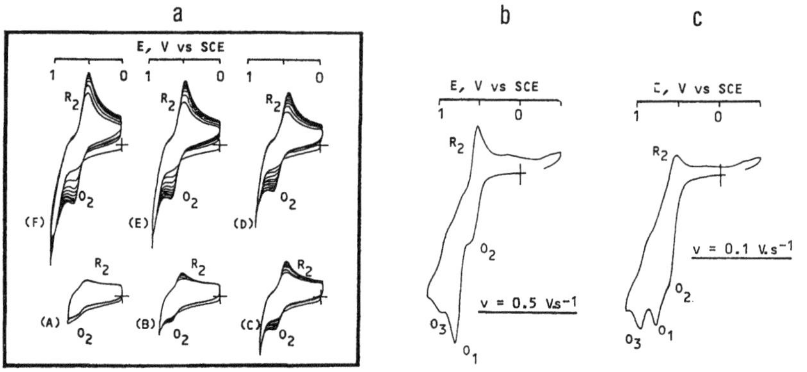

Fig. 16. (a) Repetitive cyclic voltammetry at $v = 20\ V\ s^{-1}$ (same conditions as in Fig. 15) of a 2 mM solution of $Fe(\eta^5\text{-}Me_5Cp)(CO)_2[\sigma\text{-}SC(S)NMe_2]$, indicating the progressive growing of the O_2/R_2 waves system, as a function of the upper oxidative limit of the scans $E_f = 0.75$ (A), 0.80 (B), 0.85 (C), 0.90 (D), 0.95 (E) and 1.00 (F) volts vs SCE. (b, c) Cyclic voltammetry of a 2 mM solution of $Fe(\eta^5\text{-}Me_5Cp)(CO)_2$ $[\sigma\text{-}SC(S)NMe_2]$ (same conditions as in Fig. 15a) at $v = 0.5$ (b) and 0.1 (c) $V\ s^{-1}$.

This is expected, since as discussed above, the homogeneous propagating sequence in reactions 57 and 55 is necessarily slow, the homogeneous electron transfer in Eqn. 57 being endergonic by at least 0.6 V ($\approx$12 kcal mol^{-1}).

$$\dashrightarrow Fp(\eta^2\text{-dtc})^{\cdot +} + Fp(\eta^1\text{-dtc})CO \rightleftharpoons Fp(\eta^2\text{-dtc}) + Fp(\eta^1\text{-dtc})CO^{\cdot +} \qquad (57)$$

$$Fp(\eta^1\text{-dtc})CO^{\cdot +} \rightarrow Fp(\eta^2\text{-dtc})^{\cdot +} + CO \qquad (55)$$

When longer time scales are considered the propagating sequence in Eqns. 57 and 55 is able to convert the monodentate complex into the chelate with extremely large coulombic turnovers. This is evidenced e.g. by the voltammograms at $v = 0.1$ and $0.5\ V\ s^{-1}$ in Figs. 16b, c. At these scan rates, which correspond to time scales in the range a few tenths of a second, the homogeneous chain is sufficiently efficient for effecting the catalytic conversion in the diffusion layer with only traces of $Fp(\eta^1\text{-dtc})CO^{\cdot +}$. As a result the O_2 wave of the chelate is observed on the first scan, i.e. before a potential range is attained where significant amounts of $Fp(\eta^1\text{-dtc})CO$ are electrolyzed.

From these results one may infer that catalytic conversion of the monodentate dtc to the chelate could be performed with only traces of an oxidant provided sufficient time is allowed. This would be true if the two cation radicals were stable chemically, or if the oxidant was maintained at a constant low concentration as it is the case at the electrode in Figs. 16b, c. In practice, when ferricinium salts are used as catalysts, only moderate amounts of conversion are observed [74a, 75], and a relatively high concentration of oxidant is required. This arises because ferricinium oxidizes irreversibly the chelate complex, which results in poor turnovers for the two following reasons. First the $Fp(\eta^2\text{-dtc})^{\cdot +}$ cation radical is chemically unstable on time scales exceeding a few tens of milliseconds, as evidenced in Figs. 16b, c by the progressive disappearance of the R_2 wave and the concomitant appearance of the O_3 wave, which corresponds to a follow-up product, when the scan rate decreases, i.e. when the time scale increases. Second, in the absence of large concentrations of tetraalkylammonium PF_6^-, BF_4^- or $SbCl_6^-$ salts, the chelate cation radical precipitates and therefore can not participate in the propagation of the catalytic chain in Eqns. 57 and 55.

4.4 Mechanism of Biphenyl Synthesis Catalyzed by Nickel Complexes

4.4.1 Introduction

Since the pioneering work of Semmelhack and coworkers [76], synthetic procedures for the preparation of symmetrical or unsymmetrical biaryls by the classical Ullmann reaction have been supplanted by those involving zerovalent transition metals. Thus good yields of biphenyl are obtained from bromo or iodobenzene in the presence of stoichiometric quantities of zerovalent nickel complexes generated in situ by zinc reduction of nickel(II) salts. These original methods were modified later by Kumada et al. [77], and more recently by Colon and Kelsey [78], using catalytic amounts of Ni(II) and stoichiometric quantities of a reducing metal (Zn, Mn, Mg, ...). Electrochemical synthesis of biaryls, polyparaphenylenes, or carboxylic acids, which involve catalytic amounts of nickel salt and a cathode as the reducing agent, were also reported [55, 58, 79]:

$$\text{ArX} + \text{“e”} \xrightarrow{\text{Ni(II)}} (1/2)\,\text{Ar—Ar} + \text{X}^- \qquad (58)$$

$$\text{ArX} + CO_2 + 2\text{“e”} \xrightarrow{\text{Ni(II)}} \text{Ar—}CO_2^- + \text{X}^- \qquad (59)$$

$$\text{X—Ar—X} + 2\text{“e”} \xrightarrow{\text{Ni(II)}} (1/n)\,\text{Ar}_n + 2\text{X}^- \qquad (60)$$

Several attempts to delineate the role of nickel catalysts in these synthesis have been made in the past [80]. Yet, because of the difficulty of performing kinetics in the presence of metal powders [78], most of the quantitative mechanistic work in the area corresponds to homogeneous conditions, i.e. were stoichiometric in nickel complexes [56].

There is general agreement that the catalytic cycle is initiated by oxidative addition of the aromatic halide to a zerovalent nickel complex to afford an arylnickel(II) intermediate [56, 78, 79a, 81] (see also Section 4.1), and is closed by the reductive elimination of a diarylnickel(III) species [56, 78, 82]. Indeed, although both reactions have been identified under stoichiometric conditions, they should not highly depend on the catalytic or stoichiometric nickel concentration since they are first order in nickel. However, the exact chemical sequence leading to the diarylnickel(III) from the arylnickel(II) complex under catalytic conditions and a large reductive driving force was not clear.

Tsou and Kochi, in an elegant series of papers proposed the involvement of a radical chain with paramagnetic Ni(I) and Ni(III) species as the key intermediates [56]. Although their arguments are firmly established, it must be emphasized that their experimental conditions were quite different from those involving catalytic nickel and stoichiometric reductive metals. For example, under stoichiometric conditions in ArNi(II)X, they observed that the diarylnickel(III) species is formed by the reaction (Eqn. 62) of the arylnickel(II) with an arylnickel(III). The latter intermediate was shown to result from the oxidative addition of the aromatic halide to a monovalent nickel complex Ni(I)X in Eqn. 61.

$$Ni(I)X + ArX \rightarrow ArNi(III)X_2 \quad (61)$$

$$ArNi(III)X_2 + ArNi(II)X \rightarrow Ar_2Ni(III)X + Ni(II)X_2 \quad (62)$$

$$Ar_2Ni(III)X \rightarrow Ar{-}Ar + Ni(I)X \quad (63)$$

Under catalytic conditions both nickel species involved in reaction 62 are present at trace levels. In the presence of a large reductive force, such a bimolecular step may not be suficiently rapid to overcome efficiently the direct reduction of these two intermediates to the common ArNi(I) species [25]:

$$ArNi(III)X_2 + \text{“e”} \rightarrow ArNi(II)X + X^- \quad (64)$$

$$ArNi(II)X + \text{“e”} \rightarrow ArNi(I) + X^- \quad (65)$$

This remark led Colon and Kelsey to postulate that under their catalytic and reductive conditions, reactions 64 and 65 completely overrun reaction 62, the diarylnickel(III) being then obtained by oxidative addition of the organic halide to the arylnickel(I) formed in Eqn. 65 [78]:

$$ArNi(I) + ArX \rightarrow Ar_2Ni(III)X \quad (66)$$

Although these authors presented a series of experimental kinetics, they could not establish their mechanism on quantitative grounds and concluded that "... Although a full mechanistic understanding of this complex reaction awaits further study, we believe the mechanism proposed [...] accomodates our experimental data better than any mechanism previously proposed" [83].

As pointed out above the main problem in investigating kinetically such a mechanism, which involves electron transfer reactions at metal particle surfaces, is related to the extreme difficulty in monitoring interfacial events with conventional kinetic techniques. Then one has to rely only on kinetic data relative to modification of the bulk solution composition with time. Although bearing important kinetic information (see e.g. Section 4.4.3) such data are extremely difficult to analyze in order to extract quantitative information on the exact nature of the chemical steps occurring in the immediate vicinity of a particle surface. Indeed, efficient catalytic processes as the one described here must involve rapid reactions, which then take place in the diffusion layers existing at the metal particles surface (see Section 2.3.3.a). In this respect electrochemistry, which allows a complete monitoring of such events, seems perfectly adapted to the study of this kind of reaction.

4.4.2 Electrochemical Investigation of the Mechanism of Nickel Catalysis of Biaryl Synthesis

First of all preparative electrochemistry allows investigation of the role of the reducing strength of the stoichiometric metal. Indeed from data similar to those reported in Fig. 17 one can determine easily several threshold potentials [25]. Thus, when the electrolysis potential is set under −1.2 V vs SCE, the bromobenzene concentration remains unaffected and ca. 1F charge per mole of nickel is passed, resulting in formation of unidentified Ni(I) species. When the potential is set between −1.5 V and −1.8 V vs SCE, two Faradays per mole of nickel are passed, and the bromobenzene concentration remains untouched, with the excep-

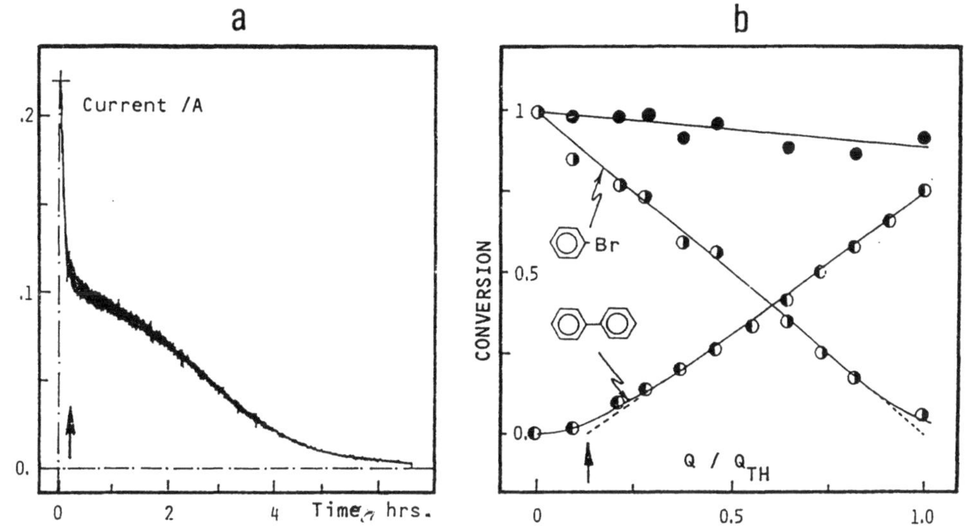

Fig. 17. Preparative electrolysis of $NiCl_2$(dppe) (1 mmol) in THF/HMPA (2/1 vols.; 0.1 M NBu_4BF_4) in the presence of 9.3 mmol of PhBr. 20°C (a) Variation of the electrolysis current with time. (b) Variation of the solution composition normalized to the initial PhBr, as a function of the charge, Q, passed. Q_{TH}: theoretical charge consumption for 1 Faraday per PhBr mole. (●): mass balance including PhBr, PhPh, and phenylnickel derivatives. The arrow in (b) corresponds to the end of the sharp initial current drop indicated by an arrow in (a).

tion of a fraction corresponding to the nickel catalytic concentration. When now the potential is made more cathodic than −2.0 V vs SCE, one observes a fast initial drop of the current (Fig. 17a) followed by a slower decrease. The charge passed during the initial current drop corresponds to 2 F per nickel, and results in the disappearance of one equivalent of bromobenzene, yet without any biphenyl formed. This is more or less comparable to what is observed when electrolysis is performed at −1.5 V vs SCE. However, following this fast initial stage, a second one corresponding to the slow current decrease in Fig. 17, amounts to an apparent electrolysis of bromobenzene (with 1 F/mole of PhBr) to afford quantitatively biphenyl. It is important to recall here that bromobenzene is not reduced under these conditions in the absence of nickel.

These phenomenological data acquire an extreme importance when compared to the voltammetry of the $Ni^{II}Cl_2$(dppe) catalyst in the presence or absence of bromobenzene. Indeed, the steady state voltammograms in Fig. 18a give a mechanistic significance to the thresold potentials identified in preparative scale electrolysis. In the absence of bromobenzene, reduction of $Ni^{II}Cl_2$(dppe) involves two successive one-electron steps with $E_{1/2} = -0.8$ and -1.35 V vs SCE. Transient cyclic voltammetry (see e.g. Section 4.1, Fig. 9) shows that these two reduction processes, in Eqns. 67 and 68, are chemically irreversible even at scan rates in the kilovolt per second range – i.e. in time scales of a few tens of microseconds [25].

$$Ni^{II}Cl_2(dppe) + e \rightarrow Ni^{I}Cl(dppe) + Cl^- \quad \text{(Wave } R_1\text{)} \tag{67}$$

$$Ni^{I}Cl(dppe) + e \rightarrow Ni^{0}(dppe) + Cl^- \quad \text{(Wave } R_2\text{)} \tag{68}$$

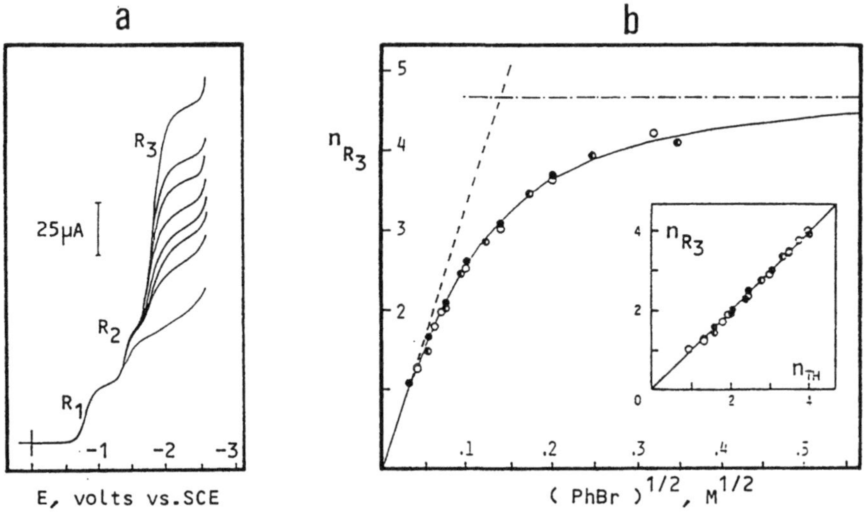

Fig. 18. (a) Steady state cyclic voltammetry at a gold Rotating Disk Electrode (ϕ: 2 mm, w: 105 rad s^{-1}, $v = 20$ mV s^{-1}) for a 2 mM solution of $NiCl_2$(dppe) solution in THF/HMPA (2/1 vols.; 0.1 M NBu_4BF_4) in the presence of different amounts of PhBr (from bottom to top: 0, 1, 2, 3, 5, 10, 20 and 50 eqs). (b) Variations of the apparent number, n_{R_3}, of electron exchanged at wave R_3 in (a), as a function of the bromobenzene concentration. Solid line: theoretical predictions for the reaction sequence in Eqns. 70–74 and Scheme III (Section 4.4.3). The same data are plotted in the insert under the form of the experimental, n_{R_3}, vs the theoretical, n_{TH}, values of the apparent number of electrons exchanged at wave R_3. "(Note that a first order dependence on PhBr concentration should correspond to $n_{R_3} \propto (PhBr)^{1/2}$. [8,25])."

As already shown in Fig. 10a, in the presence of bromobenzene a third reduction wave R_3 develops around -1.8 V vs SCE. Its limiting current depends on the bromobenzene concentration, yet without being proportional to the latter. Moreover a near linear dependence with $[PhBr]^{1/2}$ is observed for $[PhBr] < 0.01$ M, whereas a saturating effect shows up above 0.1 M. Cyclic voltammetry of an authentic sample establishes that the R_3 wave corresponds to the one-electron chemically irreversible reduction of $Ph{-}Ni^{II}Br(dppe)$, in Eqn. 69 (see Section 3.5):

$$Ph{-}Ni^{II}Br(dppe) + e \rightarrow Ph{-}Ni^{I}(dppe) + Br^- \qquad (\text{Wave } R_3) \qquad (69)$$

the latter being formed by the facile oxidative addition of bromobenzene (see Eqn. 33 in Section 4.1). to the zerovalent coordinatively unsaturated nickel generated at wave R_2 (Eqn. 68).

Several conclusions can be drawn from these observations. (i) The $Ni^{I}Cl$(dppe) complex, formed at potentials less cathodic than ca. -1.3 V vs SCE, is unreactive with bromobenzene on our time scale, as evidenced by the independence of its electrochemical wave in the presence of PhBr. (ii) Ni^0(dppe) formed on the plateau of wave R_2 (i.e. when E is more negative than -1.5 V vs SCE) reacts stoichiometrically with bromobenzene to afford a phenylnickel(II) species, which in unreactive except when the electrode potential reaches wave R_3 (i.e. $E \approx -2.0$ V vs SCE). (iii) The behavior observed at wave R_3 is indicative of redoxcatalysis (see Section 3.1), i.e. of a constant regeneration of the phenylnickel(II) species, initiated by $PhNi^{I}$(dppe) in the presence of excess PhBr. (iv) The regeneration of

$PhNi^{II}Br(dppe)$ observed at wave R_3 obeys a complex rate law with two limiting trends: first order dependence with [PhBr] at low bromobenzene concentrations; zeroth order in [PhBr] at large bromobenzene concentrations.

The simplest reaction sequence compatible with the above conclusions is presented in the series of Eqns. 70–74 [25]:

$$PhNi^{II}Br(dppe) + e \rightarrow PhNi^{I}(dppe) + Br^- \quad \text{(Wave } R_3) \tag{70}$$

$$PhNi^{I}(dppe) + PhBr \rightarrow Ph_2Ni^{III}Br(dppe) \tag{71}$$

$$Ph_2Ni^{III}Br(dppe) \rightarrow Ph{-}Ph + Ni^{I}(dppe)Br \tag{72}$$

$$Ni^{I}(dppe)Br + e \rightarrow Ni^{0}(dppe) + Br^- \quad \text{(Wave } R_2) \tag{73}$$

$$Ni^{0}(dppe) + PhBr \rightarrow PhNi^{II}Br(dppe) \tag{74}$$

All these steps have been independently characterized by previous authors (Eqn. 72) [82] or ourselves (Eqns. 70, 73, 74) [25], with the exception of step 71. Although oxidative addition of aromatic halides to monovalent nickel has been proposed several times in the literature [56, 78, 84], its involvement here may be quite critical since we have established above – point (i) – that $Ni^{I}Cl(dppe)$ was not reactive with PhBr in our time scale (i.e. from some µs to a few s). However, in our opinion this is only an apparent contradiction that arises because of the conventional but misleading notation of formal oxidation states. We believe there is no close relationship between the chemical reactivity of a "monovalent" nickel bearing a halide ligand and that of a "monovalent" nickel bearing an aryl ligand, although both are noted as Ni^{I} species [85]. Such a statement is difficult to prove and quantify for Ni^{I} species owing to their large instabilities, yet it may be transposed to the Ni^{II} level.

In this case a simple comparison of the half-wave potentials of $Ni^{II}Br_2(dppe)$ and $PhNi^{II}Br(dppe)$ shows that the latter is reduced ca. 1.1 V more cathodically than the former [25]. Such a large difference is strongly indicative of the important differences in the electronic donating properties of the two ligands. Moreover a subtle change, by comparison to the above one, from a chloride to a bromide ligand in $Ni^{I}X(dppe)$ results in an anodic shift of ca. 60 mV of the reduction potential, i.e. similar to the 160 mV anodic shift observed for the reduction of $Ni^{II}X_2(dppe)$, when X = Cl or Br [25] (compare [85c]). Although these results do not claim to demonstrate our above statement, they are direct evidence that the electronic properties of monovalent nickel strongly depend on the exact nature of the metal coordination shell. On the other hand it has been established that the rates of oxidative additions critically depend on electronic properties of the organic halide and of the metal complex: the more electron rich the metal and the more easily reducible the halide, the faster is the oxidative addition [56].

We thus considered that the sequence in Eqns. 70–74, which happens to be identical to one of the two schemes postulated by Colon and Kelsey, could be accepted on a chemical basis. In order to test for its kinetic validity we derived the theoretical law relating the apparent number of electrons, n_{R_3}, exchanged at wave R_3 to the bromobenzene concentration, on the basis of Eqns. 70–74 [25, 48]. The result of this theoretical analysis is compared to the experimental variations in Fig. 18b. From the perfect agreement observed we are inclined to accept the sequence in Eqns. 70–74 as the most likely mechanism for the catalysis of biphenyl synthesis by nickel complexes under conditions where a large reductive driving force prevails. Moreover the fitting in Fig. 18b allows us to estimate as ca. $10^3\ M^{-1}\ s^{-1}$

and 20 s^{-1} the respective rate constants for the oxidative addition of bromobenzene to a phenylnickel(I) in Eqn. 71, and for the reductive elimination of biphenyl in Eqn. 72.

4.4.3 Transposition to "Homogeneous" Conditions

In the preceding section we have shown how a likely mechanism could be established for the catalytic role of nickel species in biphenyl synthesis from halobenzenes, on the basis of electrochemical data. Yet, at this point one may wonder if this is a mechanism valid only under electrochemical conditions or if it presents some utility for "homogeneous" related cases where the electrode is replaced by a dispersed powder of a reducing metal. At various instances in this chapter we have drawn analogies between electrochemical interfacial conditions and those encountered in polyphasic reactions (see e.g. Section 2.3). Here, at the end of this chapter, we want to show that this is not a matter of subtle considerations, but is true in experimental practice.

Scheme III

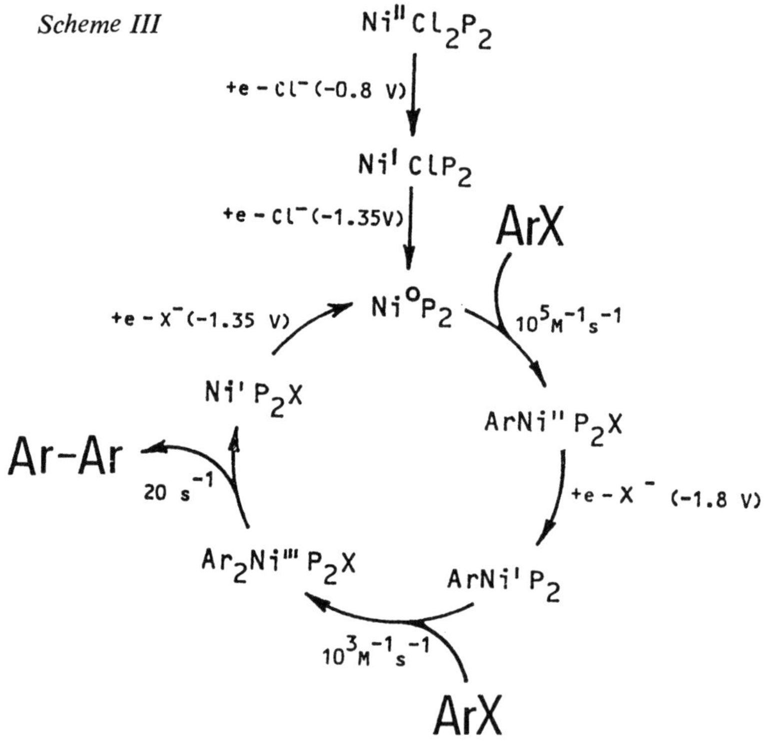

Scheme III summarizes most of the kinetic data we obtained on the mechanism [25]. It is seen from this scheme that when the electrode is replaced by a reducing metal particle such as zinc, the slower step in the chain should be the difficult reduction of the phenylnickel(II) species. Therefore most of the nickel catalyst is "stored" under the form of this intermediate, whose concentration is then close to that, $[Ni]_0$, of the nickel salt introduced in the reaction. On the other hand, as for any other chain reaction, the overall rate of the chain conversion is imposed by the slowest step of the chain, i.e. by the phenylnickel(II) reduction at the metal particle surface. As a result the rate law observed under "homogeneous" conditions ought to be as given in Eqn. 75:

$$2d[\mathrm{ArAr}]/dt = -d[\mathrm{ArX}]/dt = k_M A_M [\mathrm{Ni}]_o \quad (75)$$

where A_M is the overall active surface area of the dispersed metal particle relative to the volume of the solution, and k_M an apparent rate constant, which not only depends on the nature of the metal and on that of the phenylnickel(II), but also on the hydrodynamics of the solution and particle size [30]. Such a rate law is valid up to the point where the arylhalide concentration drops below the level where its oxidative addition to the phenylnickel(I) species becomes the rate limiting step in the chain propagation. After this point the rate law becomes as given in Eqn. 76:

$$2d[\mathrm{ArAr}]/dt = -d[\mathrm{ArX}]/dt = k^* A_M [\mathrm{Ni}]_o [\mathrm{ArX}] \quad (76)$$

where k^* is an apparent rate constant depending on the hydrodynamics of the solution [30] and on the true rate constant of the oxidative addition in reaction 71.

When both reactions compete for the control of the chain propagation the more general law in Eqn. 77 is obtained:

$$d[\mathrm{ArX}]/dt = -k_M A_M [\mathrm{Ni}]_o \{[\mathrm{ArX}]/([\mathrm{ArX}] + k_M/k^*)\} \quad (77)$$

Integration of rate law in Eqn. 77 gives the variation of the conversion, i.e. of $y = \{1 - [\mathrm{ArX}]/[\mathrm{ArX}]_o\}$, as a function of time:

$$y - \varepsilon \,\mathrm{Ln}\,(1 - y) = k_M A_M [\mathrm{Ni}]_o \, t \quad (78)$$

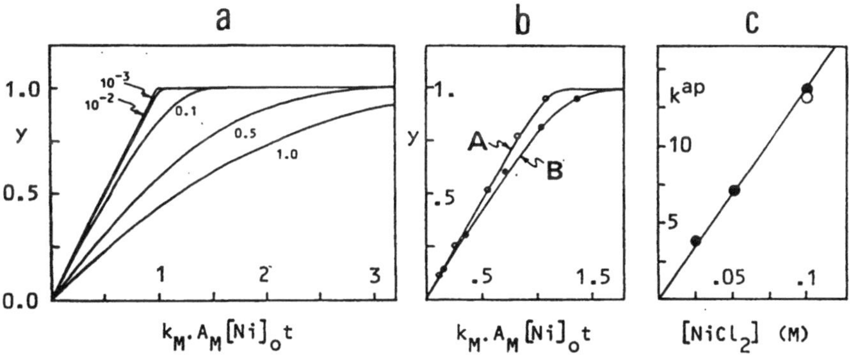

Fig. 19. "Homogeneous" catalysis of biphenyl synthesis from chlorobenzene by nickel salts in the presence of zinc powder. (a) Theoretical variation of the conversion, $y = (1 - [\mathrm{ArX}]/[\mathrm{ArX}]_o)$, as a function of time (see text and Eqns. 75–78 for definitions) for different values of ε (numbers on the curves). (b) Comparison of the theoretical predictions such as those in (a) with the experimental results obtained by Colon and Kelsey in Ref. 78. [(A) Exp: $NiCl_2$ (1 mmol), TPP (3.81 mmol), Zn (30.6 mmol), NaBr (4.86 mmol) and PhCl (9.8 mmol); Th: $k_M A_M [\mathrm{Ni}]_o = 0.55\ \mathrm{min}^{-1}$, $\varepsilon = 0.05$. (B) Exp: $NiCl_2$ (1 mmol), TPP (7.6 mmol), Zn (30.6 mmol), NaBr (9.7 mmol) and PhCl (19.7 mmol); Th: $k_M A_M [\mathrm{Ni}]_o = 0.068\ \mathrm{min}^{-1}$, $\varepsilon = 0.13$]. (c) Experimental variation of the apparent zero-order rate constant (in $\mathrm{M \cdot s^{-1}}$) deduced from the initial slope of plots such as in (a, b). (Experimental data from Table 5 in Ref. 78: [PhCl] = (○) 1 or (●) 2 M; [NaBr] = 1 M).

where $\varepsilon = k_M/(k^*[ArX]_0)$, is the parameter controlling the respective influence of the two possible rate determining steps. Fig. 19a presents the theoretical variation of the conversion with time as a function of ε. Interestingly when $\varepsilon \ll 1$, the conversion is zeroth order in organic halide, first order in nickel catalyst, first order in reducing metal (assuming that A_M is proportional to the amount of reducing metal particles). Conversely, when $\varepsilon \gg 1$ the conversion is first order in organic halide, in nickel catalyst and in reducing metal. Both conclusions could obviously be derived directly from each pertinent limiting rate law: Eqn. 75 for $\varepsilon \ll 1$; Eqn. 76 for $\varepsilon \gg 1$. These theoretical predictions, based upon the mechanism established under electrochemical conditions, are compared in Figs. 19b, c to some of the experimental "homogeneous" data reported by Colon and Kelsey [78]. As evidenced by these plots a remarkable agreement is obtained.

5. Conclusion

In this chapter we have tried to show and explain how electrochemistry may be used to gain mechanistic and quantitative informations on the reactivity of organometallic transient intermediates. The few selected examples have been chosen to illustrate various possible electrochemical approaches, either in terms of experimental strategies for solving a given problem, or in terms of the increasing complexity of the problems addressed, from a "simple" elementary act to the multistep catalytic mechanism of biphenyl synthesis. Yet, in our opinion the most important message contained in this chapter is that, in contradiction to generally received ideas, electrochemical conditions do not differ greatly from the so called "homogeneous" conditions used by most chemists. Thus, we hope this chapter will help "homogeneous" chemists to overcome a few technological activation barriers, and resort more frequently to electrochemical kinetic techniques.

To conclude this chapter we want also to mention that because of lack of space and general aim of this chapter, we have mainly insisted on the electrochemical aspects. Yet we do believe that although electrochemistry has unique potentialities for investigation of mechanisms and chemical reactivity, its use in conjunction with other chemical techniques and strategies is obviously worthwhile and desirable. For example the study of $Fe(CO)_5$ reduction mechanism would have been extremely difficult or impossible without the help of information provided by other techniques such as e.s.r. Yet, owing to the general scope of this chapter and to the fact that other authors in this book (Kochi, Rieger) develop these important aspects in their contributions, they were voluntarily made "transparent" in ours.

6. Acknowledgment

We want to express here our gratitude to those people who contributed to the work presented in this chapter and whose names can be found at several places in the reference list. We are specially indebted to Drs. Anny Jutand and Jean-Noël Verpeaux, from our research group, Dr. Paul J. Krusic from E. I. Du Pont de Nemours, Central Research Department, for their invaluable contributions and grants of knowledge.

7. References

1. See e.g. (a): N. G. Connelly and W. E. Geiger, in F. G. A. Stone and R. West (Eds.), Advances in Organometallic Chemistry, Vol. **23**, Academic Press, Orlando, 1984, pp. 1–94; (b) W. E. Geiger and N. G. Connelly, *ibid.*, Vol. **24**, 1985, pp. 87–130; (c) P. Lemoine, Coord. Chem. Rev. **47** (1982) 55; and references therein.
2. See e.g. Organic Electrochemistry, H. Lund and M. Baizer (Eds.), 3rd edn., M. Dekker, New York, 1990.
3. See e.g. A. A. Vlcek, (a) Nature **177** (1956) 1043; (b) Z. Anorg. Chem. **304** (1960) 109; (c) Prog. Inorg. Chem. **5** (1963) 211; (d) Coll. Czech. Chem. Comm. **30** (1965) 952.
4. R. E. Dessy and L. A. Bares, Acc. Chem. Res. **5** (1972) 415, and refs. therein.
5. See e.g. J. March, Advanced Organic Chemistry, 3rd edn., J. Wiley & Sons, New York, 1985.
6. A. J. Bard and R. L. Faulkner, Electrochemical Methods, J. Wiley & Sons, New York, 1980.
7. Laboratory Techniques in Electroanalytical Chemistry, P. T. Kissinger and W. R. Heineman (Eds.), M. Dekker, New York, 1984.
8. C. P. Andrieux and J. M. Savéant, in C. F. Bernasconi (Ed.), Investigation of Rates and Mechanisms of Reactions, Vol. **6**, 4/E, Part 2, Electrochemical Reactions, J. Wiley & Sons, New York, 1986, Ch. 7, pp. 305–390; and refs. therein.
9. For discussions see: (a) J. K. Kochi, Organometallic Mechanisms and Catalysis, Academic Press, New York, 1978; (b) M. Chanon and M. L. Tobe, Angew. Chem. Int. Ed. Engl. **21** (1982) 1.
10. C. Amatore, in M. Moreau and P. Turq (Eds.), Chemical Reactivity in Liquids. Fundamental Aspects, Fast Techniques in Electrochemistry: Application to the Study of Chemical Reactivity, Plenum Press, New York, 1988, pp. 73–89.
11. J. O'M. Bockris and A. K. N. Reddy, Modern Electrochemistry, Vols. **1** and **2**, Plenum Press, New York, 1977.
12. C. Amatore, in H. Lund and M. Baizer (Eds.), Organic Electrochemistry, 3rd edn., Basic Concepts, M. Dekker, New York, 1990, I. Ch. 2.
13. C. Amatore, M. Gareil and J.-M. Savéant, J. Electroanal. Chem. **176** (1984) 377, and refs. therein.
14. C. P. Andrieux and J.-M. Savéant, J. Electroanal. Chem. **205** (1986) 43, and refs. therein.
15. (a) D. Kuchynka, C. Amatore and J. K. Kochi, Inorg. Chem. **25** (1986) 4087; (b) B. A. Narayanan, C. Amatore and J. K. Kochi, Organometallics **6** (1987) 129.
16. For a simple presentation of the principles of digital simulation of see e.g. Ref. 6 pp. 675–698, and J. T. Maloy, Digital Simulation of Electrochemical Problems, in Ref. 7 pp. 417–461. For more elaborated presentation and discussions of various algorithms see e.g.: (a) S. W. Feldberg, in A. J. Bard (Ed.), Electroanalytical Chemistry, Vol. 3, Marcel Dekker, New York, 1969, ppff. 199; (b) D. Britz, Digital Simulation in Electrochemistry, Springer, Berlin, 1981; (c) D. K. Gosser and P. H. Rieger, Anal. Chem. **60** (1988) 1159.
17. Compare e.g.: (a) J. W. Hershberger and J. K. Kochi, J. Chem. Soc., Chem. Commun. (1982) 212; (b) J. W. Hershberger, R. J. Klingler and J. K. Kochi, J. Am. Chem. Soc. **105** (1983) 61; (c) J. W. Hershberger, C. Amatore and J. K. Kochi, J. Organomet. Chem. **250** (1983) 345; (c) P. M. Zizelman, C. Amatore and J. K. Kochi, J. Am. Chem. Soc. **106** (1984) 3771; (d) J. K. Kochi, J. Organomet. Chem. **300** (1986) 139.
18. Reaction of Molecules at Electrodes, N. S. Hush (Ed.), Wiley Interscience, New York, 1971.
19. (a) C. K. Mann, in A. J. Bard (Ed.), Electroanalytical Chemistry, Vol. **3**, Nonaqueous Solvents for Electrochemical Use, M. Dekker, New York, 1969, pp. 57–134; (b) H. Lund, in H. Lund and M. Baizer (Eds.), Organic Electrochemistry, 3rd edn., Basic Concepts, M. Dekker, New York, 1990, I. Ch. 6; (c) A. J. Fry and W. E. Britton, in Ref. 7, Solvents and Supporting Electrolytes, pp. 367–382.
20. (a) R. M. Wightman, Science **240** (1988) 415; (b) R. M. Wightman and D. O. Wipf, in A. J. Bard (Ed.), Electroanalytical Chemistry, Vol. **15**, Voltametry at Ultramicroelectrodes, M. Dekker, New York, 1989, pp. 267–353.
21. (a) C. Amatore, in M. Fleischmann, S. Pons, D. Rolinson and P. P. Schmidt (Eds.), Datatech Systems, Morganton, 1987, pp. 169–183; (b) S. Bruckenstein, Anal. Chem. **59** (1987) 2098.

22. M. I. Montenegro, D. Pletcher, J. Electroanal. Chem. **248** (1988) 229.
23. (a) Ref. 19a; (b) R. Lines and V. D. Parker, Acta Chem. Scand., Ser. **B**, B31 (1977) 369; (c) J. O. Howell and R. M. Wightman, Anal. Chem. **56** (1984) 524; (d) J. O. Howell and R. M. Wightman, J. Phys. Chem. **88** (1984) 3915; (e) L. Geng, A. G. Ewing, J. C. Jernigan and R. W. Murray, Anal. Chem. **58** (1986) 852; (f) C. Amatore, and F. Pflüger, submitted.
24. (a) C. Amatore, M. R. Deakin and R. M. Wightman, J. Electroanal. Chem. **225** (1987) 49; (b) C. Amatore, B. Fosset, J. Bartelt, M. R. Deakin and R. M. Wightman, J. Electroanal. Chem., **256** (1988) 255; (c) J. E. Bartelt, M. R. Deakin, C. Amatore and R. M. Wightman, Anal. Chem. **60** (1988) 2167; (d) C. Amatore, J.-N. Verpeaux and P. J. Krusic, Organometallics, **7** (1988) 2426.
25. C. Amatore and A. Jutand, Organometallics, **7** (1988) 2203.
26. See Ref. 6, pp. 500–511, as well as the refs. quoted therein for a detailed and quantified presentation of the models.
27. See e.g. (a) V. G. Levich, Physicochemical Hydrodynamics, Prentice-Hall, Englewood Cliffs, N. J., 1962; (b) J. S. Newman, Electrochemical Systems, Prentice-Hall, Englewood Cliffs, N. J., 1973, pp. 305–339; and refs. therein.
28. (a) J. Newman, Electrochemical Systems, Prentice-Hall, Inc., Englewood Cliffs (NJ), 1973, pp. 353–381 and 414–425; (b) A. M. Bond, M. Fleischmann and J. Robinson, J. Electroanal. Chem. **172** (1984) 11; (c) K. B. Oldham, J. Electroanal. Chem. **250** (1988) 1.
29. Ref. 6, pp. 128–130.
30. C. Amatore, in H. Lund and M. Baizer (Eds.), Organic Electrochemistry, 3rd edn., Relationships Between Micro and Macro Phenomena, M. Dekker, New York, 1990, I. Ch. 4.
31. (a) T. L. Brown and S. B. McCullen, J. Am. Chem. Soc. **104** (1982) 7496; (b) S. B. McCullen, H. W. Walker and T. L. Brown, J. Am. Chem. Soc. **104** (1982) 4007; (c) D. R. Kidd, C. P. Cheng and T. L. Brown, J. Am. Chem. Soc. **100** (1978) 4103.
32. (a) D. J. Kuchynka, C. Amatore and J. K. Kochi, J. Organomet. Chem. **328** (1987) 133; (b) C. Amatore, D. J. Kuchynka and J. K. Kochi, J. Electroanal. Chem. **241** (1988) 181.
33. See e.g. (a) J. Villermaux, in M. Moreau and P. Turq (Eds.), Chemical Reactivity in Liquids. Fundamental Aspects, Issues and Challenges in Chemical Engineering of Liquid Phase Reactions, Plenum Press, New York, 1988, pp. 531–541; (b) F. E. Marble, *ibid.*, Mixing, Diffusion and Chemical Reaction of Liquids in a Vortex Field, pp. 581–596.
34. (a) C. Amatore, A. Jutand and F. Pflüger, J. Electroanal. Chem. **218** (1987) 361; (b) D. O. Wipf, E. W. Kristensen, M. R. Deakin and R. M. Wightman, Anal. Chem. **60** (1988) 306; (c) C. P. Andrieux, D. Garreau, P. Hapiot, J. Pinson and J. M. Savéant, J. Electroanal. Chem. **243** (1988) 321.
35. Ref. 6, pp. 44–62.
36. Ref. 6, p. 71.
37. For example, compare the evaluation of solvation energies via the Born model [(a) W. M. Latimer, K. S. Pitzer, and C. M. Slansky, J. Chem. Phys. **7** (1939) 108; (b) R. M. Noyes, J. Am. Chem. Soc. **84** (1962) 513; (d) J. F. Coetzee and J. J. Campion, J. Am. Chem. Soc. **89** (1967) 2513; (e) M. Tanaka, Inorg. Chem. **15** (1976) 2325.] with that relative to the ionic atmosphere in the Debye-Hückel model: P. W. Atkins, Physical Chemistry, Oxford University Press, Oxford, 1978, pp. 314–321.
38. (a) For a simplified presentation see Ref. 12; (b) for an extensive review and a more specialized presentation see: N. Sutin, in S. J. Lippard (Ed.), Progress in Inorganic Chemistry, Vol. **30**, Wiley Interscience, New York, 1983, pp. 441–498.
39. (a) Z. Smoluchowsky, Physik. Chem. **92** (1917) 129; (b) P. Debye, Trans. Electrochem. Soc. **82** (1942) 265; (c) R. A. Marcus, Faraday Discuss. Chem. Soc. **74** (1982) 7.
40. (a) R. S. Nicholson and I. Shain, Anal. Chem. **36** (1964) 706; (b) L. Nadjo and J. M. Savéant J. Electroanal. Chem. **48** (1973) 113.
41. C. Amatore, M. A. Oturan, J. Pinson, J.-M. Savéant and A. Thiébault, J. Am. Chem. Soc. **107** (1985) 3451.
42. See e.g. Ref. 6, pp. 563–567, or Ref. 7, pp. 163–192, as well as Refs. 43, 44, for descriptions of related electronics.

43. J. O. Howell, W. G. Kuhr, R. E. Ensman and R. M. Wightman, J. Electroanal. Chem. **209** (1986) 77.
44. C. Amatore, C. Lefrou and F. Pflüger, J. Electroanal. Chem., in press.
45. (a) K. Farmery, M. Kilner, R. Greatex and N. N. Greenwood, J. Chem. Soc., **A** (1969) 2339; (b) H. B. Chin, M. B. Smith, R. D. Wilson and R. Bau, J. Am. Chem. Soc. **96** (1974) 5285; (c) P. J. Krusic, J. R. Morton, K. F. Preston and Y. Le Page, J. Mol. Struct., **189** (1988) 1.
46. (a) Ref. 40b; (b) J.-M. Savéant and D. Tessier, J. Electroanal. Chem. **65** (1975) 57; (c) J.-M. Savéant and D. Tessier, J. Phys. Chem. **81** (1977) 2192; **83** (1978) 1723; (d) D. Garreau, J.-M. Savéant and D. Tessier, J. Phys. Chem. **82** (1979) 3003; (e) R. J. Klingler and J. K. Kochi, J. Am. Chem. Soc. **104** (1982) 4186.
47. See e.g. (a) Ref. 32 or (b) C. Amatore and J.-M. Savéant, J. Electroanal. Chem. **107** (1980) 353.
48. See e.g. C. Amatore, A. Jutand and L. Mottier, submitted.
49. C. Amatore, J.-N. Verpeaux and P. J. Krusic, submitted.
50. See e.g. Ref. 6, pp. 280–315.
51. See Ref. 30 for discussion of this point and related ones.
52. (a) C. J. Pickett and D. Pletcher, J. Chem. Soc., Dalton Trans., (1976) 749; (b) A. Seurat, P. Lemoine and M. Gross, Electrochim. Acta **23** (1978) 1219.
53. Ref. 6, p. 143.
54. Ref. 6, p. 288.
55. (a) M. Troupel, Y. Rollin, S. Sibille, J.-F. Fauvarque and J. Perichon, J. Chem. Research (S), (1980), 26; (b) G. Schiavon, G. Bontempelli and B. Corain, J. Chem. Soc., Dalton Trans., (1981) 1074.
56. (a) T. T. Tsou and J. K. Kochi, J. Am. Chem. Soc. **101** (1979) 6319; (b) **101** (1979) 7547; (c) J. Org. Chem. **45** (1980) 1930.
57. (a) C. Amatore, M. Azzabi and A. Jutand, J. Organomet. Chem., **363** (1989) C 41.; (b) C. Amatore, M. Azzabi, and A. Jutand submitted.
58. M. Troupel, Y. Rollin, S. Sibille, J.-F. Fauvarque and J. Perichon, J. Chem. Research (S), (1980) 24.
59. (a) D. E. Dessy, F. E. Stary, R. B. King and H. Waldrop, J. Am. Chem. Soc. **88** (1966) 471; (b) C. J. Pickett and D. Pletcher, J. Chem. Soc., Dalton Trans., (1975) 879; (c) A. M. Bond, P. A. Dawson, B. M. Peake, B. H. Robinson and J. Simpson, Inorg. Chem. **16** (1977) 2199; (d) G. J. Bezems, P. H. Rieger and S. Visco, J. Chem. Soc., Chem. Commun., (1981) 265; (e) N. El Murr and A. Chaloyard, Inorg. Chem. **21** (1982) 2206; (f) B. A. Narayanan, C. Amatore and J. K. Kochi, Organometallics **5** (1986) 926.
60. H. Strong, P. J. Krusic and J. San Filippo Jr., Inorg. Synthesis **24** (1984) 157.
61. C. Amatore, M. Gareil and J.-M. Savéant, J. Electroanal. Chem. **147** (1983) 1.
62. C. Amatore and J.-M. Savéant, J. Electroanal. Chem. **125** (1981) 1.
63. J. P. Collman, R. G. Finke, P. L. Matlock, R. Wahren, R. G. Komoto and J. I. Brauman, J. Am. Chem. Soc. **100** (1978) 1119.
64. (a) J.-M. Savéant, Acc. Chem. Res. **13** (1980) 323; (b) M. Chanon, Bull. Soc. Chim. Fr. (1982) 197; (c) M. Julliard and M. Chanon, Chem. Rev. **83** (1983) 425; (d) M. Chanon, Bull. Soc. Chim. Fr., (1985) 209; (e) D. Astruc, Angew. Chem. Int. Ed. Engl. **27** (1988) 643.
65. (a) R. L. Rich and H. Taube, J. Am. Chem. Soc. **76** (1954) 2608; (b) F. Basolo, P. H. Wilks, R. G. Pearson and R. G. Wilkins, J. Inorg. Nucl. Chem. **6** (1958) 161; (c) F. Basolo, A. F. Messing, P. H. Wilks, R. G. Wilkins and R. G. Pearson, J. Inorg. Nucl. Chem. **6** (1958) 203; (d) F. Basolo, M. L. Morris and R. G. Pearson, Discuss. Faraday Soc. **29** (1960) 80.
66. (a) N. Kornblum, R. E. Michel and R. C. Kerber, J. Am. Chem. Soc. **88** (1966) 5662; (b) G. A. Russel and W. C. Danen, J. Am. Chem. Soc. **88** (1966) 5663; (c) N. Kornblum, Angew. Chem. Int. Ed. Engl. **14** (1975) 734; (d) J. F. Bunnett, Acc. Chem. Res. **11** (1978) 413.
67. A. Darchen, H. Mousser and H. Patin, J. Chem. Soc., Chem. Commun., (1988) 968.
68. C. Amatore, A. Jutand and J. N. Verpeaux, in M. Chanon, M. Julliard and J. C. Poite (Eds.) Paramagnetic Organometallic Species in Activation, Selectivity, Catalysis, Kluwer Academic Publishers, Dordrecht, 1989.
69. C. Amatore, J. Pinson, J.-M. Savéant and A. Thiébault, J. Am. Chem. Soc. **103** (1981) 6930.

70. (a) S. W. Feldberg and L. Jeftic, J. Phys. Chem. **76** (1972) 2439; (b) see Refs. 1a, b; (c) M. J. Therien, C.-L. Ni, F. C. Anson, J. Osteryoung and W. C. Trogler, J. Am. Chem. Soc. **108** (1986) 4037; (d) J. K. Kochi, J. Organomet. Chem. **300** (1986) 139; (e) in organic chemistry see e.g.: C. Amatore, C. Combellas, J. Pinson, M. A. Oturan, S. Robveille, J.-M. Savéant and A. Thiébault, J. Am. Chem. Soc. **107** (1985) 4846, and references therein on $S_{RN}1$ reactions.
71. For an early example in cluster chemistry see e.g. Ref. 59d. For non-electrochemical examples see e.g.: (a) A. Fox, J. Malito and A. Poe, J. Chem. Soc. Chem. Commun., (1981) 1052; (b) Q.-Z. Shi, T. G. Richmond, W. C. Trogler and F. Basolo, J. Am. Chem. Soc. **106** (1984) 71.
72. See e.g. F. A. Cotton and G. Wilkinson, Advanced Inorganic Chemistry, 4th edn., Wiley, New York, 1980.
73. E. Roman, D. Catheline and D. Astruc, J. Organomet. Chem. **236** (1982) 229.
74. (a) C. Amatore, J.-N. Verpeaux, A. Madonik, M.-H. Desbois and D. Astruc, J. Chem. Soc., Chem. Commun., (1988) 200; (b) compare with- M. Moran, I. Cuadrado, J. R. Masaguer and J. Losada, J. Organomet. Chem., **335** (1987) 255, and M. Moran, I. Cuadrado, C. Munoz-Reja, J. R. Masaguer and J. Losada, J. Chem. Soc., Dalton Trans., (1988) 149, for comparable and simultaneous observations with a xanthate (S_2COEt) ligand.
75. J.-N. Verpeaux, M. H. Desbois, A. Madonik, C. Amatore and D. Astruc, Organometallics, in press.
76. (a) M. F. Semmelhack, P. M. Helquist and L. D. Jones, J. Am. Chem. Soc. **93** (1971) 5908; (b) M. F. Semmelhack and L. S. Ryono, J. Am. Chem. Soc. **97** (1975) 3873; (c) M. F. Semmelhack, P. M. Helquist, L. D. Jones, L. Keller, L. Mendelson, L. S. Ryono, J. G. Smith and R. D. Stauffer, J. Am. Chem. Soc. **103** (1981) 6460.
77. M. Zembayashi, K. Tamao, J. Yoshida and M. Kumada, Tetrahedron Lett., (1977) 4089.
78. I. Colon and D. R. Kelsey, J. Org. Chem. **51** (1986) 2627.
79. (a) M. Troupel, Y. Rollin, J. Perichon and J. F. Fauvarque, Nouv. J. Chim. **5** (1981) 621; (b) J. F. Fauvarque, C. Chevrot, A. Jutand, M. François and J. Perichon, J. Organomet. Chem. **264** (1984) 273; (c) J. F. Fauvarque, M. A. Petit, F. Pflüger, A. Jutand, C. Chevrot and M. Troupel, Makromol. Chem. Rapid Commun **4** (1983) 455.
80. See e.g. Refs. 56, 78 and Refs. and discussions therein.
81. (a) G. W. Parshall, J. Am. Chem. Soc. **96** (1974) 2360; (b) M. Hidai, T. Kashiwagi, T. Ikenchi and Y. Uchida, J. Organomet. Chem. **30** (1971) 279; (c) D. R. Fahey, J. Am. Chem. Soc. **92** (1970) 402; (d) D. R. Fahey, Organomet. Chem., Rev. Ser. **A**, **7** (1972) 245; (e) D. R. Fahey and J. E. Mahan, J. Am. Chem. Soc. **99** (1977) 2501.
82. (a) M. Almemark and B. Akermark, J. Chem. Soc., Chem. Commun., (1978) 66; (b) D. G. Morrell and J. K. Kochi, J. Am. Chem. Soc. **97** (1975) 7262; for stable arylnickel(III) complexes see: (c) D. M. Grove, G. van Koten and R. Zoet, J. Am. Chem. Soc. **105** (1983) 1379; (d) D. M. Grove, G. van Koten, W. P. Mul, A. A. H. van der Zeijden, J. Terheijden, M. C. Zoutberg and C. H. Stam, Organometallics **5** (1986) 322.
83. Ref. 78, p. 2635.
84. C. G. Gosden, K. P. Healy and D. Pletcher, J. Chem. Soc., Dalton Trans., (1978) 972.
85. For dependence of "Ni^I" reactivity as a function of its ligands see e.g. (a) B. Corain, M. Bressan, P. Rigo and A. Turco, Chem. Commun., (1968) 509; (b) M. Aresta, C. F. Nobile and A. Sacco, Inorg. Chim. Acta **12** (1975) 167; (c) F. V. Lovecchio, E. S. Gore and D. H. Busch, J. Am. Chem. Soc. **96** (1974) 3109.

CHAPTER 2

Structures of Seventeen-Electron Metal-Centered Radicals

Michael C. Baird

Department of Chemistry, Queen's University, Kingston, Canada K7L 3N6

Table of Contents

1. Introduction

Organotransition metal chemistry has long been dominated by compounds containing closed-shell electron configurations and obeying the eighteen-electron rule [1]. In recent years, however, there have appeared numerous reports concerning a wide variety of seventeen-electron organotransition metal species, and it is now clear that seventeen-electron compounds form a major subdivision of organotransition metal chemistry [2].

We present here a review of publications dealing with structures of and, where possible, bonding in the major classes of known seventeen-electron organometallic complexes. Discussion of paramagnetic sandwich complexes will be omitted, as the electronic structures and relevant chemistry of this class of compounds have been amply dealt with elsewhere [3].

In order to clearly distinguish in the text seventeen-electron species from closed-shell compounds of similar structures, the chemical formulae of the former will henceforth be enclosed in brace brackets.

2. Hexacoordinate Complexes

Vanadium hexacarbonyl, $\{V(CO)_6\}$, is unique among the simple binary metal carbonyls in that it is a readily isolable, seventeen-electron compound, which is thermally stable with respect to dimerization [4]. If the structure of $\{V(CO)_6)\}$

were rigorously octahedral (O_h symmetry), the compound would have a $(t_{2g})^5$ electron configuration and a $^2T_{2g}$ electronic ground state. It should therefore be subject to a Jahn-Teller distortion, and considerable effort has been expended to detect evidence for such a distortion.

An investigation of the magnetic behavior of freshly sublimed, polycrystalline $\{V(CO)_6\}$ in the temperature range 4.2 to 300 K found that the magnetic susceptibility, χ, decreases from 4.2 to 11 K, then increases to successive maxima at 28 K and 66 K, above which χ decreases smoothly [5]. The magnetic moment, corrected for diamagnetism, was found to increase very sharply from about 0.3 BM at 4 K to about 1.8 BM, the spin-only value for one unpaired electron, at 66 K, above which transition point the increase was much more gradual. A theoretical model was proposed in which $\{V(CO)_6\}$ in its equilibrium position is octahedral above 66 K, with the $^2T_{2g}$ ground term undergoing strong vibronic coupling. Below 66 K, there ensues a static, axial distortion accompanied by an antiferromagnetic magnetic exchange interaction.

The gas phase IR spectrum of $\{V(CO)_6\}$ in a heated IR cell exhibits a single, rather broad ν(CO) at 1988 cm^{-1} [6]. Comparison with the IR spectrum of the octahedral $Cr(CO)_6$ [7] shows that the *P*, *Q* and *R* branches of the rotational structure, present in the CO stretching band structure of $Cr(CO)_6$ (at 2000 cm^{-1}) [8], are absent in the spectrum of $\{V(CO)_6\}$; this has been taken to be a clear indication of a Jahn-Teller effect. Unfortunately, the separate rotational lines could not be resolved at the resolution achieved (0.3 cm^{-1}), and no distinction could be made between a dynamic Jahn-Teller effect and, for example, a small tetragonal distortion.

The UV-visible and MCD spectra of $\{V(CO)_6\}$, diluted in inert gas matrices at 5 K [9], have been successfully assigned on the basis of a $^2B_{2g}$ ground state arising from a very substantial static Jahn-Teller distortion. Because of the large number of possible excited states, the spectra are very complicated and it was not possible to distinguish between a trigonal or a tetragonal distortion. Interestingly more concentrated solutions of $\{V(CO)_6\}$ exhibit absorption at about 17,200 cm^{-1} and are thus deeply coloured, as is the pure carbonyl in the solid state. This observation has led to speculation concerning the possibility of intermolecular electron exchange interactions, but a better rationale appears to involve low energy intermolecular ionic excited states [10], (Eq. 1).

$$2\{V(CO)_6\} \xrightarrow{h\nu} [V(CO)_6]^+ + [V(CO)_6]^- \qquad (1)$$

An electron diffraction [11] study of gaseous $\{V(CO)_6\}$ at 341 K resulted in a radial distribution curve consistent with an octahedral structure, but with generally broader peaks than expected from a comparison with $Cr(CO)_6$. The broadening was attributed to dynamic Jahn-Teller distortions, although it was suggested that the nuclear configuration of maximum probability for most of the vibrational levels populated is octahedral at the temperature of the experiment. Complementing this work, an X-ray crystallographic [12] study at 245 K showed that $\{V(CO)_6\}$ is very close to regular octahedral, with possibly a marginal tetragonal distortion. A vibrational analysis suggested that dynamic effects, if present, are also small.

In contrast, ESR studies of $\{V(CO)_6\}$ [13, 14] at 77 K in frozen cyclohexane have been interpreted in terms of a non-axial distortion, with g_x, g_y and g_z, A_x, A_y and A_z all unequal (1.9789, 2.1121, 2.1370; 69.0, 157.0, 128.3 MHz), and hence incompatible with the conclusions reached in publications cited above. Spectra of $\{V(CO)_6\}$ at 14 K doped in polycrystalline $Cr(CO)_6$ exhibit similar spectra, suggesting that the non-axial distortion is not imposed by the host. It has been suggested that the ESR data are best interpreted in terms of D_{2h} symmetry for $\{V(CO)_6\}$ [14].

It would seem that the stability of $\{V(CO)_6\}$ with respect to dimerization is to be attributed to very extensive π delocalization of the odd electron onto the ligands [10, 14]. Indeed, the relative values of the IR-active ν(CO) of $Cr(CO)_6$ and $\{V(CO)_6\}$, mentioned above (2000 cm^{-1} and 1988 cm^{-1}, respectively [6, 8]), suggest that the carbonyl groups of the latter experience the greater degree of π-bonding in spite of there being one less electron in the vanadium compound.

In accord with this hypothesis, analyses of the hyperfine structure of the ESR spectra [14], complemented by SCF-Xα-DV calculations [10], suggest that $\{V(CO)_6\}$ is a π radical, with very extensive delocalization of the singly occupied molecular orbital (SOMO) over the carbonyl ligands. In contrast to the situation with other simple seventeen-electron species (see below), the unpaired spin of $\{V(CO)_6\}$ has little directional character and the reorganization energy required for the compound to rearrange in order to interact with one-electron ligands must therefore be prohibitively high. Thus $\{V(CO)_6\}$ does not dimerize, although $[V(CO)_6]_2$, with bridging carbonyls, may have been formed on co-condensation of vanadium atoms and carbon monoxide at low temperatures [15, 16].

Consistent with the above, compounds of the type $V(CO)_6X$ (X = one-electron donor) are not known [4], although many seven-coordinated phosphine-substituted derivatives of the type $V(CO)_{6-n}L_nX$ (n = 1–6; X = H, alkyl, other metals, etc) have been reported [4]. While it is not clear why the sterically more demanding tertiary phosphines would stabilize the higher coordination number, an ESR study of *fac*-$V(CO)_3(PMe_3)_3$ [17] has shown that the SOMO of this molecule has σ rather than π character (see below). Thus the SOMO of substituted compounds may generally be hybridized in a manner more appropriate for σ bonding.

Substitution reactions of $\{V(CO)_6\}$ are generally facile, and proceed as in Eq. 2 [18].

$$\{V(CO)_6\} \xrightarrow{+L} \{V(CO)_5L\} \xrightarrow{+L} \{V(CO)_4L_2\} \xrightarrow{+L} \{V(CO)_3L_3\} \qquad (2)$$

IR data for a number of substituted derivatives have been reported [18] and are listed, with some analogous compounds of chromium(0) [19, 20], in Table 1. As can be seen, the patterns of ν(CO) are generally similar to those of the more stable and hence better characterized chromium compounds. It is interesting to note that, for the monosubstituted series, back-donation appears to be more pronounced for the d^5 vanadium compounds than for the d^6 chromium compounds, as was observed above for the hexacarbonyls. However, the differences may well arise at least in part from solvent effects, as the trend is not quite as obvious for the disubstituted compounds.

Surprisingly few ESR data are available for substituted complexes of vanadium(0). A single crystal study [21] (93 K) of $\{$*trans*-$V(CO)_4(PMe_3)_2\}$ doped into the chromium analogue resembles $\{V(CO)_6\}$ in that low temperatures in the solid state are necessary to observe a spectrum, and that g_x, g_y and g_z, A_x, A_y and A_z are all unequal (2.0002, 2.0244, 2.0419; 83.9, 218.2, 91.7 MHz). The data have been interpreted in terms of a $^2B_{2g}$ ground state in D_{2h} symmetry, with the phosphorus atoms equivalent and the single electron in a d_{xz} orbital (Z axis coincident with the P-V-P axis).

Table 1. IR Data for Selected Mono- and Disubstituted Compounds of Vanadium(0), and for Analogous Compounds of Chromium(0) [18–20]

Compound	ν_{CO} (cm^{-1})
$\{V(CO)_5PMe_3\}$	2040(w), 1922(s, br)
$\{V(CO)_5P(n\text{-}Bu)_3\}$	2042(w), 1922(s, br)
$Cr(CO)_5P(n\text{-}Bu)_3$	2061(w), 1936(s), 1944(m)
$\{V(CO)_5PMePh_2\}$	2041(w), 1930(s, br)
$\{V(CO)_5P(OMe)_3\}$	2054(w), 1930(s, br)
$Cr(CO)_5P(OMe)_3$	2075(w), 1950(s), 1965(m)
$\{V(CO)_5PPh_3\}$	2050(w), 1930(s, br)
$Cr(CO)_5PPh_3$	2066(w), 1945(s)
$\{V(CO)_5P(i\text{-}Pr)_3\}$	2040(w), 1920(s, br)
$\{V(CO)_5AsPh_3\}$	2051(w), 1934(s, br)
$Cr(CO)_5AsPh_3$	2067(w), 1946(s)
trans-$\{V(CO)_4[PBu_3]_2\}$	1981(w), 1860(vs)
trans-$Cr(CO)_4[PBu_3]_2$	1873(s)
cis-$\{V(CO)_4[PMePh_2]_2\}$	1990(m), 1895(sh), 1867(s), 1845(s)
cis-$\{V(CO)_4[P(OMe)_3]_2\}$	2007(m), 1905(sh), 1885(s), 1865(sh)
trans-$Cr(CO)_4[P(OMe)_3]_2\}$	1965(w), 1916(s)
trans-$\{V(CO)_4[P(O\text{-}i\text{-}Pr)_3]_2\}$	1983(w), 1848(s)
cis-$\{V(CO)_4[P(O\text{-}i\text{-}Pr)_3]_2\}$	2000(w), 1901(sh), 1876(s), 1849(s)
trans-$\{V(CO)_4[PPh_3]_2\}$	1990(m), 1892(sh), 1860(vs)
trans-$Cr(CO)_4[PPh_3]_2$	1884

In contrast, the well-resolved ESR spectrum of $\{fac\text{-}V(CO)_3(PMe_3)_3\}$ can be observed in liquid solution [17], and has been interpreted in terms of a $^2A_1$ ground state in C_{3v} symmetry; the three phosphorus atoms are equivalent, and the unpaired electron is in a d_{z^2} orbital. The "best fit" parameters for a frozen spectrum of $\{fac\text{-}V(CO)_3(PMe_3)_3\}$ in methylene chloride agree well with the average values of g, A_V and A_P from liquid solution, 2.013, 136 MHz and 76 MHz, respectively.

The closed-shell compound $Cr(CO)_6$ undergoes electrochemical oxidation to yield $\{Cr(CO)_6^+\}$, isoelectronic with $\{V(CO)_6\}$ [22–24] and presumably also of O_h or near O_h symmetry. This cation is stable on a time scale of seconds, in contrast to the analogous complexes of molybdenum and tungsten [22, 23], but does not as yet appear to be well characterized. Published ESR spectral data [22, 24] differ sufficiently that they may well be attributable to different species, but it has been noted that the g value reported in ref. 24 is rather more similar to the g values of other octahedral chromium(I) complexes (Table 2). However, neither spectrum appears to implicate the type of orbital degeneracy noted above for $\{V(CO)_6\}$, and a definitive ESR study of the species $\{M(CO)_6\}^+$ (M = Cr, Mo, W) is yet to be carried out.

Interestingly, observation of vibrational fine structure on photoionization bands of $Cr(CO)_6$, $Mo(CO)_6$ and $W(CO)_6$ has resulted in the reasonable conclusion that the metal-carbon bond lengths of the corresponding seventeen-electron cations are about 0.1 Å longer than in the neutral, electronically-saturated molecules [25].

Table 2. IR and ESR Data for Selected Chromium(I) Complexes and their Eighteen-Electron Analogues [24, 26, 28, 29, 31, 32]

Complexes[a]	ν_{CO} (cm^{-1})	g_{av}
$\{Cr(CO)_6\}^+$		1.982, A_{Cr} 21.6 G
$\{$*trans*-$Cr(CO)_4[PPh_3]_2\}^+$	1990	2.027, A_P 18 G
trans-$Cr(CO)_4[PPh_3]_2$	1880	
$\{$*cis*-$Cr(CO)_4[dppm]\}^{+a}$	2084, 2033, 1969	2.009, A_P 6.9 G
cis-$Cr(CO)_4[dppm]^a$	2006, 1915, 1897, 1875	
$\{$*mer*-$Cr(CO)_3[P(OMe_3]_3\}^+$	2051, 1988, 1935	2.002, A_P 40 G(*av*)
mer-$Cr(CO)_3[P(OMe_3]_3$	1981, 1900, 1878	
$\{$*mer*-$Cr(CO)_3[PMe_2Ph]_3\}^+$	2010, 1940, 1904	2.019, A_P 20 G(t) 36 G(d)[a]
$\{Cr(CO)_3(dpme)(L^{Et})\}^{+a}$	2010, 1930, 1890	2.019, A_P 22.3 G
$Cr(CO)_3(dmpe)(L^{Et})^a$	1901, 1806	

[a] dppm = $Ph_2PCH_2PPh_2$, L^{Et} = carbenoid ligand, dmpe = $Me_2PCH_2CH_2PMe_2$, d = doublet, t = triplet

Substituted seventeen-electron complexes of the types $\{M(CO)_{6-n}L_n{}^+\}$, $\{M(CO)_4(L{-}L)^+\}$ and $\{M(CO)_2(L{-}L)_2{}^+\}$ (M = Cr, Mo, W; n = 1–3; L = monodentate tertiary phosphine; L—L = bidentate tertiary phosphine) have long been known, and are much more stable thermally than are the corresponding hexacarbonyl cations [26–32], although none has been characterized by X-ray crystallography. However, many of the ions are sufficiently stable that IR and ESR spectra can been readily obtained, and cases of geometrical isomerism of octahedral structures have been inferred on the basis of carbonyl stretching modes and/or hyperfine coupling to phosphorus (Table 2). As can be seen, the g values are normally significantly larger than the free electron value, while the ν_{CO} of the seventeen-electron complexes are generally about 100 cm^{-1} higher than are the ν_{CO} of the corresponding eighteen-electron compounds.

However, as with the species $\{M(CO)_6\}^+$ (M = Cr, Mo, W), the ESR spectra of the substituted complexes listed in Table 2 are very well resolved in liquid solution, even at high temperatures. These results stand in stark contrast to the vanadium(0) complexes discussed above, the ESR spectra of which are not readily detectable in solution because of a large g anisotropy, which is incompletely averaged by tumbling. However, elucidation of the electronic structures of the seventeen-electron complexes of chromium, molybdenum and tungsten awaits further research.

As with the isoelectronic vanadium(0) complexes, the seventeen-electron complexes of chromium, molybdenum and tungsten are much more labile with respect to ligand substitution and isomerization reactions than are the parent eighteen-electron compounds. Thus it has been shown that *cis-trans* or *fac-mer* isomerization reactions are far more facile than are the same processes for the electronically saturated analogues. In agreement with theory [33], the *trans*- and *mer*-isomers are generally more stable and, as electron transfer between seventeen- and eighteen-electron analogues is often facile, *cis* → *trans and fac* → *mer* isomerization of eighteen-electron compounds can be catalyzed by traces of the seventeen-electron complex (or of an appropriate oxidant).

3. Pentacoordinate Complexes

Co-condensation of $Cr(CO)_6$, $Mo(CO)_6$ and $W(CO)_6$ with alkali metal atoms in inert gas matrices results in the formation of the species $\{M(CO)_5^-\}$ (M = Cr, Mo, W). Careful analysis of the IR spectrum of the chromium analogue, obtained with $^{13}C^{16}O/^{12}C^{16}O$ mixtures, resulted in the conclusion that the radical has a square-pyramidal (C_{4v}) structure [34], in agreement with ESR data for the chromium complex formed by UV and γ irradiation of single crystals of the hydride complex $(Ph_4P)[HCr(CO)_5]$ [35]. For the seventeen-electron species, the g tensor was essentially identical ($g_{\|} = 2.0019$, $g_{\perp} = 2.0205$) with that attributed to the same species, generated by UV irradiation of $K_2[Cr_2(CO)_{10}]$. Thus the complexes $\{M(CO)_5\}^-$ assume structures similar to the those of the isoelectronic $\{Mn(CO)_5\}$ and $\{Fe(CO)_5^+\}$ complexes, but in contrast to the trigonal bipyramidal structures of the isoelectronic iron complexes $\{Fe(CO)_3L_2^+\}$ (see below).

The earlier literature on the binary carbonyls of manganese and rhenium has been reviewed [36, 37]. The simplest stable species are of the stoichiometry $M_2(CO)_{10}$, with metal-metal bonds joining two essentially square-pyramidal $M(CO)_5$ units [38]. While it is clear that equilibrium concentrations of the monomeric species $\{M(CO)_5\}$ in solutions at room temperatures are very low [36], it has long been known that photolysis of the dimers at appropriate wavelengths can lead to homolytic dissociation and generation of the corresponding seventeen-electron monomeric species $\{M(CO)_5\}$, apparently via population of the metal-metal σ^* orbital (Eq. 3) [39–49]. Substituted analogues can be made in the same way [40, 41, 50–55] (Eq. 4).

$$M_2(CO)_{10} \xrightarrow{h\nu} 2\{M(CO)_5\} \tag{3}$$

$$M_2(CO)_8L_2 \xrightarrow{h\nu} 2\{M(CO)_4L\} \tag{4}$$

The compound $\{Mn(CO)_5\}$ has been characterized by IR [45, 56], ESR [57, 58] and electronic spectroscopy [39, 44, 45]. It appears to assume an essentially square pyramidal (C_{4v}) structure, and ν_{CO} have been reported at 1987.6 cm^{-1} (e) and 1978.4 cm^{-1} (a_1) (CO matrix at 10–20 K) [56]. On the basis of the IR intensity data, an axial-equatorial bond angle of $96 \pm 3°$ has been calculated [56]. The ESR spectra have been obtained at low temperatures and in inert matrices, and are axial, with $g_{\|} = 2.00$ and $g_{\perp} = 2.04$, and manganese hyperfine splitting constants $A_{\|} = 65$ G and $A_{\perp} = -33.5$ G [58]. The odd electron is largely localized on the C_4 axis in an orbital of a_1 symmetry ($3d_{z^2}$), as predicted by theory [59].

The rhenium analogue $\{Re(CO)_5\}$ has also been similarly characterized [43, 46, 47], as have substituted manganese and rhenium compounds [50–53], for which ESR and IR data have been obtained [50–53]. All are believed to have structures similar to $\{Mn(CO)_5\}$, disubstituted complexes $\{M(CO)_3L_2\}$ being distorted square pyramids with trans-basal phosphines (Table 3).

Thus, in contrast to $\{V(CO)_6\}$ (see above), the SOMO in these compounds is oriented appropriately for attack by a σ radical. Furthermore, and again in con-

Table 3. IR and ESR Data for Selected Manganese(0) and Rhenium(0) Complexes [50, 52, 53]

Complexes	ν_{CO} (cm^{-1})	g_{av}
$\{Mn(CO)_3[P(n\text{-}Bu)_3]_2\}$	1860(sh), 1855(s)	$g_{\parallel}$, $g_{\perp}$ 2.040; $A_{\parallel}$(Mn) 54.1, $A_{\perp}$(Mn) 35.7; $A_{\parallel}$(P) 20, $A_{\perp}$(P) 19 G g_{iso} 2.030, $A(Mn)_{iso}$ 7, $A(P)_{iso}$ 21 G
$\{Mn(CO)_3[P(i\text{-}Bu)_3]_2\}$	1860(sh), 1855(s)	
$\{Mn(CO)_3[P(O\text{-}i\text{-}Pr)_3]_2\}$	1893(s), 1875(s)	
$\{Mn(CO)_3[P(OEt)_3]_2\}$		g_{iso} 2.031, $A(Mn)_{iso}$ 14, $A(P)_{iso}$ 42 G
$\{Mn(CO)_3[P(i\text{-}Pr)_3]_2\}$	1860(sh), 1855(s)	
$\{Re(CO)_3[P(c\text{-}Hx)_3]_2\}$	1849	$g_{\parallel}$ 1.9782, $g_{\perp}$ 2.1257; $A_{\parallel}$(Re) 412, $A_{\perp}$(Re) 78.7; $A_{\parallel}$(P), $A_{\perp}$(P) 14.2 G

trast to the situation with $\{V(CO)_6\}$, little reorganization of the ligands would be required during coupling to either a radical or to another $\{M(CO)_5\}$ species to form the coupled products.

Indeed, combination reactions of two molecules of $\{Mn(CO)_5\}$, of $\{Re(CO)_5\}$, or of their phosphine-substituted derivatives at room temperature to form the corresponding eighteen-electron dimers (reverse of equations 3, 4) proceed via second order processes at rates approaching the diffusion limit [39, 40, 42–48, 53, 60] (Table 4). The apparent correlation of second order rate constants for dimerization with cone angles of L suggests that steric effects are more important than electronic effects in controlling the dimerization reactions [40].

This proclivity towards dimerization stands in direct contrast with the behaviour of $\{V(CO)_6\}$, consistent with the very significant differences in the nature of the SOMO's of the two compounds (see above), and suggests that the seventeen-electron manganese and rhenium compounds are best considered to be metal-centered radicals. As can be seen in Table 4, dimerization is faster for $\{Re(CO)_5\}$ than for

Table 4. Second Order Rate Constants k_2 for Dimerization Reactions of Selected Seventeen-Electron Compounds $\{M(CO)_4L\}$ [40]

Compound	k_2 (M^{-1} s^{-1})	Cone Angle of L
$\{Mn(CO)_5\}$	9×10^8	95°
$\{Mn(CO)_4P(n\text{-}Bu_3)\}$	1×10^8	132°
$\{Mn(CO)_4P(i\text{-}Bu_3)\}$	2×10^7	143°
$\{Mn(CO)_4P(i\text{-}Pr_3)\}$	4×10^6	160°
$\{Mn(CO)_4P(OPh)_3\}$	9×10^7	128°
$\{Mn(CO)_4PPh_3\}$	1×10^7	145°
$\{Re(CO)_5\}$	4×10^9	95°

the smaller $\{Mn(CO)_5\}$ [42], and is less facile for compounds in which carbon monoxide has been substituted by tertiary phosphines, especially sterically demanding phosphines [50, 51, 53, 62–65].

A variety of α,β-diketones has also been found to react with the manganese- and rhenium-centered radicals to form paramagnetic compounds of the type $\{M(CO)_4$ (diketone)$\}$ [66, 67]. On the basis of ESR data, however, it has been suggested that these compounds are best regarded as being eighteen-electron complexes with a radical anion ligand. Similar conclusions have been reached concerning analogous compounds containing α,β-diimines [68].

The species $\{M(CO)_5^+\}$ (M = Fe, Os) have been prepared by γ irradiation of the corresponding neutral species in various matrices at low temperatures [69–71], and appear from their ESR parameters to have C_{4v} symmetry. Thus the ESR spectrum of a magnetically active species, obtained by irradiation of $Cr(CO)_6$ doped with $Fe(^{13}CO)_5$ and identified as $\{Fe(CO)_5^+\}$, indicated clearly the presence of one unique and four equivalent carbon atoms [69]. A species having the same form for the g-tensor components but a significantly different $g_\perp$ has been obtained from irradiating a solution of $Fe(CO)_5$ in trichlorofluoromethane, and has been assigned to the same radical [70]. Thus the structure appears to be similar to that of the isoelectronic $\{Mn(CO)_5\}$ (see above).

Substituted complexes of the types $\{Fe(CO)_{5-n}L_n^+\}$ (n = 1, 2; L = tertiary phosphines) have also been synthesized by controlled oxidation of the parent eighteen-electron compounds [72–76]. Although reactive in ways not yet fully elucidated, several of the iron-centered radicals are sufficiently stable that IR and ESR spectra have been obtained and are presented in Table 5. Unlike $\{Fe(CO)_5^+\}$ and the isoelectronic manganese analogues, compounds of the type $\{Fe(CO)_3L_2^+\}$ appear to assume trigonal bipyramidal (D_{3h}) structures rather than square pyramidal (C_{2v} if the L ligands coordinate in a trans basal arrangement). Thus ESR spectra exhibit hyperfine coupling of the free electron to two equivalent phosphorus nuclei, while the IR spectra exhibit only a single ν(CO). Enrichment of $\{Fe(CO)_3(PPh_3)_2^+\}$ with ^{13}CO to the extent of 50% results in a pattern of ν(CO)

Table 5. IR and ESR Data for Selected Complexes $\{trans\text{-}Fe(CO)_3L_2^+\}$ and their Eighteen-Electron Analogues [69, 72, 76]

L	ν_{CO} (cm^{-1})	g_{av}, (A_P (G))
$\{Fe(CO)_5^+\}$		2.054
$\{Fe(CO)_3[PMe_3]_2^+\}$	1990	2.056 (18.2)
$Fe(CO)_3[PMe_3]_2$	1863	
$\{Fe(CO)_3[PBu_3]_2^+\}$	1980	2.057 (18.5)
$Fe(CO)_3[PBu_3]_2$	1855	
$\{Fe(CO)_3[PMePh_2]_2^+\}$	1993	2.052 (18.8)
$Fe(CO)_3[PMePh_2]_2$	1878	
$\{Fe(CO)_3[PPh_3]_2^+\}$	1999	2.053 (18.7)
$Fe(CO)_3[PPh_3]_2$	1878	
$\{Fe(CO)_3[P(OPh)_3]_2^+\}$	2029	2.053 (25.0)
$Fe(CO)_3[P(OPh_3]_2$	1923	
$\{Fe(CO)_3[AsPh_3]_2^+\}$	1992	2.054
$\{Fe(CO)_3[AsPh_3]_2^+\}$	1876	

very similar to that of similarly enriched $Fe(CO)_3(PPh_3)_2$ [76], which is known to be of D_{3h} symmetry.

Interestingly, SCF-Xα-DV calculations on the molecule $\{Fe(CO)_3(PH_3)_2\}^+$ in both D_{3h} and C_{2v} symmetry have been carried out, and energy levels of the frontier and near frontier orbitals been determined [76]. Comparisons of calculated and experimental electronic spectra result in a much better fit for the D_{3h} structure, but the predicted ground electronic state would then be $^2E'$, subject to a Jahn-Teller distortion. Although the IR spectrum of ^{13}CO-enriched $\{Fe(CO)_3(PPh_3)_2{}^+\}$, mentioned above, appeared to exhibit one extra $\nu(CO)$, possibly indicative of a distortion, the fact that ESR spectra of this type of complex are well-resolved in solution at room temperature [72, 76] would appear to mitigate against this family of complexes being Jahn-Teller molecules. Further research clearly needs to be done.

4. Tetra- and Tricoordinate Complexes

Although the simplest stable binary carbonyl of cobalt is $Co_2(CO)_8$ [77], monomeric $\{Co(CO)_4\}$ has been formed in a variety of matrices at low temperature. As with other metal-metal bonded systems, photolysis of $Co_2(CO)_8$ and its substituted derivatives yields the metal radicals $\{Co(CO)_3L\}$ (L = CO, tertiary phosphine) (Eq. 5) [78–80], while $\{Co(CO)_4\}$ and has also been generated via hydrogen atom abstraction from $HCo(CO)_4$ by the trityl radical (Eq. 6) [81].

$$[Co(CO)_3L]_2 \xrightarrow{h\nu} 2\{Co(CO)_3L\} \quad (5)$$

$$HCo(CO)_4 + \{Ph_3C^{\cdot}\} \longrightarrow \{Co(CO)_4\} + Ph_3CH \quad (6)$$

$\{Co(CO)_4\}$ has been characterized by IR and ESR spectroscopy [82–84]. In a CO matrix at 6–15 K, the ESR spectrum is typical of a molecule with axial symmetry, with $g_{\parallel} = 2.007$, $g_{\perp} = 2.128$, $A_{\parallel}(Co) = 58 \cdot 10^{-4}\ cm^{-1}$ and $A_{\perp}(Co) = 55 \cdot 10^{-4}\ cm^{-1}$. Interestingly, with ^{13}CO-labelled compound, a large hyperfine coupling to only one carbonyl group is observed, suggesting C_{3v} symmetry. Thus, the ground state is $^2A_1$ in C_{3v} symmetry, and the unpaired electron resides in a cobalt $3d_{z^2}$–$4p_z$ hybrid orbital. The IR spectrum is consistent with this conclusion, as three ν_{CO} are observed at 2010.7 (e), 2028.8 (a_1) and 2107.0 (a_1) cm^{-1}. The analogous tetracarbonyls of rhodium and iridium have been studied by IR spectroscopy, and probably also have C_{3v} symmetry [84].

Treatment of $Fe(CO)_5$ and $Ni(CO)_4$ with alkali metal atoms in inert gas matrices results in the species $\{Fe(CO)_4{}^-\}$ and $\{Ni(CO)_3{}^-\}$, respectively [34]. Analyses of the IR spectra suggest a C_{3v} structure for $\{Fe(CO)_4{}^-\}$, a D_{3d} structure for $\{Ni(CO)_3{}^-\}$, although ESR data for the former have also been interpreted in terms of a D_{2d} structure [70]. However, little is known of these complexes. The compounds $Cu(CO)_3$ [85–87] and $Ag(CO)_3$ [87] have also been detected at low temperatures, and have been characterized by IR and ESR spectroscopy. While the former appears to be a trigonal planar π radical, the latter may exist in both planar and pyramidal forms.

5. Complexes of Polyhapto Ligands

The fifteen-electron compound vanadocene, $(\eta^5\text{-}C_5H_5)_2V$, readily combines with a variety of π-acid ligands L (L = CO, olefins, alkynes, but not tertiary phosphines) to form seventeen-electron complexes of the type $\{(\eta^5\text{-}C_5H_5)_2VL\}$ [88–93], in which the ligand L lies on the pseudo-C_2 (z) axis of the bent $(\eta^5\text{-}C_5H_5)_2V$ moiety. Similar compounds of stoichiometry $\{(\eta^5\text{-}C_5H_7)_2VL\}$ are formed with the acyclic pentadienyl ligand [91–93].

ESR (Table 6) and SCF-Xα-DV molecular orbital calculations have been carried out for several of the compounds, and it has been shown that the SOMO for the $\eta^5\text{-}C_5H_5$ compounds is of a_1 character and is largely localized between the cyclopentadienyl planes and hence accessible for bimolecular nucleophilic attack [92]. As a result, CO exchange reactions of $\{(\eta^5\text{-}C_5H_5)_2VCO\}$ and $\{(\eta^5\text{-}C_5Me_5)_2VCO\}$ are associative; the second order rate constants for the former at various temperatures are about three times those of the more crowded $\eta^5\text{-}C_5Me_5$ complex [91].

Table 6. Data for Selected Compounds $\{(\eta^5\text{-}C_5R_5)_2VL\}$ (R = H, Me) [90, 92]

Compound	g_{av}
$\{(\eta^5\text{-}C_5H_5)_2V(CO)\}$	2.005, A_V = 28.8 G
$\{(\eta^5\text{-}C_5Me_5)_2V(CO)\}$	2.000, A_V = 18.6 G
$\{(\eta^5\text{-}C_5H_5)_2V(CF_3C_2CF_3)\}$	1.997, A_V = 47.3 G

In contrast, the SOMO's of the pentadienyl and mixed cyclopentadienyl-pentadienyl complexes, although also of a_1 symmetry, appear to be oriented towards the centers of the polyhapto ligands, and thus are much less accessible towards nucleophiles [92]. As a result, CO exchange reactions are several orders of magnitude slower, and appear to be dissociative in nature [91].

The compound $[\eta^5\text{-}C_5H_5Cr(CO)_3]_2$ has been shown to have a very long metal-metal bond [94], and the apparent weakness [95] of this bond has been invoked to rationalize the high chemical reactivity of $[\eta^5\text{-}C_5H_5Cr(CO)_3]_2$ relative to the molybdenum and tungsten analogues [96–104]. (The compound $[\eta^5\text{-}C_5H_5Mo(CO)_3]_2$ actually has a shorter metal-metal bond than does $[\eta^5\text{-}C_5H_5Cr(CO)_3]_2$ [94]). In addition, ESR, UV-visible and IR studies have shown that $[\eta^5\text{-}C_5H_5Cr(CO)_3]_2$ dissociates somewhat to the paramagnetic monomer $\{\eta^5\text{-}C_5H_5Cr(CO)_3\}$ on sublimation and dissolution [100–102], and an equilibrium constant and thermodynamic parameters for the homolytic dissociation have been measured [102]. Although $[\eta^5\text{-}C_5H_5Cr(CO)_3]_2$ dissociates only to the extent of a few percent at room temperature, homolysis is far more extensive at higher temperatures and the carbonyl stretching modes have been assigned by IR difference spectroscopy at 92 °C [102]. Interestingly, the related compound $[\eta^5\text{-}C_5Me_5Cr(CO)_3]_2$ is also dimeric in the solid state but dissociates completely in solution, presumably because of the steric requirement of the $\eta^5\text{-}C_5Me_5$ ligand [104]. The solution chemistry of this radical strongly resembles that of the $\eta^5\text{-}C_5H_5$ compound.

Cosublimation of $[\eta^5\text{-}C_5H_5Cr(CO)_3]_2$ and $\eta^5\text{-}C_5H_5Mn(CO)_3$ results in a sample of $\{\eta^5\text{-}C_5H_5Cr(CO)_3\}$ doped into the crystals of the diamagnetic host, thus permitting a single-crystal ESR study [105, 106]. In C_{3v} symmetry, the d orbitals of a $M(CO)_3$ moiety transform as a_1 $(3d_{z^2}) + e(3d_{x^2-y^2}, 3d_{xy})$ (z-axis coinciding with C_3 axis), as outlined in a theoretical discussion of $\eta^5\text{-}C_5H_5Mn(CO)_3$ [107]. However, the chromium radical is subject to a Jahn-Teller distortion, and an examination of the ESR spectrum of the doped material, of necessity carried out below 25 K because of Jahn-Teller broadening above that temperature, resulted in the tentative conclusion that the unpaired electron is in an a' orbital in C_s symmetry. Thus the ordering of d orbitals is as postulated for $\eta^5\text{-}C_5H_5Mn(CO)_3$ [107], and the SOMO of $\{\eta^5\text{-}C_5H_5Cr(CO)_3\}$ was thought to be largely of $3d_{x^2-y^2}$ character.

However, a low temperature investigation of the ESR spectrum of $\{\eta^5\text{-}C_5Me_5Cr(CO)_3\}$ doped into a single crystal of the manganese analogue by cosublimation of $[\eta^5\text{-}C_5Me_5Cr(CO)_3]_2$ and $\eta^5\text{-}C_5Me_5Mn(CO)_3$ has shown that, for this species, the unpaired electron is in an a'' orbital, of largely d_{xy} character [108]. For the latter system, a crystallographically imposed plane of symmetry removes an ambiguity concerning one of the principal axes of the g-tensor for the $\eta^5\text{-}C_5H_5$ system discussed above, and it is thought that the interpretation of the spectrum of the $\eta^5\text{-}C_5Me_5$ compound probably also applies to the $\eta^5\text{-}C_5H_5$ compound [108]. Interestingly, the paramagnetic tris(pyrazolyl)borate complex $TpMo(CO)_3$ appears to have a rather similar electronic structure, although low temperature ESR data are not yet available [109].

Consistent with spontaneous homolysis of $[\eta^5\text{-}C_5H_5Cr(CO)_3]_2$ in solution at room temperature to form the presumed substitution-labile $\{\eta^5\text{-}C_5H_5Cr(CO)_3\}$, treatment of $[\eta^5\text{-}C_5H_5Cr(CO)_3]_2$ with a variety of tertiary phosphines L results in both the facile formation of compounds of the stoichiometry $\{\eta^5\text{-}C_5H_5Cr(CO)_2L\}$ (Eq. 7), for which IR data are presented in Table 7 [98, 99, 102].

$$[\eta^5\text{-}C_5H_5Cr(CO)_3]_2 + 2L \rightarrow 2\{\eta^5\text{-}C_5H_5Cr(CO)_2L\} \qquad (7)$$

The ease of substitution decreases as the steric requirements of L increase, as does the proclivity of $\{\eta^5\text{-}C_5H_5Cr(CO)_2L\}$ to dimerize. Thus $[\eta^5\text{-}C_5H_5Cr(CO)_2P(OMe)_3]_2$ is dimeric and diamagnetic in the solid state but with an even longer chromium-chromium bond than in $[\eta^5\text{-}C_5H_5Cr(CO)_3]_2$, but is extensively dissociated in solution to substituted monomer [98]. Complexes with bulkier phosphines are completely dissociated in solution and in the solid state, and the X-ray crystal structure has been determined in one case, $\{\eta^5\text{-}C_5H_5Cr(CO)_2PPh_3\}$ [99].

Table 7. IR Data (CH_2Cl_2) for Complexes of the Type $\{\eta^5\text{-}C_5H_5Cr(CO)_2L\}$ [103]

Compound	ν_{CO} (cm^{-1})
$\{\eta^5\text{-}C_5H_5Cr(CO)_2(PEt_3)\}$	1924, 1805
$\{\eta^5\text{-}C_5H_5Cr(CO)_2(PBu_3)\}$	1921, 1803
$\{\eta^5\text{-}C_5H_5Cr(CO)_2(PCy_3)\}$	1919, 1802
$\{\eta^5\text{-}C_5H_5Cr(CO)_2(PPh_3)\}$	1920, 1800
$\{\eta^5\text{-}C_5H_5Cr(CO)_2(PMe_2Ph)\}$	1925, 1807
$\{\eta^5\text{-}C_5H_5Cr(CO)_2(PEt_2Ph)\}$	1923, 1807
$\{[\eta^5\text{-}C_5H_5Cr(CO)_2]_2(\mu\text{-dppb})\}$	1915, 1806
$\{[\eta^5\text{-}C_5H_5Cr(CO)_2]_2(\mu\text{-dpph})\}$	1915, 1795

This compound is isostructural with the eighteen-electron manganese analogue, but with one subtle difference; the OC—Cr—CO bond angle is, at 80.9°, much less than the 90–92° normally found for such "piano-stool" compounds. It was suggested, and confirmed by a single crystal ESR spectrum of $\{\eta^5\text{-}C_5H_5Cr(CO)_2PPh_3\}$ doped into the diamagnetic manganese analogue,[107, 110] that the SOMO is of largely $3d_{xy}$ character such that the "hole" is localized between the carbonyl ligands. In contrast to $\{\eta^5\text{-}C_5H_5Cr(CO)_3\}$, the substituted radical $\{\eta^5\text{-}C_5H_5Cr(CO)_2PPh_3\}$ is not a Jahn-Teller molecule and its ESR spectrum should be observable in liquid solution. Thus, although the single crystal ESR spectrum of $\{\eta^5\text{-}C_5H_5Cr(CO)_2PPh_3\}$ was best resolved at 173 K, a spectrum could also be detected in toluene at 203 K. On the basis of solution spectral parameters (g = 2.030, A_P = 99 MHz), the species in solution and in the solid were identical.

The X-ray crystal structure of the analogous compound $\{\eta^5\text{-}C_5Me_5Cr(CO)_2PMe_3\}$ has also recently been determined, and a similar distortion has been found. In this case, the OC—Cr—CO bond angle is 79.7°(3) [111], and again it would seem that the SOMO is largely localized between the carbonyl ligands although a single crystal ESR study has not yet been carried out.

These conclusions concerning the electronic structures of the chromium-centered radicals have found support [112] in LCAO-HFS calculations employing an uncontracted triple ζ-STO basis set for Cr and a double ζ-STO basis set for the ligand atoms. The electronic ground state for $\{\eta^5\text{-}C_5H_5Cr(CO)_3\}$ in C_{3v} symmetry, where OC—Cr—CO = 91° is $(1a')^2(1a'')^2(2a')^1$, of $^2A'$ symmetry; the first excited state $^2A''$ with a $(1a')^2(2a')^2(1a'')^1$ electron configuration is, however, only 12 kJ/mol higher in energy. Geometry optimization of $\{\eta^5\text{-}C_5H_5Cr(CO)_3\}$ for the two states such that the OC—Cr—CO bond angles were allowed to relax under C_s symmetry resulted in the conclusion that the $^2A'$, state, in which the OC—Cr—CO angles are 85°, 85° and 100°, is lower by 7 kJ/mol than the $^2A''$ state, in which the OC—Cr—CO angles are 84°, 92° and 92°. Furthermore, the barrier to interconversion between the two states was calculated to be only about 20 kJ/mol, consistent with the presence of the dynamic Jahn-Teller distortion implied by the ESR studies.

Interestingly, similar calculations on $\{\eta^5\text{-}C_5H_5Cr(CO)_2(PH_3)\}$ suggest that the ground state for this molecule is $^2A''$, with the OC—Cr—CO angle 83°, the P—Cr—CO angles 91° [112]. These results are in excellent agreement with the X-ray crystal structures of the two substituted compounds $\{\eta^5\text{-}C_5H_5Cr(CO)_2(PPh_3)\}$ and $\{\eta^5\text{-}C_5Me_5Cr(CO)_2(PMe_3)\}$, and with the ESR data for the former (see above).

Photolysis of $[\eta^5\text{-}C_5H_5M(CO)_3]_2$ [113–115] and $\eta^5\text{-}C_5H_5M(CO)_3H$ [116] (M = Mo, W) has been shown to yield the monomeric species $\{\eta^5\text{-}C_5H_5M(CO)_3\}$, which may also be obtained electrochemically from the dimers [117]. The IR spectra [116] of the molybdenum- and tungsten-centered radicals in CO matrices at low temperature are very similar to the solution spectra of the two chromium analogues mentioned above [102, 104] (Table 8). That all three species are similar is indicated by the observations that, for all three, the high frequency absorption is very sharp, while the lower frequency doublet is very broad.

Complexes $\{(\eta^6\text{-arene})Cr(CO)_2L^+\}$ (L = CO, tertiary phosphine) can be prepared by chemical and electrochemical oxidation of the corresponding eighteen-electron compounds and, although the tricarbonyl species appear to be too unstable to isolate, the substituted complexes are sufficiently robust that several have been isolated and characterized by IR and ESR spectroscopy (Table 9) [118–121].

Table 8. IR Data for the Compounds $\{\eta^5\text{-}C_5H_5M(CO)_3\}$ (M = Cr, Mo, W) [102, 104, 116]

Compound	$\nu(CO)$ (cm^{-1})
$\{\eta^5\text{-}C_5H_5Cr(CO)_3\}$	2011, 1901, 1894
$\{\eta^5\text{-}C_5H_5Mo(CO)_3\}$	2009, 1916, 1908
$\{\eta^5\text{-}C_5H_5W(CO)_3\}$	1999, 1900, 1897

While no crystallographic data are available for these types of compounds, cooling is necessary to obtain resolved ESR spectra, which exhibit *g* factors and phosphorus hyperfine couplings comparable to those of the above-mentioned $\{\eta^5\text{-}C_5H_5Cr(CO)_2(PPh_3)\}$. Thus very similar chemical and electronic structures seem implied for the two series of complexes. Interestingly, the cationic, diphosphine-bridged bis-arene complex $\{(\mu\text{-biphenyl})[Cr(CO)_2]_2(\mu\text{-dppm})^+\}$ displays IR and ESR spectra indicative of a charge- and spin-localized mixed valent complex [119].

Table 9. IR and ESR Data for Selected Compounds $\{(\eta^6\text{-}C_6Me_6)Cr(CO)_2L^+\}$ and their Eighteen-Electron Analogues [118, 119]

Complex	ν_{CO} (cm^{-1})	g_{av}
$\{(\eta^6\text{-}C_6Me_6)Cr(CO)_2PPh_3^+\}$	1969, 1853	2.041, A_P = 31.3 G
$(\eta^6\text{-}C_6Me_6)Cr(CO)_2PPh_3$	1857, 1796	
$\{(\eta^6\text{-}C_6Me_6)Cr(CO)_2PMePh_2^+\}$	1965, 1849	
$(\eta^6\text{-}C_6Me_6)Cr(CO)_2PMePh_2$	1853, 1793	
$\{(\eta^6\text{-}C_6Me_6)Cr(CO)_2PMe_2Ph^+\}$	1961, 1849	
$(\eta^6\text{-}C_6Me_6)Cr(CO)_2PMe_2Ph$	1850, 1787	
$\{(\eta^6\text{-biphenyl})Cr(CO)_2PPh_3^+\}$	1993, 1886	2.043, A_P = 30.6 G
$(\eta^6\text{-biphenyl})Cr(CO)_2PPh_3$	1887, 1833	

A large number of complexes of the type $\{(\eta^5\text{-}C_5H_{5-x}Me_x)M(CO)_{3-y}L_y{}^+\}$ (M = Mn, Re; L = nitrogen, phosphorus donor; x = 0, 1, 5; y = 0, 1) have been prepared via oxidation of the eighteen-electron neutral precursors [122–126]. Many of the seventeen-electron species are sufficiently stable that they have been characterized by and their reactions monitored by IR and ESR spectroscopy and electrochemistry (Table 10). In contrast to the isoelectronic $(\eta^5\text{-}C_5H_5)$chromium(I) and $(\eta^6$-arene)chromium(I) complexes mentioned above, well-resolved ESR spectra of the manganese(II) complexes discussed here are readily observed in liquid solution at room temperature. The significance of this observation in terms of electronic structures has not yet been addressed fully.

The compound $\{\eta^5\text{-}C_5H_5Fe(CO)_2\}$ has been generated in a variety of ways, especially photochemically from the corresponding dimer [127–134]. It is very unstable with respect to dimerization, but use of very fast time resolved IR spectroscopy has permitted both the observation of the carbonyl stretching modes of the short-lived iron-centered radical and the monitoring of its dimerization reaction and of its substitution reactions with various ligands L to form

Table 10. IR and ESR Data for Selected Compounds $\{(\eta^5\text{-}C_5H_{5-x}Me_x)Mn(CO)_{3-y}L_y^+\}$ and their Eighteen-Electron Analogues [122]

Complex	ν_{CO} (cm^{-1})	g_{av}
$\{(\eta^5\text{-}C_5H_5)Mn(CO)_2PPh_3^+\}$	2045, 1965	
$(\eta^5\text{-}C_5H_5)Mn(CO)_2PPh_3$	1931, 1863	
$\{(\eta^5\text{-}C_5H_5)Mn(CO)_2PEt_3^+\}$	2037, 1951	
$(\eta^5\text{-}C_5H_5)Mn(CO)_2PEt_3$	1922, 1851	
$\{(\eta^5\text{-}C_5H_4Me)Mn(CO)_2PPh_3^+\}$	2041, 1961	
$(\eta^5\text{-}C_5H_4Me)Mn(CO)_2PPh_3$	1927, 1861	
$\{(\eta^5\text{-}C_5H_4Me)Mn(CO)_2PEt_3^+\}$	2029, 1945	
$(\eta^5\text{-}C_5H_4Me)Mn(CO)_2PEt_3$	1919, 1845	
$\{(\eta^5\text{-}C_5H_4Me)Mn(CO)_2PMe_2Ph^+\}$	2029, 1953	
$(\eta^5\text{-}C_5H_4Me)Mn(CO)_2PMe_2Ph$	1923, 1852	
$\{(\eta^5\text{-}C_5H_4Me)MnCO[PPh_3]_2^+\}$	1913	2.02, A_{Mn} = 93.2 G, A_P = 24.5 G
$(\eta^5\text{-}C_5H_4Me)MnCO[PPh_3]_2$	1816	
$\{(\eta^5\text{-}C_5H_4Me)MnCO[MePPh_2]_2^+\}$	1912	2.02, A_{Mn} = 93.2 G, A_P = 24.5 G
$(\eta^5\text{-}C_5H_4Me)MnCO[MePPh_2]_2$	1815	

$\{\eta^5\text{-}C_5H_5Fe(CO)L\}$ [130–132]. The dimerization proceeds with a rate constant near the diffusion-controlled limit, while the substitution reactions are associative but less rapid.

Little is known of the structure of the species $\{\eta^5\text{-}C_5H_5Fe(CO)_2\}$, although a theoretical treatment suggested that it should be "planar", with the unpaired electron in an essentially $3d_{xy}$ orbital [135]. However, the ESR spectrum of the radical in frozen solution at 77 K suggests that the complex is axially symmetric, with $g_{\parallel}$ 2.003 and $g_{\perp}$ 2.144, and the unpaired electron in essentially a $3d_{z^2}$ orbital, orthogonal to the OC—Fe—CO plane [134]. The analogous radicals $\{\eta^5\text{-}C_5H_5Ru(CO)_2\}$ [127] and $\{\eta^5\text{-}C_5Me_5Fe(CO)_2\}$ [133] have also been studied and shown to have properties similar to those of $\{\eta^5\text{-}C_5H_5Fe(CO)_2\}$.

6. Summary

As will by now be clear, a wide variety of seventeen-electron organotransition metal complexes has been reported, but little is generally known of either chemical or electronic structures. Very few compounds have been characterized by X-ray crystallography and, in addition, very few investigations of seventeen-electron complexes have utilized single-crystal ESR spectroscopy, a technique which is in principle very powerful for the elucidation of electron structures of paramagnetic compounds.

However, as has been pointed out above, structures of many seventeen-electron complexes have only been inferred from spectroscopic data, and in several cases, complexes which should therefore be Jahn-Teller molecules appear from their ESR data not to be. In addition, in several isoelectronic series of complexes, individual members appear to have very different structures, although definitive crystallographic information is generally not yet available. Clearly more structural work is to be done on this interesting and increasingly important class of compounds.

7. Acknowledgment

Financial support from the Natural Sciences and Engineering Research Council has made much of the author's research in this area possible. Gratitude must also be expressed to T. Ziegler, K. F. Preston and J. R. Morton for information concerning unpublished results.

8. References

1. See Comprehensive Organometallic Chemistry, G. Wilkinson, F. G. A. Stone and E. W. Abel, Eds.; Pergamon: London, 1982; Chapters 19–43.
2. M. C. Baird, Chem. Rev. **88** (1988) 1217.
3. For recent publications, see (a) J. H. Ammeter, J. Magn. Res. **30** (1978) 299. (b) A. Haaland, Acc. Chem. Res. **11** (1979) 415. (c) S. P. Solodovnikov, Russ. Chem. Rev. **51** (1982) 961. (d) F. G. N. Cloke, A. N. Dix, J. C. Green, R. N. Perutz and E. A. Seddon, Organometallics **2** (1983) 1150, and references therein. (e) D. Astruc, Comments on Inorg. Chem. **6** (1987) 61. (f) D. Astruc, Acc. Chem. Res. **19** (1986) 377. (g) R. M. Nielson, M. N. Golovin, G. E. McManis and M. J. Weaver, J. Am. Chem. Soc. **110** (1988) 1745. (h) D. O'Hare, J. C. Green, T. P. Chadwick and J. S. Miller, Organometallics **7** (1988) 1335.
4. For an overview of the earlier literature, see N. G. Connelly, Comprehensive Organometallic Chemistry; G. Wilkinson, F. G. A. Stone and E. W. Abel, Eds.; Pergamon: London, 1982, Chapter 24.2.
5. J. C. Bernier and O. Kahn, Chem. Phys. Lett. **19** (1973) 414.
6. Y. Ishikawa, P. A. Hackett and D. M. Rayner, J. Am. Chem. Soc. **109** (1987) 6644.
7. H. Haas and R. K. Sheline, J. Am. Chem. Soc. **88** (1966) 3219.
8. L. H. Jones, R. S. McDowell and M. Goldblatt, Inorg. Chem. **8** (1969) 2349.
9. T. J. Barton, R. Grinter and A. J. Thomson, J. Chem. Soc., Dalton Trans. (1978) 608.
10. G. F. Holland, M. C. Manning, D. E. Ellis and W. C. Trogler, J. Am. Chem. Soc. **105** (1983) 2308.
11. D. G. Schmidling, J. Mol. Struct. **24** (1975) 1.
12. S. Bellard, K. A. Rubinson and G. M. Sheldrick, Acta Crystallogr. **B 35** (1979) 271.
13. M. P. Boyer, Y. LePage, J. R. Morton, K. F. Preston and M. J. Vuolle, Can. J. Spectrosc. **26** (1981) 181.
14. S. W. Bratt, A. Kassyk, R. N. Perutz and M. C. R. Symons, J. Am. Chem. Soc. **104** (1982) 490.
15. T. C. Devore and H. F. Franzen, Inorg. Chem. **15** (1976) 1318.
16. T. A. Ford, H. Huber, W. Klotzbücher, M. Moskovits and G. A. Ozin, Inorg. Chem. **15** (1976) 1666.
17. J. M. McCall, J. R. Morton and K. F. Preston, J. Magn. Res. **64** (1985) 414.
18. Q.-Z. Shi, T. G. Richmond, W. C. Trogler and F. Basolo, J. Am. Chem. Soc. **106** (1984) 71.
19. M. J. Wovkulich and J. D. Atwood, J. Organomet. Chem. **77** (1979) 184.
20. M. J. Wovkulich, S. J. Feinberg and J. D. Atwood, Inorg. Chem. **19** (1980) 2608.
21. J. M. McCall, J. R. Morton and K. F. Preston, Organometallics **4** (1985) 1272.
22. C. J. Pickett and D. Pletcher, J. Chem. Soc., Dalton Trans. (1975) 879.
23. H. L. Chum, D. Koran and R. A. Osteryoung, J. Organomet. Chem. **140** (1977) 349.
24. R. N. Bagchi, A. M. Bond, R. Colton, D. L. Luscombe and J. E. Moir, J. Am. Chem. Soc. **108** (1986) 3352.

25. J. L. Hubbard and D. L. Lichtenberger, J. Am. Chem. Soc. **104** (1982) 2132.
26. A. M. Bond, R. Colton, J. E. Kevekordes and P. Panagiotidou, Inorg. Chem. **26** (1987) 1430.
27. A. M. Bond, R. Colton and J. E. Kevekordes, Inorg. Chem. **25** (1986) 749.
28. R. N. Bagchi, A. M. Bond, G. Brain, R. Colton, T. L. E. Henderson and J. E. Kevekordes, Organometallics **3** (1984) 4.
29. A. M. Bond, S. W. Carr and R. Colton, Inorg. Chem. **23** (1984) 2343.
30. A. M. Bond, S. W. Carr and R. Colton, Organometallics **3** (1984) 541.
31. M. F. Lappert, R. W. McCabe, J. J. MacQuitty, P. L. Pye and P. I. Riley, J. Chem. Soc., Dalton Trans. (1980) 90.
32. P. K. Ashford, P. K. Baker, N. G. Connelly, R. L. Kelly and V. A. Woodley, J. Chem. Soc., Dalton Trans. (1982) 477.
33. D. M. P. Mingos, J. Organomet. Chem. **179** (1979) C29.
34. P. A. Breeze, J. K. Burdett and J. J. Turner, Inorg. Chem. **20** (1981) 3369.
35. J. R. Morton and K. F. Preston and M. Y. Darensbourg, Magn. Res. Chem. **26** (1988) 787.
36. P. M. Treichel, Comprehensive Organometallic Chemistry; G. Wilkinson, F. G. A. Stone and E. W. Abel, Eds.; Pergamon: London, 1982; Chapter 29.
37. N. M. Boag and H. D. Kaesz, Comprehensive Organometallic Chemistry; G. Wilkinson, F. G. A. Stone and E. W. Abel, Eds.; Pergamon: London, 1982; Chapter 30.
38. M. R. Churchill, K. N. Amoh and H. J. Wasserman, Inorg. Chem. **20** (1981) 1609.
39. T. Kobayashi, K. Yasufuku, J. Iwai, H. Yesaka, H. Noda and H. Ohtani, Coord. Chem. Rev. **64** (1985) 1.
40. R. S. Herrick, Rev. Inorg. Chem. **8** (1986) 1.
41. M. S. Wrighton and D. S. Ginley, J. Am. Chem. Soc. **97** (1975) 2065.
42. R. W. Wegman, R. J. Olsen, D. R. Gard, L. R. Faulkner and T. L. Brown, J. Am. Chem. Soc. **103** (1981) 6089.
43. W. K. Meckstroth, R. T. Walters, W. L. Waltz, A. Wojcicki and L. M. Dorfman, J. Am. Chem. Soc. **104** (1982) 1842.
44. H. Yesaka, T. Kobayashi, K. Yasufuku and S. Nagakura, J. Am. Chem. Soc. **105** (1983) 6249.
45. S. P. Church, H. Hermann, F.-W. Grevels and K. Schaffner, J. Chem. Soc., Chem. Comm. (1984) 785.
46. K. Yasufuku, H. Noda, J. Iwai, H. Ohtani, M. Hoshino and T. Kobayashi, Organometallics **4** (1985) 2174.
47. T. Kobayashi, H. Ohtani, H. Noda, S. Teratani, H. Yamazaki and K. Yasufuku, Organometallics **5** (1986) 110.
48. T. A. Seder, S. P. Church and E. Weitz, J. Am. Chem. Soc. **108** (1986) 7518.
49. D. A. Prinslow and V. Vaida, J. Am. Chem. Soc. **109** (1987) 5097.
50. D. R. Kidd, C. P. Cheng and T. L. Brown, J. Am. Chem. Soc. **100** (1978) 4103.
51. S. B. McCullen and T. L. Brown, J. Am. Chem. Soc. **104** (1982) 7496.
52. H. W. Walker, G. B. Rattinger, R. L. Belford and T. L. Brown, Organometallics **2** (1983) 775.
53. H. W. Walker, R. S. Herrick, R. J. Olsen and T. L. Brown, Inorg. Chem. **23** (1984) 3748.
54. K.-W. Lee and T. L. Brown, J. Am. Chem. Soc. **109** (1987) 3269.
55. P. Rushman and T. L. Brown, J. Am. Chem. Soc. **109** (1987) 3632.
56. S. P. Church, M. Poliakoff, J. A. Timney and J. J. Turner, J. Am. Chem. Soc. **103** (1981) 7515.
57. J. A. Howard, J. R. Morton and K. F. Preston, Chem. Phys. Lett. **83** (1981) 226.
58. M. C. R. Symons and R. L. Sweany, Organometallics **1** (1982) 834.
59. M. H. Elian and R. Hoffmann, Inorg. Chem. **14** (1975) 1058.
60. W. L. Waltz, O. Hackelberg, L. M. Dorfman and A. Wojcicki, J. Am. Chem. Soc. **100** (1978) 7259.
61. D. A. Lacombe, J. E. Anderson and K. M. Kadish, Inorg. Chem. **25** (1986) 2074.
62. D. R. Kidd and T. L. Brown, J. Am. Chem. Soc. **100** (1978) 4095.
63. S. B. McCullen, H. W. Walker and T. L. Brown, J. Am. Chem. Soc. **104** (1982) 4007.

64. D. R. Tyler, and A. S. Goldman, J. Organomet. Chem. **311** (1986) 349.
65. A. Fox, J. Malito and A. Poë, J. Chem. Soc., Chem. Commun. (1981) 1052.
66. T.-I. Ho, C.-M. Chang, S. R. Wang and C. P. Cheng, J. Chem. Soc., Dalton Trans. (1988), 123 and references therein.
67. C. P. Cheng, S. R. Wang, J. C. Lin and S.-L. Wang, J. Organomet. Chem. **249** (1988) 375.
68. R. R. Andréa, W. G. J. de Lange, T. van der Graaf, M. Rijkhoff, D. J. Stufkens and A. Oskam, Organometallics **7** (1988) 1100.
69. T. Lionel, J. R. Morton and K. F. Preston, J. Chem. Phys. **76** (1982) 234.
70. B. M. Peake, M. C. R. Symons and J. L. Wyatt, J. Chem. Soc., Dalton Trans. (1983) 1171.
71. T. Lionel, J. R. Morton and K. F. Preston, J. Magn. Res. **49** (1982) 225.
72. P. K. Baker, N. G. Connolly, B. M. R. Jones, J. P. Maher and K. R. Somers, J. Chem. Soc., Dalton Trans. (1980) 579.
73. S. W. Blanch, A. M. Bond and R. Colton, Inorg. Chem. **20** (1981) 755.
74. R. N. Bagchi, A. M. Bond, C. L. Heggie, T. L. Henderson, E. Mocellin and R. A. Seikel, Inorg. Chem. **22** (1983) 3007.
75. M. J. Therien, C.-L. Ni, F. C. Anson, J. G. Osteryoung and W. C. Trogler, J. Am. Chem. Soc. **108** (1986) 4037.
76. M. J. Therien and W. C. Trogler, J. Am. Chem. Soc. **108** (1986) 3697.
77. R. D. W. Kemmitt and D. R. Russell, Comprehensive Organometallic Chemistry; G. Wilkinson, F. G. A. Stone and E. W. Abel, Eds.; Pergamon: London, 1982, Chapter 34.
78. R. L. Sweany and T. L. Brown, Inorg. Chem. **16** (1977) 421.
79. C. L. Reichel and M. S. Wrighton, J. Am. Chem. Soc. **101** (1979) 6769.
80. R. W. Wegman and T. L. Brown, Organometallics **1** (1982) 47.
81. F. Ungváry and L. Markó, J. Organomet. Chem. **193** (1980) 383.
82. L. A. Hanlan, H. Huber, E. P. Kündig, B. R. McGarvey and G. A. Ozin, J. Am. Chem. Soc. **97** (1975) 7054, and references therein.
83. S. A. Fairhurst, J. R. Morton and K. F. Preston, J. Magn. Res. **55** (1983) 453.
84. G. A. Ozin and A. J. L. Hanlan, Inorg. Chem. **18** (1979) 2091.
85. J. A. Howard, B. Mile, J. R. Morton, K. F. Preston and R. Sutcliffe, J. Phys. Chem. **90** (1986) 1033, and references therein.
86. J. A. Howard, B. Mile, J. R. Morton, K. F. Preston and R. Sutcliffe, Chem. Phys. Lett. **117** (1985) 115.
87. J. A. Howard, B. Mile, J. R. Morton and K. F. Preston, J. Phys. Chem. **90** (1986) 2027.
88. G. Fachinetti, S. Del Nero and C. Floriani, J. Chem. Soc., Dalton Trans. (1976) 1046.
89. G. Fachinetti, C. Floriani, A. Chiesi-Villa and C. Guastini, Inorg. Chem. **18** (1979) 2282.
90. J. L. Petersen and L. Griffith, Inorg. Chem. **19** (1980) 1852.
91. R. M. Kowaleski, F. Basolo, W. C. Trogler, R. W. Gedridge, T. D. Newbound and R. D. Ernst, J. Am. Chem. Soc. **109** (1987) 4860.
92. R. M. Kowaleski, F. Basolo, J. H. Osborne and W. C. Trogler, Organometallics **7** (1988) 1425.
93. R. D. Ernst, J.-Z. Liu and D. R. Wilson, J. Organomet. Chem. **250** (1983) 257.
94. R. D. Adams, D. E. Collins and F. A. Cotton, J. Am. Chem. Soc. **96** (1974) 749.
95. J. T. Landrum and C. D. Hoff, J. Organomet. Chem. **282** (1985) 215.
96. P. Hackett, P. S. O'Neill and A. R. Manning, J. Chem. Soc., Dalton Trans. (1974) 1625.
97. T. Madach and H. Vahrenkamp, Z. Naturforsch. **34b** (1979) 573.
98. L.-Y. Goh, M. J. D'Aniello, S. Slater, E. L. Muetterties, I. Tavanaiepour, M. I. Chang, M. F. Fredrich and V. W. Day, Inorg. Chem. **18** (1979) 192.
99. N. A. Cooley, K. A. Watson, S. Fortier and M. C. Baird, Organometallics **5** (1986) 2563.
100. H. J. Keller, Z. Naturforsch. **23b** (1968) 133.
101. T. Madach and H. Vahrenkamp, Z. Naturforsch. **33b** (1978) 1301.
102. S. J. McLain, J. Am. Chem. Soc. **110** (1988) 643.
103. N. A. Cooley, P. T. F. MacConnachie and M. C. Baird, Polyhedron **7** (1988) 1965.
104. T. J. Jaeger and M. C. Baird, Organometallics **7** (1988) 2074.
105. P. J. Krusic, S. J. McLain, J. R. Morton, K. F. Preston and Y. LePage, J. Magn. Res. **74** (1987) 72.

106. J. R. Morton, K. F. Preston, N. A. Cooley, M. C. Baird, P. J. Krusic and S. J. McLain, J. Chem. Soc. Faraday Trans **1**, Vol. **83** (1987) 3535.
107. D. L. Lichtenberger, and R. F. Fenske, J. Am. Chem. Soc. **98** (1976) 60.
108. J. R. Morton and K. F. Preston, unpublished results.
109. M. D. Curtis, K.-B. Shiu, W. M. Butler and J. C. Huffman, J. Am. Chem. Soc. **108** (1986) 3335.
110. N. A. Cooley, M. C. Baird, J. R. Morton, K. F. Preston and Y. LePage, J. Magn. Res. **76** (1988) 325.
111. T. J. Jaeger, K. A. Watson, S. Fortier and M. C. Baird, unpublished results.
112. T. Ziegler, unpublished results.
113. M. S. Wrighton and D. S. Ginley, J. Am. Chem. Soc. **97** (1975) 4246.
114. J. L. Hughey, C. B. Bock and T. J. Meyer, J. Am. Chem. Soc. **97** (1975) 4440.
115. R. M. Laine and P. C. Ford, Inorg. Chem. **16** (1977) 388.
116. K. A. Mahmoud, A. J. Rest and H. G. Alt, J. Organomet. Chem. **246** (1983) C37.
117. K. M. Kadish, D. A. Lacombe and J. E. Anderson, Inorg. Chem. **25** (1986) 2246.
118. N. G. Connelly, Z. Demidowicz and R. L. Kelly, J. Chem. Soc., Dalton Trans. (1975) 2335.
119. N. van Order, W. E. Geiger, T. E. Bitterwolf and A. L. Rheingold, J. Am. Chem. Soc. **109** (1987) 5680.
120. K. M. Doxsee, R. H. Grubbs and F. C. Anson, J. Am. Chem. Soc. **106** (1984) 7819.
121. C. G. Zoski D. A. Sweigart, N. J. Stone, P. H. Rieger, E. Mocellin, T. F. Mann, D. R. Mann, D. K. Gosser, M. M. Doeff and A. M. Bond, J. Am. Chem. Soc. **110** (1988) 2109.
122. N. G. Connelly and M. D. Kitchen, J. Chem. Soc. Dalton Trans. (1977) 931.
123. J. W. Hershberger, R. J. Klingler and J. K. Kochi, J. Am. Chem. Soc. **105** (1983) 61.
124. J. W. Hershberger, C. Amatore and J. K. Kochi, J. Organomet. Chem. **250** (1983) 345.
125. P. M. Zizelman, C. Amatore and J. K. Kochi, J. Am. Chem. Soc. **106** (1984) 3771.
126. J. K. Kochi, J. Organomet. Chem. **300** (1986) 139.
127. H. B. Abrahamson, M. C. Palazotto, C. L. Reichel and M. S. Wrighton, J. Am. Chem. Soc. **101** (1979) 4123.
128. J. V. Caspar and T. J. Meyer, J. Am. Chem. Soc. **102** (1980) 7794.
129. D. R. Tyler, M. A. Schmidt and H. B. Gray, J. Am. Chem. Soc. **105** (1983) 6018.
130. B. D. Moore, M. B. Simpson, M. Poliakoff and J. J. Turner, J. Chem. Soc., Chem. Commun, (1984) 972.
131. A. J. Dixon, M. A. Healy, M. Poliakoff and J. J. Turner, J. Chem. Soc., Chem. Commun. (1986) 994.
132. A. J. Dixon, S. J. Gravelle, L. J. van de Burgt, M. Poliakoff, J. J. Turner and E. Weitz, J. Chem. Soc., Chem. Commun. (1987) 1023.
133. B. D. Moore, M. Poliakoff and J. J. Turner, J. Am. Chem. Soc. **108** (1986) 1819.

CHAPTER 3

Atom Transfer Reactions and Radical Chain Processes Involving Atom Transfer

Theodore L. Brown

School of Chemical Sciences, University of Illinois, Urbana-Champaign, 505 S. Mathews Street, Urbana, IL 61801, USA

Table of Contents

1. Introduction

During the past three decades organometallic chemistry has become a large and diverse branch of chemistry. New and unusual structural forms continue to appear as organometallic chemists apply the powerful arsenal of structure-determining tools now available to the analysis of reaction products. At the same time there has been substantial progress in developing deeper understandings of elementary processes that lead to the myriad of organometallic structures now known. In spite of this progress, however, our understandings of reactivity patterns and mechanism in organometallic reactions lags our knowledge of structural possibilities.

Organometallic compounds are, for the most part, diamagnetic, spin-paired molecules and ions, containing even numbers of electrons. With some exceptions, odd-electron organometallic radicals are generally not stable species, readily isolated and subject to some of the most popular modes of structure determination. As a result, radicals have until recently received little attention as an important subset of organometallic compounds. Because radicals are not generally stable, their existence is often evidenced initially through their effects as intermediates in reactions. These effects are especially important in radical chain processes which, under appropriate conditions, may dominate the observed chemistry. Many examples of such radical chain processes are described in this book, especially as they concern electron transfer catalysis.

Among the earliest examples of radical chain processes observed in organometallic systems were those involving the intermediacy of radicals that engage in rapid atom transfer reactions [1–3]. This chapter is about such reactions, and about the atom transfer processes that lie at the heart of them. The radicals of concern are mainly the binary metal carbonyls, $M(CO)_n^{\cdot}$, variously substituted derivatives, $M(CO)_nL_y\cdot$, and aromatic ring systems, mainly $(\eta^5\text{-}C_5H_5)M(CO)_n\cdot$ (we use the notation $Cp = \eta^5\text{-}C_5H_5$). Depending on the metal, any of these species may carry a charge. Many of the conclusions drawn from studies of these systems are applicable to other, non-carbonyl-containing organometallic radicals, and – to a lesser extent – to polynuclear systems containing an unpaired electron.

It is important to distinguish radicals in which the unpaired spin is centered largely on the metal from those in which it is resident mainly on a ligand attached to the metal. For example, in $Mn(CO)_5$, about 60% of the unpaired spin density is localized in the $3d_{z^2}$ metal orbital [4, 5]. By contrast, in the rhenium carbonyl adduct of di(t-butyl)orthoquinone (DTBQ), $Re(CO)_4(DTBQ)$, the unpaired spin is largely resident on the DTBQ, which can be thought of as a semiquinone anion ligand bound to $Re(CO)_4^+$. The esr spectra of such systems reveal only a small amount of unpaired spin density in metal-containing orbitals [6, 7]. One can think of the metal electron count as $18 + \delta$, where δ may be positive or negative, depending on the particular system [7]. The behavior of such systems with respect to reactions at the metal center, such as substitution, is only moderately affected by the presence of unpaired spin, whereas the chemical behavior of the more obviously metal-centered radicals is dramatically different from that of structurally isomorphous spin-paired systems. The chemical properties of the $18 + \delta$ systems are likely to reflect the extent of unpaired spin presence at the metal center, but there are not at present enough reliable data regarding substitution and other reactions to establish a relationship.

2. Formation and Properties of Organometallic Radicals

As described in the following section, radicals give rise to radical chain processes because of reactivity characteristics that are not typical of spin-paired organometallic systems. Before reviewing examples of chain reactions that involve atom transfer, we briefly consider the formation and properties of the radicals that participate in these reactions.

2.1 Formation

Radicals are formed in several characteristic reactions, any one of which might operate in a given reaction system to initiate a chain process.

2.1.1 Metal-metal bond homolysis

The rupture of a metal-metal bond, either photochemically or thermally, gives rise to a pair of radicals. This has been a much-studied process; only a few exemplary cases are shown here:

$$Mn_2(CO)_{10} \xrightarrow{h\nu} 2Mn(CO)_5\cdot \text{ [8–10]} \quad (1)$$

$$[CpMo(CO)_3]_2 \xrightarrow{h\nu} 2CpMo(CO)_3\cdot \text{ [11–13]} \quad (2)$$

$$[CpCr(CO)_3]_2 \xrightarrow{h\nu,\Delta} 2CpCr(CO)_3\cdot \text{ [14–16]} \quad (3)$$

2.1.2 Atom abstraction

Atom abstraction from an 18-electron metal center, such as a metal carbonyl hydride or halide, leads to a 17-electron radical. Atom transfers are discussed in detail in later sections of this chapter.

2.1.3 Oxidation or reduction

One-electron oxidation or reduction of an 18-electron metal center gives rise to a 17-electron or 19-electron metal center, respectively. Such reactions can occur via outer-sphere type electron transfer reactions involving chemical oxidants and reductants or can be effected electrochemically. These reactions are discussed in detail in other chapters of this book.

2.2 Characteristic Reactions

Organometallic radicals exhibit an impressively wide range of chemical behavior. Fig. 1 shows diagrammatically the various reactions that are possible for a prototypical $M(CO)_nL_y\cdot$ radical. All of these reactions are known for many different radicals, though not every radical necessarily exhibits high reactivity with respect to each reaction.

2.2.1 Recombination

Radicals formed via homolysis of a metal-metal bond, or capable of forming a metal-metal bonded dinuclear species, undergo recombination. Recombination

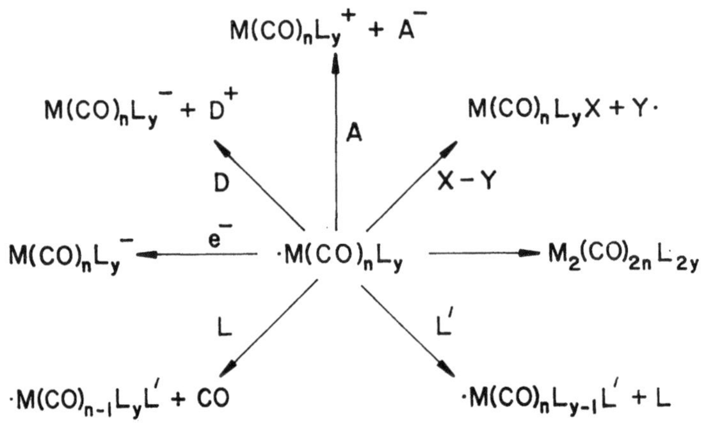

Fig. 1. Possible general reactions of a prototypical metal carbonyl radical.

reactions are generally very fast; Table 1 lists reported rate constants. Note that the fastest reactions occur at essentially diffusion-controlled rates (k(diffusion)$\sim 6\times10^{10}$ M^{-1} s^{-1}). Some recombination rate constants reflect a steric impediment to recombination produced by bulky substituents on the metal. Steric crowding at the metal leads to a weaker metal-metal bond [17]; when the steric requirement is sufficiently large, recombination is prevented, and the radicals are persistent. Examples include $Mn(CO)_3L_2\cdot$ (L = phosphine) [18, 19] and $CpCr(CO)_2PPh_3\cdot$ [16].

Table 1. Rate Constants for Recombination of Metal Carbonyl Radicals

Radical	Solvent	Temp. (°C)	k_r (M^{-1} s^{-1})	Reference
$Re(CO)_5\cdot$	C_8H_{18}	22	2.7×10^9	[20]
	C_6H_{14}	22	3.4×10^9	[21]
$Mn(CO)_5\cdot$	C_6H_{14}	20	4×10^9	[9]
	C_6H_{14}	22	9×10^8	[22]
	THF	20	2×10^9	[9]
$Mn(CO)_4[P(n\text{-}Bu)_3]\cdot$	C_6H_{14}	22	1×10^8	[22]
$Mn(CO)_4[P(i\text{-}Bu)_3]\cdot$	C_6H_{14}	22	2×10^7	[22]
$Mn(CO)_4[P(i\text{-}Pr)_3]\cdot$	C_6H_{14}	22	4×10^6	[22]
$Mn(CO)_4[P(O\text{-}Ph)_3]\cdot$	C_6H_{14}	22	9×10^7	[22]
$Mn(CO)_4[PPh_3]\cdot$	C_6H_{14}	22	1×10^7	[22]
$Co(CO)_4\cdot$	C_6H_{14}	22	4×10^8	[23]
$Co(CO)_3[P(n\text{-}Bu)_3]\cdot$	C_6H_{14}	22	9×10^7	[23]
$CpMo(CO)_3\cdot$	C_6H_{14}	20	5×10^9	[12]
	CH_3CN	20	3×10^9	[12]
	THF	20	2×10^9	[12]
$CpW(CO)_3\cdot$	THF	22	3.8×10^9	[24]

2.2.2 *Atom transfer*

Atom transfer reactions of the radicals, of central importance in their chemistry, are discussed in Sections 8.1 and 9.

2.2.3 *Electron transfer*

Wrighton and Hepp have pointed out that the unpaired electron in a 17-electron metal carbonyl radical lies at a higher energy than the electron pair that forms the metal-metal bond in the dinuclear compound from which the radical may be thought to derive, as illustrated in Fig. 2 [25]. For this reason, the radical is a stronger reducing agent than the parent dinuclear species. Many examples of such behavior are known, e.g.:

$$(\eta^5\text{-}C_5H_5)W(CO)_3\cdot + (\eta^5\text{-}C_5H_5)_2Fe^+ \rightarrow (\eta^5\text{-}C_5H_5)W(CO)_3^+ + (\eta^5\text{-}C_5H_5)_2Fe \quad [25] \qquad (4)$$

$$Re(CO)_4P(O\text{-}i\text{-}Pr)_3\cdot + CH_3NC_5H_4(4\text{-}CN)^+ \rightarrow Re(CO)_4P(O\text{-}iPr)_3^+ + CH_3NC_5H_4(4\text{-}CN)\cdot \quad [26] \qquad (5)$$

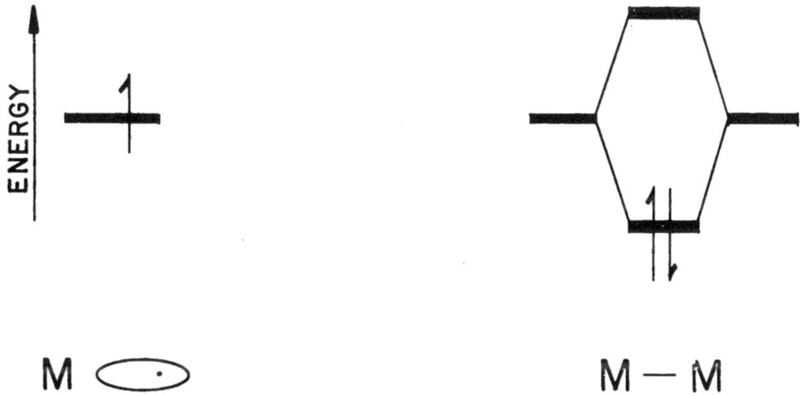

Fig. 2. Energies of the SOMO and HOMO in a metal-centered radical and the corresponding dinuclear species, respectively.

In both these cases the parent dinuclear compound and the oxidant are stable together in solution. However, upon homolysis of the metal-metal bond, a fast electron transfer process occurs. Radicals can also act as electron acceptors, e.g.:

$$Mn(CO)_3(PPh_3)_2\cdot + e^- \rightarrow Mn(CO)_3(PPh_3)_2^- \quad [27] \qquad (6)$$

$$(\eta^5\text{-}C_5H_4Me)Mn(CO)_2(PPh_2Me)^+ + PPh_2Me \rightarrow$$
$$(\eta^5\text{-}C_5H_4Me)Mn(CO)_2(PPh_2Me) + PPh_2Me^+ \quad [28]$$

Because radicals can behave as either oxidants or reductants, it is not surprising that disproportionation reactions are widely observed. Many reactions leading to formation of ionic disproportionation products are radical chain in character, and are driven by substitution at the metal center, which shifts redox potentials to render the electron transfer process energetically feasible [29–31]. The role of 19-electron species in such reactions has been the subject of considerable recent discussion [32, 33].

2.2.4 Substitution

Perhaps the most extraordinary property of many organometallic radicals is their exceptional lability toward substitution. This high lability was first apparent from studies of photochemical substitution into dinuclear compounds such as $Mn_2(CO)_{10}$ [8, 10]. Work in our laboratory led us to conclude initially that substitution in $Mn(CO)_5\cdot$ proceeded via a CO dissociative process, although in substituted Mn radicals the evidence suggested an associative pathway [10]. It is now clear, however, that radical substitution processes typically occur via an associative pathway [34–36]. Table 2 lists representative values for bimolecular substitution rate constants for a selection of radicals. The major point to note is that the rate constants are generally very high; they approach the diffusion limit for substitution of the least sterically hindered radicals such as $Re(CO)_5\cdot$ or $Mn(CO)_5\cdot$ by good nucleophiles such as PBu_3.

Table 2. Rate Constants for Substitution of Organometallic Radicals

Radical	Substituting Ligand	Solvent	Temp. (°C)	k_s (M^{-1} s^{-1})	Ref.
$V(CO)_6\cdot$	PMe_3	C_6H_{14}	25	132	[35]
$V(CO)_5P(n\text{-}Bu)_3\cdot$ [a]	$P(n\text{-}Bu)_3$	C_6H_{14}	25	5.4×10^{-4}	[35]
$Re(CO)_5\cdot$ [b]	$P(n\text{-}Bu)_3$	C_6H_{14}	22	1.7×10^{9}	[34a]
$Mn(CO)_5\cdot$	PBu_3	C_6H_{14}	24	1.0×10^{9}	[36]
	$P(i\text{-}Pr)_3$	C_6H_{14}	24	6.7×10^{7}	[36]
	PPh_3	C_6H_{14}	24	1.7×10^{7}	[36]
	$AsPh_3$	C_6H_{14}	24	6.5×10^{4}	[36]
	$P(O\text{-}i\text{-}Pr)_3$	C_6H_{14}	24	3.1×10^{7}	[36]
$Mn(CO)_4PPh_3\cdot$ [c]	$P(OEt)_3$	decalin	49.9	7.4×10^{2}	[38]
$Mn(CO)_3[P(i\text{-}Bu)_3]_2\cdot$ [d]	CO	C_6H_{14}	20	0.32	[39]
$(\eta^5\text{-}C_5H_4Me)(Mn(CO)_2(NCMe)\cdot^{+}$	$P(OMe)Ph_2$	CH_3CN	25	2.5×10^{4}	[28]
$(\eta^5\text{-}C_5H_4Me)Mn(CO)_2(py)\cdot^{+}$	$PMePh_2$	CH_3CN	25	4×10^{2}	[28]
$(\eta^5\text{-}C_5H_4Me)(Mn(CO)_2\text{-}$	$PPhEt_2$	CH_2Cl_2	22	4.5×10^{2}	[40]
$(4\text{-}NC_5H_4OCH_3)^{+\cdot}$	PPh_2Et	CH_2Cl_2	22	39	[40]
	PPh_3	CH_2Cl_2	22	2.8	[40]

[a] Reaction affords trans-$V(CO)_4[P(n\text{-}Bu)_3]_2\cdot$
[b] k_s based upon reported value of 4.9×10^7 M^{-1} s^{-1} for reaction of $Re(CO)_5\cdot$ with CCl_4 [37].
[c] Reaction affords $Mn(CO)_4P(OEt)_3\cdot$
[d] Reaction affords $Mn(CO)_4P(i\text{-}Bu)_3\cdot$, then $Mn_2(CO)_8[P(i\text{-}Bu)_3]_2$.

As we shall see later, fast substitution is an essential feature of a radical chain pathway for substitution involving atom transfer as a chain propagating step. Fast substitution is also essential in substitution processes in which the propagating step is electron transfer; i.e., in electron-transfer-chain (ETC) catalyzed substitutions [28].

The data in Table 2, and other available data permit a few generalizations regarding rates of substitution into radicals:

i) Rate constants for substitution vary in the order

$M(CO)_n\cdot > M(CO)_{n-1}L\cdot > M(CO)_{n-2}L_2\cdot$

where L is a strongly donor ligand such as a phosphine, of significantly larger steric requirement than CO.

ii) Substitution at metal centers to which an η^5-C_5H_5 ligand is attached is slower than at a metal center containing only CO ligands. For example, $Mn(CO)_5\cdot$ is more labile than $CpMn(CO)_2L^+$. It is likely that $Mn(CO)_5\cdot$ is more labile than $CpMo(CO)_3\cdot$.

iii) Even in the most labile systems, the rate of substitution is sensitive to the nucleophilicity of the entering ligand. This suggests a considerable degree of bond formation between the metal and the entering ligand, even though the free energy barrier to attaining the transition state may be quite low.

3. Radical Chain Processes Involving Atom Transfer

In this section we review those reactions for which the mechanism can be understood in terms of a radical chain process that involves atom transfer as one of the propagation steps. The overall result of the reaction may be substitution, reduction or something else. Not included here are electron-transfer-chain (ETC) catalyzed processes that might have the same overall result, but in which a key propagation step is electron transfer rather than atom transfer. By analogy with the ETC designation, reactions of interest in this chapter might be called atom-transfer-chain (ATC) catalyzed processes.

Since the first examples of radical chain processes involving atom transfer to an organometallic species [1, 2], many conceptually analogous processes have been reported. Although the various mechanisms described share certain common features, they differ in important and interesting ways that determine the conditions under which a particular kind of radical chain process may occur.

3.1 Alkyl Halide Reduction

In 1972 Booth and Shaw proposed a radical chain mechanism for the reaction between $HMn(CO)_4PPh_3$ and haloalkanes [1]:

$$HMn(CO)_4PPh_3 + CF_3I \rightarrow IMn(CO)_4PPh_3 + HCF_3 \quad (6)$$

The proposed mechanism is:

$$R\cdot + HMn(CO)_4PPh_3 \rightarrow RH + Mn(CO)_4PPh_3\cdot \quad (7)$$

$$CF_3I + Mn(CO)_4PPh_3\cdot \rightarrow IMn(CO)_4PPh_3 + CF_3\cdot \quad (8)$$

$$CF_3\cdot + HMn(CO)_4PPh_3 \rightarrow CF_3H + Mn(CO)_4PPh_3\cdot \quad (9)$$

where R· is any initiator. This mechanism is closely analogous that proposed for radical chain reaction of trialkyltin hydrides with haloalkanes [41]. In keeping with the proposed radical character of the reaction, the rate was observed to be retarded by radical scavengers.

Chain termination steps in this system involve coupling of the various radicals, e.g.,

$$CF_3\cdot + Mn(CO)_4PPh_3\cdot \rightarrow CF_3Mn(CO)_4PPh_3 \quad (10)$$

$$2\,Mn(CO)_4PPh_3\cdot \rightarrow Mn_2(CO)_8[PPh_3]_2 \quad (11)$$

For the chain process to compete effectively with these bimolecular termination steps, the hydrogen atom transfer process, Eq. (9), must be rapid.

Reactions of $Os(CO)_4H(R)$ compounds and related di- and trinuclear Os compounds, e.g., $HOs(CO)_4Os(CO)_4R$, with CCl_4 are assigned an analogous reaction pathway [42]. These reactions are usually rapid at room temperature, but proceed at nonreproducible rates.

Alkyl halide reduction via a radical chain reaction was demonstrated by Kinney et al. for $CpV(CO)_3H^-$ ($Cp = \eta^5\text{-}C_5H_5$) [43]. The mechanism in best accord with the observations is entirely analogous to equations (7)–(9). When the reaction is carried out in the presence of a large excess of PPh_3 the initial product is still $CpV(CO)_3X^-$, rather than $CpV(CO)_3PPh_3$. An alternative electron transfer process of the form

$$CpV(CO)_3H^- + RX \rightarrow CpV(CO)_3H\cdot + R\cdot + X^- \quad (12)$$

$$CpV(CO)_3H\cdot + R\cdot \rightarrow RH + CpV(CO)_3 \quad (13)$$

$$CpV(CO)_3 + X^- \rightarrow CpV(CO)_3X^- \quad (14)$$

is thus ruled out; the 16-electron $CpV(CO)_3$ intermediate, if formed, should have been trapped by excess PPh_3.

It is noteworthy that although electron transfer is ruled out in the reactions with alkyl halides, it is invoked by Bergman et al. to account for reactions of the same anion with transition metal alkyls and acyls, and for the catalytic effect of $CpV(CO)_3H^-$ on ligand substitution reactions of metal carbonyl compounds [44]. The evidences for radical chain character in these reactions are, however, limited, and alternative non-chain, non-radical pathways are not ruled out. For example, the reactions of $CpV(CO)_3H^-$ with transition metal carbonyl compounds might occur via nucleophilic interaction of the hydride at a CO of the substrate molecule. In several other cases, such a pathway accounts for the labilization of substrate carbonyl systems toward substitution [45–48]:

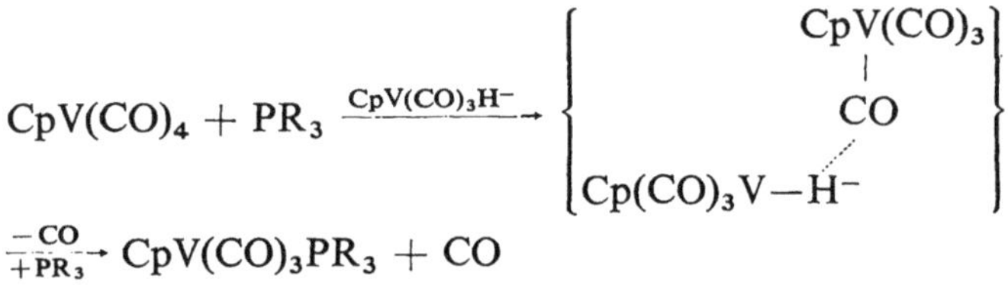

(15)

This same mode of interaction could lead to hydride transfer between $CpV(CO)_3H^-$ and $CpV(CO)_4$, albeit at a much slower rate than the catalyzed substitution at $CpV(CO)_4$.

Use of organic radicals capable of rearrangement or cyclization at known rates permits an estimate of about $2 \times 10^7\ M^{-1}\ s^{-1}$ for the rate constant k_H for hydrogen atom transfer from $CpV(CO)_3H^-$ to alkyl radicals [43]. This is more than an order of magnitude larger than the measured value for $HSnBu_3$.

Alternative pathways to the radical chain mechanism outlined above are available to anionic metal carbonyl hydrides [49]. These include oxidative addition of the X—C bond to the metal center, or nucleophilic displacement of halide at carbon by the anionic hydride in an S_{N2} process. Darensbourg, Newcomb et al. have evaluated the competing nucleophilic displacement and radical chain reductions of alkyl bromides by use of a two-radical probe methodology [50]. The neopentyl-like bromide, 1, is inherently unreactive toward nucleophilic halide displacement. On the other hand, 6-bromo-1-hexene, 2, is susceptible to reduction via either halogen atom transfer (assuming, consistent with Bergman's work that an SET pathway is not operative) or S_{N2} displacement.

1

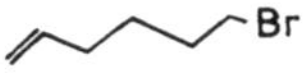

2

The rate of rearrangement of the radical derived from 1, Eq. (16), is known ($k_r = 6.0 \times 10^6\ s^{-1}$):

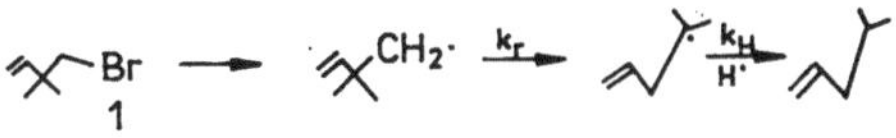

(16)

Measurement of the ratio of concentrations of unrearranged to rearranged alkane in reaction of 1 with anionic carbonyl hydrides at known concentrations of the latter permits estimation of the rate constants k_H for hydrogen atom transfer, Eq. (16); these are on the order of $10^7\ M^{-1}\ s^{-1}$ (see Section 9). The *apparent* values of k_H for reaction of 2 with the same anionic carbonyl hydrides are 10 to 10^2 times larger, whereas it would be reasonable to expect them to be somewhat less than for the corresponding reactions with 1. It is possible to conclude on this basis that the predominant reaction with 2 involves the nucleophilic displacement pathway. The percentage of reduction by the S_{N2} pathway increases in the order $HCr(CO)_5^{-\cdot}$ (90%) < $HW(CO)_5^{-\cdot}$ (99.7%) < $HW(CO)_4P(OMe)_3^{-\cdot}$ (99.97%).

The reaction of $HOs_2(CO)_8R$ with CCl_4 is rapid at room temperature [42]; it probably proceeds via a radical chain pathway:

$$In\cdot + HOs_2(CO)_8R \rightarrow InH + \cdot Os_2(CO)_8R \qquad (17)$$

$$\cdot Os_2(CO)_8R + CCl_4 \rightarrow ClOs_2(CO)_8R + \cdot CCl_3 \qquad (18)$$

$$\cdot CCl_3 + HOs_2(CO)_8R \rightarrow CHCl_3 + \cdot Os_2(CO)_8R \qquad (19)$$

Reactions of this kind are likely to be generally facile for all metal carbonyl hydrides.

Metal carbonyl radicals have been shown to facilitate the reduction of $RCCl_3$ to $RCCl_2H$ by several reducing agents, including $HSiR_3$, $CH_3CH(OH)CH_3$, and DMF [51]. The reactions involving dinuclear metal carbonyls proceed under irradiation. They are clearly radical in character, as evidenced by EPR results, spin trapping studies and the chemical characteristics of the system.

3.2 Substitution

3.2.1 Carbonyl hydrides

The first example of a radical chain pathway for substitution of an organometallic compound, involving the substitution of $HRe(CO)_5$ by phosphines, was reported by Byers and Brown [2]. The initial observations were that $HRe(CO)_5$ appeared to undergo substitution by phosphines at highly irreproducible rates. The prior history of the sample, character of the container for the solution, exposure to light and other factors played a role in determining the initial rate of reaction in a given sample. However, with carefully prepared hydrocarbon solutions in the dark, thermal substitution of $HRe(CO)_5$ by PBu_3 was shown to proceed to a negligible extent after several hours. Addition of $Re_2(CO)_{10}$ at a concentration no more than ten percent that of $HRe(CO)_5$, followed by irradiation at 311 nm, caused the reaction to proceed slowly for a few minutes, then rapidly to completion.

The following radical chain pathway accounts for the observations:

$$Re_2(CO)_{10} \xrightarrow{h\nu} 2\,Re(CO)_5\cdot \tag{20}$$

$$Re(CO)_5\cdot + PBu_3 \rightarrow Re(CO)_4PBu_3 + CO \tag{21}$$

$$Re(CO)_4PBu_3\cdot + PBu_3 \rightarrow Re(CO)_3(PBu_3)_2\cdot + CO \tag{22}$$

$$Re(CO)_4PBu_3\cdot + HRe(CO)_5 \rightarrow HRe(CO)_4PBu_3 + Re(CO)_5\cdot \tag{23}$$

$$Re(CO)_3(PBu_3)_2\cdot + HRe(CO)_5 \rightarrow HRe(CO)_3(PBu_3)_2 + Re(CO)_5\cdot \tag{24}$$

Homolysis of the Re—Re bond in $Re_2(CO)_{10}$, Eq. (20), is effected by irradiation at 311 nm; the absorption maximum corresponding to the $\sigma - \sigma^*$ transition of the metal-metal bond in $Re_2(CO)_{10}$ occurs at 313 nm.

The extent to which the second substitution occurs depends on the concentrations of phosphine and hydride. For $PBu_3 \sim 10^{-2}$ M and $HRe(CO)_5 \sim 10^{-3}$ M, $[HRe(CO)_4PBu_3]/[HRe(CO)_3(PBu_3)_2] \sim 3$. When the substituting ligand is PPh_3, only the mono-substituted product hydride is observed. In the initial slow phase of the reaction it is likely that various impurities that can act as radical traps are active. Once these have been scavenged by reaction with the metal carbonyl radicals first formed, the reaction proceeds rapidly. Termination steps are likely to involve mainly dimerization of the radicals:

$$2\,Re(CO)_5\cdot \rightarrow Re_2(CO)_{10} \tag{25}$$

$$Re(CO)_5\cdot + Re(CO)_4PBu_3\cdot \rightarrow Re_2(CO)_9PBu_3 \tag{26}$$

$$2\,Re(CO)_4PBu_3\cdot \rightarrow Re_2(CO)_8(PBu_3)_2 \tag{27}$$

These chain termination steps are not strongly competitive with the chain propagation cycle, as evidenced by the fact that only very small absorbances due to substituted dinuclear species or reductions in absorbance due to $Re_2(CO)_{10}$ are seen when reaction is complete.

It seems likely that the most important adventitious inhibitor present at the beginning of the reaction is O_2. When known quantities of O_2 were added to the solutions before reaction, the induction period was observed to lengthen in proportion to the amount added [3]. It seems likely that O_2, which does not react with $HRe(CO)_5$, forms an adduct of the form $(CO)_5ReO_2\cdot$, analogous to the dioxygen adducts observed to form with other metal carbonyl radicals [52].

Although reaction is rapid under irradiation once the induction period is past, interruption of the irradiation causes the reaction to slow considerably. Thus, the chain termination steps, Eqs. (25)–(27), or scavenging of the radicals, compete with the propagation steps. Nevertheless, it was qualitatively clear that the quantum yield for formation of substituted hydride, based on the number of $Re_2(CO)_{10}$ molecules homolyzed, was far greater than one. This implies in turn that all of the steps in the propagation chain must be rapid with respect to chain termination. (It was already known that the recombination of $Mn(CO)_5\cdot$ or $CpMo(CO)_3\cdot$ radicals to form the respective dinuclear compounds are fast, with rate constants at or near the diffusion limit.) Thus, both the substitution step, Eq. (21), and atom transfer step, Eq. (23), must be fast relative to radical recombination.

The propensity of the metal carbonyl radical to undergo hydrogen atom transfer with a suitable hydrogen atom source might be thought to be roughly comparable to that of a main group metal-centered radical such as $Bu_3Sn\cdot$, so it is not surprising that step (23), should be fast. However, the fact that it is fast suggests that the reaction is exergonic, or at worst ergoneutral. There is an implication that substitution of phosphines for CO on the metal center results in a stronger M—H bond. This suggestion is consistent with qualitative observations of a generally greater thermal stability and lower pK_a in water of the substituted as compared with the unsubstituted hydrides (see Section 9).

There is no precedent in the chemistry of main group element radicals for the observed lability of the metal carbonyl radicals toward substitution. On the other hand, Wrighton and Ginley [8], and then Kidd and Brown [10], had shown that substitution of $Mn(CO)_5\cdot$ by phosphines is facile. An associative mechanism has been established for substitution processes at such 17-electron metal carbonyl radical centers such as $Re(CO)_5\cdot$ [34a], $Mn(CO)_5\cdot$ [36], $CpW(CO)_3\cdot$ [53] $V(CO)_6\cdot$ [35], and $(MeCp)Mn(CO)_2L^+$ [28]. Because substitution is bimolecular, ligand nucleophilicity and concentration are important in determining chain length in the overall process.

Additional observations of significance for characterizing the radical chain process were made by Hoffman and Brown in their study of the substitution reactions of $CpW(CO)_3H$ [54]. The ligand substitution reactions of this hydride proceed slowly at room temperature in hydrocarbon solutions, to yield a *cis-trans* mixture of monosubstituted hydride complexes, $CpW(CO)_2LH$. However, the observed rates varied widely under nominally identical conditions of temperature and concentration. Addition of galvinoxyl or nitrosodurene in concentrations less than 20 mol % that of the hydride caused the reaction to proceed at rates at least 10^3 faster. (In the absence of PBu_3, galvinoxyl reacts with the hydride to generate the dimer, $Cp_2Mo_2(CO)_6$.) However, in contrast to the observation in the $HRe(CO)_5$ substitution, traces of O_2 do not significantly affect the rate of phosphine substitution into $CpW(CO)_3H$.

Irradiation of solutions of $CpW(CO)_3H$ and PBu_3 with light of 500 nm wavelength for periods of 1 h or so produced no singificant extent of substitution. Analogous irradiation of solutions containing small amounts (<10 mol % relative to the hydride) of either $Cp_2W_2(CO)_6$ or $Cp_2Mo_2(CO)_6$ resulted in rapid reaction. These results are consistent with a radical chain process of the form:

$$Cp_2W_2(CO)_6 \xrightarrow{h\nu} 2CpW(CO)_3\cdot \tag{28}$$

$$CpW(CO)_3\cdot + L \rightarrow CpW(CO)_2L\cdot + CO \tag{29}$$

$$CpW(CO)_2L\cdot + CpW(CO)_3H \rightarrow CpW(CO)_2LH + CpW(CO)_3\cdot \tag{30}$$

It has been established that irradiation of $Cp_2W_2(CO)_6$ in the 500 nm wavelength region leads to quantum yields on the order of 0.4 to 0.6 for disappearance of $Cp_2W_2(CO)_6$ due to reaction with halogen atom donors [13, 55]. On the basis of actinometric measurements the quantum yield for formation of $CpW(CO)_2LH$ ($L = PBu_3$) in the radical chain substitution process with small amounts of $Cp_2W_2(CO)_6$ present was estimated to be on the order of 1000, suggesting that chain lengths in the radical chain process are on the order of 2000. A much smaller quantum yield of about 50 was obtained for $Cp_2Mo_2(CO)_6$ as photoinitiator. Given that the W—H bond energy is likely to be higher than the Mo—H bond energy [56], the rate constant for the initiating step,

$$CpMo(CO)_2PBu_3\cdot + CpW(CO)_3H \rightarrow CpMo(CO)_2PBu_3H + CpW(CO)_3\cdot \quad (31)$$

should be much slower, even though the subsequent chain processes should proceed as when $Cp_2W_2(CO)_6$ is employed as initiator. Evidence regarding the relative energies of the Mo—H and W—H bonds is found in the observation that 500 nm irradiation of a solution of $CpW(CO)_3H$ and $Cp_2Mo_2(CO)_6$ does not lead to formation of $CpMo(CO)_3H$, whereas irradiation of a solution of $CpMo(CO)_3H$ and $Cp_2W_2(CO)_6$ leads to formation of $CpW(CO)_3H$ [54].

The analogous radical chain process is not readily demonstrated for CpMo—$(CO)_3H$ because it reacts too rapidly with PBu_3 under room temperature conditions (reaction complete in less than 10 s at 30 °C) to permit study of photoinitiation. It reacts completely over a period of about 15 minutes with $P(OEt)_3$. Remarkably, however, $CpMo(CO)_3H$ fails to react at all with $AsPh_3$ over a period of 15 h. Addition of a small quantity of $Cp_2Mo_2(CO)_6$ followed by irradiation at 500 nm for 35 min also failed to produce reaction.

It seems likely that the substitution reactions of $CpMo(CO)_3H$ proceed via a radical chain pathway. The lack of reactivity of $AsPh_3$ as substituting ligand can be ascribed to its comparatively low nucleophilicity. The bimolecular rate constants for substitution of $Mn(CO)_5\cdot$ by several nucleophiles are listed in Table II. It is evident that $AsPh_3$ reacts much less rapidly than the phosphorus ligands listed. If the ligand substitution process in the overall chain process is too slow, the chain propagation cycle does not compete well with recombination or other loss processes for the radicals and a rapid chain process is not observed.

Whereas irradiation of a solution of $CpW(CO)_3H$ and PBu_3 in hexane with 500 nm light does not significantly affect the rate of substitution of the hydride, irradiation with light of wavelength shorter than 400 nm materially speeds the substitution [54]. Quantum yields for substitution varied considerably from one solution to another, but values on the order of 30 could be obtained with light of 311 nm wavelength. Substitution under these conditions is proposed to proceed via the radical chain pathway outlined above, but with initiation of the reaction occasioned by homolysis of the W—H bond. Assuming chain lengths of 2000, this suggests a quantum yield for W—H bond homolysis of about 0.015. Consistent with the proposed mechanism is the observation that irradiation of solutions of $CpW(CO)_3H$ with light of 358 nm wavelength in the presence of trityl chloride leads to complete conversion to $CpW(CO)_3Cl$. Evidence for homolysis of the W—H bond under irradiation is also available from matrix isolation studies [57].

The mononuclear osmium complex, $Os(CO)_4(H)CH_3$, does not undergo substitution under thermal solution reaction conditions in the absence of air. It does undergo slow thermal decomposition to form CH_4 and predominantly $HOs_2(CO)_8CH_3$, plus small amounts of trinuclear Os compounds [42]. When oxygen is present in small amounts the substitution reaction proceeds more rapidly than the conversion to dinuclear compounds, presumably via a radical chain reaction:

$$In + Os(CO)_4(H)CH_3 \rightarrow HIn + Os(CO)_4CH_3\cdot \quad (32)$$

$$Os(CO)_4CH_3\cdot + PPh_3 \rightarrow Os(CO)_3(PPh_3)CH_3\cdot + CO \quad (33)$$

$$Os(CO)_3(PPh_3)CH_3\cdot + Os(CO)_4(H)CH_3 \rightarrow Os(CO)_3(PPh_3)CH_3(H) + Os(CO)_4CH_3\cdot \quad (34)$$

The reaction is retarded by larger amounts of oxygen, possibly because the O_2 combines with the metal-centered radicals to form peroxy radicals, which are not sufficiently reactive to maintain the chain process.

Substitution of $Os(CO)_4(H)CH_3$ occurs readily on photolysis. However, it is not clear whether the increased rate is due to homolysis of the Os—H bond, initiating a radical chain process, or to CO dissociation.

Similarly, the cis-dihydride, $Os(CO)_4H_2$, undergoes rapid substitution to form the trans-disubstituted product:

$$Os(CO)_4H_2 + 2PEt_3 \rightarrow Os(CO)_2(PEt_3)_2H_2 + 2CO \quad (35)$$

The reaction is reported to be extremely rapid after an induction period of variable length, indicating a radical chain process.

The substitution of $Os(CO)_4H_2$ by PPh_3 in benzene yields the monosubstituted product, $Os(CO)_3(PPh_3)H_2$ [58]. Addition of AIBN (2,2-azobis(2-methyl-propionitrile) to the benzene solutions causes the reaction to proceed more rapidly, even at 35 °C.

Metal carbonyl hydrides of the first transition series are generally quite labile. It is likely that this lability is due to the existence of radical chain pathways for substitution, with possibly subsequent reactions such as electron transfer. However, alternative reaction pathways could operate. The existence of a comparatively weak M—H bond, which might occasion rapid radical chain reactions under appropriate conditions, also might lead to hydride migration. There are few unambiguous examples of hydride migration as an avenue for substitution. Attempts to eliminate radical chain pathways for substitution of $HMn(CO)_5$ were not entirely successful [59]. Although the reactions were in some respects well-behaved, and the rate law suggestive of hydride migration as a pre-equilibrium step, it proved impossible to obtain a degree of reproducibility that would entirely rule out a radical chain contribution.

In general, the comparatively more rapid substitutions of metal carbonyl hydrides as compared with corresponding metal carbonyl halides is likely to originate in more facile radical chain reactions for the former. For example, CpRu $(CO)_2H$ reacts much more readily than the corresponding chloride with bisphosphines, such as $Ph_2P(CH_2)_nPPh_2$, to form quantitatively $CpRu(Ph_2P(CH_2)_n PPh_2)H$ [60]. The reaction probably proceeds via a radical chain pathway. On the other hand, as related next, carbonyl halides are also capable of undergoing radical chain substitution under appropriate conditions.

3.2.2 Carbonyl halides

The first example of substitution in metal carbonyl halides via a radical chain pathway was provided by Alway and Barnett as part of their study of the photosubstitution reactions of $CpM(CO)_3X$ (M = Mo, X = Br or I; M = W, X = Cl, Br or I) [61]. Addition of $[CpMo(CO)_3]_2$ to benzene solutions of $CpMo(CO)_3X$ and PPh_3 causes a marked increase in the quantum yield for photosubstitution on irradiation with 436-nm light. No loss in absorption due to $[CpMo(CO)_3]_2$ was noted at the end of the reaction. Whereas in the absence of the added dimer photosubstitution occurs via CO loss, a radical chain pathway driven by homolysis of the dimer operates when the dimer is present:

$$[CpMo(CO)_3]_2 \xrightarrow{h\nu} 2CpMo(CO)_3\cdot \quad (36)$$

$$CpMo(CO)_3\cdot + PR_3 \rightarrow CpMo(CO)_2PR_3\cdot + CO \quad (37)$$

$$CpMo(CO)_2PR_3\cdot + CpMo(CO)_3X \rightarrow CpMo(CO)_2PR_3X + CpMo(CO)_3\cdot \quad (38)$$

The authors showed that homolysis of the M—X bond is not significant as a source of radicals for the chain process. The termination steps presumably involve recombination of radicals to form dimers.

The quantum yields in these photo-initiated reactions, while greater than one, are not nearly so large as observed in the substitution of $CpW(CO)_3H$ in the presence of $[CpW(CO)_3]_2$ [54]. In part this may be due to absorption of much of the incident light by the carbonyl halide. It may also result, however, from shorter chain lengths due to slower halogen atom transfer, Eq. (38), as compared with hydrogen atom transfer. No rate constant data are available for halogen atom transfer between metal carbonyl centers.

Independently of the Alway and Barnett work, Coville showed that the thermal substitution of $CpMo(CO)_3I$ by isonitriles is catalyzed by $[CpMo(CO)_3]_2$ [62]. He reported that the substitution rates are increased by light, and that the reactions are inhibited by hydroquinone or galvinoxyl. The free radical chain process expressed in Eqs. (36)–(38), was proposed, with the added feature that homolysis of the dimer might occur via a thermal process. Coville, Singleton and co-workers have gone on to study the catalytic effects of metal dimers on a variety of substitution processes. They have shown that $[CpFe(CO)_2]_2$ is effective in promoting substitution of CO by RNC (R = t-Bu, Ch_2Ph) or PPh_3 into $CpMo(CO)_3I$ or $CpFe(CO)_2X$ (X = Br, I) [63, 64]. At 80 °C the reaction clearly has a radical chain component analogous to that described above:

$$[CpFe(CO)_2]_2 \xrightarrow{h\nu,\Delta} 2\,CpFe(CO)_2\cdot \quad (39)$$

$$CpFe(CO)_2\cdot + RNC \rightarrow CpFe(CO)(CNR)\cdot + CO \quad (40)$$

$$CpFe(CO)(CNR)\cdot + CpFe(CO)_2I \rightarrow CpFe(CO)(CNR)I + CpFe(CO)_2\cdot \quad (41)$$

Reaction (40) was confirmed by the reaction of $CpFe(CO)_2^-Na^+$ with tropylium tetrafluoroborate. In the absence of added RNC, $[CpFe(CO)_2]_2$ was formed; presumably the 17-electron radical is formed via electron transfer, and then dimerizes. When RNC is present, $Cp_2Fe_2(CO)_3(CNR)$ and $Cp_2Fe_2(CO)_2(CNR)_2$ are formed.

At 42 °C in benzene, the reactions

$$[CpFe(CO)_2]_2 + t\text{-}BuNC \rightarrow Cp_2Fe_2(CO)_3(CN\text{-}t\text{-}Bu) \quad (42)$$

$$CpFe(CO)_2I + Cp_2Fe_2(CO)_3(CN\text{-}t\text{-}Bu) \rightarrow CpFe(CO)(CN\text{-}t\text{-}Bu)I \quad (43)$$

were found to occur much more slowly than the catalyzed reaction under the same reaction conditions. On these grounds the authors conclude that the radical chain pathway cannot be the dominant one for product formation at 42 °C [64]. However, neither reaction (42) nor (43) is a chain process. The slowness of these reactions at 42 °C is presumably due to the slow rate of thermal homolysis of the dimers. On the other hand, the overall reaction is a chain process. If the chain lengths are reasonably long, one expects that the overall rate of the chain process will be much higher than the rates of initiating events such as metal-metal bond homolysis.

The reported crossover experiments confirm this view (product yields >80%):

$$CpFe(CO)_2I + (MeCp)_2Fe_2(CO)_4 + t\text{-}BuNC \rightarrow CpFe(CO)(CNR)I \quad (44)$$

$$(MeCp)Fe(CO)_2I + [CpFe(CO)_2]_2 + t\text{-}BuNC \rightarrow (MeCp)Fe(CO)(CN\text{-}t\text{-}Bu)I \quad (45)$$

The fact that in these versions of the catalyzed substitution reaction there is little crossover of the label between dimer and carbonyl halide product is evidence that the chain lengths are long; that is, the number of molecules undergoing substitution in the chain process is large compared with the number of metal-metal bond homolysis events that initiate the chains. It is worthy of note that irradiation of the reaction mixtures at 42 °C causes very rapid reactions, consistent with an efficient chain process initiated via photochemical metal-metal bond homolysis.

Under certain conditions the reactions give rise to ionic products. For example, $CpFe(CN\text{-t-Bu})_3I$ is formed from reaction of CpFe(CO)(CN-t-Bu)I with t-BuNC, with $[CpFe(CO)_2]_2$ as catalyst. In such cases it is evident that an electron transfer pathway intervenes in the course of the simple radical chain process.

The potential for synthetic utility of radical chain pathways is illustrated in a recent report of the synthesis of $(\eta^5\text{-}C_5(CH_3)_5)Ru(CO)(PMe_2Ph)I$ [65]. No reaction between $(\eta^5\text{-}C_5(CH_3)_5)Ru(CO)_2I$ and PMe_2Ph in toluene at 80 °C is observed to take place. However, upon addition of a small amount of AIBN an immediate reaction occurs, resulting in a 77% yield of the desired product.

3.2.3 *Metal carbonyl alkyls*

Fabian and Labinger first reported a facile substitution via a radical chain pathway involving transfer of an alkyl group [66]:

$$CpFe(CO)_2(\eta^1\text{-}Cp) + P(OPh)_3 \rightarrow CpFe(CO)(P(OPh)_3)(\eta^1\text{-}Cp) + CO \qquad (46)$$

Whereas the thermal reaction is slow the reaction proceeds very quickly upon irradiation of a solution containing small amounts of $[CpFe(CO)_2]_2$. A radical chain mechanism was proposed:

$$[CpFe(CO)_2]_2 \xrightarrow{h\nu} 2\,CpFe(CO)_2\cdot \qquad (47)$$

$$CpFe(CO)_2\cdot + L \rightarrow CpFe(CO)(L)\cdot + CO \qquad (48)$$

$$CpFe(CO)(L)\cdot + CpFe(CO)_2(\eta^1\text{-}Cp) \rightarrow CpFe(CO)(L)(\eta^1\text{-}Cp) + CpFe(CO)_2\cdot \qquad (49)$$

The novel feature in this scheme is transfer of the η^1-Cp ligand, postulated to occur via an addition-elimination process:

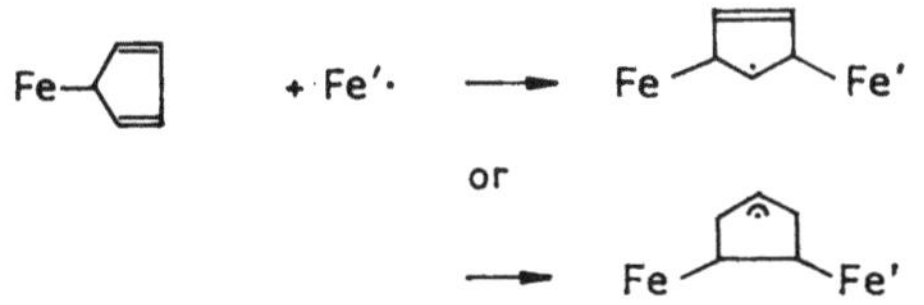

(50)

Crossover experiments were performed to establish that the alkyl group transfers between metal centers. A similar reaction pathway applies to substitution of $CpFe(CO)_2(\eta^1\text{-}C_3H_5)$. However, it does not apply to methyl or other alkyl groups incapable of undergoing the addition-elimination process. The nature of the substituting ligand is important in determining the outcome of the reaction with $CpFe(CO)_2(\eta^1\text{-}Cp)$. Substitution by bulky PPh_3 does not occur under the reaction conditions employed by Fabian and Labinger. Presumably PPh_3 substitutes into the radial too slowly to maintain the radical chain process. On the other hand, more strongly donating ligands, such as PPh_2Me or PMe_3 lead to ionic products. In this case an electron transfer chain process, probably involving the substituted radical, appears to intervene:

$$CpFe(CO)_2\cdot + PMe_3 \rightarrow CpFe(CO)(PMe_3)\cdot + CO \quad (51)$$

$$CpFe(CO)(PMe_3)\cdot + CpFe(CO)_2(\eta^1\text{-}Cp) \rightarrow$$
$$CpFe(CO)(PMe_3)^+ + CpFe(CO)_2(\eta^1\text{-}Cp)^{-}\cdot \quad (52)$$

$$CpFe(CO)(PMe_3)^+ + PMe_3 \rightarrow CpFe(CO)(PMe_3)_2^+ \quad (53)$$

$$CpFe(CO)_2(\eta^1\text{-}Cp)^{-}\cdot \rightarrow CpFe(CO)_2\cdot + Cp^- \quad (54)$$

In a closely related study Rosenblum and Waterman demonstrated facile substitution of CO by the bicyclic phosphite, $P(OCH_2)_3CCH_3$, into $CpFe(CO)_2$-(η^1-C_3H_5) and $CpFe(CO)_2(\eta^1\text{-}C_3H_4CH_3)$, using $[CpFe(CO)_2]_2$ as photoinitiator of the chain process [67]. They also postulate the addition-elimination pathway involving the alkyl group, noting that the reaction is not facile for η^1-2-butenyl and η^1-cyclopentyl as ligands, neither of which should be capable of undergoing the addition-elimination process.

4. Metal-Metal Bond Formation

Radical intermediates appear to be involved in a class of reactions leading to metal-metal bond formation, with elimination of alkane. Edidin and Norton have demonstrated that the following reactions proceed via radical chain pathways [58]:

$$2Ph_3PAuCH_3 + H_2Os(CO)_4 \rightarrow 2CH_4 + (Ph_3PAu)_2Os(CO)_4 \quad (55)$$

$$Ph_3PAuCH_3 + HMn(CO)_5 \rightarrow Ph_3PAuMn(CO)_5 + CH_4 \quad (56)$$

Reaction (55) is extremely slow when oxygen is rigorously excluded from this system. It is greatly accelerated by addition of $H_2Os_2(CO)_8$ at 10 mol % of $H_2Os(CO)_4$ with excess Ph_3PAuCH_3 in benzene, followed by irradiation with Pyrex-filtered light. The irradiation presumably results in homolysis of the Os—Os bond, forming $HOs(CO)_4\cdot$ radicals.

When Ph_3PAuCH_3 and Ph_3P are reacted together with $H_2Os(CO)_4$ in benzene at 35 °C, with AIBN added as initiator, $(Ph_3PAu)_2Os(CO)_4$ and $H_2Os(CO)_3PPh_3$ are formed concurrently, indicating a common intermediate. A radical chain pathway involing $HOs(CO)_4\cdot$ as intermediate was independently established for $H_2Os(CO)_4$ substitution. Based on these evidences, the authors conclude that the reaction is radical chain in character, with $HOs(CO)_4\cdot$ as chain carrier.

The favored mechanism for reaction is as follows:

$$In\cdot + H_2Os(CO)_4 \rightarrow InH + HOs(CO)_4\cdot \quad (57)$$

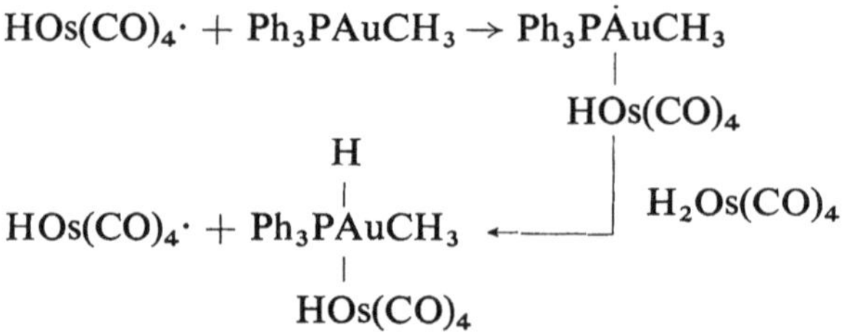

$$Ph_3PAu(H)CH_3 \rightarrow Ph_3PAuOs(H)(CO)_4 + CH_4 \qquad (58)$$
$$\quad |$$
$$\quad HOs(CO)_4$$

$$Ph_3PAuOs(H)(CO)_4 + Ph_3PAuCH_3 \xrightarrow{\text{fast}} (Ph_3PAu)_2Os(CO)_4 + CH_4 \qquad (59)$$

An analogous reaction leads to product in the reaction of $HMn(CO)_5$, Eq. (56). The authors found $HRe(CO)_5$ totally unreactive with Ph_3PAuCH_3 under analogous reaction conditions, even when $Re_2(CO)_{10}$ was added and the solution irradiated to form $Re(CO)_5\cdot$ radicals. It is likely that H atom transfer from Re to Au is slow because the Re—H bond energy is substantially higher than the Au—H bond energy.

Other reactions involving M—M bond formation with alkane elimination may also proceed via the same general radical chain pathway, e.g. [68]:

$$(CH_3)_2Pt(\text{cyclooctadiene}) + 2HMn(CO)_5 \xrightarrow{CO} 2CH_4 + Mn_2Pt(CO)_{12} \qquad (60)$$

Radical intermediates are likely to be involved also in formation and isomerization of dinuclear compounds containing donor-acceptor metal-metal bonds [69]. As an example, formation of $(H)(Me_3P)(OC)_3Os—Re(CO)_5$ from $HRe(CO)_5$ and $Os(CO)_4(PMe_3)$ might proceed as follows:

$$HRe(CO)_5 + In\cdot \rightarrow HIn + Re(CO)_5\cdot \qquad (61)$$

$$Re(CO)_5\cdot + Os(CO)_4(PMe_3) \rightarrow \cdot Re(CO)_4 + CO \qquad (62)$$
$$\qquad\qquad |$$
$$\qquad\qquad Os(CO)_4(PMe_3)$$
$$Re(CO)_5 \leftarrow$$
$$|$$
$$\cdot Os(CO)_3(PMe_3)$$

$$(Me_3P)(OC)_3\dot{O}s—Re(CO)_5 + HRe(CO)_5 \rightarrow (H)(Me_3P)(OC)_3OsRe(CO)_5 + Re(CO)_5\cdot \qquad (63)$$

The second step in this scheme is based on the known high substitution lability of the $Re(CO)_5\cdot$ radical [34a] and the established propensity of certain neutral 18-electron organometallic compounds to act as two-electron donor ligands [70]. The proposed mechanism is as yet unsupported by kinetic or mechanistic studies establishing the radical character of the process.

5. Decarbonylation of Formylmetals

Kochi and coworkers were first to propose a radical chain mechanism for decarbonylation of formylmetal complexes [71]. Their first report of this reaction involved a study of the decomposition of cis-$(OC)_5ReRe(CO)_4CHO^-Bu_4N^+$, I. An oxygen-free solution of this salt in either THF or acetonitrile remains unchanged for several days at 25 °C in the dark. Addition of 10 mol % AIBN caused a transformation to form cis-$(OC)_5ReRe(CO)_4H^-Bu_4N^+$, II, in 90% yield within 4 h. Based on the known rate of homolysis of AIBN a turnover number of >500 was estimated for the chain process leading to decarbonylation. Further, the rate of decarbonylation under these conditions is markedly retarded in the presence of one equivalent of dihydroanthracene, a good hydrogen donor.

Irradiation of THF solutions of I with 400 nm light affords II in 75% yield in 3 h. From actinometry measurements, the quantum yield for decarbonylation is estimated to be >400 [71c]. The proposed reaction mechanism is as follows:

$$(OC)_5ReRe(CO)_4CHO^- + In\cdot \rightarrow HIn + (OC)_5ReRe(CO)_5^{-\cdot} \qquad (64)$$

$$(OC)_5ReRe(CO)_5^{-\cdot} \rightarrow (OC)_5ReRe(CO)_4^{-\cdot} + CO \qquad (65)$$

$$(OC)_5ReRe(CO)_4^{-\cdot} + (OC)_5ReRe(CO)_4CHO^- \rightarrow (OC)ReRe(CO)_4H^- + (OC)_5ReRe(CO)_5^- \qquad (66)$$

Abstraction of hydrogen, Eq. (64), leads to a 19-electron metal center, which extrudes a CO group, Eq. (65). Hydrogen atom abstraction initially could occur via $Re(CO)_5\cdot$ radicals produced photochemically from metal-metal bond homolysis, via $(CH_3)_2(NC)C\cdot$ radicals formed from homolysis of AIBN, or via other initiators.

The radical chain decarbonylation can also be initiated via electrochemical or chemical reduction. In either case radical intermediates are formed to act as initiators:

$$(OC)_5ReRe(CO)_4CHO^- + e^- \rightarrow Re(CO)_5\cdot + Re(CO)_4CHO^{2-} \qquad (67)$$

or

$$\rightarrow Re(CO)_5^- + Re(CO)_4CHO^-$$

Propagation of the chain in these circumstances is limited by further reduction of the chain propagating species:

$$(OC)_5ReRe(CO)_4^- + e^- \rightarrow (OC)_5ReRe(CO)_4^{2-} \xrightarrow{H^+} (OC)_5ReRe(CO)_4H^- \qquad (68)$$

Kochi et al. have demonstrated the formation of formylmetal complexes via reduction of metal carbonyl compounds in the presence of a suitable H atom donor [71c]. For example, the reduction of $Cr(CO)_6$ in THF containing 0.1 M tetraethylammonium perchlorate as supporting electrolyte, in the presence of 0.25 M $HSnBu_3$ leads to about 20% conversion to $Cr(CO)_5CHO^-$. The authors suggest that the initially formed 19-electron species, $Cr(CO)_6^-$, is trapped by hydrogen atom transfer instead of dissociating CO to form $Cr_2(CO)_{10}^{2-}$, a principal product in the absence of $HSnBu_3$. Similarly, upon reduction of $Fe(CO)_5$ in the presence of $HSnBu_3$, $Fe(CO)_4CHO^-$ is obtained in 15–30% yield.

The decay of these formylmetal systems, which are inherently less stable than $(OC)_5ReRe(CO)_4CHO^-$, proceeds at least in part via the same radical chain pathway as outlined in Eqs. (64)–(66). The radical chain can be interrupted by addition of hydrogen atom donors. For example, $Cr(CO)_5CHO^-$, formed from reduction of $Cr(CO)_6$ by $K(i\text{-}PrO)_3BH$, decomposes in THF at 25 °C with a half-life of about 40 min, to afford $Cr(CO)_5H^-$ and $(OC)_5CrHCr(CO)_5^-$ in 10% and 17% yields, respectively. Upon addition of a hydrogen donor such as γ-terpenine, dihydroanthracene or $HSnBu_3$, the half-life is markedly extended. The yield of $Cr(CO)_5H^-$ is increased and the bridging hydrido species is absent. These results are consistent with the radical chain pathway for decarbonylation; the hydrogen atom donor interrupts the chain process by trapping the chain propagating $Cr(CO)_5^-$ species to form $Cr(CO)_5H^-$. Other formylmetal species stabilized by the presence of $HSnBu_3$ include $CH_3CORe(CO)_4CHO^-$, $W(CO)_5CHO^-$, $BrRe(CO)_4CHO^-$ and $Mn(CO)_3(PPh_3)_2CHO$ [71c].

Kochi et al. noted that the half-life for decomposition of $Cr(CO)_5CHO^-$ reaches a limiting value of about 170 min upon addition of hydrogen atom donors, independent of the nature or concentration of the retarder. Pathways other than the radical chain decarbonylation are available for decomposition of this and the other formylmetal species studied.

A closely analogous study of Cp'Ru(CO)(L)CHO (Cp' = η^5-$C_5(CH_3)_5$; L = CO, PMe_2Ph) has been reported by Nelson and Sumner [65]. Cp'Ru(CO)(PMe_2Ph)CHO is stable in degassed benzene at room temperature, but decomposes at 40 °C or above to yield an approximately 1 : 1 ratio of Cp'Ru$(CO)_2$H and Cp'Ru(CO)(PMe_2Ph)H. Addition of 2 equivalents of dihydroanthracene extends the decomposition time from about 1 h to up to 14 days. A radical chain pathway analogous to that proposed by Kochi et al. was proposed by the authors. The mixture of products arises because either CO or phosphine can dissociate from the 19-electron intermediate.

When decomposition of the formyl complex Cp'Ru(CO)(PMe_2Ph)CHO is carried out in the presence of Cp'Ru$(CO)_2$I the opportunity arises for iodine atom transfer to the radical intermediate in the decomposition chain. Accordingly, Cp'Ru(CO)(PMe_2Ph)I is seen as a product. In addition, however, the ionic compound Cp'Ru$(CO)_2$(PMe_2Ph)$^+$I$^-$ is also observed. Indeed, in a polar solvent such as acetone it is the predominant phosphine-containing product. The ionic product is thought to arise via electron transfer to the carbonyl iodide from the 19-electron intermediate formed in the radical chain process.

Finally, it should be noted that where the thermodynamics of the system permit, the microscopic reverse of this radical chain decarbonylation leads to carbonyl insertion into a metal-hydrogen bond, as in the insertion of CO into the Rh—H bond of (octaethylporphyrin)rhodium hydride [72a]:

$$Rh_2(OEP)_2 \rightleftarrows 2(OEP)Rh\cdot \quad (69)$$

$$(OEP)Rh\cdot + CO \rightarrow (OEP)RhCO \quad (70)$$

$$(OEP)RhCO\cdot + (OEP)RhH \rightarrow (OEP)RhCHO + (OEP)Rh\cdot \quad (71)$$

6. Insertion Reactions

Insertions of unsaturated functional groups into M—H or M—M bonds of transition organometallic compounds are not generally observed to occur via radical chain pathways. However, a few examples are known.

Insertion of styrene or certain other olefins into Rh—H or Rh—Rh bonds of (OEP)RhH and $(OEP)_2Rh_2$, (OEP = octaethylporphyrin) occurs with facility [72a]. The kinetics evidence suggests radical chain pathways. For the hydride:

$$(OEP)RhH + In\cdot \rightarrow (OEP)Rh\cdot + HIn \quad (72)$$

$$(OEP)Rh\cdot + PhCH{=}CH_2 \rightarrow (OEP)RhCH_2CHPh\cdot \quad (73)$$

$$(OEP)RhCH_2CHPh\cdot + (OEP)RhH \rightarrow (OEP)RhCH_2CH_2Ph + (OEP)Rh\cdot \quad (74)$$

This reaction was found to be catalyzed by $(OEP)_2Rh_2$, which serves as a source of (OEP)Rh· radicals. The reaction is analogous to the homolytic chain mechanism observed in reactions of main group hydrides such as trialkytin hydrides with unsaturated functional groups [73]. Reaction of the rhodium dimer with styrene is similar:

$$(OEP)_2Rh_2 \rightleftarrows 2(OEP)Rh\cdot \qquad (75)$$

$$(OEP)Rh\cdot + PhCH{=}CH_2 \rightarrow (OEP)RhCH_2CHPh\cdot \qquad (76)$$

$$(OEP)RhCH_2CHPh\cdot + (OEP)_2Rh_2 \rightarrow (OEP)RhCH_2CH(Ph)Rh(OEP) + (OEP)Rh\cdot \qquad (77)$$

This reaction is unique in involving a homolytic displacement on a metal-metal bond. It is made feasible by a weak Rh—Rh bond, and by the absence of an axial ligand, which renders the metal center accessible to the attacking radical. The dimer also undergoes oxidative addition to benzyl bromide,

$$(OEP)_2Rh_2 + PhCH_2Br \rightarrow (OEP)RhCH_2Ph + (OEP)RhBr \qquad (78)$$

and to hydrogen donors such as $HSnBu_3$, probably by analogous radical chain processes.

The reaction of $CpMo(CO)_2LH$ ($L = CO, PMe_3$) with hexamethylsilacyclopropane, HMS, provides an interesting example of an apparently radical chain insertion [74]:

$CpMo(CO)_2(L)H +$  $\rightarrow CpMo(CO)_2(L)SiMe_2H + C_2(CH_3)_4$ (79)

The proposed mechanism is as follows:

$$CpMo(CO)_2(L)\cdot + HMS \rightarrow CpMo(CO)_2(L)SiMe_2CMe_2CMe_2\cdot \qquad (80)$$

$$CpMo(CO)_2(L)SiMe_2CMe_2CMe_2\cdot \rightarrow CpMo(CO)_2(L)SiMe_2\cdot + Me_2C{=}CMe_2 \qquad (81)$$

$$CpMo(CO)_2(L)SiMe_2\cdot + CpMo(CO)_2(L)H \rightarrow CpMo(CO)_2(L)SiMe_2H + CpMo(CO)_2(L)\cdot \qquad (82)$$

Elimination of tetramethylethylene is fast compared with the rate of hydrogen atom transfer to the initially-formed product of ring-opening. However, some $CpMo(CO)_2(L)SiMe_2CMe_2CMe_2H$ was observed when the starting hydride was reacted with neat HMS.

In general, the radical chain pathway for addition of transition metal hydrides to unsaturated functional groups is not observed. The reaction is more likely to proceed via direct hydrogen atom transfer to the unsaturated functionality, followed by hydrogen abstraction from a second molecule of hydride. The overall result is hydrogenation via a non-chain radical process, as demonstrated elegantly by Halpern and co-workers [72].

7. Miscellaneous Reactions

In this brief section we mention three reports of what appear to be radical chain reactions that require or could involve atom transfer steps. In neither case, however, is the evidence completely clear. The reaction of cyclopentadiene with $Fe(CO)_5$ to yield $[CpFe(CO)_2]_2$ has been studied [75]. One component of the set of reactions involved in the overall process, the migratory decarbonylation of

(η^4-C_5H_6)$Fe(CO)_3$ to yield $CpFe(CO)_2H$ and CO, is catalyzed by $[CpFe(CO)_2]_2$ and appears to be radical chain in character:

$$(\eta^4\text{-}C_5H_6)Fe(CO)_3 + CpFe(CO)_2\cdot \rightarrow CpFe(CO)_2H + CpFe(CO)_2\cdot + CO \quad (83)$$

Angelici et al. have observed an apparently radical chain reaction of PPh_3 or PPh_2Me with $Os_3(CO)_{12}Br_2$ in THF at room temperature, to give a mixture of products [76]:

$$Os_3(CO)_{12}Br_2 + L \rightarrow Br_2Os_3(CO)_{10}L_2\ (12\text{–}20\%) + Br_2Os(CO)_2L_2 + Os(CO)_4L(60\%) \quad (84)$$

The reaction is inhibited by addition of galvinoxyl or excess CCl_4. When conducted in CH_3CN the reaction with PPh_3 is complete within 5 h; however, it requires less than 20 min to go to completion when 5 equivalents of NBu_4Br have been added. The authors do not suggest specific radical pathways. The following steps seem reasonable possibilities:

$$Os_3(CO)_{12}Br_2 \rightarrow Br(OC)_4Os{-}Os(CO)_4\cdot + Br(OC)_4Os\cdot \quad (85)$$

$$Br(OC)_4Os\cdot + 2L \rightarrow Br(OC)_2(L)_2Os\cdot + 2CO \quad (86)$$

$$Br(OC)_2(L)_2Os\cdot + Os_3(CO)_{12}Br_2 \rightarrow Br_2(OC)_2(L)_2Os + \cdot Os(CO)_4Os_2(CO)_8Br \quad (87)$$

$$\cdot Os(CO)_4Os_2(CO)_8Br + 2L \rightarrow \cdot Os(CO)_2(L)_2Os_2(CO)_8Br + 2CO \quad (88)$$

$$\cdot Os(CO)_2(L)_2Os(CO)_8Br + Os_3(CO)_{12}Br_2 \rightarrow BrOs(CO)_2(L)_2Os_2(CO)_8Br\ \cdot Os(CO)_4Os_2(CO)_8Br \quad (89)$$

The comparatively high yield of mononuclear product suggests that elimination processes involving the radicals may be important, e.g.:

$$\cdot Os(CO)_4Os_2(CO)_8Br \rightarrow Os(CO)_4 + \cdot Os(CO)_4Os(CO)_4Br \quad (90)$$

$$Os(CO)_4 + L \rightarrow Os(CO)_4L$$

$$\cdot Os(CO)_4Os(CO)_4Br + Os_3(CO)_{12}Br_2 \rightarrow Os_2(CO)_8Br_2 + \cdot Os(CO)_4Os_2(CO)_8Br \quad (91)$$

$$Os_2(CO)_8Br_2 \rightarrow Br_2 + Os_2(CO)_8 \quad (92)$$

$$Os_2(CO)_8 + 2L \rightarrow 2Os(CO)_4L \quad (93)$$

Detty and Jones have observed that oxidation of $CpRe(L)_2H_2$ ($L{=}P(aryl)_3$) leads to 17-electron cations stable in CH_3CN [77]. However, further oxidation to form a small quantity of $CpRe(L)_2H_2{}^{2+}$, leads to rapid loss of the 17-electron cation from the solution. The authors propose that the reaction could be driven by either a proton or hydrogen atom transfer. For a hydrogen atom transfer, the reaction would proceed as follows:

$$CpRe(L)_2H_2{}^{2+} + CpRe(L)_2H_2{}^{+\cdot} \xrightarrow[CH_3CN]{} CpRe(L)_2(NCCH_3)H^{\cdot 2+} + CpRe(L)_2H_3{}^{+} \quad (94)$$

This atom transfer step could serve as the initial process in a radical chain disproportionation to form $CpRe(L)_2H_3{}^+$ and $CpRe(L)_2(NCCH_3)H^+$, through the intermediacy of an electron transfer step:

$$CpRe(L)_2(NCCH_3)H^{\cdot 2+} + CpRe(L)_2H_2{}^{\cdot +} \rightarrow CpRe(L)_2(NCCH_3)H^{+} + CpRe(L)_2H_2{}^{2+} \quad (95)$$

8. Requirements for Radical Chain Behavior

Before reviewing what is known of the rate constants for atom transfer reactions of organometallic radicals, it is useful to consider briefly the magnitudes of rate constants for atom transfer and substitution required to sustain a radical chain process of considerable length. For this purpose, consider the following simple reaction scheme:

$$M(CO)_n\cdot + L \xrightarrow{k_s} M(CO)_{n-1}L\cdot + CO \qquad (96)$$

$$M(CO)_{n-1}L\cdot + HM(CO)_n \xrightarrow{k_a} HM(CO)_{n-1}L + M(CO)_n\cdot \qquad (97)$$

$$2M(CO)_n\cdot \xrightarrow{k_r} M_2(CO)_{2n} \qquad (98)$$

By some means, either through thermal or photochemical homolysis of the M—M bond in $M_2(CO)_{2n}$ or via hydrogen atom abstraction from $HM(CO)_n$ by an initiator, there arises a concentration of $M(CO)_n\cdot$ radicals. Maintenance of the chain process requires that propagation steps compete effectively with termination steps. For simplicity we consider only the one termination process shown in Eq. (98). The pseudo-first-order rate term for loss of $M(CO)_n\cdot$ radicals via recombination is

$$\frac{1}{[M(CO)_n\cdot]}\,\frac{-\,d[M(CO)_n\cdot]}{dt} = 2k_r[M(CO)_n\cdot] \qquad (99)$$

Steps (96) and (97) must be fast with respect to this loss term:

$$\frac{1}{[M(CO)_n\cdot]}\,\frac{-d[M(CO)_n\cdot]}{dt} = k_s[L] \gg 2k_r[M(CO)_n\cdot] \qquad (100)$$

$$\frac{1}{[M(CO)_n\cdot]}\,\frac{-d[M(CO)_n\cdot]}{dt} = k_a[HM(CO)_n] \gg 2k_r[M(CO)_n\cdot] \qquad (101)$$

If, for example, the chain length is to be on the order of 1000, then roughly,

$$k_s[L] \text{ or } k_a[HM(CO)_n] > 10^3 k_r[M(CO)_n\cdot]$$

Assume that k_r is 10^8 M^{-1} s^{-1}. If $[L] \sim 10^{-3}$ M, $[M(CO)_n\cdot] \sim 10^{-8}$ M, then k_s should be on the order of 10^6 M^{-1} s^{-1} or larger. If it is as low as 10^4 M^{-1} s^{-1}, the chain length will be too short to sustain the process. Similarly, if $[HM(CO)_n] \sim 10^{-4}$ M, then k_a must be on the order of 10^7 M^{-1} s^{-1} for chain lengths to be on the order of 1000. Thus, the existence of an efficient homolytic chain process for substitution requires under the usual solution concentration conditions large values for the rate constants for *both* atom transfer and substitution.

Clearly, the requirements on the rate constants for the propagation steps in any given radical chain process depend on the rates of termination processes. For example, the radical chain substitution reactions of $(\eta^5\text{-}C_5H_4CH_3)Mn(CO)_2L^+$ radicals, studied by Kochi and coworkers [40, 71d], proceed via ETC catalysis. Chain lengths are observed to be long, although substitution rate constants are on the order of 10–10^4 M^{-1} s^{-1}. In this case the termination steps do not involve recombination of radicals; the radicals, once formed, are fairly long lived.

9. Halogen Atom Transfer

Metal carbonyl radicals react readily with halogen atom sources to form metal carbonyl halides. Wrighton and Ginley [8, 13] showed that such halogen atom transfer processes can compete with recombination of radicals formed via homolysis of the metal-metal bond in dinuclear metal carbonyls. Inasmuch as the rate constants of the recombination reactions were known to approach the diffusion limit, it was clear that the halogen atom transfer reactions are also fast.

Wrighton and Abrahamson studied the formation of metal carbonyl halides in competition with metal-metal bond formation upon photolysis of heterodinuclear metal carbonyl compounds [78]. From the product distributions they were able to infer the following order of reactivities of the radicals toward typical halogen atom donors such as CCl_4: $Re(CO)_5 > Mn(CO)_5 > CpW(CO)_3 > CpMo(CO)_3 > CpFe(CO)_2 > Co(CO)_4$. Although some rough quantitative estimates of relative reactivities were made, the nature of the experiment does not admit of much precision, nor does it permit quantitative comparisons over a large range of relative reactivities.

The first quantitative assessment of halogen atom transfer rate constants was made by Laine and Ford [24], who studied the quantum yields for photochemical reaction of $[CpW(CO)_3]_2$ with several halogen donors. The quantum yield for disappearance of the starting dinuclear compound in THF solution was measured as a function of halogen donor concentration. The rate constant for recombination of the radicals is known. Loss of dinuclear starting compound due to reaction of the CO-loss product or other pathway was assessed by observing the limiting disappearance quantum yield in the absence of halogen donor. The method lacks quantitative precision because the disappearance quantum yield is a function of the intensity of absorbed light, and this is not constant throughout the sample. Nevertheless, the rate constants derived from the analysis are reasonably good, particularly as regards the relative values for different halogen donors.

The first direct observations of halogen atom transfer to metal carbonyl radicals were made by Waltz and coworkers, who employed pulse radiolysis of $Mn_2(CO)_{10}$ [79] or $Re_2(CO)_{10}$ [20] to form the radicals. They were able to observe the optical absorption spectra of the radicals in ethanol, and note the rates of disappearance of the absorptions as a function of CCl_4 concentration.

Flash photolysis provides a convenient and quantitatively accurate means of evaluating the rate constants for atom transfer reactions of many metal carbonyl radicals, particularly those that can be formed via homolysis of the metal-metal bond in dinuclear metal carbonyl compounds. Fig. 3 shows schematically the absorptions due to $Mn_2(CO)_{10}$, $Mn_2(CO)_9$, and $Mn(CO)_5\cdot$. When a solution containing the metal carbonyl is flashed with an intense pulse of short duration (for example, a 10 ns pulse of wavelength 337 nm from an N_2 laser), the dinuclear compound undergoes CO loss or homolysis to generate $Mn_2(CO)_9$ or $Mn(CO)_5\cdot$, respectively. Because in this and many other related systems absorption of the radical is distinct from those of other components in the solution, it is possible to probe the solution at longer wavelengths to observe the time-dependence of the concentration of the radical. In the absence of a halogen or other atom transfer source the absorption due to the radical decays as the radicals recombine to form the dinuclear compound. When a halogen atom source is present in sufficiently high concentration, the pseudo first order atom transfer process is faster than the bimolecular recombination. Thus, observation of the transient decay permits

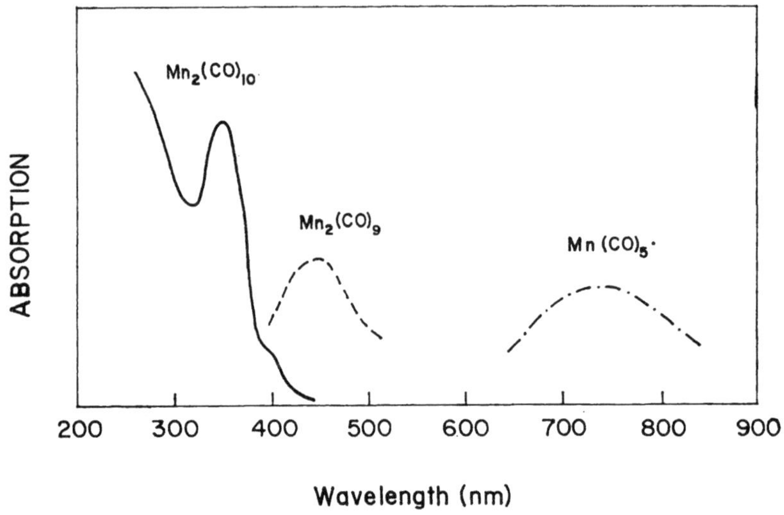

Fig. 3. Major absorptions due to $Mn_2(CO)_{10}$, $Mn_2(CO)_9$, and $Mn(CO)_5$·

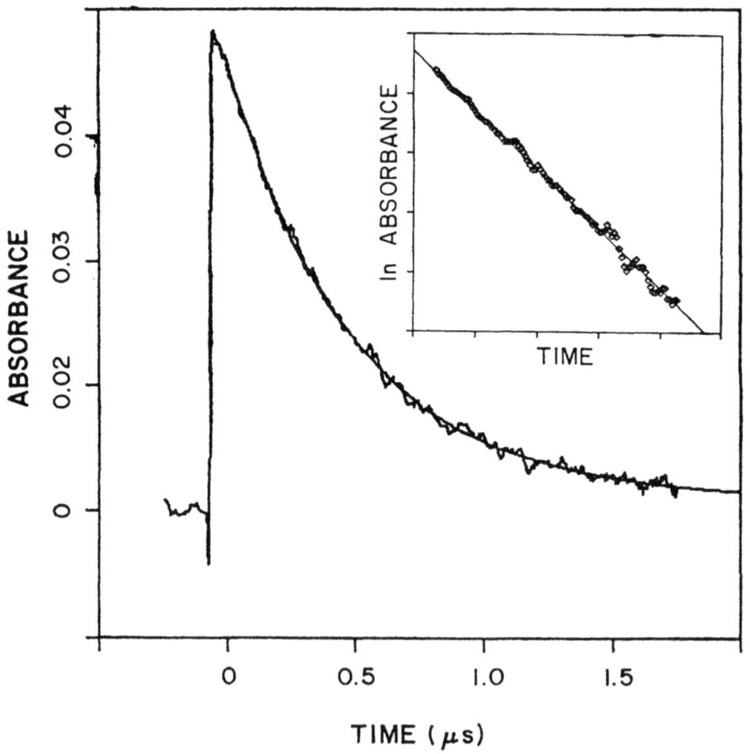

Fig. 4. Transient absorption at 550 nm due to $Re(CO)_4[P(O\text{-}i\text{-}Pr)_3]$· after flash photolysis in C_6H_{14}, in the presence of 0.010 M CCl_4.

evaluation of the atom transfer rate constant. Fig. 4 shows the decay of transient absorption of the $Re(CO)_4[P(O\text{-}i\text{-}Pr)_3]\cdot$ radical where the rate determining process is atom transfer from CCl_4. Also shown is the corresponding pseudo first order plot of ln $[Re(CO)_4L\cdot]$ vs. time.

It is possible to form substituted metal carbonyl radicals that are persistent in solution at room temperature [18, 19]. Formation of the metal-metal bond in these cases is precluded by the presence of two bulky ligands on the metal. Atom transfer reactions of such radicals have been studied by stopped-flow IR [19] and electrochemical methods [80].

Table 3. Rate Constants for Halogen Atom Transfer to Metal Carbonyl Radicals

Radical	Substrate	Temp. (°C)	Solvent	k_T (M^{-1} s^{-1})	Ref.
$Mn(CO)_5\cdot$	CCl_4	20	C_6H_6	1.0×10^6	[81]
	CCl_4	22	CCl_4	9.1×10^5	[82]
	CCl_4	22	C_2H_5OH	6.1×10^5	[20]
	CCl_4	21	C_6H_{14}	1.4×10^6	[83]
	CBr_4	21	C_6H_{14}	1.5×10^9	[83]
	$CHBr_3$	21	C_6H_{14}	1.0×10^7	[83]
	$PhCH_2Br$	21	C_6H_{14}	4.8×10^5	[83]
	CH_2Br_2	21	CH_2Br_2	7×10^3	[83]
	$PhCH_2Br$	20	C_6H_6	4.6×10^6	[81]
	CH_2CHCH_2Br	20	C_6H_6	2.8×10^6	[81]
$Mn(CO)_4P(n\text{-}Bu)_3\cdot$	CCl_4	21	C_6H_{14}	1.8×10^6	[83]
$Mn(CO)_4P(i\text{-}Bu)_3\cdot$	CCl_4	24	C_6H_{14}	8.9×10^4	[83]
$Mn(CO)_4P(C_6H_{11})_3\cdot$	CCl_4	24	C_6H_{14}	2.0×10^4	[83]
	$CHCl_3$	24	$CHCl_3$	7.3	[83]
$Mn(CO)_4P(i\text{-}Pr)_3\cdot$	CCl_4	24	C_6H_{14}	2.8×10^4	[83]
	CH_2Br_2	24	C_6H_{14}	8.6×10^2	[83]
	$CHCl_3$	24	$CHCl_3$	1.6×10^1	[83]
$Mn(CO)_4P(O\text{-}i\text{-}Pr)_3\cdot$	CCl_4	21	C_6H_{14}	2.2×10^4	[83]
$Mn(CO)_3[P(i\text{-}Bu)_3]_2\cdot$	CCl_4	24	C_6H_{14}	3.0×10^4	[83]
	$PhCH_2Cl$	24	C_6H_{14}	0.80	[83]
	CH_2Cl_2	24	C_6H_{14}	4.0×10^{-2}	[83]
$Mn(CO)_3[P(i\text{-}Pr)_3]_2\cdot$	CCl_4	24	C_6H_{14}	1.0×10^3	[83]
$Mn(CO)_3[P(n\text{-}Bu)_3]_2\cdot$	$PhCH_2Cl$	24	C_6H_{14}	1.0×10^3	[83]
$Re(CO)_5\cdot$	CCl_4	20	C_6H_6	3.0×10^7	[84]
	CCl_4	22	C_6H_{14}	4.9×10^7	[37]
	CCl_4	22	C_2H_5OH	3.9×10^7	[20]
	$CHCl_3$	20	C_6H_6	3.8×10^4	[84]
	$CHCl_3$	22	C_6H_{14}	4.4×10^5	[37]
	$PhCH_2Br$	20	C_6H_6	7.7×10^7	[84]
	CH_2CHCH_2Br	20	C_6H_6	1.2×10^7	[84]
	$PhCH_2Cl$	20	C_6H_6	1.2×10^5	[84]
	CH_2Br_2	22	C_6H_{14}	4.8×10^7	[37]
	CH_2Br_2	22	CH_3CN	5.1×10^7	[37]
$Re(CO)_4PMe_3\cdot$	CCl_4	22	C_6H_{14}	2.5×10^{9} [a]	[37]
	CH_2Br_2	22	C_6H_{14}	1.8×10^{9} [a]	[37]

Table 3. continued

Radical	Substrate	Temp. (°C)	Solvent	k_T (M^{-1} s^{-1})	Ref.
	CH_2Br_2	22	CH_3CN	8.7×10^{8a}	[37]
	CH_2Br_2	22	toluene	8.5×10^{8a}	[37]
	CH_2Br_2	22	THF	5.1×10^8	[37]
	CH_2Br_2	22	pyridine	6.1×10^8	[37]
	$CHCl_3$	22	C_6H_{14}	2.1×10^7	[37]
$Re(CO)_4P(OMe)_3\cdot$	CCl_4	22	C_6H_{14}	3.3×10^8	[37]
	CH_2Br_2	22	C_6H_{14}	1.5×10^8	[37]
	CH_2Br_2	22	CH_3CN	8.9×10^7	[37]
	$CHCl_3$	22	C_6H_{14}	1.5×10^6	[37]
$Re(CO)_4P(O\text{-}i\text{-}Pr)_3\cdot$	CCl_4	22	C_6H_{14}	2.0×10^8	[37]
	CH_2Br_2	22	C_6H_{14}	8.1×10^7	[37]
	CH_2Br_2	22	CH_3CN	3.4×10^7	[37]
	$CHCl_3$	22	C_6H_{14}	9.3×10^5	[37]
$Re(CO)_4P(n\text{-}Bu)_3\cdot$	CCl_4	22	C_6H_{14}	7.6×10^8	[37]
	CH_2Br_2	22	C_6H_{24}	3.9×10^8	]37]
	CH_2Br_2	22	CH_3CN	2.2×10^8	[37]
	$CHCl_3$	22	C_6H_{14}	5.4×10^6	[37]
$Re(CO)_4P(i\text{-}Bu)_3\cdot$	CCl_4	22	C_6H_{14}	8×10^7	[37]
	CH_2Br_2	22	C_6H_{14}	2.7×10^7	[37]
	CH_2Br_2	22	CH_3CN	1.6×10^7	[37]
	$CHCl_3$	22	C_6H_{14}	4.5×10^5	[37]
$Re(CO)_4P(i\text{-}Pr)_3\cdot$	CCl_4	22	C_6H_{14}	9.3×10^7	[37]
	CH_2Br_2	22	C_6H_{14}	1.4×10^7	[37]
	CH_2Br_2	22	CH_3CN	8.5×10^6	[37]
	$CHCl_3$	22	C_6H_{14}	4.0×10^5	[37]
$Re(CO)_4P(C_6H_{11})_3\cdot$	CCl_4	22	C_6H_{14}	7.4×10^7	[37]
	CH_2Br_2	22	C_6H_{14}	1.5×10^7	[37]
	$CHCl_3$	22	C_6H_{14}	3.0×10^5	[37]
$Re(CO)_4As(C_2H_5)_3\cdot$	CCl_4	22	C_6H_{14}	1.3×10^{9a}	[37]
	CH_2Br_2	22	C_6H_{14}	7.4×10^8	[37]
	CH_2Br_2	22	CH_3CN	3.7×10^8	[37]
	$CHCl_3$	22	C_6H_{11}	8.7×10^6	[37]
$\cdot(OC)_4Re(dcype)Re(CO)_4\cdot^b$	CCl_4	22	toluene	1.6×10^8 6.5×10^7	[85]
	CH_2Br_2	22	toluene	1.8×10^7 3.3×10^7	[85]
$\cdot(OC)_4Re(dppb)Re(CO)_4\cdot^b$	CCl_4	22	toluene	9.3×10^{8a} 2.1×10^8	[85]
	CH_2Br_2	22	toluene	2.9×10^8 2.0×10^8	[85]
$\cdot(OC)_4Re(dppp)Re(CO)_4\cdot^b$	CCl_4	22	toluene	5.4×10^8	[85]
	CH_2Br_2	22	toluene	2.2×10^8	[85]
$CpMo(CO)_3\cdot$	CCl_4	20	C_6H_6	2.2×10^4	[81]
	$PhCH_2Br$	20	C_6H_6	2.0×10^5	[81]
	CH_2CHCH_2Br	20	C_6H_6	1.0×10^5	[81]

Table 3. continued

Radical	Substrate	Temp. (°C)	Solvent	k_T (M^{-1} s^{-1})	Ref.
$CpW(CO)_3\cdot$	CCl_4	22	THF	1.3×10^4	[24]
	$CHCl_3$	22	THF	2.1×10^1	[24]
	$PhCH_2Cl$	22	THF	1.4×10^1	[24]
	CH_2Cl_2	22	THF	$<6\times10^{-1}$	[24]
	$PhCH_2Br$	20	C_6H_6	1.3×10^5	[81]
	CH_2CHCH_2Br	20	C_6H_6	6.5×10^4	[81]

[a] Because these values for k_T approach the diffusion limit, a correction should be applied to obtain the activated rate constant. See reference listed.
[b] (dcype) = bis(dicyclohexylphosphino)ethane; (dppb) = bis(diphenylphosphino)butane; (dppp) = bis(diphenylphosphino)propane. Where two rate constants are listed, they correspond to successive halogen atom transfers to the two metal centers of the biradical, where these could be distinguished.

Table 4. Rate Constants for Halogen Atom Transfer to $Re(CO)_4L\cdot$ Radicals in Toluene at 22°C [86a]

RX	L = PMe_3 $k_T\times10^{-7}$ (M^{-1} s^{-1})	L = $P(O\text{-}i\text{-}Pr)_3$ $k_T\times10^{-7}$ (M^{-1} s^{-1})	$-E^{1/2}$ of RX· [86b]
allyl	390 ± 10	200 ± 10	0.23
CBr_4	460 ± 10	410 ± 20	0.30
C_2Cl_6	200 ± 10	23 ± 1	0.62
$CHBr_3$	470 ± 20	220 ± 20	0.64
α,α,α-trichlorotoluene		4.3 ± 0.1	0.68
CCl_4	160 ± 10	16 ± 1	0.78
4-nitrobenzyl bromide		71 ± 4	0.80
4-cyanobenzyl bromide	140 ± 10	26 ± 1	0.96
4-(trifluoromethyl)benzyl bromide		20 ± 1	
3-bromobenzyl bromide	100 ± 10	17 ± 1	1.16
4-bromobenzyl bromide		18 ± 1	1.26
4-methylbenzyl bromide		8.3 ± 0.3	1.34
benzyl bromide	110 ± 10	12 ± 1	1.36
4-fluorobenzyl bromide		13 ± 1	1.38
allyl bromide	93 ± 4	9.4 ± 0.2	1.21
CH_2Br_2	85 ± 4	4.7 ± 0.1	1.48
$BrCH_2CH_2Br$	11 ± 1	0.51 ± 0.03	1.52
CH_3CHBr_2	51 ± 3	0.86 ± 0.03	1.62
$CHCl_3$	1.4 ± 0.1	0.073 ± 0.002	1.67
α,α-dichlorotoluene	1.9 ± 0.1	0.11 ± 0.01	1.80
4-bromobenzonitrile	0.058 ± 0.005		1.80
allyl chloride	0.038 ± 0.002		1.91
benzyl chloride	0.14 ± 0.01	0.0070 ± 0.001	1.94
1-bromonaphthalene	0.021 ± 0.001		1.96

Tables 3 and 4 list quantitative data regarding halogen atom transfer rate constants of metal carbonyl radicals. Unfortunately, although some metal carbonyl radical systems are well-represented, reliable data based on flash photolysis or other direct observation methods are lacking for many radicals of importance. Nevertheless, several generalizations can be drawn from the available results:

1. The "parent" radicals clearly vary in reactivity, with relative reactivities in the order given by Abrahamson and Wrighton [78]. In quantitative terms, the relative rate constants for reaction with a given halogen donor will depend on the reactivity of the donor toward halogen atom abstraction; for highly reactive donors the selectivity is predictably lower than for reactions with less reactive substrates. There are too few points for reliable quantitative comparisons, but as a rough rule, the relative reactivities are $Re(CO)_5\cdot$ (10^4) > $Mn(CO)_5\cdot$(10^2) > $CpMo(CO_3)\cdot$ $\approx$ $CpW(CO)_3\cdot$ (1).

2. For a given atom transfer reaction, the rate constant is generally smaller in more polar solvents, but the effects are small. For example, the rate constant for reaction of $Re(CO)_4PMe_3\cdot$ with CH_2Br_2 varies over a range of only three in solvents of widely differing polarity and nucleophilicity. This suggests that neither solvent reorganization nor interaction of the solvent with the metal carbonyl radical plays a large role in affecting the reaction rate.

3. Ligands on the metal in replacement of CO can exert major effects on the rate constants. In general, strongly donor ligands enhance the rate of atom transfer; steric crowding by bulky ligands retards it. Ligand effects can be expressed in terms of a two-parameter linear free energy relationship [40, 87, 88]. For a radical $M(CO)_nL\cdot$ the measured rate constant can be expressed as

$$\ln k_T = a\delta + b\theta + c \qquad (102)$$

where δ is a measure of the electronic charge at the metal center and θ is a measure of the steric effects of the ligand L. For the former, Brown and coworkers have employed the ^{13}C chemical shift for the CO carbons in $Ni(CO)_3L$ compounds [37], measured by Bodner [89]. The steric parameter is represented by the cone angle as defined and estimated by Tolman [90]. The coefficients a and b measure the sensitivity of the reaction to changes in each parameter. An example of such a correlation, for the reaction

$$Re(CO)_4L\cdot + CH_2Br_2 \rightarrow Re(CO)_4LBr + CH_2Br\cdot \qquad (103)$$

is shown in Fig. 5.

The evaluation of ligand effects via the two-parameter free energy relationship affords a convenient means of summarizing data. More importantly, it provides a means of evaluating the electronic and steric characteristics of the reaction. The positive value of 1.4 for a in the correlation of Fig. 5 indicates that the rate constant increases with increased electronic charge buildup at the metal center. The negative value of -0.14 for b indicates that the reaction is impeded by crowding at the metal center, as reflected in the cone angle of the ligand.

9.1 Systematics of Halogen Atom Transfer Reactions

Chemists are perennially concerned with models and theoretical constructs that can form the basis of a systematic understanding of the many reactions they ob-

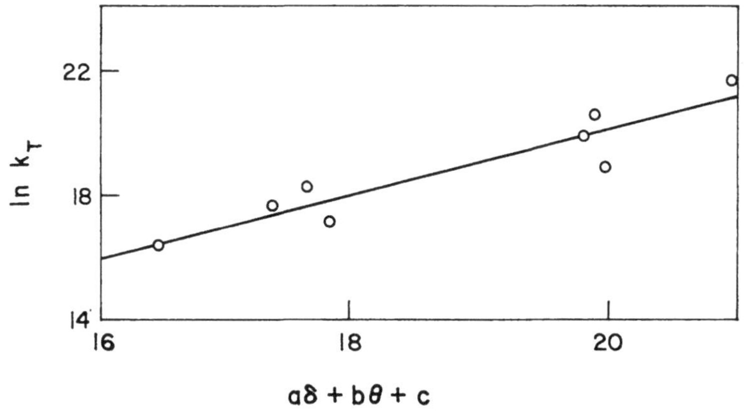

Fig. 5. Least squares correlation of observed rate constants for reaction (103) in hexane with those calculated from equation (102). The values for *a* and *b* are 1.4 and −0.14, respectively [37]. The ligands represented by the points are, from left to right: $P(i\text{-}Pr)_3$; CO; $P(O\text{-}i\text{-}Pr)_3$; $P(i\text{-}Bu)_3$; $P(n\text{-}Bu)_3$; $AsEt_3$; $P(OMe)_3$; PMe_3.

serve. Structure-reactivity correlations, which relate rate constants to the properties of reactants and products, are an important tradition in both inorganic and organic chemistry. Linear free energy relationships, which relate the free energy barrier of a reaction or the equilibrium constant for an equilibrium to parameters that characterize the components of a reaction series, have proved very useful. However, such relationships, as well as various "principles" or postulates related to reactivity, such as the Hammond postulate [91] or the Selectivity-Reactivity Principle [92], are recognized as having limited applicability. In recent years the focus of research has shifted to the application of more general structure-reactivity relationships [93]. Chief among these are the family of so-called Marcus type formulations, inspired by Marcus's original formulations of the free energy barrier for weak overlap electron transfer on the one hand [94] and for strong-overlap electron transfer or atom, proton, or group transfer reactions to the other [95].

In the Marcus model for atom transfer, which concerns us here, the free energy barrier is formulated in terms of a thermodynamic component, expressed as the net free energy change in the rate-determining process, and a kinetic component, expressed as the intrinsic barrier to reaction. The intrinsic barrier, $\Delta G^{\neq}(0)$, is defined as the free energy barrier for the hypothetical reaction of overall zero free energy change.

The prior and post-equilibria associated with a rate-determining process of interest will vary from one reaction system to another. It is useful to treat the atom transfer process in terms of the equilibria shown in equation (104).

$$\begin{aligned} RX + \cdot Re(CO)_4L &\underset{k_{21}}{\overset{k_{12}}{\rightleftharpoons}} \{RX \ldots \cdot Re(CO)_4L\} \underset{k_{32}}{\overset{k_{23}}{\rightleftharpoons}} \{RX^{-+}Re(CO)_4L\} \\ &\xrightarrow{k_{30}} R\cdot + XRe(CO)_4L \end{aligned} \qquad (104)$$

In this formulation we have assumed that the rate-determining process is an electron transfer, with rate constant k_{23}. The characteristics of the overall formulation do not, however, require electron transfer as the rate-determining step; the reaction could as well be visualized as a simple atom transfer from carbon to metal. The rationale for choosing the electron transfer formulation is discussed below.

The rate constant for the overall atom transfer process, k_T, is given by:

$$k_T = \frac{k_{12}}{1 + (k_{21}/k_{23}) + (k_{32}/k_{23})(k_{21}/k_{30})} \tag{105}$$

The constant k_{12} is the diffusion rate constant for the solvent in which the reaction occurs. The rate constant k_{23} for electron transfer and the ratio k_{32}/k_{23} can be expressed as

$$k_{23} = k_{23}^{\circ} \exp(-\Delta G_{23}^{\neq}/RT) \tag{106}$$

$$k_{23}/k_{23} = \exp(\Delta G_{23}/RT)$$

Here k_{23}°, $\Delta G_{23}^{\neq}$ and ΔG_{23} are the frequency factor, standard free energy of activation and standard free energy change for the electron transfer step. Employing these expressions we obtain

$$k_T = \frac{k_{12}}{1 + \frac{k_{21}}{k_{23}^{\circ}} \exp(\Delta G_{23}^{\neq}/RT) + \frac{k_{21}}{k_{30}} \exp(\Delta G_{23}/RT)} \tag{107}$$

Marcus derived an expression for the free energy of activation, $\Delta G_{23}^{\neq}$, in terms of the overall free energy change and the intrinsic barrier, by employing Johnston's bond energy-bond order model [96] for the potential energy surface of the reaction. Agmon and Levine showed later that an equivalent expression arises from more general considerations of the nature of the free energy profile [97].

$$\Delta G_{23}^{\neq} = \Delta G_{23} + \frac{\Delta G^{\neq}(0)}{\ln 2} \ln \left\{1 + \exp\left[\frac{-\Delta G_{23} \ln 2}{\Delta G^{\neq}(0)}\right]\right\} \tag{108}$$

From this expression for $\Delta G_{23}^{\neq}$ one can obtain the desired formulation of the rate constant for atom transfer, k_T:

$$\log k_T = \log k_{12} - \log \left[1 + A \exp(\Delta G_{23}/RT) \exp \left\{\frac{\Delta G^{\neq}(0)}{RT \ln 2}\right.\right.$$

$$\left.\left. \ln [1 + \exp(-\Delta G_{23} \ln 2/\Delta G^{\neq}(0))]\right\} + B \exp(\Delta G_{23}/RT)\right] \tag{109}$$

The constant $A = k_{21}/k_{23}^{\circ}$ is a measure of the efficiency with which the precursor complex undergoes atom transfer as opposed to diffusively dissociating. The constant $B = k_{21}/k_{30}$ is a measure of the relative rates of diffusive separation of the precursor and successor complexes. The value of A determines the asymptotic limit for k_T in the highly exergonic region, whereas B affects the behavior of k_T only in the strongly endergonic regime.

It is often desirable to apply the Marcus/Agmon Levine (MAL) equation to a body of kinetics data in which a single atom acceptor is reacted with a series of atom donors, or a single donor with a series of atom acceptors. A *reaction series* is one in which A, B and $\Delta G^{\neq}(0)$ are effectively constant [86, 95]. The intrinsic barrier for a given reaction can be expressed as the average value of the intrinsic barriers for the two degenerate reactions. That is, $\Delta G^{\neq}(0)$ for the reaction.

$$\text{M}\cdot + \text{XR} \rightarrow \text{MX} + \text{R}\cdot \tag{110}$$

is the average of the intrinsic barriers for the reactions

$$M\cdot + XM' \rightarrow MX + M' \tag{111}$$

$$R\cdot + XR' \rightarrow RX + R' \tag{112}$$

In a reaction series in which RX varies, the intrinsic barrier to the degenerate reaction (112) could vary. Unfortunately we know very little about the intrinsic barriers to halogen atom transfer reactions. The assumption of constant $\Delta G^{\neq}(0)$ in a series in which the overall energy change varies considerably is not necessarily a very good one. A series of plots of log k_T vs. ΔG_{23} is shown in Fig. 6 for various values of $\Delta G^{\neq}(0)$, and fixed values of A and B. It is evident that the shape of the curves is critically dependent on $\Delta G^{\neq}(0)$.

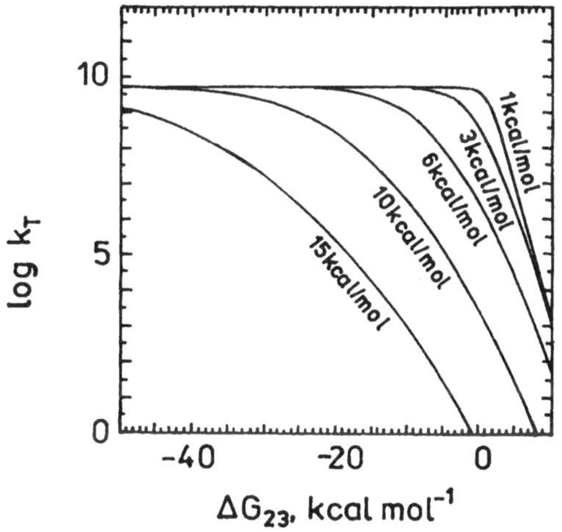

Fig. 6. Log K_T vs. ΔG_{23} for various values of the intrinsic barrier $\Delta G^{\neq}(0)$ [85]. Preprinted by permission of the American Chemical Society.

Lee and Brown [86] applied equation (109), modified for use of electrochemical reduction potentials, to the data for halogen atom transfer reactions of $Re(CO)_4PMe_3\cdot$ and $Re(CO)_4P(O\text{-}i\text{-}Pr)_3\cdot$ radicals shown in Table 4, choosing as a measure of ΔG_{23} the irreversible reduction potentials of the organic halogen donors (Table 4). The results are illustrated in Fig. 7. The fit is generally good for both series, especially considering that chlorides, bromides and iodides are all included in the correlation. However, there is evidence of steric effects in the scatter in the data for the phosphite-substituted radical. The points denoted as solids correspond to the most hindered organic donors, CCl_4, C_2Cl_6 and $C_6H_5CCl_3$. Such departures from the predicted relationship could arise because $\Delta G^{\neq}(0)$ is larger as a result of a steric impediment to close approach in the transition state, or from a larger value of A, which could also result from increased steric repulsions.

Both of the reaction series depicted are characterized by low values for the intrinsic barriers; 3.5 kcal mol^{-1} and 5.0 kcal mol^{-1} for the $Re(CO)_4PMe_3\cdot$ and $Re(CO)_4P(O\text{-}i\text{-}Pr)_3$ radicals, respectively. These are about the same as the intrinsic barriers to more obviously outer-sphere electron transfers from the same two radi-

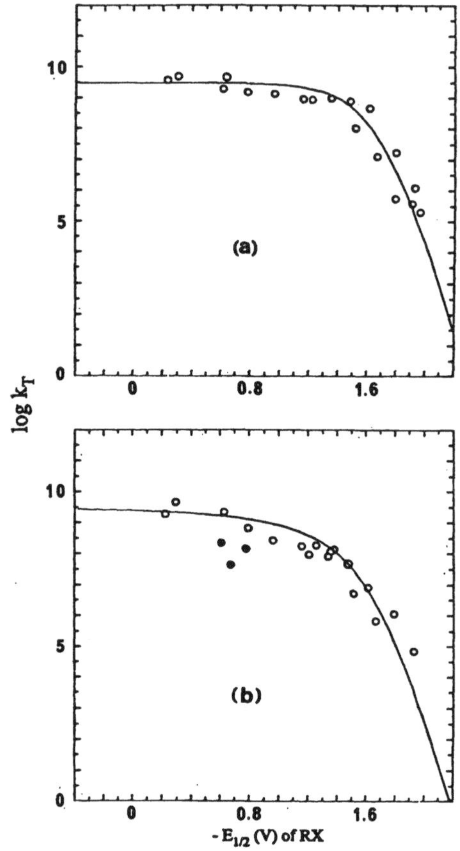

Fig. 7. Log k_T for reactions of $Re(CO)_4PMe_3\cdot$ and $Re(CO)_4P(O\text{-}i\text{-}Pr)_3\cdot$ with organic halides vs. the $E_{1/2}$ values for the halides in 75% dioxane/25% water (Table 4) [85]. Reprinted by permission of the American Chemical Society.

cals to a series of N-methylpyridinium salts in CH_3CN [98]; in these cases $\Delta G^{\neq}(0)$ is 3 kcal mol^{-1}. The low values suggest that there is probably little contribution from solvent reorganization to the barrier. Inasmuch as the reactants are neutral, in a hydrocarbon solvent, this is an expected result. There is apparently little variation in the intrinsic barrier in the series; the fits are generally quite good, especially for the $Re(CO)_4PMe_3\cdot$ series. The configurational changes that occur in the approach to the transition state must in each series be largely compensating.

Marcus suggested, on the basis of the approximate BEBO-based model he employed, that $\Delta G^{\neq}(0)$ values on the order of 5 to 10% of the bond dissociation energy might be expected in atom transfers [95, footnote 15]. Assuming an average dissociation energy of about 60 kcal mol^{-1} for the C—X and M—X bonds, values from 3 to 6 kcal mol^{-1} are expected, in agreement with the range of values obtained. However, this estimate ignores all contributions to $\Delta G^{\neq}(0)$ other than the bond stretching and bond formation associated with the transferred atom. As additional data accumulate on intrinsic barriers to atom transfer reactions in a variety of contexts, a more detailed general model will be required to account for variations in this important quantity.

In the examples cited, use of the irreversible reduction potential as the measure of ΔG_{23} leads to good correlations. This suggests, but does not prove, that electron transfer plays an important role in halogen atom transfer to the metal carbonyl radicals. The parameter C in the expression $\Delta G_{23} = -23.06E_{1/2} + C$, employed to transform Equation (109) to one relating k_T to $E_{1/2}$ [86], embodies the irreversible oxidation potential of the metal carbonyl in the reaction complex, and reflects the strong irreversibility of the halide reduction. The relationship between log k for electron transfer from $Re(CO)_4PMe_3$ and $Re(CO)_4P(O\text{-}i\text{-}Pr)_3$ and the cathodic peak potentials for reduction of a series of N-methylpyridinium salts, while yielding closely similar intrinsic barriers, leads to substantially less negative values for C for a given metal carbonyl than for the atom transfer reaction series. This suggests that the same driving force may operate in the two series of reactions, one of which is quite clearly an outer-sphere electron transfer process, while the other is more inner-sphere in character.

10. Hydrogen Atom Transfer

10.1 Bond Energy Considerations

By analogy with the reactions of main group organometallic radicals such as $\cdot SnR_3$, transition element organometallic radicals in which the unpaired electron is centered largely on the metal can be expected to undergo direct hydrogen atom transfer reactions with appropriate donors, forming metal-hydrogen bonds. This expectation must be tempered, however, by the realization that transition element-hydrogen bonds are generally weak. There is an unfortunate lack of reliable thermochemical data regarding M—H bond dissociation energies for neutral metal carbonyl hydrides and other organometallic hydrides. However, the limited data available, a few of which are listed in Table V, serve to illustrate the point that transition element-hydrogen bond energies are weaker than most other element-hydrogen bonds.

It is difficult to draw reliable conclusions regarding the factors determining M—H bond energies from the limited data and observations available, but two trends appear operative:

(a) For equivalent structures, M—H bonds are stronger for third transition row elements than for first. There are weak thermochemical evidences for this [103]; in addition, M—H stretching frequencies are generally higher for third row M—H bonds than for first row; $Re_2(CO)_{10}$ reacts photochemically with H_2 to form products containing Re—H bonds much more readily than $Mn_2(CO)_{10}$ [3], and so on.

(b) Among carbonyl hydrides, replacement of CO by phosphines or phosphites appears to lead to stronger M—H bonds. One notes that such replacement leads to lower M—H bond acidity [107]; in addition, kinetics evidence such as that referenced in Table 5 suggests stronger M—H bonds.

10.2 Alternatives to Direct Atom Transfer

Because M—H bonds are generally quite weak, other pathways compete with direct H atom transfer for formation of M—H bonds. For example:

(a) Oxidative addition of X—H can occur to the metal carbonyl radical. Given the extraordinary lability of radicals toward substitution, an associative displacement of CO by X—H may be quite facile where steric hindrance is not excessive:

$$\cdot M(CO)_n + X{-}H \rightarrow \cdot(CO)_{n-1}MX(H) + CO \tag{113}$$

Table 5. Comparative energies of bonds to hydrogen

Bond	Dissoc. Energy (kcal mol^{-1})	Ref.
$C_6H_5CH_2{-}H$	85	[99]
$Cl_3C{-}H$	90	[99]
Fluorene(C—H)	80	[100]
$Et_3Si{-}H$	90	[101, 102]
$Bu_3Sn{-}H$	74	[101]
$(CO)_5Mn{-}H^a$	60	[103]
$(CO)_3[P(n\text{-}Bu)_3]_2Mn{-}H$	≥ 64	[19]
$(CO)_3[P(OPh)_3]_2Mn{-}H$	≥ 65	[104]
$(CO)_4Co{-}H$	58	[105]
$Cp(CO)_3Mo{-}H$	66 ± 7	[106]

^{a} Based on an assumed Mn—Mn bond energy in $Mn_2(CO)_{10}$ of 22.5 kcal mol^{-1}.

In many such reactions a subsequent hydrogen atom transfer and reductive elimination of H_2 may occur, forming the X—M product, and H_2, as in Scheme I [108].

Scheme I

$$\cdot M(CO)_n + X{-}H \rightarrow \cdot(CO)_{n-1}MX(H) + CO \tag{114}$$

$$\cdot(CO)_{n-1}MX(H) + X{-}H \rightarrow (CO)_{n-1}MX(H)(H) + X\cdot \tag{115}$$

$$(CO)_{n-1}MX(H)(H) \xrightarrow{CO} XM(CO)_n + H_2 \tag{116}$$

However, an alternative last step would involve elimination of HX, leaving the M—H bond product, as in Eq. (117).

$$(CO)_{n-1}MX(H)(H) \xrightarrow{CO} HM(CO)_n + X{-}H \tag{117}$$

In this event, X—H would have acted as a catalyst for hydrogen atom transfer; the initial oxidative addition of X—H would have had the effect of forming a radical capable of forming a stronger M—H bond. To date, no clear examples of such a process have been identified.
(b) In photochemical reactions involving dinuclear or polynuclear metal carbonyl compounds and X—H, oxidative addition may occur at the metal center in the CO-loss, product, forming intermediates that go on to form products not readily distinguished from the product of a direct hydrogen atom transfer. Examples of oxidative addition to a 16-electron coordinatively unsaturated metal center have been established, or can be inferred from the nature of photochemical reactions products [109].

(c) Hydrogen atom transfer may occur to other than the metal center of the radical. Examples of hydride ion as well as hydrogen atom transfer to ligands are known, [110, 111]. In at least some cases an apparent hydride ion transfer may proceed via a single electron transfer followed by hydrogen atom transfer. Hydrogen atom transfer to the carbon of coordinated CO can occur in electron-rich radicals. Kochi and coworkers have provided convincing electrochemical and spectroscopic evidence for this type of reaction [27, 71]. The metal carbonyl substrate for such a reaction is in all cases a formally 19-electron radical. Indeed, it is just the presence of an electron in a comparatively high energy orbital on the metal center that promotes a delocalization of charge to coordinated CO, creating the potential for formation of a C—H bond.

The immediate product of H atom transfer to carbon is a formyl species, which generally undergoes rapid ligand loss and migration of hydrogen to metal to form a metal hydride. Thus, for example [27],

$$Mn(CO)_3(PPh_3)_2(NCCH_3) + HSnBu_3 \rightarrow Mn(CO)_2(PPh_3)_2(NCCH_3)CO(H) \quad (118)$$

$$Mn(CO)_2(PPh_3)_2(NCCH_3)CO(H) \rightarrow HMn(CO)_3(PPh_3)_2 + CH_3CN \quad (119)$$

10.3 Hydrogen Atom Transfer from Metal Hydrides

Hydrogen atom transfer from metal to substrate is important in many reactions of metal carbonyl hydrides. The overall reaction process in which atom transfer is an elementary step is often non-chain in character. Here are a few examples:

$$2CpCr(CO)_3H + CH_2{=}C(CH_3)CH{=}CH_2 \rightarrow Cp_2Cr_2(CO)_6 + CH_3C(CH_3){=}CHCH_3 \text{ [112]} \quad (120)$$

$$2Mn(CO)_5H + C_6H_5C(CH_3){=}CH_2 \rightarrow C_6H_5CH(CH_3)_2 + Mn_2(CO)_{10} \text{ [72b]} \quad (121)$$

$$CpFe(CO)_2H + CH_2{=}CHCH{=}CH_2 \rightarrow CpFe(CO)_2CH_2CH{=}CHCH_3 \text{ [113]} \quad (122)$$

Although such reactions are commonplace, few rate constants have been measured. A few reported values for H atom transfer to alkyl radicals are listed in Table 6.

Table 6. Rate Constants for Hydrogen Atom Transfer from M—H

M—H Species	Substrate	Solvent	Temp. (°C)	k_H ($M^{-1}\,s^{-1}$)	Ref.
$PPN^+HCr(CO)_5^-$		THF	25°	1.8×10^7	[50]
$PPN^+HW(CO)_5^-$	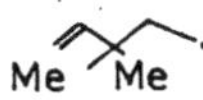	THF	25°	0.96×10^7	[50]
$HCpCr(CO)_3$	$CH_3\text{-}\dot{C}(\Delta)(Ph)$	Toluene	RT	4×10^7	[114]
$HCpFe(CO)_2$	$CH_3\text{-}\dot{C}(\Delta)(Ph)$	Hexane	RT	$\sim 10^9$	[114]
$PPN^+HCpV(CO)_3^-$	$CH_2{=}CHCH_2OCH_2CH_2\cdot$	None	25	2×10^7	[43]

10.4 Hydrogen Atom Transfer to Metal Centers

Reactions in which hydrogen is transferred *to* a metal center, forming an M—H bond, are comparatively less common, because M—H bonds are weak. Most commonly, such reactions occur when hydrogen is transferred from one metal to another, as in the radical chain processes described in 3.2.1.

Table 7 lists reported rate constant for H atom transfer from $HSnBu_3$ to several manganese and rhenium carbonyl radicals. The rate constants for the manganese radicals are comparatively small. Atom transfer from $HSnBu_3$ to $\cdot Mn(CO)_3[PR_3]_2$ is likely to be slightly endothermic or, at best, ergoneutral. Depending on the magnitude of the intrinsic barrier, the rate constants should be several orders of magnitude below the diffusion limit. Steric factors obviously also play a role; the rate constant for transfer to $\cdot Mn(CO)_3[P(i\text{-}Pr)_3]_2$, which contains the bulkiest ligand in the series, is 10^4 times slower than for $\cdot Mn(CO)_3[P(n\text{-}Bu)_3]_2$.

Table 7. Rate Constants for Hydrogen Atom Transfer from $HSnBu_3$·

Metal Radical	Solvent	Temp. (°C)	k_H (M^{-1} s^{-1})	Ref.
$Mn(CO)_3[P(n\text{-}Bu)_3]_2\cdot$	C_6H_{14}	20	10.7	[19]
	THF	25	13.2	[80]
$Mn(CO)_3[P(i\text{-}O\text{-}Pr)_3]_2\cdot$	C_6H_{14}	20	0.11	[19]
	THF	25	5.0	[80]
$Mn(CO)_3[P(i\text{-}Bu)_3]_2\cdot$	C_6H_{14}	20	0.786	[19]
$Mn(CO)_3[P(i\text{-}Pr)_3]_2\cdot$	C_6H_{14}	20	2.0×10^{-3}	[19]
$Mn(CO)_3(DPPE)\cdot$	THF	25	25.1	[80]
$Mn(CO)_3(PPh_3)_2\cdot$	THF	25	0.5	[80]
$Mn(CO)_3[P(OPh)_3]_2\cdot$	THF	25	7.7	[104]
$Mn(CO)_3[P(O\text{-}(o\text{-}CH_3)C_6H_4)_3]_2\cdot$	THF	25	2.1	[80]
$Re(CO)_5\cdot$	C_6H_{14}	22	5.1×10^7	[37]
$Re(CO)_4PMe_3\cdot$	C_6H_{14}	22	1.7×10^8	[37]
$Re(CO)_4P(n\text{-}Bu)_3\cdot$	C_6H_{14}	22	5.1×10^7	[37]
$Re(CO)_4P(O\text{-}i\text{-}Pr)_3\cdot$	C_6H_{14}	22	1.5×10^7	[37]
$Re(CO)_4P(i\text{-}Bu)_3\cdot$	C_6H_{14}	22	5.3×10^6	[37]
$Re(CO)_4P(i\text{-}Pr)_3\cdot$	C_6H_{14}	22	1.3×10^6	[37]
$Re(CO)_4AsEt_3\cdot$	C_6H_{14}	22	7.3×10^7	[37]
$Re(CO)_4P(OMe)_3\cdot$	C_6H_{14}	22	6.4×10^7	[37]
$Re(CO)_4P(C_6H_{11})_3\cdot$ [a]	C_6H_{14}	22	4.1×10^6	[37]

[a] C_6H_{11} = cyclohexyl

By comparison, the rate constants for H atom transfer to the rhenium radicals are much greater. The largest values are within two orders of magnitude of the diffusion limit, suggesting that the overall reaction is exergonic or at least thermoneutral with a small intrinsic barrier. Direct comparisons between the Mn and Re systems are unfortunately not possible, because data for reactions of radicals with equivalent structures are not available. The rhenium radicals are monosubstituted, and thus less crowded than the disubstituted manganese radicals. The effect of changing ligand steric requirement in the rhenium series (e.g., $\cdot Re(CO)_4P(n\text{-}Bu)_3$ vs. $\cdot Re(CO)_4P(i\text{-}Pr)_3$) is smaller than the aforementioned change in the manganese systems, where two ligands are bound to the metal and the metal radius is smaller.

Gasanov and Dotdaev have obtained rate constants for H atom transfer from $HSiEt_3$ to $\cdot Mn(CO)_5$, $\cdot CpFe(CO)_2$ and $\cdot CpW(CO)_3$,based upon an EPR analysis of photolyzed solutions and the use of spin traps [115]. The rate constants for transfer to $\cdot Mn(CO)_5$, $\cdot CpFe(CO)_2$ and $CpW(CO)_3$ are reported as 3×10^4, 7.8×10^3 and 4.8×10^4 M^{-1} s^{-1}, respectively. Given the substantially endergonic character of these reactions, as suggested by the data in Table 5, it is unlikely that these values represent the direct atom transfer. Rather, the reactions probably proceed by one of the other possible pathways suggested above.

11. Summary and Conclusions

On the basis of what is presently known regarding rate constants for atom transfer processes and the substitution and electron transfer properties of metal carbonyl radicals, it would appear that radical chain mechanisms for substitution and other characteristic reactions of metal carbonyl compounds should be quite common. That they are not more frequently observed may be due to trapping of radical intermediates by impurities or the solvent. Molecular oxygen may be especially important in this regard, because many organometallic compounds are not sufficiently reactive with O_2 to scavenge it from solution. The O_2 is thereby present to react with radical intermediates, forming peroxy complexes that do not propagate the chain process [52]. Careful exclusion of O_2, initiation of radical chain processes via photochemical generation of an initiator, or use of an initiator source such as AIBN are all useful steps to promote a radical chain process.

Radical chain processes have the potential for forming products distinct from those attainable via other pathways. In favorable cases the reactions proceed rapidly under mild conditions to yield a single product. There is therefore some incentive to further explore the use of radical chain processes in organometallic synthesis, just as Barton has advocated in organic synthesis [116].

Finally, it should be noted that the term atom transfer can be taken more generally to mean also the transfer of characteristic functional groups such as —SCN or —SR, in which the steric requirements of the group and the presence of one or more unshared electron pairs permits group transfer in the same manner as atom transfer, and with comparable facility [117].

12. Acknowledgment

This contribution and the work from our laboratories described in this Chapter, have been supported by the National Science Foundation via Research Grant NSF-CHE86-08839.

13. References

1. B. L. Booth and B. L. Shaw, J. Organomet. Chem. **43** (1972) 369.
2. B. H. Byers and T. L. Brown, J. Am. Chem. Soc. **97** (1975) 947.
3. B. H. Byers and T. L. Brown. J. Am. Chem. Soc. **99** (1977) 2527.

4. J. A. Howard, J. R. Morton and K. F. Preston, Chem. Phys. Lett. **83** (1981) 226.
5. G. B. Rattinger, R. L. Belford, H. Walker and T. L. Brown, Inorg. Chem. **28** (1989) 1059.
6. (a) K. A. M. Creber and J. K. S. Wan, Trans. Met. Chem. **8** (1983) 253.
(b) K. A. M. Creber and J. K. S. Wan, J. Am. Chem. Soc. **103** (1981) 2101.
(c) C. P. Cheng, S. R. Wang and T. C. Chou, J. Clin. Chem. Soc. **33** (1986) 99.
(d) T.-I. Ho, C.-M. Chang, S. R. Wang and C. P. Cheng, J. Chem. Soc., Dalton Trans. **1988**, 123.
7. (a) W. Kaim, Coord. Chem. Rev. **76** (1987) 187.
(b) R. R. Andrēa, W. G. J. de Lange, T. van der Graaf, M. Rijkhoff, D. J. Stufkens and A. Oskam, Organometallics **7** (1988) 1100 and references therein.
8. M. S. Wrighton and D. S. Ginley, J. Am. Chem. Soc. **97** (1975) 2065.
9. J. L. Hughey, IV, C. P. Anderson and T. J. Meyer, J. Organomet. Chem. **125** (1977) C49.
10. D. R. Kidd and T. L. Brown, J. Am. Chem. Soc. **100** (1978) 4095.
11. A. R. Burkett, T. J. Meyer and D. G. Whitten, J. Organomet. Chem. **67** (1974) 67.
12. J. L. Hughey IV, C. R. Bock and T. J. Meyer, J. Am. Chem. Soc. **97** (1975) 4440.
13. M. S. Wrighton and D. S. Ginley, J. Am. Chem. Soc. **97** (1975) 4246.
14. R. D. Adams, D. E. Collins and F. A. Cotton, J. Am. Chem. Soc. **96** (1974) 749.
15. T. Madach and H. Vahrenkamp, Z. Naturforsch. **33b** (1978) 1301.
16. N. A. Cooley, K. A. Watson, S. Fortier and M. C. Baird, Organometallics **5** (1986) 2563.
17. (a) A. J. Poë and R. A. Jackson, Inorg. Chem. **17** (1978) 2330.
(b) R. A. Jackson and A. J. Poë, Inorg. Chem. **17** (1978) 997.
18. D. R. Kidd, C. P. Cheng and T. L. Brown, J. Am. Chem. Soc. **100** (1978) 4103.
19. S. B. McCullen and T. L. Brown, J. Am. Chem. Soc. **104** (1982) 7496.
20. W. K. Meckstroth, R. T. Walters, W. L. Waltz, A. Wojcicki and L. M. Dorfman, J. Am. Chem. Soc. **104** (1982) 1842.
21. R. W. Wegman, R. J. Olsen, D. R. Gard, L. R. Faulkner and T. L. Brown, J. Am. Chem. Soc. **103** (1981) 6089.
22. H. W. Walker, R. S. Herrick, R. J. Olsen and T. L. Brown, Inorg. Chem. **23** (1984) 3748.
23. R. J. Olsen, R. W. Wegman and T. L. Brown, unpublished observations.
24. R. M. Laine and P. C. Ford, Inorg. Chem. **16** (1977) 388.
25. A. F. Hepp and M. S. Wrighton, J. Am. Chem. Soc. **103** (1981) 1258.
26. P. Rushman and T. L. Brown, J. Am. Chem. Soc. **109** (1987) 3632.
27. B. A. Narayanan, C. Amatore and J. K. Kochi, Organometallics **6** (1987) 129.
28. J. W. Hershberger, C. Amatore and J. K. Kochi, J. Organomet. Chem. **250** (1983) 345.
29. M. Absi-Halabi and T. L. Brown, J. Am. Chem. Soc. **99** (1977) 2982.
30. S. B. McCullen and T. L. Brown, Inorg. Chem. **20** (1981) 3528.
31. A. E. Stiegman, M. Stieglitz and D. R. Tyler, J. Am. Chem. Soc. **105** (1983) 6032.
32. A. E. Stiegman and D. R. Tyler, Comments Inorg. Chem. **5** (1986) 215.
33. W. C. Trogler, Int. J. Chem. Kinetics **19** (1987) 1025.
34. (a) A. Fox, J. Malito and A. Poë, J. Chem. Soc., Chem. Comm. (1981) 1052.
(b) A. J. Poë, J. Trans. Metal Chem. (Weinheim Ger.) **7** (1982) 65.
35. Q.-Z. Shi, T. G. Richmond, W. C. Trogler and F. Basolo, J. Am. Chem. Soc. **104** (1982) 4032.
36. T. R. Herrinton and T. L. Brown, J. Am. Chem. Soc. **107** (1985) 5700.
37. J. M. Hanckel, K.-W. Lee, P. Rushman and T. L. Brown, Inorg. Chem. **25** (1986) 1852.
38. A. Poë and C. V. Sekhar, J. Am. Chem. Soc. **107** (1985) 4874.
39. S. B. McCullen, H. W. Walker and T. L. Brown, J. Am. Chem. Soc. **104** (1982) 4007.
40. P. M. Zizelman, C. Amatore and J. K. Kochi, J. Am. Chem. Soc. **106** (1984) 3771.
41. H. G. Kuivila, Acc. Chem. Res. **1** (1968) 299.
42. W. J. Carter, J. W. Kelland, S. J. Okrasinski, K. E. Warner and J. R. Norton, Inorg. Chem. **21** (1982) 3955.
43. R. J. Kinney, W. D. Jones and R. G. Bergman, J. Am. Chem. Soc. **100** (1978) 7902.
44. W. D. Jones, J. M. Huggins and R. G. Bergman, J. Am. Chem. Soc. **103** (1981) 4415.
45. (a) T. L. Brown and P. A. Bellus, Inorg. Chem. **17** (1978) 3727.
(b) P. A. Bellus and T. L. Brown, J. Am. Chem. Soc. **102** (1980) 6020.

46. F. Basolo, Inorg. Chim. Acta. **100** (1985) 33.
47. D. J. Darensbourg, N. Walker and M. Y. Darensbourg, J. Am. Chem. Soc. **102** (1980) 1213.
48. S. L. Webb, C. M. Giandomenico and J. Halpern, J. Am. Chem. Soc. **108** (1986) 345.
49. M. Y. Darensbourg and C. E. Ash, Adv. in Organomet. Chem. **27** (1987) 1.
50. C. E. Ash, P. W. Hurd, M. Y. Darensbourg and M. Newcomb, J. Am. Chem. Soc. **109** (1987) 3313.
51. R. Kh. Freidlina, R. G. Gasanov, N. A. Kuz'mina and E. Ts. Chukovskaya, Uspekhi Khim. **54** (1985) 1127. (Russ. Chem. Revs. **54** (1985) 662).
52. (a) S. A. Fieldhouse, B. W. Fullam, G. W. Neilson and M. C. R. Symons, J. Chem. Soc., Dalton Trans (1974) 567.
    (b) W. E. Lindsell and P. N. Preston, J. Chem. Soc. Dalton (1979) 1105.
53. N. N. Turaki and J. M. Huggins, Organometallics **5** (1986) 1703.
54. N. W. Hoffman and T. L. Brown, Inorg. Chem. **17** (1978) 613.
55. (a) M. S. Wrighton and D. S. Ginley, J. Am. Chem. Soc. **97** (1975) 3533.
    (b) D. S. Ginley, C. R. Bock and M. S. Wrighton, Inorg. Chim. Acta. **23** (1977) 85.
56. E. O. Fischer, W. Hafner and H. O. Stahl, Z. anorg. allg. Chem. **282** (1955) 47. The tungsten hydride is much more stable than the molybdenum analogue.
57. K. A. Mahmoud, A. J. Rest and H. G. Alt, J. Chem. Soc., Dalton Trans., (1984) 187.
58. R. T. Edidin and J. R. Norton, J. Am. Chem. Soc. **108** (1986) 948.
59. B. H. Byers and T. L. Brown, J. Organomet. Chem. **127** (1977) 181.
60. C. White and E. Cesarotti, J. Organomet. Chem. **287** (1985) 123.
61. D. G. Alway and K. W. Barnett, Inorg. Chem. **19** (1980) 1533.
62. N. J. Coville, J. Organomet. Chem. **190** (1980) C84.
63. N. J. Coville, M. O. Albers, T. V. Ashworth and E. Singleton, J. Chem. Soc., Chem. Comm. (1981) 408.
64. N. J. Coville, M. O. Albers and E. Singleton, J. Chem. Soc., Dalton Trans. (1983) 947.
65. G. O. Nelson and C. E. Sumner, Organometallics **5** (1986) 1983.
66. (a) B. D. Fabian and J. A. Labinger, J. Am. Chem. Soc. **101** (1979) 2239.
    (b) B. D. Fabian and J. A. Labinger, Organometallics **2** (1983) 659.
67. M. Rosenblum and P. S. Waterman, J. Organomet. Chem. **187** (1980) 267.
68. M. A. Urbancic, S. R. Wilson and J. R. Shapley, Inorg. Chem. **23** (1984) 2954.
69. F. W. B. Einstein, M. C. Jennings, R. Krentz, R. K. Pomeroy, P. Rushman and A. C. Willis, Inorg. Chem. **26** (1987) 1341.
70. (a) F. W. B. Einstein, R. K. Pomeroy, P. Rushman and A. C. Willis, J. Chem. Soc., Chem. Comm. (1983) 854.
    (b) F. W. B. Einstein, R. K. Pomeroy, P. Rushman and A. C. Willis, Organometallics **3** (1985) 250.
71. (a) B. A. Narayanan, C. Amatore, C. P. Casey and J. K. Kochi, J. Am. Chem. Soc. **105** (1983) 6351.
    (b) B. A. Narayanan, C. Amatore and J. K. Kochi, Organometallics **3** (1984) 802.
    (c) B. A. Narayanan, C. Amatore and J. K. Kochi, Organometallics **5** (1986) 926.
    (d) J. K. Kochi, J. Organomet. Chem. **300** (1986) 139.
72. (a) J. Halpern, Pure Appl. Chem. **58** (1986) 575.
    (b) R. L. Sweany and J. Halpern, J. Am. Chem. Soc. **99** (1977) 8335.
73. R. C. Poller, The Chemistry of Organotin Compounds, Academic Press, New York, 1970.
74. D. H. Berry and J. H. Mitstifer, J. Am. Chem. Soc. **109** (1987) 3777.
75. T. H. Whitesides and J. Shelly, J. Organomet. Chem. **92** (1975) 215.
76. Y.-S. Chen, S.-L. Wang, R. A. Jacobsen and R. J. Angelici, Inorg. Chem. **25** (1986) 1118.
77. M. R. Detty and W. D. Jones, J. Am. Chem. Soc. (1987) 5666.
78. H. B. Abrahamson and M. S. Wrighton, Inorg. Chem. **17** (1978) 1003.
79. W. L. Waltz, O. Hackelberg, L. M. Dorfman and A. J. Wojcicki, J. Am. Chem. Soc. **100** (1978) 7259.
80. D. J. Kuchynka, C. Amatore and J. K. Kochi, J. Organomet. Chem. **328** (1987) 133.
81. R. G. Gasanov and E. G. Sadykhov, Izvest. Akad. Nauk SSSR Ser. Khim. (1987) 993.

82. H. Yesaka, T. Kobayashi, K. Yasufuku and S. Nagakura, J. Am. Chem. Soc. **105** (1983) 6249.
83. R. S. Herrick, T. R. Herrinton, H. W. Walker and T. L. Brown, Organometallics **4** (1985) 42.
84. R. G. Gasanov, E. G. Sadykhov and R. Kh. Freidlina, Izvest. Akad. Nauk. Ser. Khim. (1986) 1031.
85. K.-W. Lee, J. M. Hanckel and T. L. Brown, J. Am. Chem. Soc. **108** (1986) 2266.
86. (a) K.-W. Lee and T. L. Brown, J. Am. Chem. Soc. **109** (1987) 3269.
(b) C. K. Mann and K. K. Barnes, "Electrochemical Reactions in Nonaqueous Systems", Marcel Dekker, Inc., New York, 1970. Half-wave potentials were measured in 75% dioxane/25% water against SCE. The half-wave potentials of substituted benzyl bromides, measured in DMF against Ag/AgBr, were normalized to the other halide values by adding the shift value (−0.54 V) that is based on the comparisons of five organic halides in the two different standard electrodes and media [85].
87. (a) P. Heimback, J. Kluth, H. Schenkluhn, B. Weimann, Angew. Chem., Int'l. Ed. Engl. **19** (1980) 569.
(b) H. Schenkluhn, R. Berger, B. Pittel and M. Zahres, Trans. Metal. Chem. (Weinheim, Ger.) **6** (1981) 277.
88. M. N. Golovin, M. M. Rahman, J. E. Belmonte and W. P. Giering, Organometallics **4** (1985) 1981.
89. G. M. Bodner, M. P. May, L. E. McKinney, Inorg. Chem. **19** (1980) 1951.
90. C. A. Tolman, Chem. Rev. **77** (1977) 313.
91. G. S. Hammond, J. Am. Chem. Soc. **77** (1955) 334.
92. (a) A. Pross, Adv. Phys. Org. Chem. **14** (1977) 69.
(b) S. S. Shaik, Prog. Phys. Org. Chem. **15** (1985) 197.
93. (a) W. J. Albery, Ann. Rev. Phys. Chem. **31** (1980) 227.
(b) E. S. Lewis, J. Phys. Chem. **90** (1986) 3756.
94. (a) R. A. Marcus, J. Chem. Phys. **24** (1956) 966.
(b) R. A. Marcus, Ann. Rev. Phys. Chem. **15** (1964) 155.
(c) R. A. Marcus, Far. Discuss. Chem. Soc. **74** (1982) 7.
95. R. A. Marcus, J. Phys. Chem. **72** (1968) 891.
96. H. S. Johnston, Gas Phase Reaction Rate Theory, The Ronald Press Co., New York, 1966, pp. 179, 209.
97. (a) N. Agmon and R. D. Levine, Chem. Phys. Lett. **52** (1977) 197.
(b) R. D. Levine, J. Phys. Chem. **83** (1979) 159.
(c) N. Agmon, Int. J. Chem. Kinetics **13** (1981) 333.
98. M. R. Burke and T. L. Brown, J. Am. Chem Soc., **111** (1989) 5185.
99. J. A. Kerr, Chem. Rev. **66** (1966) 465.
100. F. G. Bordwell, J.-P. Cheng and J. A. Harrelson, Jr. J. Am. Chem. Soc. **110** (1988) 1229.
101. R. A. Jackson, J. Organomet. Chem. **166** (1979) 17.
102. J. M. Kanabus-Kaminska, J. Hawari and D. Griller, J. Am. Chem. Soc. **109** (1987) 5267.
103. G. Pilcher in Thermochemistry and its Applications to Chemical and Biochemical Systems, M. A. V. Riberio da Silva, Ed., NATO ASI Series **119**, D. Reidel Publ. Co., Dordrecht, Holland, 1984, p. 353.
104. C. Amatore, D. J. Kuchynka and J. K. Kochi, J. Electroanal. Chem. **241** (1988) 181.
105. F. J. Ungvary, J. Organomet. Chem. **36** (1972) 363.
106. S. P. Nolan, R. Lopez de la Vega and C. D. Hoff, J. Organomet. Chem. **315** (1986) 187.
107. (a) R. G. Pearson, Chem. Rev. **85** (1985) 41.
(b) R. T. Edidin, J. M. Sullivan and J. R. Norton, J. Am. Chem. Soc. **109** (1987), 3945.
108. R. W. Wegman and T. L. Brown, Organometallics **1** (1982) 47.
109. (a) M. Wrighton and M. A. Schroeder, J. Am. Chem. Soc. **95** (1973) 5765.
(b) P. J. Krusic, W. J. Cote and A. Grand, J. Am. Chem. Soc. **106** (1984) 4642.
(c) H. Nagorski and M. J. Mirbach, J. Organomet. Chem. **291** (1985) 199.
110. S. G. Davies, M. L. H. Green and D. M. P. Mingos, Tetrahedron **34** (1978) 20.
111. P. Michaud, D. Astruc and J. H. Ammeter, J. Am. Chem. Soc. **104** (1982) 3755.

112. A. Miyake and H. Kondo, Angew. Chem., Int'l. Ed. **7** (1968) 631.
113. M. J. Thomas, T. A. Shackleton, S. C. Wright, D. J. Gillis, J. P. Colpa and M. C. Baird, J. Chem. Soc., Chem. Comm. **1986**, 312.
114. R. M. Bullock and E. G. Samsel, J. Am. Chem. Soc. **109** (1987) 6542.
115. R. G. Gasanov and S. Kh. Dotdaev, Ivest. Akad Nauk. SSSR, Ser. Khim. **1986**, 1981.
116. D. H. R. Barton and S. R. Zard, Substituent Effects in Radical Chemistry, H. G. Vieke, Ed., (1986) 443.
117. K.-W. Lee and T. L. Brown, Inorg. Chem. **26** (1987) 1852.

CHAPTER 4

Synthetic Applications of Electron Transfer Catalysis: The Ligand Substitution Reaction

Neil J. Coville

Department of Chemistry, University of the Witwatersrand,
P.O. Wits 2050, Johannesburg, South Africa

Contents

List of Abbreviations

bipy – 2,2'-bipyridyl
bpym – 2,2'-bipyrimidine
Cp – C_5H_5
Cp* – C_5Me_5
MeCp – MeC_5H_4
DAD – diazadiene
dba – dibenzylideneacetone
dppe – $Ph_2PCH_2CH_2PPh_2$·
dppm – $Ph_2PCH_2PPh_2$
py – pyridine
PPN – bis(triphenylphosphine)nitrogen (+1)
ttas – $[o\text{-}C_6H_4(AsMe_2)]_2AsMe$
tripod – $CH(PPh_2)_3$
tBu – $(CH_3)_3C$
Xy – 2,6-dimethylphenyl (xylyl)

1. Introduction

One of the most well-studied organometallic reaction types is the ligand substitution reaction which stoichiometrically can be written as

$$ML_x + L' \rightarrow ML_{x-1}L' + L \qquad (1)$$

i.e. a reaction in which a ligand L, initially attached to a metal atom M, is replaced by another ligand L'. Notwithstanding the simplicity of the stoichiometry of the reaction, mechanistic studies have revealed a variety of processes by which the substitution reaction can take place [1]. Since the rewards of a correct mechanistic interpretation of a reaction can be substantial and lead to better control of a reaction sequence, much time and effort has been spent on the elucidation of reaction mechanisms of the substitution reaction. Determination of a reaction mechanism is not however a trivial process and it must always be remembered that a mechanism is an interpretation of information, usually kinetic, and as such is open to correction and modification as new techniques and new data become available.

This chapter will review the information presently at hand on one class of ligand substitution reactions, the catalysed substitution reactions involving odd electron intermediates. The emphasis in this chapter will focus on the use of the technique in synthetic studies rather than on a detailed consideration of reaction mechanisms, which will be discussed elsewhere in this book. It is also to be noted that numerous of the reactions to be discussed in this chapter are catalyzed by heterogeneous catalysts and are hence not readily amenable to the mechanistic probes generally used to characterize homogeneous (solution) reactions. Indeed, it may turn out that as more information becomes available that some of the catalyzed reactions described below may occur by alternative mechanisms to those presently proposed. This will not however detract from the experimental data, namely, that certain classes of compounds catalyze the ligand substitution reaction.

As will become apparent, most of the work to be described in this chapter relates to substitution reactions of metal carbonyl or substituted metal carbonyl complexes. Although catalyzed ligand substitution reactions in inorganic chemistry are well documented [2], very few inorganic complexes containing a metal-carbon bond and lacking CO ligands have been investigated. It can be confidently predicted that as the ground rules for the catalyzed substitution reaction become known an extension to a wider range of reactants will be undertaken.

2. Ligand Substitution Reactions

As mentioned above, the simplicity of the stoichiometry of a reaction does not necessarily give an indication of the detail of a reaction mechanism. The mechanism is typically determined by kinetic studies accompanied by other physical measurements for ascertaining the composition and stereochemistry of the intermediates (or transition states) and products in a chemical reaction. The kinetic studies allow for the elucidation of the rate determining step and from the many studies performed in both inorganic and organic chemistry two *limiting* types of reactions have been proposed

$$ML_x \rightarrow ML_{x-1} + L \xrightarrow{L'} ML_{x-1}L' \quad (S_N1 \text{ or } D) \tag{2}$$

$$ML_x + L' \rightarrow L \ldots ML_{x-1} \ldots L' \rightarrow ML_{x-1}L' + L \quad (S_N2 \text{ or } A) \tag{3}$$

ie. a first order nucleophilic substitution (S_N1) or dissociative (D) process or a second order nucleophilic substitution (S_N2) or associative (A) process.

Intermediate cases can be envisaged which do not neatly fit into the above categories and these alternative possibilities such as the I_A and I_D interchange mechanisms have been thoroughly discussed elsewhere [3]. The above reaction types hold whether the reaction steps involve one or two electron transfer processes, even though further categorization of the reaction can be envisaged which is dependent on the number of electrons involved in the reaction [4].

Activation of the reactant ML_x by means of the addition or removal of an electron provides a means of inducing increased reactivity at the metal center

$$ML_x \xrightarrow{\pm e^-} ML_x\cdot \rightarrow \text{products} \tag{4}$$

Depending on the reactant environment, the products can arise from a substitution, oxidative-addition, isomerization, decomposition reaction, etc. The redox process can be induced by chemical, photochemical or electrochemical techniques, and the reaction rate to product formation can be enhanced by many orders of magnitude relative to the classical thermal technique. Our interest in this chapter will be on the electron induced substitution reaction, which involves organometallic radicals, and the attempts that have been made to make the reaction catalytic. Where appropriate, mention will be made of the alternative pathways that sometimes compete with the substitution reaction (eg. isomerization) during the redox process.

Numerous mechanistic pathways can be written for the electron induced substitution reaction of a typical 18 electron complex ML_x by L′ to give $ML_{x-1}L'$ (M = transition metal; L, L′ = 2 electron neutral or anionic donor ligand).

$$ML_x \xrightarrow{-e^-} (ML_x)o \begin{cases} \rightarrow (ML_{x-1})o + L \xrightarrow{L'} (ML_{x-1}L')o \xrightarrow{+e^-} ML_{x-1}L' & (5) \\ \rightarrow (L \ldots ML_{x-1} \ldots L')o \rightarrow (ML_{x-1}L')o + L \xrightarrow{+e^-} ML_{x-1}L' & (6) \end{cases}$$

$$ML_x \xrightarrow{+e^-} (ML_x)r \begin{cases} \rightarrow (ML_{x-1})r + L \xrightarrow{L'} (ML_{x-1}L')r \xrightarrow{-e^-} ML_{x-1}L' & (7) \\ \rightarrow (L \ldots ML_{x-1} \ldots L')r \rightarrow (ML_{x-1}L')r + L \xrightarrow{-e^-} ML_{x-1}L' & (8) \end{cases}$$

where (ML_x)o refers to an oxidized reagent molecule and (ML_x)r refers to a reduced reagent molecule etc. The electron charge could be associated with either the metal or the ligand (see below).

Recent studies have indicated that the 17–19 electron cycle for radicals (similar to the 16–18 electron cycle for non-radical complexes [5]) provides a guideline for predicting a reaction mechanism [6]. Reactions (6) and (7) are hence the suggested pathways for the substitution reactions of the radical species (ML_x)o and (ML_x)r – in these reactions the electron count of the radicals never drops below 17 or exceeds 19 electrons.

3. Catalyzed Substitution Reactions

In the above reaction scheme (5)–(8), the overall reaction involves $ML_x + L' \rightarrow ML_{x-1}L'$ occurring via the intervention (gain/loss) of an electron but it is to be noted that the *overall* reaction does not however involve a change in the metal oxidation state (ML_x, $ML_{x-1}L'$). Two possible methods for making the reaction catalytic are discussed below.

3.1 Electrocatalysis* [7]

The electrocatalyzed substitution reaction requires that the reactant ML_x gain (or lose) an electron from an initiator which could be a catalyst molecule, electrode or photon, prior to ligand exchange. To make the reaction catalytic, it is imperative that the $(ML_{x-1}L')$o or $(ML_{x-1}L')$r species be able to transfer an electron rapidly to (or from) the initiator or another species (e.g. another reactant molecule, ML_x) to complete the reaction sequence and form the product $ML_{x-1}L'$. This could occur in a simple process involving catalyst (cat) via oxidation (9) or reduction (10) of ML_x.

$$\text{cat} + ML_x \rightarrow (\text{cat})o(ML_x)r \xrightarrow{+L', -L} (\text{cat})o(ML_{x-1}L')r \rightarrow \text{cat} + ML_{x-1}L' \quad (9)$$

$$\text{cat} + ML_x \rightarrow (\text{cat})r(ML_x)o \xrightarrow{+L', -L} (\text{cat})r(ML_{x-1}L')o \rightarrow \text{cat} + ML_{x-1}L' \quad (10)$$

* Electrocatalysis: "... concerns the catalysis of a reaction by an 'electron' or an 'electron hole' which, overall, does not result in a redox change ..." (definition taken from Ref. [7]).

No mechanistic detail of the actual substitution process (A or D mechanism) is indicated in the above schemes. The reactions (9) and (10) show the intervention of an electron transfer process which induces a more rapid substitution reaction relative to the thermal substitution reaction.

An alternative mechanism, for which there is much evidence (see later) involves a reaction scheme as shown below

$$\text{initiator} + ML_x \rightarrow (\text{initiator})o + (ML_x)r \quad (11)$$

$$(ML_x)r + L' \rightarrow (ML_{x-1}L')r + L \quad (12)$$

$$(ML_{x-1}L')r + ML_x \rightarrow ML_{x-1}L' + (ML_x)r \quad \text{etc.} \quad (13)$$

Equations (12) and (13) constitute a chain process which could continue until all the reactant ML_x was consumed or an alternative chain termination reagent intervened in the cycle. The rate of the reaction will be determined by the substitution and/or the electron transfer reaction (reactions (12) and (13)). A similar reaction scheme can be envisaged in which the initiator acts as a reducing agent. The initial reaction (11) can be viewed as an outer sphere electron transfer process in inorganic chemistry terminology.

3.2 Radical Chain Catalysis

The reaction sequence described in Equations (11)–(13) is similar to an alternative scheme used to describe activation/catalysis in ligand substitution reactions, namely the radical chain process [8]. In this process, a free radical is formed, typically by homolytic metal-metal or metal-hydride bond cleavage, and this metal centered radical enters a catalytic cycle e.g.

$$L_xM - ML_x \xrightarrow{\text{initiator}} 2\, ML_x\cdot \quad (14)$$

$$ML_x\cdot + L' \rightarrow ML_{x-1}L'\cdot + L \quad (15)$$

$$ML_{x-1}L'\cdot + RML_x \rightarrow RML_{x-1}L' + ML_x\cdot \quad \text{etc.} \quad (16)$$

This reaction sequence generally involves noncharged radicals and requires atom transfer (of R) i.e. it can be regarded as an inner sphere electron transfer process. However, similarities in the chain propogation steps (15) and (16) with the electrocatalysed reaction scheme ((12) and (13)) are apparent. Indeed, it appears that some catalysts e.g. $[CpFe(CO)_2]_2$ can act as both electrocatalysts and radical chain catalysts [9]. Similar explanations of a dual role for an electrochemically induced reaction, namely the decomposition of a formyl group

$$Re(CO)_5Re(CO)_4(CHO) \rightarrow Re(CO)_5Re(CO)_4H + CO \quad (17)$$

have been suggested [10].

3.3 Radical Anions

In general, addition of an electron to a ML_x species leads to $ML_x\cdot$ in which the electron is regarded as being associated with the metal. This is an approximation to the real situation in which partial delocalization of the electron onto the ligand can occur – a situation which can in principle, be evaluated by ESR spectroscopy. By judicious choice of ligand and low oxidation state metal, the generation of paramagnetic species in which the electron resides almost exclusively on the ligand is also possible. This type of radical can be written as $ML_x(L'\cdot)$ (rather than

$ML_xL'\cdot$) to indicate the position of the unpaired electron on the ligand. The radical anion could be synthesized by two distinct routes

$$ML_xL' \xrightarrow{+e^-} ML_x(L'\cdot) \quad (18)$$

$$ML_x + L'\cdot \rightarrow ML_x(L'\cdot) \quad (19)$$

From a number of studies, it appears that $L'\cdot$ is a better nucleophile than L′ [11] and that (19) provides a facile route to odd electron species. The $ML_x(L'\cdot)$ species also shows enhanced substitution reactivity relative to ML_xL' species [12] e.g. the enhanced reactivity of the radical relative to the diamagnetic precursor in (20) below

$$[Ru(bipy)_2(CO)Cl]^- \xrightarrow{2e^-} [Ru(bipy^{-\cdot})_2(CO)Cl]^-$$

$$\xrightarrow{CH_3CN} [Ru(bipy^{-\cdot})_2(CO)(CH_3CN)]^+Cl^- \quad (20)$$

In this chapter, we will not differentiate between the $ML_xL'\cdot$ and $ML_x(L'\cdot)$ species unless appropriate to the discussion. It must also be recognized that the addition of an electron to an ML_x species can result in ligand reorientation processes (isomerization) and this process will be influenced by the location of the electron on the radical.

4. Synthetic Aspects of the Catalyzed Substitution Reaction

Classical methods for the ligand substitution reaction (1) are via thermal and photochemical procedures. However, other reagent induced methods (e.g. electrochemical [13], Me_3NO[14, 15]) as well as the use of phase transfer catalysts [16] have now become common place and can easily lead to a wide range of substituted complexes under mild conditions. Alternative catalytic techniques, to be discussed below, are also readily available for the ligand substitution reaction.

In seeking a route to a substitution reaction (catalytic or other), a number of factors need to be considered in establishing the usefulness of the process. These include the following

(i) the reaction should be carried out under mild conditions. The reactants ML_x, L′ and the product $ML_{x-1}L'$ need to be stable under the reaction conditions and this criterion will often be met by lower reaction temperatures.

(ii) the reaction should be rapid.

(iii) the reaction should preferably lead to only one product e.g. exclusive monosubstitution. The reaction conditions should thus enhance the discrimination between formation of potential reaction products

(iv) when reagents or catalysts are used to improve on a substitution reaction (yield, time), the reagents, or catalysts should preferably be stable towards O_2 and H_2O, traces of which are commonly found in *most* solvents.

(v) in some instances, sequential multi-substituted products may be required and the generalization of the technique to incorporate this possibility may be useful. Similarly, the possibility of extending the reaction conditions to a wide range of L′ is an important objective in this area of chemistry.

(vi) if reagents or catalysts are used they should be cheap, easily obtained, and in some instances readily regenerated. It is also important that the required product be easily separated from the additives or catalysts (e.g. an electrode) at the end of the reaction.

It is to be noted that the ready access to electrochemical equipment (e.g. coulometers) has made electrochemical synthesis a potentially routine procedure in many laboratories.

In the discussion to follow a survey and assessment of the chemical literature pertaining to the catalyzed ligand substitution reaction in which odd electron species have been implicated has been undertaken. Further exploration of the use of electrocatalysis and free radical catalysis in synthetic strategies both for the reactions discussed and also for other substitution reactions, which presently occur thermally with difficulty, is to be expected in the ensuing years.

5. The Use of Catalysts in Synthetic Organometallic Chemistry

Much of the work that has been reported in the literature on the use of catalysts in enhancing the ligand substitution reaction in organometallic chemistry has related to a study of metal carbonyl and substituted metal carbonyl complexes. For this reason, the discussion that follows will arbitrarily break down the information according to whether the catalyst displaces CO groups or not and secondly according to the complexity of the reactant molecules i.e. to the number of metal atoms in the complex (monometallic, bimetallic, cluster).

5.1 Substitution Reactions of Monometallic Homoleptic Carbonyl Complexes

There are only a limited number of complexes of the type $M(CO)_x^{n+/-}$ (e.g. x = 4–6; n = 0–2; x and n determined by the metal, M) that are permissible by either the 16–18 or 17–19 electron counting rules. These complexes provide a cheap, readily available source of transition metal complexes in low oxidation states. Reactivity studies have been well documented on the $M(CO)_x^{n+/-}$ complexes and in particular the CO substitution reaction [17]

$$M(CO)_x + L \rightarrow M(CO)_{x-1}L + CO \qquad (21)$$

has been exhaustively studied and continues to be used as a model reaction for mechanistic studies of more complex molecules [18].

Numerous redox studies, following from the original reports of Dessy and coworkers [19] have been performed on the homoleptic metal carbonyl complexes and provide valuable information for predicting the possible effect of electrocatalysts on the CO substitution reaction. [10, 20–23] Non-catalytic electrochemical studies have also provided a route to the synthesis of substituted transition metal carbonyl complexes [24].

It appears that the first report of an electrocatalyzed substitution reaction of a homoleptic metal carbonyl complex was the reaction between $Fe(CO)_5$ and group 15 donor ligands, L, to form $Fe(CO)_4L$ [25a, b]. Reaction conditions were mild and the electrocatalyst was generated in situ from iron cluster anions. Unfortunately, an electrochemical investigation of the reaction of $Fe(CO)_5$ in the presence of L (e.g. L = PPh_3), although indicating a small reduction in the d.c. polarogram (10%) and the presence of a wave corresponding to the presence of $Fe(CO)_4L$, showed no catalytic synthetic possibilities [26].

This has to be contrasted with the extensive use of a wide range of supported transition metals as well as transition metal complexes that have been shown to bring about the substitution reactions shown in (22) and (23) below (see Table 1; Refs. [9, 27–40, 48]).

$$Fe(CO)_5 + L \rightarrow Fe(CO)_4L \quad (L = RNC, \text{group 15 donor ligand}) \tag{22}$$

$$M(CO)_6 + L \rightarrow M(CO)_5L \quad (M = Cr, Mo, W; L = RNC) \tag{23}$$

Table 1. Survey of catalysts used in the substitution reaction $M(CO)_x + L \rightarrow M(CO)_{x-1}L + CO$[a]

Catalyst used	Comment	Ref.
$Fe(CO)_5 + RNC \rightarrow Fe(CO)_4(CNR)$		
$CoCl_2$, $NiCl_2$, other salts	Range of R, yields 79–95%	[27]
$RhCl(PPh_3)_3$, $RuCl_2(PPh_3)_3$, $Ir(CO)Cl(PPh_3)_2$	R = Xy, benzene 45 °C	[30a]
catalyst supports, zeolites		[29]
$Fe(CO)_5 + PR_3 \rightarrow Fe(CO)_4(PR_3)$[b,c]		
CoX_2 (X = Cl, Br, I etc.)	d, yields >60% for phosphines	[27,30a, 31, 32]
$[Cp'Fe(CO)_2]_2$ (Cp′ = Cp, MeCp, Cp*)	d, e, yields >85%	[33, 34]
Pd/C, $Pd/CaCO_3$, PdO, PtO_2	R = Ph, yields >90% (Pd catalysts)	[29]
$[Fe_2(CO)_8]^{2-}$	yields 25–70%	[25]
$Pt(PPh_3)_4$	$PR_3 = PPh_3$	[48]
electrochemical	poor catalysis (<10% reaction)	[26]
$[Fe(CO)_3(\mu\text{-}SMe)]_2$	$PR_3 = PPh_3$, yields >90%	[28]
$M(CO)_6 + RNC \rightarrow M(CO)_5(CNR)$[f]		
PdO, PtO_2, Pd/C	R = $PhCH_2$, tBu, Xy, Me	[29, 35–38]
$RhCl(PPh_3)_3$	g	[30a]
$CoCl_2$	R = tBu, p-$C_6H_4CH_3$	[27a, 36, 38, 86]
zeolites, catalyst supports	variable catalysis	[29]
$Cp_2Fe_2(CO)_{4-x}(CNR)_x$ (x = 0–2)		[39]
metal carbonyl dimers, clusters	many active catalysts	[39]
$M(CO)_6 + PPh_3 \rightarrow Mo(CO)_4(PPh_3)_2$		
$[Fe(CO)_3(\mu\text{-}SMe)]_2$	80% yield	[28]
$Pt(PPh_3)_4$	M = W, 92% yield	[48]

[a] Specific details of the reaction conditions, range of catalysts and types of ligands used are to be found in the references listed.
[b] ^{13}CO exchange also catalyzed by Pd/C [40].
[c] PR_3 can be part of a polystyrene support [34].
[d] PR_3 = range of phosphines, phosphites. $SbPh_3$, $AsPh_3$ and $AsMe_2Ph$ also used.
[e] A photochemically catalyzed reaction has also been reported between dppe and $Fe(CO)_5$ to give $Fe(CO)_4(\eta^1\text{-dppe})$ [33].
[f] M = Cr, Mo, W; All the reactions give near quantitative yields; reactions in benzene (25–80 °C).
[g] Also used as a polymer support catalyst.

The data derived from the various studies on the use of catalysts in reactions (22) and (23) can be summarised as follows:

(i) A wide range of isocyanide and phosphine complexes have been prepared in high yield (typically >90%) and in high purity (>90%) under mild reaction conditions. In one instance, a chiral phosphine, L, has been used in the synthesis of $Fe(CO)_4L$ [31]. The methodology is superior to the alternative thermal and photochemical procedures.

(ii) A wide range of catalysts has been used to catalyze the reactions. Modification of the molecular catalysts is possible e.g. $[CpFe(CO)_2]_2$ or $[Fe(CO)_3(\mu\text{-SMe})]_2$ may be modified by chemical reaction to $Cp_2Fe_2(CO)_3L$ or $[Fe(CO)_2L(\mu\text{-SMe})]_2$.

(iii) The catalyst can be attached to polymer supports to assist with the ease of separation of catalyst from reagents [30a].

(iv) Polymerization of RNC can occur in the presence of some catalysts e.g. $NiCl_2$ but correct experimental procedures can overcome this competitive reaction [27].

(v) Studies have shown that some catalyst supports including zeolites and activated carbon can catalyze the substitution reactions, e.g. (22) [29].

(vi) Attempts have been made to establish the mechanism of the ligand substitution reaction and it appears that different mechanisms are operating for different classes of catalysts. Both free radical and electrocatalytic pathways are consistent with experimental observations for different catalysts [27] and this aspect will be discussed more fully below.

(vii) The reaction between $Fe(CO)_5$ and L (L = dppe) can be photochemically catalysed ($\lambda > 560$ nm) by $[CpFe(CO)_2]_2$ [33].

(viii) An extension of the methodology to other homoleptic metal carbonyls e.g. $Os(CO)_5$ has still to be undertaken.

5.2 Substitution Reactions of $ML_x(CO)_y$: Replacement of CO

In complexes of the type $ML_x(CO)_y$ substitution of either L or CO is possible. In this section, catalytic reactions involving CO loss from $ML_x(CO)_y$ complexes will be discussed while in Section 5.3 reactions involving loss of L will be considered.

A survey of reactions that have been carried out to date on the CO substitution of $ML_x(CO)_y$ complexes is given in Table 2 (Refs. [9, 27–30, 32, 36–39, 41–71]) and specific aspects of these reactions are discussed below. As can be seen in Table 2, a range of commonly used carbonyl complexes have been investigated and the CO catalyzed displacement reaction with two exceptions have been achieved only with RNC and group 15 donor ligands. The exceptions are shown in Eqn. 24 and 25.

$$\text{(1)}\quad Cp'Fe(CO)_2(\eta'\text{-}SCSNR_2) \rightarrow Cp'Fe(CO)(\eta\text{-}SCSNR_2) + CO\ (Cp' = Cp, Cp^*) \qquad (24)$$

achieved for Cp′ = Cp* with the ferrocene cation [41a]. Interestingly, when Cp′ = Cp catalysis has been achieved with $Mo(CO)_6$, $CpTiCl_2$ and PPh_3/CS_2 [41b]. The driving force for the reaction is the irreversibility of the reaction induced by the chelate effect [7b, 41]. The other exception

$$\text{(2)}\quad [CpFe(CO)_2L]I \rightarrow CpFe(CO)LI + CO \qquad (25)$$

is catalyzed by $[CpFe(CO)_2]_2$ [42].

Table 2. Survey of catalytic carbonyl substitution reactions of $ML_x(CO)_y$ complexes[a]

Reactant/Ligand	Catalyst[b]	Product	Ref.
$CpV(CO)_4$/RNC, PR_3	PdO, $CpV(CO)_3H$	$CpV(CO)_3L$	[50a, 50b]
$M(CO)_5X$/RNC (M = Mn, Re)	PdO	$M(CO)_{5-x}(CNR)_x$ (x = 2–5) + salt	[53]
$M(CO)_5H$/RNC (M = Mn, Re)	$M_2(CO)_{10}$	$M(CO)_{5-x}(PR_3)_xH$ (x = 1, 2)	[43, 45, 60b]
$CpMn(CO)_3$/RNC	PdO/$[CpFe(CO)_2]_2$	$CpMn(CO)_2(CNR)$	[54, 56]
$Re(CO)_3(CNR)$acac/RNC	PdO	$Re(CO)_{3-x}(CNR)_{1+x}$(acac) (x = 1, 2)	[52]
$Re(CO)_5CH_3/PPh_3$	$Pt(PPh_3)_4$	$Re(CO)_4(PPh_3)CH_3$	[48]
$M(CO)_5(CNR)$/RNC (M = Cr, Mo, W)	various	$M(CO)_{6-x}(CNR)_x$ (x = 2–6)	[29, 30a, 35–39, 88]
$M(CO)_4$bipy/PR_3 (M = Cr, Mo, W)	Na, elec	$M(CO)_3(PR_3)$bipy	[63, 71]
arene$Cr(CO)_3$/RNC	PdO/$[CpFe(CO)_2]_2$	arene $Cr(CO)_2(CNR)$	[55, 56]
$CpM(CO)_3H/PR_3$ (M = Mo, W)	$[CpM(CO)_3]_2$, photo	$CpM(CO)_2(PR_3)H$	[44]
$CpM(CO)_3I$/RNC, PR_3 (M = Mo, W)	dimers, photo	$CpM(CO)_{3-x}L_xI$ (x = 1, 2) + salt	[9, 46, 49]
$W(CO)_5(CH_3CN)$/RNC	PdO	$W(CO)_4(CNR)(CH_3CN)$	[57]
$Fe(CO)_4PR_3/PR_3$	$CoCl_2$, met. anions	$Fe(CO)_3(PR_3)_2$	[25a, 32]
$Fe(CO)_4(CNR)$/RNC	various[d]	$Fe(CO)_{5-x}(CNR)_x$ (x = 2–5)	[9, 27, 29, 30a, 39, 5]
$Fe(CO)_4$olefin/RNC	$Rh(CO)Cl(PPh_3)_2$	$Fe(CO)_{4-x}(CNR)_x$olefin (x = 1, 3)	[30b]
$Cp'Fe(CO)_2I$[e]/PR_3, RNC	$[Fe(CO)_3(\mu\text{-SMe})]_2$ $Pt(dba)_2$, $[CpFe(CO)_2]_2$	$Cp'Fe(CO)_{2-x}L_xI$ (x = 1, 2) + salt	[9, 28, 42, 48]
$Cp'Fe(CO)_2(\eta^1\text{-SCSNR}_2)$[f]	g	$Cp'Fe(CO)(\eta^2\text{-SCSNR}_2)$	[41]
$Cp'Fe(CO)_2R'$[h]/PR_3	$[CpFe(CO)_2]_2$	$Cp'Fe(CO)(PR_3)R'$	[47]
$CpRu(CO)_2I$/RNC, PR_3	various	$CpRu(CO)_{2-x}L_xI$ (x = 1, 2) + salt	[58, 59, 61]
$Os(CO)_4(Me)H/PR_3$	$H_2Os_2(CO)_8$, O_2	$Os(CO)_3(PR_3)(Me)H$	[60a]
$SnCl_3Co(CO)_4/PR_3$	photo	$[Co(CO)_3(PR_3)_2]SnCl_4$	[62]

[a] Specific details of the reaction conditions, range of catalysts and types of ligands used are to be found in the references listed.
[b] electro = electrochemical catalysis, photo = photocatalysis.
[c] See also refs. [69, 70].
[d] See also Table 3.
[e] Cp′ = Cp, MeCp.
[f] Cp′ = Cp, Cp*.
[g] catalysts used include Cp_2Fe^+ and $Mo(CO)_6$
[h] R′ = η^1-C_5H_5, η^1-allyl; Cp′ = MeCp, Cp.

A logical extension of the CO substitution reaction is to achieve multiple CO substitution. Indeed this has been achieved and complete substitution of all CO groups by RNC has been accomplished from $Fe(CO)_5$ [27], $M(CO)_6$ [37] (M = Mo, W) and $Re(CO)_5Br$ [53]. Whereas the NMe_3O and thermal routes may be competitive alternative options to the electrocatalyzed routes in the synthesis of say $Fe(CO)_4(CNR)$ and $Fe(CO)_3(CNR)_2$ these routes do not yield the higher substituted derivatives e.g. $Fe(CO)(CNR)_4$. The possibility that the NMe_3O reactions proceed by electron transfer also should be considered.

Another feature of the catalyzed reaction is the stepwise nature of the CO substitution process. This allows for the synthesis of *mixed ligand* derivatives e.g. the synthesis of $Fe(CO)_3(CNBut)(CNXy)$ [27] and $Mo(CO)_4(CNMe)(CNBut)$ [36].

5.2.1 Chemical Reagents

A wide range of catalysts have been used to achieve catalytic CO replacement and a number of mechanisms have been proposed to rationalize the data. The earliest reported study on these catalyzed reactions appeared in 1975 when Byers and Brown described the unusual substitution reactivity of $HRe(CO)_5$ [43a]. The authors observed that the reaction between $HRe(CO)_5$ and L (L = PPh_3, PBu_3) was catalyzed by $Re(CO)_5\cdot$, generated photolytically from $Re_2(CO)_{10}$ or from other radical species [43b]. A free radical chain mechanism to rationalize the results was proposed.

$$Re_2(CO)_{10} \rightarrow 2Re(CO)_5\cdot \quad (26)$$

$$Re(CO)_5\cdot + L \rightarrow Re(CO)_4L\cdot \quad (27)$$

$$Re(CO)_4L\cdot + HRe(CO)_5 \rightarrow HRe(CO)_4L + Re(CO)_5 \quad \text{etc.} \quad (28)$$

Adventious O_2 was also found to initiate the reaction via production of $Re(CO)_5\cdot$ from $HRe(CO)_5$. This free radical process was later extended to explain the substitution behaviour of $CpM(CO)_3H$ (M = Mo, W) [44], $Mn(CO)_5H$ [45], $CpMo(CO)_3I$ [46] and more recently $H_2Os(CO)_4$ [60b] with similar group 15 donor ligands. Similar reactivity was found and reaction schemes proposed for the substitution reactions of $MeCpFe(CO)_2(\eta^1\text{-allyl})$ [47a] and $CpFe(CO)_2(\eta^1\text{-}C_5H_5)$ [47b]. A logical extension to the photochemical production of metal radicals from metal dimers is via thermal M—M bond cleavage. Thus initial results obtained on addition of $[CpMo(CO)_3]_2$ as catalyst to a refluxing solution of $CpMo(CO)_3I$ and RNC to give $CpMo(CO)_2(CNR)I$ could be rationalized mechanistically in terms of a free radical sequence similar to Equations (26)–(28). [49] Further studies on the related $CpFe(CO)_2I$/RNC reaction in refluxing benzene is the presence of $[CpFe(CO)_2]_2$ as catalyst viz.

$$CpFe(CO)_2I + RNC \rightarrow CpFe(CO)_{2-x}(CNR)_xI \quad (x = 1, 2) \quad (29)$$

could also be rationalized by a free radical process for x = 1, viz.

$$[CpFe(CO)_2]_2 \rightarrow 2CpFe(CO)_2\cdot \quad (30)$$

$$CpFe(CO)_2\cdot + L \rightarrow CpFe(CO)L\cdot + CO \quad (31)$$

$$CpFe(CO)L\cdot + CpFe(CO)_2I \rightarrow CpFe(CO)LI + CpFe(CO)_2\cdot \quad \text{etc.} \quad (32)$$

However, exchange reactions performed at lower temperatures using labelled catalyst and/or reactant and the production of higher substituted derivatives in the catalyzed reaction (29) could also be explained by an electron transfer pathway [9]:

$$\text{catalyst} + CpFe(CO)_2I \rightarrow \text{catalyst}^- + CpFe(CO)_2I^+ \quad (33)$$

$$CpFe(CO)_2I^+ + L \rightarrow CpFe(CO)LI^+ + CO \quad (34)$$

$$CpFe(CO)LI^+ + CpFe(CO)_2I \rightarrow CpFe(CO)LI + CpFe(CO)_2I^+ \text{ etc.} \quad (35)$$

(Other similar schemes in which the catalyst acts as a reducing agent are also possible). Importantly, the above mechanism suggests that other reactant molecules, which did not contain the '$CpFe(CO)_2$' unit, should also be catalyzed by the dimer catalyst and this was indeed shown to be true [9].

Since the original studies on reaction (29) numerous publications have appeared which have attempted to delineate the reaction type further. In particular, the reaction

$$Fe(CO)_4(CNR) + RNC \rightarrow Fe(CO)_3(CNR)_2 + CO \quad (36)$$

has been used as a probe reaction to explore catalysts and reaction mechanisms (see Table 2). This reaction was chosen since the starting material $Fe(CO)_4(CNR)$ is easily synthesized in high yield from readily available $Fe(CO)_5$ with a range of R groups (e.g. R = tBu, Xy, etc.), is air stable, and the reaction can readily be monitored by IR spectroscopy. Below are described some pertinent data on specific catalyst systems that have been obtained from studies on reactions (36).

5.2.1.1 Transition Metal Salts The use of transition metal salts, such as $CoCl_2$, have been comprehensively investigated. It has been observed that with $CoCl_2$ as a catalyst that a range of solvents (wet or dry) can be used for reaction (36) – high boiling solvents which solubilize the catalyst give more rapid reactions. A wide range of CoX_2 salts (X = PF_6, $1/2SO_4$, NO_3, CH_3COO) catalyze the reaction and the data suggest that the counterion has little influence on the reaction mechanism. In solution the Co^{2+} coordinates to the ligand e.g. $Co(CNR)_x^{2+}$ is formed, but a bridging mechanism involving transfer of the coordinated RNC ligand to Fe has been ruled out by exchange reactions [27]. No direct evidence for an electron transfer mechanism was obtained, although a mechanism involving electron transfer to or from Co(II) is possible. Reactions are rapid, yields are nearly quantitative, and products obtained after catalyst removal are pure.

5.2.1.2 Metal-Metal Bonded Complexes A second class of catalysts that brings about reaction (36) are metal dimer and cluster complexes. As mentioned above $[CpFe(CO)_2]_2$ was found to act as a catalyst for reagents that did not contain the $CpFe(CO)_2$ unit e.g. such as $Fe(CO)_4CNR$. In this instance only a marginal catalytic effect was initially observed (rate increase <50%). However, modifications of the catalyst by changing the ring substituent, the metal, or replacement of CO by an RNC group resulted in dramatic increases in the reaction rate (Table 3). Also, irradiation of the reaction mixture in the presence of a catalyst resulted in an increase in the reaction rate. Radical scavengers and hydroquinone were also observed to *increase* the reaction rate (see also Ref. [44]). The ability of $[CpFe(CO)_2]_2$, which contains a metal-metal bond, to catalyze reaction (36) suggested other metal-metal bonded species should also catalyze the reaction. Provided that the dimer or cluster e.g. $H_4Ru_4(CO)_{12}$, $Mn_2(CO)_{10}$, or $Co_4(CO)_{12}$ did not undergo a deactivating competing reaction with the ligand (e.g. RNC reacted

Table 3. Effect of iron dimer catalysts on the reaction $Fe(CO)_4(CNBut)$ + tBuNC → $Fe(CO)_3(CNBut)_2$[a,b]

Catalyst	Reaction time/min[c]
blank	120
$[CpFe(CO)_2]_2$	75
$Cp_2Fe_2(CO)_3(CNBut)$	75
$[CpFe(CO)(CNBut)]_2$	2
$[CpFe(CO)_2]_2/PtO_2$	2
$[MeCpFe(CO)_2]_2$	30
$[CpCO_2MeFe(CO)_2]_2$	10
$[CpOMeFe(CO)_2]_2$	4
$[CpRu(CO)_2]_2$	18
$[Cp^*Fe(CO)_2]_2$	nc[d]
$[Cp^*Fe(CO)_2]_2$	40[e]

[a] Reaction conditions: $Fe(CO)_4(CNBut)$ – 1 mmol; tBuNC – 1.1 mmol; catalyst – 0.1 mmol; benzene – 10 ml.

[b] Data taken from Ref. [39].

[c] Time for reaction to go to completion as determined by IR spectroscopy.

[d] nc = no catalysis.

[e] Room temperature; 500 W irradiation.

with $[CpMo(CO)_3]_2$ to yield $[CpMo(RNC)_4]^+$ and $Fe_3(CO)_{12}$ reacted with RNC to yield $Fe(CO)_{5-x}(CNR)_x$ – both monometallic species are catalytically inactive) catalysis was observed. Some cluster reagents were totally inactive e.g. $Os_3(CO)_{12}$ and $Re_2(CO)_{10}$.

Aime and co-workers [28] have extended the use of metal dimers as catalysts in ligand substitution reactions. Thus $[Fe(CO)_3(\mu\text{-SMe})]_2$ has been found to catalyze the reaction of PPh_3 with $CpFe(CO)_2I$ to yield $CpFe(CO)(PPh_3)I$ rapidly and in excellent yield. A metal-metal bond cleavage reaction of the dimer has been proposed to rationalize the data.

5.2.1.3 Supported Metals The possibility that the catalyzed CO substitution reaction could be rationalized by an electron transfer mechanism suggested that metal oxides and supported metals might also function as catalysts. After a screening of a wide range of possible materials PdO and Pd/C were found to be excellent catalysts for bringing about the CO substitution reaction of a wide range of $ML_x(CO)_y$ reagents including $Fe(CO)_4(CNR)$. It is worth noting that the catalyst support does influence catalytic activity and that the support can itself catalyze CO substitution reactions. It appears that adventious radicals trapped in the support or zeolite may be responsible for this effect.

5.2.1.4 Combination Catalysts It has been found that a catalyst consisting of a mixture of a metal-metal bonded complex and a supported metal complex can have unexpected catalytic properties. This is highlighted by an example. $Os_3(CO)_{12}$ does not catalyze reaction (36) and PdO does not catalyze reaction (36) at room temperature to any significant extent. However, addition of $PdO/Os_3(CO)_{12}$ to

$Fe(CO)_4(CNBut)$ and tBuNC at room temperature gives a rapid reaction to yield $Fe(CO)_3(CNBut)_2$ quantitatively (35 min). Replacement of PdO by Pd/C (12 min) and $Pd/CaCO_3$ (23 min) also results in rapid catalysis. The reaction of $Os_3(CO)_{12}$ and RNC is catalyzed by PdO to give $Os_3(CO)_{12-x}(CNR)_x$ ($x > 1$) at *elevated* temperatures and whether trace amounts of these complexes are the "real" catalysts must await further investigation [39].

5.2.1.5 Transition Metal Complexes Examples of this class of catalyst include $Rh(CO)Cl(PPh_3)_2$ [30] and more recently $Pt(PPh_3)_4$ [48]. An electron transfer mechanism may also be operating for these catalysts. An analysis of the CO force constants of the reactants suggests that $Pt(PPh_3)_4$ is only effective as a catalyst on electron poor metal carbonyl complexes [48].

5.2.1.6 Group I Metals Recently, the use of substoichiometric amounts (10–20%) of potassium metal to induce electron transfer catalysis for the reaction

$$Mo(CO)_4(bipy) + PPh_3 \rightarrow Mo(CO)_3(PPh_3)(bipy) + CO \quad (37)$$

and related reactions has been reported [63]. The catalytic effect is not dramatic but the thorough investigation has revealed the role of $Mo(CO)_4(bipy)^{\cdot -}$ (in which the electron resides predominantly on the ligand) in the catalytic reaction. The catalytic technique may be quite widespread and useful in other organometallic synthesis reactions [15, 64].

5.2.2 Electrochemical Studies

The use of electrochemical procedures to induce *non-catalytic* CO substitution in substituted metal carbonyl complexes has been investigated (see e.g. [15]) and includes recent investigations on arene$Cr(CO)_3$ and $CpMn(CO)_3$ [65], $Cr(CO)_4(PPh_3)_2$ [66], $DADMo(CO)_4$ [67] and $[Mo(CO)_5]_2$bpym [68]. In general the reactions which involve initial electrolytic conversion of the 18 to a 17 or a 19 electron system, occur rapidly to yield the substituted carbonyl radical which then undergoes a further redox reaction to give the stable 18 electron substituted product. A suggestion of a light catalyzed substitution reaction accompanying the catalyzed (redox) reaction

$$Cr(CO)_4(PPh_3)_2 + CO \rightarrow Cr(CO)_5PPh_3 \quad (38)$$

has been made [66]. A recent report on the electrochemically activated reaction

$$\text{arene}Cr(CO)_3 + L \rightarrow \text{arene}Cr(CO)_2L + CO \quad (39)$$

suggests that catalysis may also be occurring in this reaction [69, 70].

Finally, it is to be noted that an example of an electrochemically induced catalytic CO displacement reaction has been reported. In this reaction

$$Mo(CO)_4\text{bipy} + PPh_3 \rightarrow Mo(CO)_3(PPh_3)\text{bipy} \; (M = Cr, Mo, W) \quad (40)$$

product yields of 85–95% and facile product isolation procedures were reported [71]. A mechanism similar to that of the potassium metal catalyzed reaction must exist [63].

To summarize: Although mechanistic details of the above reactions are sparse, it is apparent that catalytic CO displacement from $ML_x(CO)_y$ can be achieved by a wide range of reagents. In general, the reactions are quite specific and selective e.g. catalyzed arene$Cr(CO)_3$ reactions yield only CO substituted products and not arene displacement products as found for the uncatalyzed thermal reactions [55, 56]. Further developments must relate to the extension of the concept to the replacement of CO by other types of incoming ligands e.g. alkenes, alkynes, etc., and to a study of other metal carbonyl reagents.

5.3 Substitution Reactions of $ML_x(CO)_y$: Replacement of L

A common method of achieving CO ligand substitution in ML_xCO_y complexes is to proceed via an intermediate complex which contains a readily displaceable ligand L, which is a better leaving group than CO.

$$ML_x(CO)_y + L' \rightarrow ML_xL'(CO)_{y-1} + CO \quad (41)$$

$$ML_xL''(CO)_{y-1} + L' \rightarrow ML_xL'(CO)_{y-1} + L'' \quad (42)$$

The function of L″ is to replace a CO ligand usually under forcing conditions and L″ is then later replaced by L′ under relatively mild conditions. This indirect replacement of CO by L′ results in cleaner reactions (non-multiple replacement reactions) and is useful when the product $ML_xL'(CO)_{y-1}$ is thermally and/or photolytically unstable.

The technique has been widely used in non-catalytic substitution reactions but more recently Kochi and co-workers have studied the use of catalysts on the second reaction, (42) [72–82]. Typically, electrochemical techniques are employed (Table 4), although chemical reagents such as $MeCpMn(CO)_2(CH_3CN)$ (electrochemically generated, [75]) and O_2 [75] have also been used to catalyze the substitution reaction. Some specific examples of the reactions studied are shown below

Table 4. Catalyzed displacement reactions of non-CO ligands, L, from $ML_x(CO)_y$ [a]

Reaction	Comments [b]	Ref.
$MeCpMn(CO)_2L + L' \rightarrow MeCpMn(CO)_2L'$	electro., yields >80%	[72, 75]
$CpRe(CO)_2L + L' \rightarrow CpRe(CO)_2L'$	electro.	[72]
$M(CO)_{6-x}(CH_3CN)_x + L \rightarrow M(CO)_{6-x}(CH_3CN)_{x-y}L_y$	electro.; L = PR_3, $P(OR)_3$; for M = Mo, W yields >80%; for M = Cr yields <10%	[81]
$W(CO)_5(CH_3CN) + tBuNC \rightarrow W(CO)_5(CNBut)$	mixed products, PdO catalyst	[57]
$CpM(CO)_2I + L \rightarrow CpM(CO)_2L^+$, $CpM(CO)L_2^+$	M = Fe, Ru, dimer catalysts	[42, 58]
$CpM(CO)_2I + RNC \rightarrow CpM(CNR)_3^+$	M = Fe, Ru, dimer catalysts	[9, 59]
$Re(CO)_3(CNR)_2Br + RNC \rightarrow Re(CO)(CNR)_5^+$	mixed products, PdO catalyst	[53]
$Cp_2CrCO + CO \rightarrow CpCr(CO)_2(\eta^3\text{-}C_5H_5)$	catalysts: air, I_2	[89]
$SnCl_3Co(CO)_4 + L \rightarrow [Co(CO)_3L_2]^+$	L = PR_3	[62]
$Os(CO)_4H_2 + PPh_3AuMe \rightarrow (PPh_3Au)_2Os(CO)_4$	photo.-$H_2Os_2(CO)_8$	[60b]
$Mn(CO)_5H + PPh_3AuMe \rightarrow PPh_3AuMn(CO)_5$	photo.-$Mn_2(CO)_{10}$	[60b]
$CpFe(CO)(PPh_3)COMe + CO \rightarrow CpFe(CO)_2Me$	catalyst: $CpV(CO)_3H$	[50b]

[a] Specific details of the reaction conditions, range of catalysts and types of ligands used can be found in the references listed.
[b] electro = electrochemical catalysis; photo. = photocatalysis.

(Table 4):

$$MeCpMn(CO)_2(CH_3CN) + PPh_3 \rightarrow MeCpMn(CO)_2PPh_3 + CH_3CN \quad [74] \quad (43)$$

$$CpRe(CO)_2py + PPh_3 \rightarrow CpRe(CO)_2PPh_3 + py \quad [72] \quad (44)$$

$$W(CO)_3(CH_3CN)_3 + tBuNC \rightarrow W(CO)_3(CNBut)_3 + CH_3CN \quad [81] \quad (45)$$

$$Mn(CO)_3(CH_3CN)_3{}^+ + PPh_3 \rightarrow Mn(CO)_3(PPh_3)_2H \quad [78\text{–}80] \quad (46)$$

Reactions have generally been carried out on a small scale under anaerobic conditions. Reactions (43)–(45) occur by electron addition to create a 19 electron system which rapidly loses the 'intermediate' ligand (e.g. py, CH_3CN). A catalytic cycle then follows [74]

$$MeCpMn(CO)_2L' + e^- \rightarrow MeCpMn(CO)_2L'\cdot \quad (47)$$

$$MeCpMn(CO)_2L'\cdot + L \rightarrow MeCpMn(CO)_2L\cdot + L' \quad (48)$$

$$MeCpMn(CO)_2L\cdot + MeCpMn(CO)_2L' \rightarrow MeCpMn(CO)_2L + MeCpMn(CO)_2L'\cdot \text{ etc.} \quad (49)$$

The combined steric and electronic effects of both the leaving (L) and entering (E) group for reactions of type (43) have been incorporated into a single Equation [74]

$$\log k_1 = A + BpK_E + CpK_L + D\theta_E$$

where k_1 = associative rate constant for the substitution reaction; A, B, C and D are constants; pK_E and pK_L relate to the electronic properties and θ_E to the steric effects of the leaving (L) and entering (E) group respectively. (The pK and θ values for the ligands are known measurable quantities). Recent kinetic studies on related *uncatalyzed* substitution reactions have also been described in terms of rate constants made up of additive components and this has led to the creation of the concept of steric profiles (see Refs. [83] and [84] for further details). Whether similar electrocatalyzed substitution reactions of synthetic utility can also be analyzed to reveal similar ligand influences must await further experimentation.

Some general comments can be made relating to the electrocatalyzed reactions represented by Equations (43)–(46).

(i) Reactions occur under mild conditions (25 °C) and product yields are high (80–100%).

(ii) Most of the complexes can be produced by alternative procedures. However, multiple substitution (e.g. (45)) and partial substitution provides a route to the more difficult to obtain mixed ligand derivatives [81] e.g.

$$W(CO)_4(CH_3CN)_2 + PPh_3 \rightarrow W(CO)_4(PPh_3)(CH_3CN) + CH_3CN \quad (50)$$

$$W(CO)_4(py)_2 + PPh_3 \rightarrow W(CO)_4(PPh_3)(py) + py \quad (51)$$

(iii) Reactions are catalytic with turnover numbers varying between 2–1000.

(iv) Rate increases of 10^6 relative to the thermal uncatalyzed reaction rate have been measured.

(v) Products are stable under the conditions of the electrochemical procedures.

(vi) The synthesis of $Mn(CO)_3(PPh_3)_2H$ occurs via an electron removal process and provides a facile high yield route to the metal hydride (Equation (46)). The catalytic cycle is proposed to occur via a dissociative process and to involve the formation of a formyl group.

Some other reactions are known in which non CO groups have been catalytically displaced by electrocatalysts. For example reactions (52) and (53) are catalyzed by $[CpFe(CO)_2]_2$

$$CpFe(CO)_2I + L \rightarrow [CpFe(CO)_2L]I \ (L = PR_3, \text{diphosphines}) \quad (52)$$

$$CpFe(CO)(L)I + L \rightarrow [CpFe(CO)(L)_2]I \ (L = PMePh_2) \quad (53)$$

while reaction (54) is catalysed by Pd/C.

$$Re(CO)_3(CNR)_2Br + RNC \rightarrow [Re(CO)(CNR)_5]Br \quad (54)$$

Wrighton and co-workers [51 a] have reported a novel method for inducing a catalytic substitution reaction. Thus irradiation of $Re(CH_3CN)(CO)_3phen^+$ in the presence of a suitable electron donating species yields the 19 electron complex $Re(CH_3CN)(CO)_3phen\cdot$ which is substitutionally labile and leads to $ReL(CO)_3$-phen· in the presence of L (e.g. $L = PPh_3$) and finally $ReL(CO)_3phen^+$ on completion of a catalytic cycle. The catalyzed reaction can also be performed electrochemically. The use of this approach in synthetic organometallic chemistry has yet to be exploited further. The alternative process of electrochemical generation of a 'photochemical' reaction may also provide future catalytic synthetic possibilities [51 b].

Norton and co-workers have established that a radical chain mechanism explains the effect of catalysts on the reactions

$$HMn(CO)_5 + PPh_3AuMe \rightarrow PPh_3AuMn(CO)_5 \quad (55)$$

$$H_2Os(CO)_4 + PPh_3AuMe \rightarrow (PPh_3Au)_2Os(CO)_4 \quad (56)$$

the reactions being catalysed by $Mn_2(CO)_{10}$ and $H_2Os_2(CO)_8$, respectively [60 b]. These reactions can be formally regarded as substitution reactions of H by the isolobal $AuPPh_3$ group.

5.4 Substitution Reactions of Non-CO Containing Complexes

The classic work on redox properties and electrocatalysts in inorganic chemistry dates from the early work of Taube and co-workers [2]. These reactions were performed on classical inorganic compounds. More recent examples of non-carbonyl containing organometallic complexes which undergo electrocatalysis include

$$CpFe(\text{arene}) + L \rightarrow CpFeL_3 + \text{arene} \quad (57)$$

$$CpM(CNR)_2I + RNC \rightarrow [CpM(CNR)_3]I \ (M = Fe, Ru) \quad (58)$$

A final example which may involve electrocatalysis is the reported activation of a Re hydride complex by Pt(0) e.g. $Pt[PPh(Bu^t)_2]_2$ [87]

$$ReH_7(PR_3)_2 + C_2H_4 \rightarrow ReH_3(C_2H_4)_2(PR_3)_2 \quad (59)$$

a reaction which may have similarities with the Pt(0) catalyzed CO substitution reaction reported by Jang and Angelici [48].

5.5 Substitution Reactions: Metal Dimers

This section can be regarded as an extension of the studies on complexes of the type $ML_x(CO)_y$ in which one of the L groups is a metal fragment. It is convenient to treat this class of complexes separately however, since on cleavage of a metal-

metal bond *two* metal based radicals are formed which may have quite different chemical and physical properties from each other if the metals or their ligand sets are different. Further, in this section, metal dimers will be discussed in terms of whether they contain bridging or capping ligands which hold the two radicals together once M—M bond cleavage has occurred.

Table 5. Catalyzed substitution reactions of metal dimer complexes[a]

Dimer/ligand	Product/comment[b]	Ref.
$Mn_2(CO)_{10}/PR_3$, RNC, NR_3	various catalysts, cleavage when $L = NR_3$; $Mn_2(CO)_{10-x}L_x$ ($x = 1, 4$)	[35b, 39, 91, 92, 99, 110–112]
$Re_2(CO)_{10}/PR_3$, RNC	various catalysts, $Re_2(CO)_{10-x}L_x$	[39, 48, 93, 94]
$MnRe(CO)_{10}$/RNC	substitution on Re, PdO	[97, 98]
$[CpRu(CO)_2]_2$/RNC	$Cp_2Ru_2(CO)_{4-x}(CNR)_x$ ($x = 1$–4), PdO	[95]
$[CpFe(CO)_2]_2/PR_3$, RNC	poor cat. with PtO_2, PdO (L = RNC)	[28, 39]
$[Fe(CO)_3(\mu\text{-SR})]_2/P(OMe)_3$	electro., range of R	[104]
$Fe_2(CO)_6[C(OR)SSMe]$ (*1*, *2*)	electro., $L = P(OMe)_3$	[106, 107, 115, 116]
$Cp_2Fe_2(CO)_2(\mu\text{-CO})(\mu\text{-CSMe})^+/PR_3$		[126]
$Fe_2(CO)_8{}^{2-}/Co_2(CO)_8$	prep of $CoFe(CO)_8{}^{2-}$	[125]
$Cp'Fe(CO)_2M(CO)_5$[c]/RNC	subst. on M (M = Mn, Re)	[100, 101]
$CpM'(CO)_3M(CO)_5$/RNC	M′ = Mo, W; M = Mn, Re; subst. on M	[103]
$CpM(CO)_3Co(CO)_3(\mu\text{-RCCR})$[d]/RNC	M = Mo, W; subst. on Co	[102]
$[CpMo(CO)_2(\mu\text{-SR})]_2{}^{2+}/CH_3CN$	electro.	[105]
$[Cp'M(CO)_3]_2$[e]/py, PR_3	cleavage products; photo.	[113, 114]
$Co_2(CO)_8/PR_3$	cleavage products; $Co(CO)_3(PR_3)_2{}^+$	[108, 109]
$Co_2(CO)_6(\mu\text{-RCCR})/PR_3$, CH_3CN	BPK, electro., $[CpFe(CO)_2]_2$	[26, 117–123]
$Co_2(CO)_8(\mu\text{-RCCR})/CpMo(CO)_3$	$CoMo(CO)_5(\mu\text{-RCCR})$	[124]

[a] Specific details of the reaction conditions, range of catalysts and types of ligands used can be found in the references listed.
[b] electro = electrochemical catalysis, photo = photocatalysis.
[c] Cp′ = Cp, MeCp and indenyl.
[d] R = COOMe, COOEt.
[e] Cp′ = Cp, MeCp.

It is worth noting that numerous studies have suggested that the M—M and M—CO bonds in many cluster and dimer complexes have comparable bond energies [90a]. Consequently direct thermal substitution reactions of metal dimers with ligands can result in mechanisms in which both M—M and/or M—CO bond cleavage processes occur. This duality of reaction mechanism has also been observed in photochemical substitution reactions [90b]. Catalytic processes thus have the potential for reducing kinetic barriers to either the M—M or M—CO cleavage reactions and could thus influence the final product distribution. The data in this section is summarized in Tables 5 and 6 (also Refs. [28, 39, 48, 91–126]).

Table 6. Catalysts used for the substitution reaction $Re_2(CO)_{10}$ + tBuNC → $Re(CO)_9(CNBut)$[a,b]

Catalyst	Reaction time/min[c]
d	nr[e]
f	210
PdO	1
Pd/C (10%)	105
$Pd/CaCO_3$ (5%)	1
Pd/Al_2O_3 (5%)	70
$[CpFe(CO)_2]_2$	40
$Mn_2(CO)_{10}$	10
$Ru_3(CO)_{12}$[g]	10
$Ir_4(CO)_{12}$[g]	15
$Co_4(CO)_{12}$[g]	5
$Co_2(CO)_8$[g]	5

[a] Reaction conditions: benzene, 55 °C; 0.2 mmol reagents; 4 mg catalyst.
[b] Data taken from Refs. [39] and [94].
[c] Time for completion of reaction as determined by IR spectroscopy.
[d] no catalyst; flask covered with foil.
[e] nr = no reaction
[f] no catalyst; laboratory light.
[g] Reaction conditions: benzene 80 °C; 1.0 mmol reagents; 0.1 mmol catalyst. In the absence of catalyst no reaction was observed in 60 min.

5.5.1 *Unbridged (Non-capped) Metal Dimers*

In an attempt to extend the concept of the use of metals and metal complexes as catalysts to CO substitution reactions of metal dimer complexes the reaction between $Mn_2(CO)_{10}$ and tBuNC in the presence of potential catalysts was screened for catalytic behaviour.

$$Mn_2(CO)_{10} + tBuNC \rightarrow Mn_2(CO)_9(CNBut) + CO \qquad (60)$$

This reaction was chosen since the starting material is moderately air stable, contains no bridging CO groups and the products can readily be characterized by IR spectroscopy. The reaction was found to be catalyzed by a wide range of catalysts and the best results were obtained for a variety of Pd catalysts [91]. Using these catalysts room temperature replacement of CO by a range of RNC gave the required $Mn_2(CO)_9(CNR)$ cleanly and in high yield. The reaction could be extended to give higher substituted derivatives, $Mn_2(CO)_{10-x}(CNR)_x$ ($x = 1$–4). A kinetic investigation of reaction (61)

$$Mn_2(CO)_{10} + tBuNC \rightarrow Mn_2(CO)_{10-x}(CNBut)_x \ (x = 1, 2) \qquad (61)$$

indicated a 500 fold rate increase on the use of PdO as a catalyst for the reaction when $x = 1$. The study also revealed that the kinetic product from the catalyzed reaction between $Mn_2(CO)_9(CNBut)$ and tBuNC was the 2,6 isomer

2,3 isomer   2,6 isomer

On raising the temperature to *ca.* 50 °C the 2,3 isomer could also be formed, the final product ratio being *ca.* 4 : 1 for the 2,6 and 2,3 isomers respectively [92]. Since the direct thermal reaction for (59) (x = 2) requires temperatures greater than 50 °C, no data are available as to the kinetic product of the direct thermal reactions. Labelling studies in which mixtures of $Mn_2(CO)_{10}$ and $Mn_2(^{13}CO)_{10}$ were reacted with tBuNC in both the presence and absence of PdO revealed that in *both* reactions, substitution proceeded without metal-metal bond cleavage i.e. reaction occurred via a classical Mn—CO bond cleavage process [99]. The reaction between $Mn_2(CO)_{10}$ and PPh_3 to yield $Mn_2(CO)_8(PPh_3)_2$ was found to be catalyzed by $[Cp'Fe(CO)_2]_2$ (Cp′ = Cp, MeCp, *Cp) [39].

Catalysis was also observed on extension of the catalytic procedures to reactions of $Re_2(CO)_{10}$ with L (L = RNC [35b, 39, 94] and PPh_3 [48, 93]) in the presence of Pd or Pt catalysts.

$$Re_2(CO)_{10} + L \rightarrow Re_2(CO)_{10-x}L_x + CO \quad (x = 1\text{–}4) \qquad (62)$$

Reaction (62) (L = RNC) was also catalyzed by a range of metal dimer and cluster complexes [39] (Table 6). This is a remarkable result in that a metal dimer complex is being catalyzed by another metal dimer. The implications for dimer and cluster chemistry are apparent. As yet very little information on the reaction mechanism is available e.g. does the catalyst undergo substitution?, are metal radicals implicated in the reaction? etc.

The effect of photochemical irradiation of metal dimer complexes to yield substituted products in which radical chain mechanisms have been implicated has been reported. In the reaction ϕ, the quantum yield, is greater than 1; the effect of the irradiation is to produce heterolytic metal-metal bond cleavage

$$M(CO)_xM(CO)_x + L \rightarrow [M(CO)_{x+1-n}L_n][M(CO)_x] \qquad (63)$$

These are not substitution reactions in the classical sense in that the integrity of the reactant has been destroyed via M—M bond cleavage. However, the mechanistic studies have bearing on all metal dimer substitution reactions. Studies have been performed on $Co_2(CO)_8$, $Mn_2(CO)_{10}$ and $[CpMo(CO)_3]_2$ by Brown and co-workers [108–110] and by Tyler and co-workers [111–114]. The mechanisms proposed by both groups are similar, differing in the importance of a 17 or 19 electron intermediate in the reaction sequence. The mechanism proposed by Tyler involving the 19 electron radical is given below

$$[CpMo(CO)_3]_2 \rightarrow 2CpMo(CO)_3\cdot \qquad (64)$$

$$CpMo(CO)_3\cdot + 2L \rightarrow CpMo(CO)_2L_2\cdot + 2CO \qquad (65)$$

$$CpMo(CO)_2L_2\cdot + [CpMo(CO)_3]_2 \rightarrow CpMo(CO)_2L_2^+ + [CpMo(CO)_3]_2^- \qquad (66)$$

$$[CpMo(CO)_3]_2^- \rightarrow CpMo(CO)_3^- + CpMo(CO)_3\cdot \qquad (67)$$

This mechanism readily explains the salt formation observed; the key step is the electron transfer process (66) in which the radical acts as a reducing agent. The 19 electron $CpMo(CO)_2L_2\cdot$ radical is believed to be more readily oxidized than a $CpMo(CO)_2L\cdot$ (17 electron) radical. The above mechanism also rationalizes the photochemical behavior of $CpMo(CO)_3X$ and other $ML_x(CO)_yX$ species [127].

Studies have also been undertaken on the use of catalysts in the substitution reactions of heterobimetallic dimers. A key question, still unanswered, in the study of the substitution chemistry of heterobimetallic (and cluster) complexes is the site selectivity question – which metal will preferably lose a ligand and will the incoming ligand bind to the same metal that lost the leaving group? This problem can be highlighted by the reaction of $MnRe(CO)_{10}$, a simple heterobimetallic dimer which contains no bridging CO groups. From the reactivity behaviour of $Mn_2(CO)_{10}$ and $Re_2(CO)_{10}$ substitution of CO from Mn would be predicted. However, the isolated product from the reaction

$$MnRe(CO)_{10} + L \rightarrow MnRe(CO)_9L + CO \quad (L = PR_3\ [128],\ RNC\ [97]) \qquad (68)$$

contains the incoming ligand attached to the Re atom. Two possible mechanisms can account for this observation

(i) initial loss of CO from Mn, transfer of a CO group from Re to Mn, addition of L to Re.

(ii) direct replacement of CO by L on Re.

For the PdO catalyzed reaction (68) (L = tBuNC), labelling studies have resolved this problem. Reaction of $Mn(CO)_5Re(^{13}CO)_5$ with tBuNC gave $Mn(CO)_5Re(^{13}CO)_4L$ (direct replacement of CO on Re) and not $Mn(CO)_4(^{13}CO)Re(^{13}CO)_4L$ as determined by mass spectroscopy [98]. Presumably the reaction direction is determined by cis-labilization effects [129]. Unfortunately, the direct thermal reaction between $MnRe(CO)_{10}$ and L cannot differentiate between the two mechanisms as $^{12}CO/^{13}CO$ scrambling on the dimer occurs at a lower temperature than required for the substitution reaction [130].

Attempts to modify the site of reactant attack in mixed metal dimers by the use of electrocatalysts have been attempted on a number of systems (e.g. $CpFe(CO)_2M(CO)_5$, M = Mn, Re). However, to date the substituted metal dimers which have been synthesized from the parent dimer via either catalytic or non-catalytic routes have always contained the substituent on the same metal atom [100–103].

5.5.2 Bridged (Capped) Metal Dimers

A series of studies have been performed on capped metal dimers of cobalt in which an acetylene moiety spans the metal-metal bond (Refs. [26, 117–124]). An early electrochemical investigation of $Co_2(CO)_6(\mu\text{-}RCCR)$

$$Co_2(CO)_6(\mu\text{-}RCCR) + L \rightarrow Co_2(CO)_5L(\mu\text{-}RCCR) + CO \qquad (69)$$

indicated the potential of the electrocatalyzed substitution of the cobalt dimer reactions [26]. This has since been substantiated by studies on a wide range of $Co_2(CO)_6(\mu\text{-}RCCR)$ complexes. In particular, when R = CF_3, a radical chain process becomes possible as the presence of the electron withdrawing group results in radicals with extended lifetimes. Comparative studies with the $Ph_2CO^{-\cdot}$ radical catalyzed reaction have been made. (A discussion on the use of the $Ph_2CO^{\cdot-}$ radical as catalyst is found in Section 5.6.1) and excellent yields were observed using both procedures ((69); L = PR_3, CH_3CN etc.) The products are found with L in the axial position on the dimer except when L = CH_3CN; presumably steric effects determine the stereochemistry of the product. Studies have also been performed on a Co complex containing a 'flyover' alkyne ligand, $[(CF_3)C_6]$. In these reactions stereochemical changes as well as the substitution reaction are catalysed by electrochemical procedures [121, 122]. Substitution reactions of $Co_2(CO)_6(\mu\text{-}CF_3CCCF_3)$ by chelating di and tri phosphines and arsines have extablished that different products can be observed as compared to the direct thermal reaction. For example, the thermal reaction between $Co_2(CO)_6(\mu\text{-}CF_3CCCF_3)$ and dppm yields $Co_2(CO)_4(\mu\text{-dppm})(\mu\text{-}CF_3CCCF_3)$ in excellent yield (>80%) whereas the $Ph_2CO^{\cdot-}$ or electrochemically catalyzed reaction gives $Co_2(CO)_7(\mu\text{-dppm})(\mu\text{-}CF_3CCCF_3)$ as the only product (97% yield). Similar results are found using dppe and ttas in place of dppm. A variation of the above involves the use of metal anions as ligands in the substitution reaction and $Ph_2CO^{\cdot-}$ has been found to initiate reaction (70) below [124]

$$Co_2(CO)_6(\mu\text{-}CF_3CCCF_3) + CpMo(CO)_3 \rightarrow Co(CO)_3CpMo(CO)_2(\mu\text{-}CF_3CCCF_3) \quad (70)$$

A related reaction in which $Fe_2(CO)_8{}^{2-}$ reacts with $Co_2(CO)_8$ to yield $FeCo(CO)_8{}^-$ via an electron transfer chain reaction has also been reported [125].

$$Fe_2(CO)_8{}^{2-} + Co_2(CO)_8 \rightarrow Fe_2(CO)_8{}^{\cdot-} + Co_2(CO)_8{}^{-\cdot} \quad (71)$$

$$Co_2(CO)_8{}^{-\cdot} \rightarrow Co(CO)_4{}^- + Co(CO)_4{}^{\cdot} \quad (72)$$

$$Fe_2(CO)_8{}^{2-} + Co(CO)_4{}^{\cdot} \rightarrow Fe_2Co(CO)_{12}{}^{2-\cdot} \quad (73)$$

$$Fe_2Co(CO)_{12}{}^{2-\cdot} \rightarrow FeCo(CO)_8{}^{2-\cdot} + Fe(CO)_4 \quad (74)$$

$$FeCo(CO)_8{}^{2-\cdot} + Co_2(CO)_8 \rightarrow FeCo(CO)_8{}^- + Co_2(CO)_8{}^{-\cdot} \quad \text{etc.} \quad (75)$$

Finally, it is to be noted that $[CpFe(CO)_2]_2$ has also been reported to catalyze the substitution reactions of the $Co_2(CO)_6(\mu\text{-RCCR})$ species – indeed, it appears to be a superior catalyst to the $Ph_2CO^{-\cdot}$ and electrochemically catalyzed reaction described above when PhCCPh is the bridging alkyne [123].

A series of studies have been performed on the catalyzed CO substitution reaction of some capped iron dimer complexes in which two metal atoms are in different chemical environments [106, 107, 117] e.g. 1 and 2 below

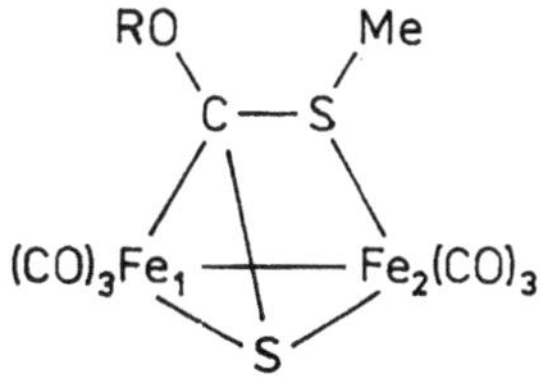

1

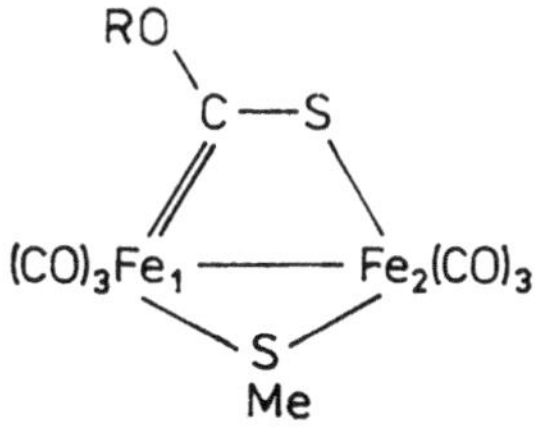

2

Electrochemical procedures have been used to study the substitution reactions of both *1* and *2* by L (e.g. L = $P(OMe)_3$). In complex *1*, the first CO replaced is on Fe2 and in *2*, it is on Fe1. The reaction has been studied in some detail and conversion of derivatives of *1* to *2* has also been catalyzed electrochemically [115]. An intermediate *3* consistent with labelling studies has been proposed to explain

the data

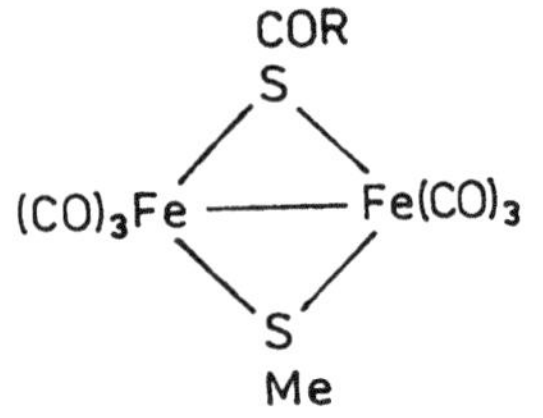

3

A carbene bridged metal dimer $Cp_2Fe_2(CO)_2(\mu\text{-}CO)(\mu\text{-}CSMe)^+$ has also been found to undergo catalytic CO replacement via electrochemical procedures. The intermediate radical $Cp_2Fe_2(\mu\text{-}CO)(\mu\text{-}CSMe)\cdot$ has been found to readily undergo CO replacement reactions with ^{13}CO and the catalytic cycle can be completed with less than 1 equivalent of electrons [126].

5.6 Substitution Reactions: Metal Trimers

This section will include a discussion of open (*4*) and closed (*5*) as well as bridged (capped) (*6*) and non-bridged (non-capped) (*4*, *5*) metal trimers, examples of which are given below

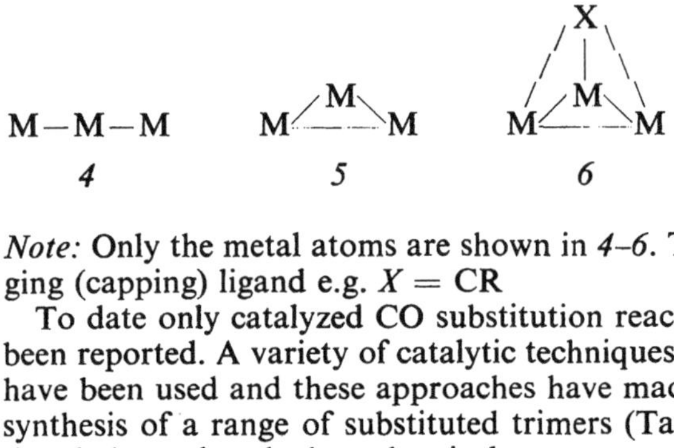

4 *5* *6*

Note: Only the metal atoms are shown in *4–6*. The X group in *6* represents a bridging (capping) ligand e.g. $X = CR$

To date only catalyzed CO substitution reactions of the above complexes have been reported. A variety of catalytic techniques (electrochemical, reagent induced) have been used and these approaches have made possible the stepwise, high yield synthesis of a range of substituted trimers (Table 7, Refs. [131–162]). The traditional thermal and photochemical routes generally lead to reactions dominated by low yielding multiproduct mixtures, which also contain products derived from cluster fragmentation.

5.6.1 Reagent Catalyzed Reactions

One of the most important reagents that has been exmployed in this area of chemistry is the in situ generated Na/Benzophenone (BPK) radical reagent, $Ph_2CO^{-\cdot}$. This reagent has been particularly useful in the substitution chemistry of ruthenium clusters, especially $Ru_3(CO)_{12}$ [147].

Addition of catalytic amounts of the blue BPK solution, which contains the $Ph_2CO^{-\cdot}$ radical to $Ru_3(CO)_{12}$ and ligand, L, at room temperature, rapidly leads to the synthesis of a wide range of $Ru_3(CO)_{12-x}L_x$ (x = 1–3) derivatives. Much of the synthetic work has been summarized in a recent review by Bruce (also Table 7) [137]. Key issues and some more recent reports on the use of the BPK reagent will be discussed here.

Table 7. Catalyzed CO substitution reactions of metal trimer complexes[a,b]

Reactant/ligand	Catalyst/comment[c]	Ref.
$Fe_3(CO)_{12}/PR_3$, RNC	$[CpFe(CO)_2]_2$, $[Fe(CO)_3(\mu\text{-}SMe)]_2$, BPK	[28, 121, 123]
$Fe_3(CO)_8(PhCCPh)/P(OMe)_3$	$[CpFe(CO)_2]_2$	[123]
$Fe_3(CO)_{11}^{2-}/Mn(CO)_5CH_3CN$	prep. of $Fe_2Mn(CO)_{12}$	[125]
$Fe_3(CO)_9(\mu\text{-}R)_2/L$	L, R variable, electro.	[150–153]
$Fe_3(CO)_8L(\mu\text{-}S)_2$[d]/L′	electro.	[155]
$Fe_3(CO)_9(N{-}N)$[e]$/PR_3$	electro.	[154]
$Fe_3(CO)_9(\mu\text{-}CMe)(\mu\text{-}CNEt)/PR_3$	BPK	[156]
$Fe_2Ru(CO)_{12}/PR_3$	BPK, subst. on Ru	[146]
$FeRu_2(CO)_{12}/PR_3$	BPK, subst. on Ru	[146]
$FeCo_2(CO)_9(EtCCEt)/PR_3$	$[Fe(CO)_3(\mu\text{-}SMe)]_2$, subst. on Co	[161]
$Ru_3(CO)_{12}/PR_3$, RNC, ROH	BPK, Pd/C, $[CpFe(CO)_2]_2$,[g]	[28, 29, 33, 39, 121, 123, 137, 157–160]
$Ru_3(CO)_9(\mu\text{-}NPh)_2/RNC$, PPh_3	BPK	[149]
$Ru_3(CO)_9(\mu\text{-}H)_x(\mu\text{-}L)$[f]$/PR_3$	BPK	[137]
$Os_3(CO)_{12}/PR_3$, RNC	BPK, Pd/C, $Pt(PPh_3)_4$, $[CpFe(CO)_2]_2$,[g]	[29, 33, 35b, 48, 121, 137]
$Os_3(CO)_{10}(CH_3CN)_2$	$PdCl_2$-Os hexamers formed	[162]
μ_3-$CRCo_3(CO)_9$[h]/L	BPK, electro.	[26, 120, 121, 131–137]

[a] Specific details of the reaction conditions, range of catalysts and types of ligand used are listed in the references.
[b] In some of the examples listed an electron induced rather than an electrocatalyzed reaction has been reported.
[c] electro. = electrochemical catalysis, photo = photocatalysis.
[d] L = $S_2C_3H_2$ (carbene)
[e] N–N = diazanobornene
[f] $x = 1$, L = CCtBu; $x = 2$, L = S
[g] Also catalyzed by $[Fe(CO)_3(\mu\text{-}SMe)]_2$
[h] R variable, L = metal anion, PR'_3 etc.

A recent investigation of the electrochemical behaviour of $Ru_3(CO)_{12}$ and related species has shed further light on the mechanism of the BPK initiated substitution reaction [138]. It has been shown that addition of an electron to $Ru_3(CO)_{12}$ yields the shortlived $Ru_3(CO)_{12}^{-\cdot}$ radical ($t_{1/2} \sim 10^{-6}$ s) [138, 139], which rapidly isomerises to an open structure $Ru_3(CO)_{12}^{-\cdot *}$. This latter species also has a short lifetime ($\sim 10^{-6}$ s) and the open structure can either dimerize or undergo a redox reaction

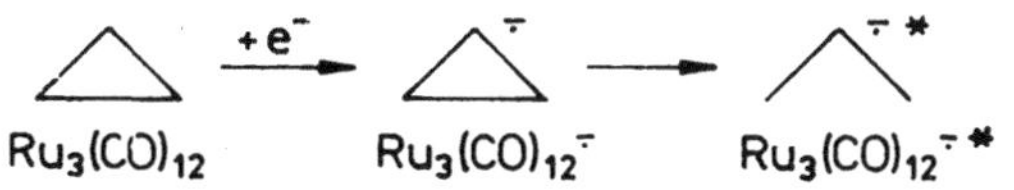

It appears that reaction of ligand L with $Ru_3(CO)_{12}^{-\cdot}$ (or $Ru_3(CO)_{12}^{-\cdot *}$) generated using BPK in THF can compete with the above processes to give the CO

substituted product, $Ru_3(CO)_{11}L$. Intermediate species such as $[\eta'\text{-}C(O)XRu_3(CO)_{11}]$ (formed by attack of X e.g. H_2O or OH^- on a bonded CO ligand) could also lead to fast and efficient nucleophilic CO substitution processes. This suggests that a common mechanism for both the PPN^+ catalyzed reaction and the BPK catalyzed reaction of $Ru_3(CO)_{12}$ may exist [140–143].

Some significant results that have been obtained from the BPK initiated reaction include

(i) The reaction occurs with stepwise loss of CO; mixed ligand derivatives are thus possible on multi CO substitution of a trimer e.g. $Ru_3(CO)_9(PMe_3)[P(OMe)_3](PPh_3)$ (overall 41% yield) has been synthesized with each ligand being attached to a different Ru atom [137]. The reaction rate decreases with the degree of substitution.

(ii) The reaction occurs in a regiospecific manner on mixed metal clusters e.g.

$$Fe_2Ru(CO)_{12} + P(OMe)_3 \rightarrow Fe_2Ru(CO)_{11}P(OMe)_3 + CO \qquad (76)$$

with exclusive addition of $P(OMe)_3$ to the Ru atom [146].

(iii) $Ru_3(CO)_{12}$ has been attached to polymer supports via the BPK method [144, 145]

$$Ru_3(CO)_{12} + PPh_2CH_2CH_2Si(OEt)_3 \rightarrow Ru_3(CO)_{11}PPh_2CH_2CH_2Si(OEt)_3 \rightarrow$$

polymerisation of the Si chain. (77)

(iv) The accessibility of a wide range of $Ru_3(CO)_{12-x}L_x$ (x = 1–4) derivatives has permitted an extensive study of the chemical and physical properties of the trimer system e.g. extensive X-ray crystallographic studies have recently been reported and analyzed on the above complexes [157–159].

(v) Substitution reactions have not been limited to $Ru_3(CO)_{12}$ (see Table 7).

(vi) Substitution by a metal anion can lead to metal fragment exchange [125, 132–134] via a mechanism similar to that depicted in Equations (71)–(75). i.e. a catalyzed chain reaction is proposed. The reaction is favored by the presence of a capping group and the ability of $Co(CO)_4$ to act as a good leaving group.

Alternative reagents to BPK have been used to induce substitution reactions on metal trimers. Thus, reaction of $M_3(CO)_{12}$ with L (L = PR_3, $P(OR)_3$, RNC) is catalyzed by $[CpFe(CO)_2]_2$ [39, 123]. A comparative study on the thermal, BPK, and $[CpFe(CO)_2]_2$ catalyzed routes suggests that the $[CpFe(CO)_2]_2$ route is superior for those reactions reported. An electrocatalyzed pathway involving iron carbonyl anions as chain initiators has been proposed from studies on the light induced substitution of CO on $M_3(CO)_{12}$ (M = Ru, Os) by $PMePh_2$ in the presence of $[CpFe(CO)_2]_2$ as catalyst [33]. It is to be noted that the reaction between $M_3(CO)_{12}$ and RNC is catalyzed by Pd/C [29] and Cp_2Co has been reported to catalyze the reaction between $Fe_3(CO)_9(\mu_3\text{-}PR)$ and group 15 donor ligands [150–153].

The dimers $[Fe(CO)_2L(\mu\text{-}SMe)]_2$ (L = CO, PPh_3) also catalyze the reaction between $M_3(CO)_{12}$ (M = Fe, Ru) [28] or $FeCo_2(CO)_9(EtCCEt)$ [161] and various nucleophiles. It was suggested that the superior ability of the substituted catalyst (L = PPh_3) results from steric factors; the PPh_3 group makes cleavage of the Fe—Fe bond in the dimer facile leading to more ready availability of the capped dimer biradical, the active catalyst [28].

5.6.2 *Electrochemical Catalysis* [148]

The first report of an electrochemically induced catalytic reaction was made in 1981 [26]. In this study $XCCo_3(CO)_9$ (X = Ph, Cl) underwent a one electron transfer process to yield $XCCo_3(CO)_8PPh_3$ in the presence of PPh_3. The cobalt trimers have continued to attract attention and reactions with a range of capping groups (CR), and ligands have since been reported (Table 7).

A wide range of capped open metal trimers have been studied, mainly by Kochi and co-workers. In particular, mechanistic studies on a series of trimers $Fe_3(CO)_9(\mu_3\text{-PPh})_2$ with a range of different ligands, and $Fe_3(CO)_9(\mu_3\text{-X})$ (X = S, Se, NR) with $P(OMe)_3$ have been reported [150–153]. The use of ESR spectroscopy has revealed that for the μ_3-Ph capped clusters substitution requires a pathway involving Fe—P bond cleavage to generate a 17 electron metal centre. Reactions occur with stepwise replacement of up to three CO ligands by adjustment of the electrode potential of the electrochemical cell.

The ability to catalytically induce selective, stepwise ligand replacement on metal trimer complexes has made the substitution chemistry of this class of complexes tractable. It can be anticipated that future studies will attempt to further extend and delineate the scope of the reactions.

5.7 Substitution Reactions: Large Metal Clusters

The limited availability of cheap sources of the large carbonyl clusters has limited studies on their general chemistry and to date very few studies of catalyzed substitution reactions of clusters containing four or more metal atoms have been reported (Table 8). Similar procedures as outlined in Sections 5.1–5.6 have been used (chemical reagents such as BPK, $[CpFe(CO)_2]_2$; electrochemical techniques) to induce substitution reactivity. In the larger cluster systems, the 16–18 electron rule

Table 8. Catalyzed CO substitution reactions of large metal clusters[a,b,c]

Reactant	Catalyst	Ref.
$M_3Co(CO)_{13}^-$ (M = Fe, Ru)	radicals	[125]
$CpFeRh(CO)_9(\mu\text{-PPh}_2)$	non-catalytic	[168]
$RuH_4(CO)_{12}$[d]	BPK, $[CpFe(CO)_2]_2$	[123, 137, 147]
$RuM_3(\mu\text{-H})(CO)_{12}$ (M = Rh, Co)	BPK	[171]
$Co_4(CO)_{10}(\mu\text{-PPh})_2$	BPK, electro.	[164–166]
$Co_4(CO)_4$(tripod)	BPK, electro.	[169]
$Cp_2Co_4(CO)_4(\mu\text{-CO})(\mu\text{-CF}_3CCCF_3)$	non-catalytic	[170]
$Rh_6(CO)_{16}$	BPK	[137]
$Ir_4(CO)_{12}$	BPK	[121]

[a] Specific details of the reaction conditions and types of ligands used are to be found in the references listed.
[b] In some of the examples listed an electron induced rather than an electrocatalyzed reaction was observed.
[c] CO replaced by a group 15 donor ligand.
[d] CO also replaced by RNC.

fails, but alternative rules have been developed to ascertain the relationship between electron count and cluster stability [163]. It can be anticipated that the addition or removal of an electron to or from a metal cluster will modify the reactivity of the cluster. Simultaneous rearrangement of the cluster geometry by a change in electron count of the cluster could also provide alternative pathways to products different from the direct thermal substitution reaction.

A mechanistic study of the BPK catalyzed reaction between $Co_4(CO)_{10}(\mu_4\text{-}PPh_2)_2$ and ligands L has been carried out [164–166]. Studies have revealed that the CO substitution reaction occurs via a dissociative mechanism with multiple CO displacement occurring at different Co atoms. Differences in reactivity between the catalyzed and uncatalyzed reactions were observed.

Other reports on the substitution of Co, Rh, Rh and mixed metal clusters have also been reported (Table 8). A recent report [167] has described the electron induced substitution behaviour of clusters with up to 10 metal atoms. Although the reactions are non-catalytic, the increased reactivity of the cluster on addition of an electron suggests that catalytic procedures for this class of complexes would be effective.

6. Other Catalyzed Reactions

As mentioned in the introductory section, addition or removal of electrons to or from a reactant molecule can result in formation of radical species, which are more susceptible to chemical reactions than their diamagnetic starting materials. Reviews by Chanon [172] have comprehensively assessed the catalytic features of a range of reaction types using this concept and in some instances this has led to a reinterpretation of earlier mechanistic work involving the redox processes of some reactions.

6.1 Catalyzed Insertion Reactions

The insertion reaction is a variant of the classical substitution reaction in which the leaving group is incorporated into the reactant molecule

$$L_xM(CO)R \rightarrow L_xM(CO)\cdot \xrightarrow{L'} L_xM(COR)L' \quad (78)$$

The classic work by Giering and co-workers [173] using electrochemical and reactant catalyzed procedures on the reaction

$$MeCpFe(CO)(PPh_3)CH_3 + CO \rightarrow MeCpFe(CO)(PPh_3)COCH_3 \quad (79)$$

resulted in reactions with rate increases of 10^6 relative to the noncatalyzed insertion reactions. The utility of this approach has been exploited in synthetic organic chemistry using chemical oxidants e.g. Cp_2Fe^+, Ag^+, Ce(IV) as catalysts [174]

$$CpFe(CO)[P(OPh)_3](^1\eta\text{-alkenyl}) + CO \rightarrow CpFe(CO)[P(OPh)_3](\text{alkenylacyl}) \quad (80)$$

In the above catalyzed reactions partial or complete isomerization of the alkenyl double bond can occur. Importantly isomerization of the alkenyl group is catalyzed in the absence of CO [174a].

6.2 Catalyzed Isomerization Reactions

Addition or elimination of an electron to or from a reactant can create radical species, which will undergo rearrangement to an alternative isomer. After the isomerization process, the further loss/gain of the electron will produce a diamagnetic species, which could be an isomer of the original reactant. The process can be made catalytic. This reaction type can be regarded as a variant of the substitution reaction in which the overall effect is the formation of a molecule in which a ligand is replaced by another ligand already attached to the central metal atom.

The early work in this area of chemistry was first reported by Rieke, Kojima, and Olefe [175], in which group 6 metal carbonyl dicarbene complexes underwent cis/trans isomerisation. This work has since been extended to other carbene ligands [176]. Mention has also been made in the text of other redox catalyzed isomerization reactions which were discovered during investigations of catalyzed substitution reactions. A review by Gieger [177] can be consulted for further details of these catalyzed reactions.

6.3 Odds and Ends

A number of reports that have bearing on the subject material are to be noted:

(i) An electron transfer catalyzed pathway has been suggested for the conversion of $Fe_3(CO)_{12}$ to $Fe(CO)_5$. In this reaction, CO generated from cluster decomposition adds to an open trimer radical $Fe_3(CO)_{12}^{-\cdot}$, resulting ultimately in $Fe(CO)_5$ formation [179].

(ii) Reaction of $CpW(CO)_2CR$ in the presence of catalytic amounts of $[CpCr(CO)_2]_2$ affords the alkyne bridged complex $[CpW(CO)_2]_2(\mu\text{-}RCCR)$. The mechanism is proposed to occur via the intervention of a $CpCr(CO)_2\cdot$ radical [180].

(iii) The reaction between $Fe_3(CO)_{12}$ and benzalazine is catalyzed by $M(CO)_6$ (M = Cr, Mo, W) to yield bimetallic ligand bridged species [181].

(iv) The reaction between $Ru_3(CO)_{12}$ and nitrobenzene to yield $Ru_3(CO)_7(\mu\text{-}NPh)(C_6H_6)$ requires the presence of $Co_2(CO)_8$.

Further investigations may reveal the presence of free radical species and/or redox processes in reactions (ii) to (iv).

7. Conclusion

The use of redox catalysts in the substitution chemistry of organometallic complexes and the role of 17–19 electron species in the catalyzed reaction have now been well-documented. It is apparent that catalysis can occur by addition or removal of electrons to reactant molecules and a variety of mechanisms all involving chain processes can be envisaged to rationalize the data. The many examples listing the types of catalysts, substrates, and substituting ligands have been summarized in the Tables listed in this chapter. The procedure of using catalysts is simple and the advantages of clean high yielding reactions performed under mild conditions has now made available many new complexes not previously accessible by conventional synthetic procedures. The future of the technique will lie in the ability to tailor the redox catalysts to bring about highly stereospecific substitution reactions in which discrimination of ligands of the same ($A = B$) or different ($A \neq B$) type attached to the same metal centre can be achieved.

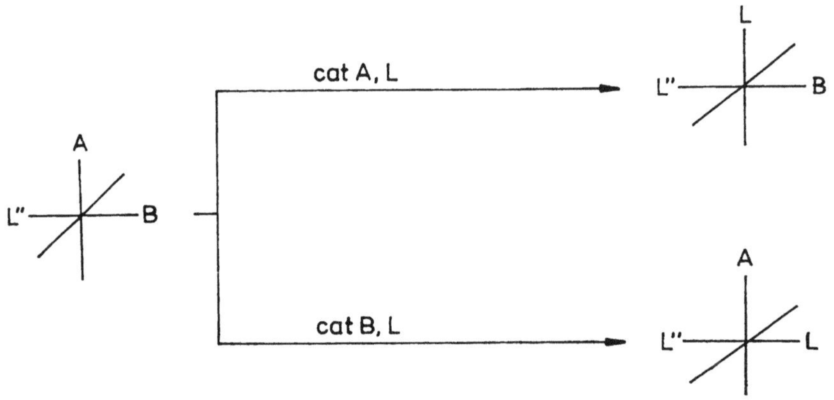

The ability to modify catalysts by chemical changes has already been demonstrated and further refinement can also be expected to lead to superior catalytic properties in the future.

8. Acknowledgment

I wish to thank my many students and collaborators who have contributed to some of the work described in this review. I also wish to thank Prof. Jack Faller, Yale University, for his hospitality during the writing of this manuscript. Finally, my thanks to the FRD, Pretoria, and the University of the Witwatersrand for their financial assistance without which this work would not have been possible.

9. References

1. J. P. Collman, L. S. Hegedus, J. R. Norton and R. G. Finke, Principles and Applications of Organotransition Metal Chemistry, University Science Books, Mill Valley, California, 1987.
2. H. Taube, Electron Transfer Reactions of Complex Ions in Solution, Academic Press, London, 1970; H. Taube, Angew. Chem., Int. Ed., Engl. **23** (1984) 329.
3. C. H. Langford and H. B. Gray, Ligand Substitution Processes, W. A. Benjamin, New York, 1965.
4. R. W. Alder, J. Chem. Soc., Chem. Comm. (1980) 1184.
5. C. A. Tolman, Chem. Soc. Rev. **1** (1972) 337.
6. (a) A. E. Steigman and D. R. Tyler, Comments Inorg. Chem. **5** (1986) 215.
   (b) See however: R. M. Kowaleski, F. Basolo, W. C. Trogler, R. W. Gedridge, T. D. Newbound and R. D. Ernst, J. Am. Chem. Soc. **109** (1987) 4860.
7. (a) D. Astruc, Comments Inorg. Chem. **6** (1987) 61.
   (b) D. Astruc, Angew. Chem., Int. Ed., Engl. **27** (1988) 643 and references cited therein.
8. T. L. Brown, Ann. N. Y. Acad. Sci. **331** (1980) 80.
9. N. J. Coville, M. O. Albers and E. Singleton, J. Chem. Soc., Dalton Trans. (1983) 947.
10. B. A. Narayanan, C. Amatore and J. K. Kochi, Organometallics **5** (1986) 926.
11. W. Kaim, Coord. Chem. Rev. **76** (1987) 187.
12. B. P. Sullivan, D. Conrad, and T. J. Meyer, Inorg. Chem. **24** (1985) 3640.
13. P. Lemoine, Coord. Chem. Rev. **83** (1988) 169.
14. T-Y. Luh, Coord. Chem. Rev. **60** (1984) 255.

15. M. O. Albers and N. J. Coville, Coord. Chem. Rev. **53** (1984) 227.
16. H. Alper, Adv. Organomet. Chem. **19** (1981) 183.
17. F. A. Cotton and G. Wilkinson, Advanced Inorganic Chemistry, 4th Ed., Wiley, New York, 1980.
18. J. D. Atwood, Inorganic and Organometallic Reaction Mechanisms, Brooks/Cole, Monterey. California, 1985.
19. R. E. Dessy, F. E. Stary, R. B. King and M. Waldrop, J. Am. Chem. Soc. **88** (1966) 471
20. C. J. Pickett and D. Pletcher, J. Chem. Soc., Dalton Trans. (1975) 879; (1976) 749.
21. A. Seurat, P. Lemoine and M. Gross, Electrochim. Acta **23** (1978) 1219.
22. A. M. Bond, P. A. Dawson, B. M. Peake, B. H. Robinson and J. Simpson, Inorg. Chem. **16** (1977) 2199.
23. R. N. Bagchi, A. M. Bond, R. Colton, D. L. Luscombe and J. E. Moir, J. Am. Chem. Soc. **108** (1986) 3352 and references cited therein.
24. See for example Y. Mugnier, P. Reeb, C. Moise and E. Lavrion, J. Organomet. Chem. **254** (1983) 111.
25. (a) S. B. Butts and D. F. Shriver, J. Organomet. Chem. **169** (1979) 191. (b) See however Ref. [40] for reports of early reactions with labelled CO.
26. G. J. Bezems, P. H. Rieger and S. Visco, J. Chem. Soc., Chem. Commun. (1981) 265.
27. (a) M. O. Albers, N. J. Coville, T. V. Ashworth, E. Singleton and H. E. Swanepoel, J. Chem. Soc., Chem. Commun. (1980) 489.
(b) M. O. Albers, N. J. Coville and E. Singleton, J. Chem. Soc., Dalton Trans., (1982) 1069.
(c) M. O. Albers, Ph. D. thesis, 1982.
28. S. Aime, M. Botta, T. Gobetto, and D. Osella, Organometallics **4** (1985) 1475.
29. M. O. Albers, E. Singleton and N. J. Coville, J. Organomet. Chem. **323** (1987) 37.
30. (a) M. O. Albers, N. J. Coville, C. P. Nicolaides, R. A. Webber, T. V. Ashworth and E. Singleton, J. Organomet. Chem. **217** (1981) 247.
(b) M. O. Albers, N. J. Coville, P. ten Doeschate and E. Singleton, S. Afr. J. Chem. **34** (1981) 81.
31. J. A. S. Howell, J. C. Kola, D. T. Dixon, P. M. Burkinshaw and M. J. Thomas, J. Organomet. Chem. **266** (1984) 83.
32. M. O. Albers, N. J. Coville, T. V. Ashworth and E. Singleton, J. Organomet. Chem. **217** (1981) 385.
33. (a) A. S. Goldman and D. R. Tyler, Inorg. Chim. Acta **98** (1985) L47.
(b) A. S. Goldman and D. R. Tyler, Inorg. Chem. **26** (1987) 253.
34. M. O. Albers, N. J. Coville and E. Singleton, J. Organomet. Chem. **232** (1982) 261.
35. (a) N. J. Coville, R. Ghirlando and T. A. Ford, S. Afr. J. Chem. **39** (1986) 224.
(b) M. O. Albers, N. J. Coville and E. Singleton, J. Chem. Soc., Chem. Comm. (1982) 96.
36. M. O. Albers, N. J. Coville, T. V. Ashworth, E. Singleton and H. E. Swanepoel, J. Organomet. Chem. **199** (1980) 55.
37. N. J. Coville and M. O. Albers, Inorg. Chim. Acta **65** (1982) L7.
38. M. O. Albers, E. Singleton and N. J. Coville, J. Chem. Ed. **63** (1986) 444.
39. M. O. Albers, E. Singleton and N. J. Coville, J. Organomet. Chem. **326** (1987) 229.
40. (a) K. Noack and M. Ruch, J. Organomet. Chem. **17** (1969) 309.
(b) A. N. Webb and J. J. Mitchell, J. Phys. Chem. **63** (1959) 1878.
(c) F. Basolo, A. T. Brault and A. J. Poë, J. Chem. Soc. (1964) 676.
41. (a) C. Amatore, A. Verpaus, A. Madonix, M.-H. Desbois and D. Astruc, J. Chem. Soc., Chem. Commun. (1988) 200.
(b) D. Catheline, E. Román E and D. Astruc, Inorg. Chem. **23** (1984) 4508.
42. (a) N. J. Coville, E. A. Darling, A. W. Hearn and P. Johnston, J. Organomet. Chem. **328** (1987) 375.
(b) P. Johnston, M. S. Loonat, W. L. Ingham, L. Carlton, and N. J. Coville, Organometallics **6** (1987) 2121.
43. (a) B. H. Byers and T. L. Brown, J. Am. Chem. Soc. **97** (1975) 947.
(b) B. H. Byers and T. L. Brown, J. Am. Chem. Soc. **99** (1977) 2527.
44. N. W. Hoffman and T. L. Brown, Inorg. Chem. **17** (1978) 613.

45. B. H. Byers and T. L. Brown, J. Organomet. Chem. **127** (1977) 181.
46. D. G. Alway and K. W. Barnett, Inorg. Chem. **19** (1980) 1533.
47. (a) B. D. Fabian and J. A. Labinger, J. Am. Chem. Soc. **101** (1979) 2239.
(b) M. Rosenblum and P. S. Waterman, J. Organomet. Chem. **187** (1980) 267; **206** (1981) 197.
48. S.-Y. Jang and R. J. Angelici, Inorg. Chem. **27** (1988) 3233.
49. N. J. Coville, J. Organomet. Chem. **190** (1980) C84, **218** (1981) 337.
50. (a) N. J. Coville, G. W. Harris and D. Rehder, J. Organomet. Chem. **293** (1985) 365.
(b) W. D. Jones, J. M. Huggins and R. G. Bergman, J. Am. Chem. Soc. **103** (1981) 4415.
51. (a) D. P. Summers, J. C. Luong and M. S. Wrighton, J. Am. Chem. Soc. **103** (1981) 5238.
(b) H. Kunkely, A. Merz and A. Vogler, J. Am. Chem. Soc. **105** (1983) 7241.
52. R. S. Wright, R. Bhavan and N. J. Coville, S. Afr. J. Chem. **39** (1986) 184.
53. P. Johnston, unpublished work.
54. G. W. Harris, J. C. A. Boeyens and N. J. Coville, J. Organomet. Chem. **255** (1983) 87.
55. G. W. Harris, M. O. Albers, J. C. A. Boeyens and N. J. Coville, Organometallics **2** (1983) 609.
56. M. O. Albers, N. J. Coville and E. Singleton, J. Organomet. Chem. **234** (1982) C13.
57. C. A. Dickson, A. McFarlane and N. J. Coville, Inorg. Chim. Acta. **158** (1989) 205.
58. (a) N. J. Coville and E. A. Darling, J. Organomet. Chem. **277** (1984) 105.
(b) M. S. Loonat, L. Carlton and N. J. Coville, S. Afr. J. Chem. **41** (1988) 72. .
59. J. C. A. Boeyens, N. J. Coville, and K. Soldenhoff, S. Afr. J. Chem. **37** (1984) 153.
60. (a) W. J. Carter, J. W. Kelland, S. J. Okrasinski, K. E. Warner and J. R. Norton, Inorg. Chem. **21** (1982) 3955.
(b) R. T. Edidin and J. R. Norton, J. Am. Chem. Soc. **108** (1986) 948.
61. K. Tabatabaian and C. White, Inorg. Chem. **20** (1981) 2020.
62. M. Absi-Halabi and T. L. Brown, J. Am. Chem. Soc. **99** (1977) 2982.
63. B. Olbrich-Deussner and W. Kaim, J. Organomet. Chem. **340** (1988) 71.
64. See Ref. [15]. Also J. Gröbe and H. Zimmermann, Z. Naturforsch. **39B** (1984) 962 and references cited therein.
65. M. G. Peterleitner, M. V. Tolstaya, V. V. Krirykh, L. I. Denisovitch, and M. I. Rybinskaya, J. Organomet. Chem. **254** (1983) 313.
66. R. N. Bagchi, A. M. Bond, G. Brain, R. Colton, T. L. E. Henderson, and J. E. Kevekordes, Organometallics **3** (1984) 4.
67. H. tom Dieck and E. Kuhl, Z. Naturforsch. **37B** (1982) 324.
68. W. Kaim, Inorg. Chem. **23** (1984) 3365.
69. C. G. Zoski, D. A. Sweigart, N. J. Stone, P. H. Rieger, E. Mocellin, T. F. Mann, D. R. Mann, D. K. Gosser, M. M. Doeff and A. M. Bond, J. Am. Chem. Soc. **110** (1988) 2109.
70. See also K. M. Doxsee, R. H. Grubbs and F. C. Anson, J. Am. Chem. Soc. **106** (1984) 7819, for an earlier report on related areneCr$(CO)_3$ complexes.
71. D. Miholova and A. A. Vlček, J. Organomet. Chem. **279** (1985) 317.
72. J. W. Hershberger, C. Amatore and J. K. Kochi, J. Organomet. Chem. **250** (1983) 345.
73. (a) J. W. Hershberger and J. K. Kochi, J. Chem. Soc., Chem. Commun. (1982) 212.
(b) J. K. Kochi, J. Organomet. Chem. **300** (1986) 139.
74. P. M. Zizelman, C. Amatore and J. K. Kochi, J. Am. Chem. Soc. **106** (1984) 3771.
75. J. W. Hershberger, R. J. Klinger and J. K. Kochi, J. Am. Chem. Soc. **105** (1983) 61.
76. J. W. Hershberger and J. K. Kochi, Polyhedron **2** (1983) 929.
77. N. G. Connelly and M. O. Kitchen, J. Chem. Soc., Dalton Trans. (1977) 931.
78. B. A. Narayanan, C. Amatore and J. K. Kochi, J. Chem. Soc., Chem. Commun. (1983) 397.
79. B. A. Narayanan, C. Amatore and J. K. Kochi, Organometallics **6** (1987) 129.
80. D. J. Kuchynka, C. Amatore and J. K. Kochi, Inorg. Chem. **25** (1985) 4087.
81. J. W. Hershberger, R. J. Klinger and J. K. Kochi, J. Am. Chem. Soc. **1982** (104) 3034.
82. C. Amatore, D. J. Kuchynka and J. K. Kochi, J. Electroanal. Chem. **241** (1988) 181.
83. A. J. Poë, Pure Appl. Chem. **60** (1988) 1209 and references cited therein.
84. Md. M. Rahman, H. Y. Liu, A. Prock and W. P. Giering, Organometallics **6** (1987) 650, and references cited therein.

85. (a) A. Darchen, J. Chem. Soc., Chem. Commun. (1983) 768.
(b) A. Darchen, J. Organomet. Chem. **302** (1986) 389.
86. C. Moinet, E. Roman and D. Astruc, J. Electroanal. Chem., Interfacial Electrochem. **241** (1981) 121.
87. N. J. Hazel, J. A. K. Howard and J. L. Spencer, J. Chem. Soc., Chem. Commun. (1984) 1663.
88. M. P. Guy, J. L. Coffer, J. S. Rommel and D. W. Bennett, Inorg. Chem. **27** (1988) 2942.
89. Reported in Ref. [79] of Ref. [7b].
90. (a) Ref. 1, pp. 239–240.
(b) T. J. Meyer and C. V. Caspar, Chem. Rev. **85** (1985) 187.
91. M. O. Albers and N. J. Coville, S. Afr. J. Chem. **35** (1982) 139.
92. D. J. Robinson, G. W. Harris, J. C. A. Boeyens and N. J. Coville, J. Chem. Soc., Chem. Commun. (1984) 1307.
93. G. W. Harris, J. C. A. Boeyens and N. J. Coville, J. Chem. Soc., Dalton Trans. (1985) 2277.
94. G. W. Harris and N. J. Coville, Organometallics **4** (1985) 908.
95. N. J. Coville, unpublished results.
96. N. J. Coville, M. O. Albers and E. Singleton, J. Chem. Soc., Dalton Trans (1982) 1389.
97. D. J. Robinson, E. A. Darling, and N. J. Coville, J. Organomet. Chem. **310** (1986) 203.
98. P. Johnston, G. J. Hutchings and N. J. Coville, J. Am. Chem. Soc., **111** (1989) 1902.
99. N. J. Coville, A. M. Stolzenberg and E. L. Muetterties, J. Am. Chem. Soc. **105** (1983) 2499.
100. (a) P. Johnston, G. J. Hutchings and N. J. Coville, Inorg. Chem. Acta. **117** (1986) L11.
(b) P. Johnston, C. A. Dickson, A. J. Markwell, L. Denner, J. C. A. Boeyens and N. J. Coville, Inorg. Chem. Acta **144** (1988) 185.
101. P. Johnston, G. J. Hutchings, L. Denner, J. C. A. Boeyens and N. J. Coville, Organometallics **6** (1987) 1292.
102. C. A. Dickson, unpublished results.
103. W. L. Ingham, unpublished results.
104. A. Darchen, H. Mousser and H. Patin, J. Chem. Soc., Chem. Commun. (1988) 968.
105. M. Guéguen, J. E. Guerchais, F. Y. Pétillon and J. Talarmin, J. Chem. Soc., Chem. Commun. (1987) 557.
106. A. Darchen, E. K. Lhadi and H. Patin, J. Organomet. Chem. **259** (1983) 189.
107. (a) E. K. Lhadi, C. Mahe, H. Patin and A. Darchen, J. Organomet. Chem. **246** (1983) C61.
(b) A. Darchen, E. K. Lahdi and H. Patin, J. Organomet. Chem. **327** (1987) C37.
(c) A. Darchen, E. K. Lhadi, D. Grandjean, A. Mousser and H. Patin, J. Organomet. Chem. **342** (1988) C15.
108. M. Absi-Halabi, J. D. Atwood, N. P. Forbus and T. L. Brown, J. Am. Chem. Soc. **102** (1980) 6248.
109. N. P. Forbus, R. Otieza, S. G. Smith and T. L. Brown, J. Organomet. Chem. **193** (1980) C71.
110. S. B. McCullen and T. L. Brown, Inorg. Chem. **20** (1981) 3528.
111. A. E. Stiegman, A. G. Goldman, D. B. Leslie and D. R. Tyler, J. Chem. Soc., Chem. Commun. (1984) 632.
112. A. E. Stiegman and D. R. Tyler, Inorg. Chem. **23** (1984) 527.
113. A. S. Goldman and D. R. Tyler, J. Am. Chem. Soc. **106** (1984) 4066.
114. A. E. Stiegman, M. Stieglitz, and D. R. Tyler, J. Am. Chem. Soc. **105** (1983) 6032.
115. E. K. Lahdi, H. Patin, and A. Darchen, Organometallics **3** (1984) 1128.
116. A. Lagadec, B. Misterkiewicz, H. Patin, A. Mousser and J.-Y. Le Marouille, J. Organomet. Chem. **315** (1986) 201.
117. R. G. Cunninghame, L. R. Hanton, S. D. Jensen, B. H. Robinson and J. Simpson, Organometallics **6** (1987) 1470.
118. S. D. Jensen, B. H. Robinson and J. Simpson, Organometallics **6** (1987) 1479.
119. C. M. Arewgoda, P. H. Rieger, B. H. Robinson, J. Simpson and S. J. Visco, J. Am. Chem. Soc. **104** (1982) 5633.
120. R. J. Cunninghame, A. J. Downard, L. R. Hanton, S. D. Jensen, B. H. Robinson and J. Simpson, Organometallics **3** (1984) 180.

121. C. M. Arewgoda, B. H. Robinson and J. Simpson, J. Am. Chem. Soc. **105** (1983) 1893.
122. C. M. Arewgoda, B. H. Robinson and J. Simpson, J. Chem. Soc., Chem. Commun. (1982) 284.
123. S. Aime, M. Botta, R. Gobetto, and D. Osella, Inorg. Chim. Acta. **115** (1986) 129.
124. S. D. Jensen, B. H. Robinson and J. Simpson, Organometallics **5** (1986) 1690.
125. C. P. Horwitz, E. M. Holt and D. F. Shriver, Organometallics **4** (1985) 1117.
126. N. C. Schroeder and R. J. Angelici, J. Am. Chem. Soc. **108** (1986) 3688.
127. X. Pan, C. E. Philbin, M. P. Castellani and D. R. Tyler, Inorg. Chem. **27** (1988) 671.
128. D. C. Sonnenberger and J. D. Atwood, J. Am. Chem. Soc. **102** (1980) 3484.
129. J. D. Atwood and T. L. Brown, J. Am. Chem. Soc. **98** (1976) 3160.
130. S. P. Schmidt, F. Basolo, C. M. Jensen and W. C. Trogler, J. Am. Chem. Soc. **108** (1986) 1894.
131. A. J. Downard, B. H. Robinson and J. Simpson, Organometallics **5** (1986) 1122, 1132, 1140.
132. S. D. Jensen, B. H. Robinson, and J. Simpson, J. Chem. Soc., Chem. Commun. (1983) 1081.
133. U. Honrath and H. Vahrenkamp, Z. Naturforsch. **39B** (1984) 545, 555, 559.
134. P. N. Lindsay, B. M. Peake, B. H. Robinson, J. Simpson, U. Honrath, H. Vahrenkamp and A. M. Bond, Organometallics **3** (1984) 413.
135. M. I. Bruce, J. G. Matisons, B. K. Nicholson, and M. L. Williams, J. Organomet. Chem. **236** (1982) C57.
136. K. Hinkelmann, F. Mahlendorf, J. Heinze, H. T. Schacht, J. S. Field and H. Vahrenkamp, Angew. Chem., Int. Ed., Engl. **26** (1987) 352.
137. M. I. Bruce, Coord. Chem. Rev. **76** (1987) 1.
138. A. J. Downard, B. H. Robinson, J. Simpson and A. M. Bond, J. Organometal. Chem. **320** (1987) 363.
139. J. E. Cyr, J. A. DeGray, D. K. Gosser, E. S. Lee and P. H. Rieger, Organometallics **4** (1985) 950.
140. G. Lavigne and H. D. Kaesz, J. Am. Chem. Soc. **106** (1984) 4647.
141. G. Lavigne, N. Lugan, J.-J. Bonnet, J. Chem. Soc., Chem. Commun. (1987) 957 and references cited therein.
142. D. A. Brandes and R. J. Puddephatt, Inorg. Chim. Acta **113** (1986) 17.
143. For a recent synthesis of [PPN]Br, see F. J. Lalor and S. Chaona, J. Organomet. Chem. **344** (1988) 163.
144. J. Evans and B. P. Gracey, J. Chem., Soc., Chem. Commun. (1983) 247.
145. S. L. Cook and J. Evans, J. Chem. Soc., Chem. Commun. (1983) 713.
146. T. Venalainen and T. Pakkanen, J. Organomet. Chem. **266** (1984) 269.
147. M. I. Bruce, J. G. Matisons and B. K. Nicholson, J. Organomet. Chem. **247** (1983) 321.
148. P. Lemoine, Coord. Chem. Rev. **83** (1988) 169; **47** (1982) 56.
149. M. I. Bruce, M. G. Humphrey, O. B. Shawkataly, M. R. Snow and E. R. T. Tiekink, J. Organomet. Chem. **315** (1986) C51.
150. H. H. Ohst and J. K. Kochi, J. Am. Chem. Soc. **108** (1986) 2897.
151. H. H. Ohst and J. K. Kochi, Inorg. Chem. **25** (1986) 2066.
152. H. H. Ohst and J. K. Kochi, J. Chem. Soc., Chem. Commun. (1986) 121.
153. T. M. Bockman and J. K. Kochi, J. Am. Chem. Soc. **109** (1987) 7725.
154. P. A. Lahuerta, J. Latorre, M. Sanau, and H. Kisch, J. Organomet. Chem. **286** (1985) C27.
155. A. Darchen, C. Mahe and H. Patin, J. Chem. Soc., Chem. Commun., (1982) 243; Nouv. J. Chim. **6** (1982) 539.
156. E. Cabrera, J. C. Daran and Y. Jeannin, J. Chem. Soc., Chem. Commun. (1988) 607.
157. M. I. Bruce, M. J. Liddell, C. A. Hughes, B. W. Skelton and A. H. White, J. Organomet. Chem. **347** (1988) 157.
158. M. I. Bruce, M. J. Liddell, C. A. Hughes, J. M. Patrick, B. W. Skelton and A. H. White, J. Organomet. Chem. **347** (1988) 181.
159. M. I. Bruce, M. J. Liddell, O. bin Shawkataly, C. A. Hughes, B. W. Skelton and A. H. White, J. Organomet. Chem. **347** (1988) 207.
160. S. Aime, M. Botta, R. Gobetto, D. Osella and F. Padovan, J. Chem. Soc., Dalton Trans. (1987) 253.

161. D. Boccardo. M. Botta, R. Gobetto, D. Osella, A. Tiripicchio and M. Tiripicchio Camellini, J. Chem. Soc., Dalton Trans. (1988) 1249.
162. R. J. Goudsmit, J. G. Jeffrey, B. F. G. Johnson, J. Lewis, R. C. S. McQueen, A. J. Sanders and J.-C. Liu, J. Chem. Soc., Chem. Commun. (1986) 24.
163. See for example: D. M. P. Mingos, Acc. Chem. Res. **17** (1984) 311.
164. M. G. Richmond and J. K. Kochi, Organometallics **6** (1987) 254.
165. M. G. Richmond and J. K. Kochi, Inorg. Chem. **25** (1986) 656.
166. M. G. Richmond and J. K. Kochi, Inorg. Chem. **26** (1987) 541.
167. S. R. Drake, B. F. G. Johnson and J. Lewis, J. Chem. Soc., Chem. Commun. (1988) 1033.
168. H. H. Ohst and J. K. Kochi, Organometallics **5** (1986) 1359.
169. J. Rimmelin, P. Lemoine, M. Gross, A. A. Bahsoun and J. A. Osborn, Nouv. J. Chim. **9** (1985) 181.
170. F. Y. Petillon, R. Rumin and J. Talarmin, J. Organomet. Chem. **346** (1988) 111.
171. J. Pursiainen, T. A. Pakkanen and J. Jaaskelainen, J. Organomet. Chem. **290** (1985) 85.
172. (a) M. Chanon and M. L. Tobe, Angew. Chem., Int. Ed., Engl. **21** (1982) 1.
(b) M. Chanon, Bull. Soc. Chim. Fr. (1982) 197; (1985) 209.
(c) M. Julliard and M. Chanon, Chem. Rev. **83** (1983) 425.
(d) M. Julliard and M. Chanon, Chem. Brit. **18** (1982) 558.
(e) M. Chanon, Asc. Chem. Res. **20** (1987) 214.
173. (a) R. H. Magnuson, S. Zulu, W. M. T'sai and W. P. Giering, J. Am. Chem. Soc. **102** (1980) 6887.
(b) R. H. Magnuson, R. Meirowitz, S. J. Zulu and W. P. Giering, J. Am. Chem. Soc. **104** (1982) 5790.
(c) R. H. Magnuson, R. Mierowitz, S. J. Zulu and W. P. Giering, Organometallics **2** (1983) 460.
(d) M. N. Golovin, R. Mierowitz, Md. M. Rohman, H. Y. Liu, A. Prock and W. P. Giering, Organometallics **6** (1987) 2285.
(e) See also M. J. Therien and W. C. Trogler, J. Am. Chem. Soc. **109** (1987) 5127.
(f) See also D. Miholova and A. A. Vlček, J. Organomet. Chem. **240** (1982) 413.
174. (a) D. L. Reger and E. Mintz, Organometallics **3** (1984) 1759.
(b) D. L. Reger, E. Mintz, and L. Lebioda, J. Am. Chem. Soc. **108** (1986) 1940.
(c) D. L. Reger, S. A. Klaeren, J. E. Babin and R. D. Adams, Organometallics **7** (1988) 181.
(d) D. L. Reger and S. A. Klaeren, J. Organomet. Chem. **342** (1988) 77.
175. R. D. Rieke, H. Kojima and K. Ofele, J. Am. Chem. Soc. **98** (1976) 6735.
176. (a) R. D. Rieke, H. Kojima, T. Saji, P. Rechberger and K. Ofele, Organometallics **7** (1988) 749 and references cited therein.
(b) See also N. G. Connelly, S. J. Raven, G. A. Carriedo and V. Riera, J. Chem. Soc., Chem. Commun. (1986) 992.
177. (a) A. M. Bond, R. Colton, and K. McGregor, Inorg. Chem. **25** (1986) 2378.
(b) A. M. Bond, R. Colton and J. E. Kevekordes, Inorg. Chem. **25** (1986) 749.
(c) A. M. Bond, S. W. Carr and R. Colton, Organometallics **3** (1984) 541.
178. W. E. Gieger, Prog. Inorg. Chem. **33** (1985) 275.
179. S. L. Yang, C. S. Li and C. H. Cheng, J. Chem. Soc., Chem. Commun. (1987) 1872.
180. M. Green, S. J. Porter and F. G. A. Stone, J. Chem. Soc., Dalton Trans. (1983) 513.
181. N. S. Nametkin, V. D. Tyurin, V. V. Trusov, A. S. Batsanov and Yu. T. Struchkov, J. Organomet. Chem. **219** (1981) C26.

CHAPTER 5

Redox Reactions of Symmetrical Metal Sandwich Compounds

William E. Geiger

Department of Chemistry, University of Vermont, Burlington, Vermont 05405, U.S.A.

Table of Contents

1. Introduction

The redox properties of transition metal sandwich complexes have been probed often since the discovery of the ferrocene/ferrocenium couple (Fc/Fc^+) in an early paper [1]. Reviews in the 1970's covered early developments in the area [2], and a chapter by Kotz treated metallocene redox chemistry through 1983 [3]. The chief objective of this Chapter is the presentation of a systematic treatment of the electron-transfer reactions of metal sandwich complexes, with emphasis on more recent work. Included will be complexes in which a single transition metal is pi-bonded to two cyclopentadienyl ligands, or two arene ligands, or their analogues (see structures below). In the interest of limiting the scope of the treatment, mixed sandwich complexes containing ligands of different size [e.g., CpM(arene)] are not covered. Particular attention will be placed on the various ions and

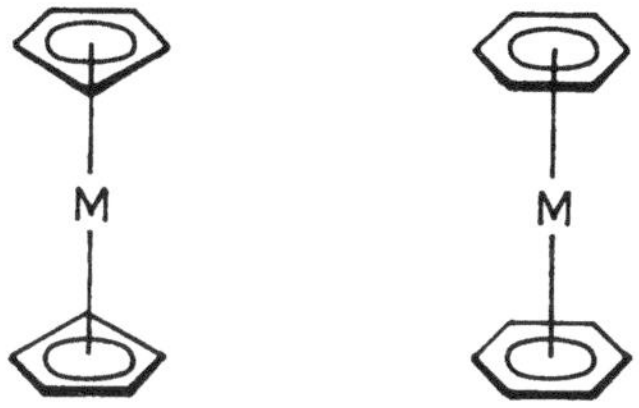

molecules that can be generated from a particular sandwich complex by oxidation or reduction. Given the central role of electrochemistry in the development of this field, a separate section is included evaluating the concept wherein the ferrocene/ferrocenium ion couple is employed as a reversible one-electron standard.

Synthetic reactions of these complexes will be covered mostly in cases wherein the reaction is induced by an electron-transfer process. For example, the reactivity of the anions derived from cobaltocene and vanadocene will be discussed, but the reactions of the corresponding neutral complexes [4] will not.

The redox chemistry of ferrocene is vast, and we will not discuss applications in which this complex acts only as an electron-transfer mediator. The reader is referred to other sources for information about the interesting areas of electron-transfer mediated photoelectrochemistry, polymer electrodes, etc. [5].

A major problem in surveying and reporting on electron-transfer reactions is the use of different reference electrodes. It has been recommended that each paper report potentials versus the ferrocene/ferrocenium couple in the electrolyte employed [6]. If potentials are reported versus another couple (e.g., the saturated calomel electrode), the potential for ferrocene under identical conditions should be reported in the same paper. It should be noted that the measured E^0 for $Fc^{0/+}$ varies significantly from one nonaqueous solvent/supporting electrolyte medium to another, although most if not all of the changes may be ascribed to differences in liquid junction potentials of the reference electrode employed. Table 1 gives representative values for the Fc/Fc^+ couple measured against the saturated calomel electrode, SCE.

In this review, wherever possible, we have converted potentials to the SCE reference. In some cases, potentials are referenced to the Fc/Fc^+ couple, and specific mention is made when this is the case.

A list of abbreviations used in this Chapter follows.

Cp	η^5-cyclopentadienyl
Cp*	η^5-pentamethylcyclopentadienyl

diolef	diolefin ligand
DMF	dimethylformamide
DMSO	dimethylsulfoxide
$E°$	formal potential (rigorously, $E°'$)
E_p	voltammetric peak potential
ΔE_p	difference in anodic and cathodic peak potentials in cyclic voltammetric wave
ETS	electron transmission spectroscopy
Fc	ferrocene
hfs	ESR hyperfine splitting
HOMO	highest occupied molecular orbital
k_s	standard heterogeneous electron-transfer rate constant
LUMO	lowest unoccupied molecular orbital
μ	micrometer
olef	olefin ligand
SCE	saturated calomel electrode
SCF	self consistent field
THF	tetrahydrofuran

Table 1. Some Reported Potentials for the Ferrocene/Ferrocenium Couple *versus* the Saturated Calomel Electrode for Various Solvent/Supporting Electrolyte Systems

Solvent	0.1 M $[Bu_4N][ClO_4]$[a]	0.1 M $[Bu_4N][PF_6]$[b]
THF	0.53 V	0.56 V
acetone	0.50 V	0.48 V
pyridine	0.50 V	
CH_2Cl_2	0.48 V	0.46 V
DMF	0.47 V	
DMSO	0.45 V	
CH_3CN	0.40 V	0.40 V
propylene carbonate	0.36 V	
CH_3NO_2	0.34 V	

[a] From D. Chang, T. Malinski, A. Ulman, and K. M. Kadish, Inorg. Chem. **23** (1984) 817.
[b] From the author's laboratory.

2. Electronic Structures of Metal Sandwich Complexes

Every level of *d*-orbital occupancy, d^1–d^9, has been observed for transition metal sandwich compounds. Tables 2 and 3 summarize the existence of molecules and ions of bis-cyclopentadienyl and bis-arene complexes, by categorizing them according to their *d*-orbital occupancy. Only species which have either been isolated or characterized *in situ* by electrochemistry or spectroscopy are included. Access to such a wide variety of compounds differing in *d*-orbital occupancy, but not in gross structure, has been profitable for the study of electronic structures of organometallic pi-complexes.

Table 2. Isolated or Electrochemically Characterized Metallocenes, Arranged by *d*-Orbital Occupancy. M denotes Cp_2M. M* denotes Cp^*_2M, and Signifies that no Example of the Cp Analogue is Known. Literature Through 1988

d^1	d^2	d^3	d^4	d^5	d^6	d^7	d^8	d^9
	V(+)	V	V(−)	Cr(−)	Mn*(−)	Fe(−)	Co(−)	Ni(−)
		Cr(+)	Cr	Mn	Fe	Co	Rh(−)	
			Mn*(+)	Re*	Ru	Rh	Ni	
			Fe*(2+)	Fe(+)	Os	Ir		
				Ru*(+)	Co(+)	Ni(+)		
				Os*(+)	Rh(+)			
					Ir(+)			
					Ni(2+)			

Table 3. Isolated or Electrochemically Characterized Bis-Arene Metal Complexes, Arranged by *d*-Orbital Occupancy. M Denotes $(benzene)_2M$. M* Denotes $(C_6Me_6)_2M$, and Signifies that no Benzene Analogue is Known. Literature Through 1988

d^1	d^2	d^3	d^4	d^5	d^6	d^7	d^8	d^9
			Ti	Ti(−)	V(−)	Cr(−)	Fe	Co*
			V(+)	V	Nb(−)	Fe*(+)	Ru*	
			Nb(+)	Nb	Cr	Ru*(+)	Co*(+)	
				Ta	Mo	Co*(2+)	Ni*(2+)	
				Cr(+)	W			
				Mo(+)	Mn(+)			
				W(+)	Tc(+)			
					Re(+)			
					Fe(2+)			
					Ru(2+)			
					Os(2+)			

The most commonly accepted molecular orbital diagram for bis-cyclopentadienyl and bis-arene complexes is reproduced from the work of Mingos [7] as Fig. 1, specifically in reference to the d^6, 18-electron, complexes bis (cyclopentadienyl)iron and bis(arene)chromium.

There is general [7–9] but not universal [10], agreement that the LUMO of ferrocene is a degenerate e_{1g^*} set composed of a combination of metal d_{xz}, d_{yz}, and cyclopentadienyl e_1 orbitals. It is the relative ordering of the $a_{1g}(d_{z^2})$ and e_{2g} $(d_{x^2-y^2}, d_{xy})$ orbitals in d^6 systems that is open to question. Xα calculations and optical spectral analysis on ferrocene place the a_{1g} level above the e_{2g}, but the opposite ordering is implied from uv-photoelectron spectroscopy and from esr data on the ferrocenium ion [7, 11]. That the d^5 ferrocenium ion has an $^2E_{2g}$ ground state, $(e_{2g})^3(a_{1g})^2$, appears to be on firmer grounds [12], but that fact does not fix the relative ordering of the two orbitals in the d^6 complex, since orbital relaxation may invert the a_{1g} and e_{2g} levels when electron loss occurs.

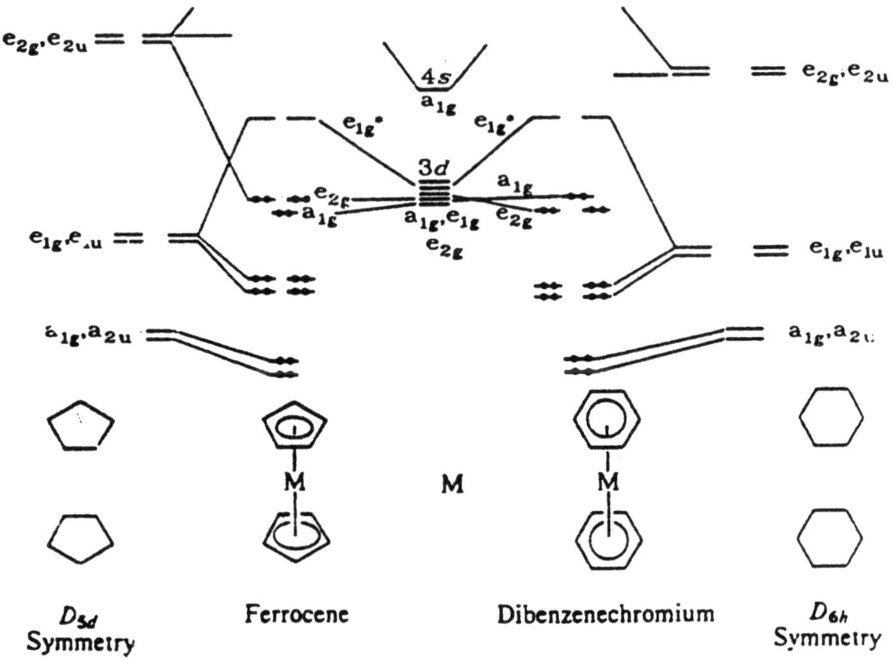

Fig. 1. Molecular orbital scheme for bis-cyclopentadienyl metal complexes and bis-arene metal complexes, assuming the symmetry shown. Occupancy shown for d^6 metals. Redrawn from D. M. P. Mingos in G. Wilkinson and F. G. A. Stone (eds.), Comprehensive Organometallic Chemistry, Pergamon Press, Oxford, 1982, Vol. 3, p. 28.

Apparently the e_{2g} level becomes more bonding with larger carbocyclic ligands [13], so that it drops below the a_{1g} in bis-arene complexes. Therefore, d^5 (arene)$_2$M complexes have the $^2A_{1g}$ ground state, $(e_{2g})^4(a_{1g})^1$.

INDO-SCF calculations indicate that for ferrocene the two possible HOMO orbitals, a_{1g} and e_{2g}, are about 90% metal d in character, but that the LUMO, e_{1g*}, is only about 65% metal. In simple terms, therefore, one may think about metallocene redox reactions involving 18 or fewer metal valence electrons as involving non-bonding, metal-based, orbitals. On the other hand, redox reactions involving metals with 19 or more valence electrons involve orbitals significantly delocalized over the metal and the cyclopentadienyl or arene ligand.

3. Ferrocene

The one-electron oxidation of bis(cyclopentadienyl)iron is a paradigm for metal sandwich redox processes, and is therefore covered first. The other metallocenes will be treated subsequently, arranged according to the positions of the metals in the periodic table.

3.1 The d^5—d^6 Couple

The blue color of the ferrocenium ion, Cp_2Fe^+, was first noticed by Kealy and Pauson when ferrocene was treated with oxidizing acids [14]. The deliberate oxidation of Fc by one-electron oxidizing agents in organic solvents led to isolation of the Fc^+ ion as the tetragallate salt [15]. A potential of +0.31 V *vs* SCE was reported for the Fc/Fc^+ couple in acidic ethanol [1, 15]. The stoichiometry and reversibility of the couple was established by chemical oxidation and reduction, dc polarography, and controlled potential coulometry. Starting from that surprisingly sophisticated beginning, there have been scores of papers written on this "simple" one-electron transfer reaction, so that Fc/Fc^+ is certainly the most highly studied organometallic redox couple.

Substituents bonded to the cyclopentadienyl ring affect the E^0 of the couple. Kotz has discussed the results of studies correlating the observed shift in potentials with substituent parameters [3], so only cursory comments will be made here. Data on substituted phenylferrocenes indicated that the E^0 varied linearly with Hammett σ constants [16]. Recent extensions of this work correlating E^0 variations with C-13 chemical shifts and with optical absorption wavelengths have confirmed that the substituent effect is transferred from the phenyl ring to the ferrocenyl group by an inductive mechanism [17].

Good correlations are also reported for E^0 variations with substituent parameters when the group is bonded directly to the cyclopentadienyl ring. Again, the transmission mechanism was found to be primarily inductive [18], but the most extensive study found that the best fit required a hybrid substituent constant which was 1/3 σ_m and 2/3 σ_p [18c].

Two more recent studies are worth mention. The first ionization potentials of ferrocene derivatives measured by photoelectron spectroscopy correlate well with the formal potentials of the derivatives. This appears to be the most extensive study to date on metal pi complexes indicating a linear relationship between the solution E^0 value and the gas-phase ionization potential [19]. A linear correlation is also found between E^0 values and ^{57}Fe Mossbauer quadrupole splittings (QS) for a series of mono- and di-substituted ferrocenes [20]. Taken together, these studies indicate a high degree of covalency in ferrocene, with significant interplay between the metal and the cyclopentadienyl ligands.

The formal potentials of amino- and dimethylamino-ferrocene were found to be about 200 mV more negative than expected based on their Hammett σ parameters [21, 22]. This is apparently the first good evidence for a ferrocene ring substituent interaction being governed by resonance rather than by induction.

Table 4 gives formal potentials for a number of ferrocene derivatives. In this table, we reference the potentials to that of Fc/Fc^+ for easier comparison of the substituent effect. The most negative value we found for oxidation of a substituted ferrocene was −0.63 V *vs.* Fc/Fc^+ (−0.23 V *vs.* SCE) [21]. The most positive value observed for a reversible system appears to be +0.74 V *vs.* Fc/Fc^+ for the phosphazene-annelated ferrocene 1,5-$N_4P_4F_6(C_5H_4)_2Fe$ [23]. Thus, a spread of almost 1.4 volt in the potentials of the oxidation of ferrocene derivatives is known.

The Fc/Fc^+ couple has been frequently used as a test system to investigate new types of electrolyte media. Recent examples involve the use of micellar media [24], very low temperature nonaqueous solvents [25], and low polarity solvents [26]. In the last study, the ferrocene oxidation was investigated in benzene, toluene, and xylene solutions with tetrahexylammonium perchlorate as the supporting

Table 4. Selected E° Values for Subtituted Ferrocene Derivatives. Each Couple Recorded in Acetonitrile. Potentials are Referenced to the Ferrocene/Ferrocenium Couple

Substituent(s)	E° vs. Fc/Fc^+	Reference
1,5-$N_4P_4F_6$	+0.74 V	[23]
1,1'-$(CF_3)_2$	+0.64 V	[82]
1,1'-diacetyl	+0.49 V	[77]
F_5-phosphazene	+0.39 V	[23]
CN	+0.31 V	[18d]
acetyl	+0.27 V	[77]
1,1'-Br_2	+0.25 V	[18d]
p-nitrophenyl	+0.12 V	[18a]
p-bromophenyl	+0.09 V	[77]
phenyl	+0.03 V	[18a]
octaphenyl	+0.03 V	[68]
vinyl	+0.02 V	[77]
Me_3Si	0.00 V	[18a]
H	0	
benzyl	−0.03 V	[18a]
CH_3	−0.06 V	[18a]
1,1'-$(CH_3)_2$	−0.10 V	[18a]
OCH_3	−0.12 V	[18d]
NH_2	−0.37 V	[22]
$(CH_3)_{10}$	−0.57 V	[67]
1,1'-$(NMe_2)_2$	−0.63 V	[21]

electrolyte, utilizing ultramicroelectrodes to minimize ohmic (iR) loss. Quite useful semiquantitative data was obtained at slow scan rates using electrodes with radii below 12 μ.

Hillman and co-workers have studied the oxidation of a number of ferrocene compounds with two or more polymethylene bridges between the cyclopentadienyl rings (see generalized structure **1**). In a study of 26 such complexes, it was shown that systems with shorter metal-to-cyclopentadienyl ring distances had relatively positive E° values, indicative of a thermodynamic destabilization of the oxidation product [27]. When three bridges of $(CH_2)_3$ were present, a comparison of the crystal structures of the redox partners showed that elongation of the iron-cyclopentadienyl bond by 4.4 pm occured upon oxidation [28]. This is in the normal range for ferrocene/ferrocenium pairs (see leading refs in Ref. 28). The $(CH_2)_3$ bridges accomodate the expansion of the sandwich structure by changing the angle between the α-carbons and the cyclopentadienyl rings.

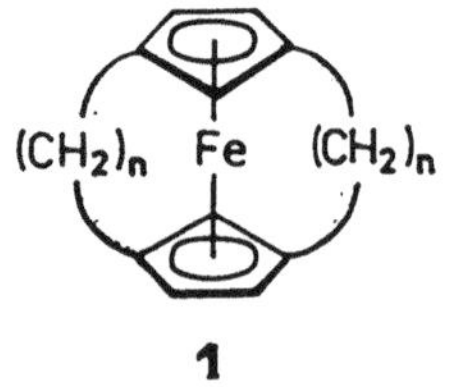

1

3.2 The d^4—d^5 and d^6—d^7 Couples

The decamethylferrocence dication has been generated through electrolysis of the corresponding monocation in an acidic ionic melt consisting of $AlCl_3$ and n-butylpyridinium chloride. Unsubstituted ferrocenium is also oxidizable, but the d^4 dication is unstable [29]. The dication of ferrocene, as well as those of Co and Ni, was produced in the gas phase by electron impact ionization [30] and its reactions were studied by tandem mass spectrometry. The favored reaction route involved extrusion of M^+ and formation of $C_{10}H_{10}^+$, but another route, involving formation of $CpFe^+$ and Cp^+, was also identified.

The d^7 configuration is better characterized than is the d^4 configuration for ferrocene. The first indication of possible existence of ferrocene anions involved one-electron reductions of ferrocenes substituted with groups that might themselves have been responsible for the electron uptake [31, 32]. However, reduction of the unsubstituted molecule has been demonstrated, and the d^7 configuration for ferrocene, as Cp_2Fe^-, is now on firm grounds [33, 34]. Section 5 is devoted to the properties and reactions of the anions of the metallocenes.

3.3 Ferrocene/Ferrocenium as a Reference Redox Couple

In 1984, IUPAC recommended that Fc/Fc^+ and $(benzene)_2Cr^{0/+}$ be employed as reference redox systems (not to be confused with reference electrodes) for non-aqueous electrolytes [6a]. In practice, the potentials of test redox couples are measured against a stable reference electrode (such as the SCE) under conditions identical to those used for measurement of one of the reference redox couples. Then the potential of the test couple is related to that of the reference couple. Often, ferrocene can used as an internal standard in investigations of new compounds [6b], but the possibility of ferrocene or ferrocenium interacting with the unknown substrate should not be ignored, since Fc/Fc^+ may catalyze electron-transfer reactions normally occuring at more positive potentials. All newly-reported potentials should be referenced to ferrocene either directly or indirectly.

The concept of a reference redox system is an attempt to overcome the problem of apparent potential variations due to changes in liquid junction potentials between the analyte solution and the reference electrode. The ideal reference redox system should have the following attributes [35]: (a) identical differences in solvation energies between the two redox forms when changing solvents (favored for large spherical molecules of low charge); (b) rapid and reversible electron-transfer kinetics at a variety of electrode surfaces; (c) stable and soluble redox partners; (d) no appreciable geometric changes during electron transfer; and (e) an $E°$ value in an accessible potential range. Investigations of the Fc/Fc^+ couple show that it has close to ideal behavior from several points of view, but some limitations which have become apparent will be discussed below (Sec 3.4).

The *difference* between the formal potentials of Fc/Fc^+ and dibenzenechromium$^{0/+}$ was shown to be $1.124\ V \pm 0.012\ V$ for 23 solvents [6a, 36]. This supports criterion (a) above, although it does not rule out the possibility that the formal potentials of the iron and chromium compounds shift identically in different solvents. Further investigations of solvation effects on these reference systems would be welcome. A recent study has shown that adsorption of ferrocene limits the use of that couple in aqueous media, especially at Hg electrodes [37].

3.4 How Reversible is the Ferrocene/Ferrocenium Couple?

Referring to Eqs. 1 and 2, there are two aspects of reversibility of the Fc/Fc^+ couple that may be discussed [38]. A *chemically* reversible system requires that

$$Cp_2Fe \rightleftharpoons Cp_2Fe^+ + e^- \text{ rate at } E^\circ = k_s \tag{1}$$

$$Cp_2Fe^+ \rightarrow \text{products} \tag{2}$$

there be no side reaction of either redox partner during the electrochemical experiment. This criterion is often, but not always, met with the ferrocenium ion (*vide infra*). An *electrochemically* reversible system is one in which the concentrations of the redox partners at a given potential are always those predicted by the Nernst equation. It is only when the standard heterogeneous electron transfer rate, k_s of Eq. 1, is "large" that the system behaves in a Nernstian fashion. In this context "large" is relative, depending on the time scale of the electrochemical experiment. In a cyclic voltammetry experiment at a scan rate of 0.5 V/s, a separation of peak potentials of no more than 65 mV, within normal experimental error of the Nernstian value, requires a standard heterogeneous electron transfer rate for ferrocene of $>$ *ca.* 0.8 cm/s. This value is affected in a relatively minor way by the diffusion coefficient of ferrocene in the solvent used.

The Fc/Fc^+ system has often been reported to have a k_s value below 1 cm/s [39–45], especially when measured at platinum electrodes. By way of contrast, aromatic hydrocarbons and many other organic redox systems have k_s values in the 1–5 cm/s range [46]. The data in references 39–45 would seem to indicate that the Fc/Fc^+ couple is not suitable to be a Nernstian standard at solid electrodes, but the range of values reported for closely analogous conditions ($k_s = 0.007$ to 0.09 cm/s), and the possibility that some of the values were affected by experimental error, suggested further studies were in order.

A number of more recent studies, using mercury electrodes or solid ultramicroelectrodes (the latter having a diameter of 10 μ or less), have reported k_s values of 1–3 cm/s for the ferrocene couple [47–51]. These values are more consistent with expectations, since structural differences between Cp_2Fe and Cp_2Fe^+ are small, leading to a very low inner-shell barrier to electron transfer. The fact that the self-exchange rate for Cp_2Fe with Cp_2Fe^+ is very fast in the gas phase [52] and in solution [53] implicates the electrode surface as playing a role in slowing down the rate of the heterogeneous electron transfer in voltammetric experiments.

Pons and co-workers have reviewed some possible reasons for the above inconsistencies in reported heterogeneous charge-transfer rates [51]. It is most likely that mildly insulating films are present on Pt electrodes, given the high tendency of platinum to adsorb impurities from solution. These films would tend to slow down the charge-transfer process. The authors of Ref. 51 found that the measured k_s value was higher ($>$2 cm/s) on a 1 μ diameter electrode than on a 25 μ electrode, and that the measured value also decreased with increasing ferrocene concentration. It was concluded that when the current flux is relatively high, the Pt electrode suffers deposition of a passivating film arising from the electrode process itself. Formation of an insoluble salt of Fc^+ or a decomposition product may be responsible. From the point of view of the heterogeneous charge-transfer, the voltammetry of ferrocene is far from ideal, especially when employing noble metal electrodes. It may be worthwhile seeking an organometallic Nernstian standard from among other pi complexes.

The chemical reversibility of the Fc/Fc$^+$ couple is also incomplete in some media owing to the tendency of the ferrocenium ion to react with nucleophiles. Reagents such as ethanol [45], Cl^- [54, 55], and DMSO [55] react with Cp_2Fe^+ to regenerate, in part, neutral Cp_2Fe. In ethanol [45, 56] and in water [37], the reaction of the ferrocenium ion with the solvent is fast enough to be noticable in cyclic voltammetry experiments. Ferrocenium also reacts slowly with DMSO and DMF, regenerating Cp_2Fe in the process, along with $Fe(solv)_6{}^{2+}$ [55]. In the majority of nonaqueous solvents, however, the ferrocenium ion is stable on the cyclic voltammetric time scale (*ca.* 20 s), and the Cp_2Fe/Cp_2Fe^+ couple may be considered to be chemically reversible.

3.5 Ferrocene Analogues with Heterocyclic or Acyclic Ligands

Sandwich complexes of Fe with 5-membered heterocyclic rings containing either phosphorus or arsenic in place of a CH moiety have been prepared, and electrochemical studies have been reported. The mixed sandwich complex **2** and the bisphosphaferrocene **3** both undergo reversible one-electron oxidations to monocations which are stable in CH_2Cl_2. Introduction of the P atom into the cyclopentadienyl ring leads to a positive $E°$ shift of 185 mV for each phosphorus [57]. A similar study of 1,1′ Diarsaferrocene gave a shift of +120 mV per As [58].

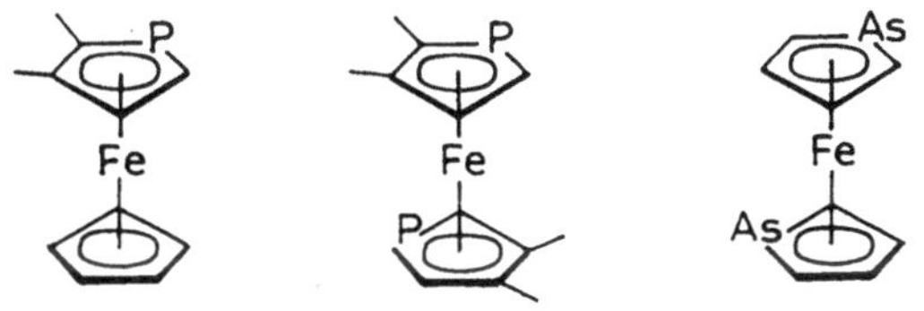

2 **3** **4**

The ESR spectrum of **3**$^+$ supports a ground state $^2E_{2g}$ formulation, the same as the ferrocenium ion [57]. A second, irreversible, oxidation of **2** and **3** may be due to ligand oxidation. Interestingly, though, **2**–**4** all exhibit one-electron reductions which are at least partly chemically reversible. At −38 °C in glyme, the $E°$ of **4**/**4**$^-$ was −2.19 V *vs.* SCE, representing a shift of +370 mV per As, compared to unsubstituted Fc/Fc$^-$. It was concluded that compared with the cyclopentadienyl ligand, the arsacyclopentadienyl group is more stabilizing towards the LUMO than the HOMO [58]. A similar effect is seen with the phosphacyclopentadienyl complexes, for which a shift of *ca.* +400 mV per *P* is noted for stabilization of the monoanion [57]. Calculations on the *P*-subtituted complexes suggest that the metal orbital contributions are about 90% for the HOMO, but only 35% for the LUMO, perhaps explaining why the heteroatom substitution has such a large effect on the potential of the reduction as compared to the oxidation. An extension of the work on phosphaferrocenes to include permethylated analogues has also appeared [59].

The "open metallocene" analogues in which one or two 2,4-dimethylpentadienyl ligands replace cyclopentadienyl ligands deserve inclusion here. The iron complexes **5** and **6** were oxidized at glassy carbon electrodes in dimethoxyethane [60]. The oxidation of the "half-open" complex **5** was reversible at room temperature (+0.29 V *vs.* SCE), but reversibility was only observed for **6**/**6**$^+$ when the temperature was lowered to −45 °C. Replacing one cyclopentadienyl ring with its pentadienyl analogue results in a negative potential shift of only about 100 mV,

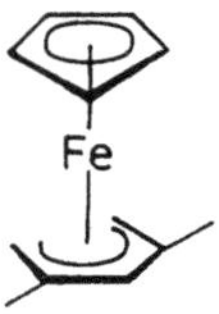

5

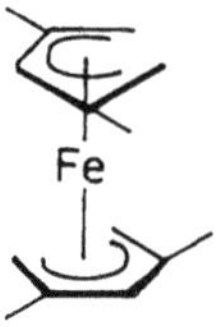

6

roughly equivalent to replacing two cyclopentadienyl hydrogens with methyl groups. In going from the bis-cyclopentadienyl complex to the half-open complex to the fully open complex, the monocations are progressively stabilized in the thermodynamic sense (more negative E° values), but destabilized in the kinetic sense (cations have shorter lifetimes for more open structures). One very interesting observation is that the gas-phase ionization potentials of **6** and bis(dimethylcyclopentadienyl)iron are identical, in contrast with the difference in E° values. The implication is that a larger gain in solvation energy of the open metallocene stabilizes the oxidation product, and that the open pentadienyl ligand does not actually alter the metal electron density, compared with its cyclic analogue.

This picture is complicated by ESR studies of $\mathbf{5}^+$ and $\mathbf{6}^+$. Compared with the spectra of the bis(dimethylcyclopentadienyl) iron cation, those of $\mathbf{5}^+$ and $\mathbf{6}^+$ show very little *g*-value anisotropy, consistent with a $^2A_1$ ground state, rather than the usual $^2E_2$ state. SCF calculations confirm that the ground state in the cations is $^2A_1$ (96% Fe d_{z^2}), but argue that the HOMO of the neutral complexes is a *ligand*-based orbital of E type. If so, this is an example of inversion of orbital levels as a consequence of one-electron oxidation [60]. UV-PES studies on the fully open **6** and its Ru and Os analogues confirm the ^2A_g state for the iron cation, but suggest a ^2B_u ground state for the two heavier metals [61].

7

Analogous to the open cyclopentadienyl complexes is the bicyclic complex **7**, which has allyl and olefin systems, both pi-bonded to iron, but electronically separated. It displays a one-electron reversible oxidation in THF at about the same potential observed for decamethylferrocene, suggesting that the interrupted conjugation in **7** stabilizes the Fe(III) cation [62].

Finally, small carboranes like $B_4C_2H_6^{2-}$ pi-bond to metals with a 5-membered B_3C_2 ring which is isoelectronic with the cyclopentadienyl ring. The dinegative charge on the carborane ring has an enormously large stabilizing effect on higher metal oxidation states, and the Fe(II)/Fe(III) potential for the complex CpFe ($B_4C_2H_6$), **8**, is −0.52 V *vs.* SCE, almost one volt negative of ferrocene itself [63]. Oxidation to a formal Fe(IV) complex is observed at +1.20 V *vs.* SCE [63].

4. Other Metallocenes

4.1 Titanium Group

Titanocene or its ions do not exist in solution. The Cp_2Ti^+ ion has been detected in the gas phase by ion cyclotron resonance spectroscopy during the reaction of Ti^+ with ferrocene [64]. Metallocenes with very few *d* electrons are rare, and to our knowledge no examples of d^1 metallocenes have been reported.

4.2 Vanadium Group

The d^2 vanadocenium ion, Cp_2V^+, is apparently the most electron deficient sandwich complex known, but it has not been established with certainty that this monocation retains the basic sandwich structure with parallel cyclopentadienyl rings.

The one-electron oxidation of Cp_2V was reported at −0.55 V *vs.* SCE in THF [65] and −0.70 V in CH_2Cl_2 [66]. The resulting cation is reactive and was not studied by spectroscopy. An analysis of the potentials of Cp_2M/Cp_2M^+ for M = V, Cr, Fe, Co, Ni showed that the $E°$ values of the last four metals correlated well with the gas-phase ionization potentials of the metallocenes. However, the $E°$ of Cp_2V/Cp_2V^+ was about 750 mV negative of that predicted on the basis of the reported ionization potential of Cp_2V [65]. It is likely that the oxidation is facilitated by coordination of a solvent molecule in the cation, with the overall redox reaction being given by Eq. 3:

$$Cp_2V + solv \rightarrow Cp_2V(solv)^+ + e^- \qquad (3)$$

The reported oxidation of the permethylated analogue, $Cp^*{}_2V$, is also consistent with Eq. 3. No electrochemistry was reported, but oxidation of the neutral complex by Cp_2Fe^+ in acetonitrile led to isolation of $Cp^*{}_2V(CH_3CN)^+$ [67].

The oxidation of bis(tetraphenylcyclopentadienyl)vanadium is irreversible in THF, the oxidation peak coming at +0.39 V *vs.* SCE [68]. This value would appear to be more in line with that expected for the d^2–d^3 vanadocene couple, especially since phenylation of the cyclopentadienyl ring does not result in large potential shifts. Thus, solvent coordination probably does not occur concomitant with electron transfer for the system $(C_5Ph_4H)_2V^{0/+}$. Detailed voltammetric measurements would be required to support this interpretation.

Unsubstituted vanadocene may also be reduced by one electron in THF, but the potential is very negative (values of −2.74 V and −3.0 V are reported [65, 66]), and the resulting anion loses a cyclopentadienyl ligand. This has led to useful appoaches to cyclopentadienyl vanadium chemistry (Sec. 5.1).

4.3 Chromium Group

Unlike the situation with the ferrocenium ion, for which the $E°$ of the Fc/Fc^+ couple was published in the same paper reporting isolation of the monocation [15], the $E°$ of Cp_2Cr/Cp_2Cr^+ was not reported until 22 years after preparation [69] of both redox partners. Both Cp_2Cr and Cp_2Cr^+ are quite air sensitive, and a vacuum electrochemical cell was used to study the couple, which has an $E°$ of −0.67 V *vs.* SCE in CH_3CN [70, 71] and −0.61 V in CH_2Cl_2 [66].

As expected, the permethylated analogue, Cp^*_2Cr, has a Cr(II)/Cr(III) potential considerably negative of that of the bis-cyclopentadienyl complex, −1.04 V *vs.* SCE in CH_3CN [67]. The 15 electron, d^3, $Cp^*_2Cr^+$ cation was isolated after oxidation of the neutral complex by Cp_2Fe^+ and shown to have the $^3E_{2g}$ ground state, $(e_{2g})^3\,(a_{1g})^1$. $(C_5Ph_4H)_2Cr$ is oxidized to the monocation at −0.77 V *vs.* SCE in THF [68]. The monocation, prepared by oxidation of the neutral complex with $AgPF_6$, has much higher kinetic stability than does the cation of the cyclopentadienyl analogue. ESR and magnetic susceptibility measurements are also consistent with the $^3E_{2g}$ ground state for the cation [68].

Chromocene also undergoes a reversible one-electron reduction to Cp_2Cr^-. A potential of −2.3 V *vs.* SCE has been reported for CH_3CN or THF solutions [65, 66, 70, 71]. The chromocene anion has not been spectroscopically characterized, and it would be interesting to see whether this d^5 complex has an A or E ground state. The anion has been reported to have a lifetime of about 10 s at room temperature [65]. A recent cyclic voltammetric study showed that the Cp_2Cr/Cp_2Cr^- wave was completely reversible at 0 °C, but that the anion undergoes slow elimination of the cyclopentadienyl anion [72] (see Sec. 5.1).

4.4 Manganese Group

Redox studies have not been reported on unsubstituted manganocene, probably since it would suffer displacement of the cyclopentadienyl anion under the polar solvent conditions normally used for electrochemical work. However, the related bis(pentamethylcyclopentadienyl)manganese compound, Cp^*_2Mn, has been studied in depth [67, 73, 74]. The neutral complex undergoes both oxidation (−0.56 V *vs.* SCE) and reduction (−2.50 V) in CH_3CN, establishing the electron transfer series $Cp^*_2Mn^+/Cp^*_2Mn/Cp^*_2Mn^-$. The anion, prepared as the orange pyrophoric sodium salt, was the first isolated metallocene anion [73, 74]. It is of special interest because it is isoelectronic with ferrocene.

The d^5 neutral complex Cp^*_2Mn has the $^2E_{2g}$ ground state [75]. The red monocation can be prepared through oxidation of Cp^*_2Mn by Cp_2Fe^+ or ferric ion, and is sufficiently stable to be handled in air for brief periods [74].

The analogous rhenium complex, Cp^*_2Re, has only recently been prepared. It displays matrix-dependent variations between high- and low-spin ground states, but no redox chemistry has yet been reported [76].

4.5 Ruthenium and Osmium

In spite of a number of investigations [77–81], the oxidation of ruthenocene is still not well understood. With the exception of experiments in molten salts, in which a quasireversible one-electron oxidation was indicated [81], the Cp_2Ru oxidation has been characterized as an irreversible two-electron process. The re-reduction of the oxidation product regenerates the original neutral metallocene, but there is little definitive information about the structure of the oxidized form. At a mercury electrode the oxidation product is $[(Cp_2Ru)_2Hg]^{2+}$ [80].

A number of unsymmetrical ruthenocenes, with one pentamethylcyclopentadienyl ligand and one Cp or substituted Cp ligand, have recently been studied [82]. All the complexes gave irreversible oxidations, but no mechanistic details were given. In spite of the irreversible nature of the oxidations, the E_p values correlated with the Ru binding energies measured by X-ray photoelectron spectroscopy [82].

The only bis(cyclopentadienyl)ruthenium complex to undergo a reversible one-electron oxidation under synthetically meaningful conditions is Cp^*_2Ru. Its E° was reported as +0.55 V *vs.* SCE in CH_2Cl_2/Bu_4NPF_6 [83, 84] or +0.42 V *vs.* SCE in CH_2Cl_2/Bu_4NClO_4 [82]. The monocation has only limited stability in acetonitrile, but persists in CH_2Cl_2 and may be spectrally characterized. The reaction with CH_3CN apparently results in **8** [85]. The ESR spectrum of $Cp^*_2Ru^+$ was said to favor on ^2E ground state for the d^5 complex, but the *g*-value anisotropy is very small, compared to Fe complexes possessing the same electronic ground state [83]. A second oxidation wave at +1.25 V was irreversible.

In a further study, Koelle and Grub studied the reaction of the monocation $Cp^*_2Ru^+$ with weakly coordinating Lewis bases, such as chloride [84]. Hydrogen-atom abstraction was commonly observed, yielding the tetramethylfulvalene cation **8** in addition to neutral decamethylruthenocene. Complex **8** can also be prepared by the reaction of Cp^*_2Ru with trityl ion, and its reactions with nucleophiles (OH^-, ROH, R_2NH) give $Cp^*Ru(C_5Me_4CH_2Nu)$ [84]. The monocation $Cp^*_2Ru^+$ is reported in another study to be unstable in CH_2Cl_2, even at −25 °C [85].

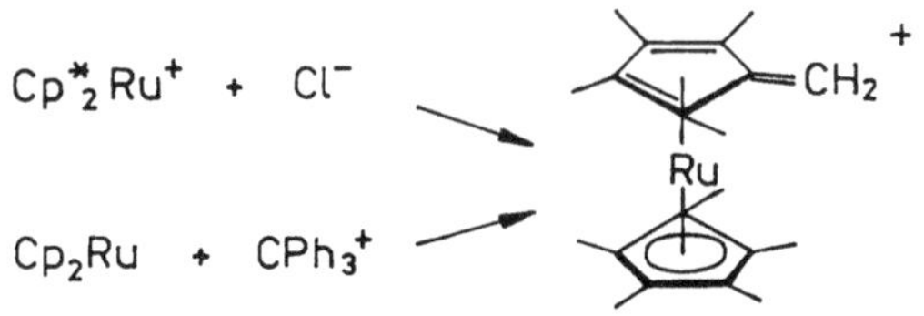

8

Bis(cyclopentadienyl)osmium is oxidized in an irreversible one-electron process at *ca.* +0.7 V *vs.* SCE [77–79, 84]. The salt of composition $[Cp_2Os][BF_4]$, isolated after controlled potential coulometry [78], was most likely the osmocenium dimer, $[Cp_2Os]_2$, **9**. The latter was prepared by oxidation of Cp_2Os with Ce^{4+} in acetonitrile [86]. It consists of two osmocenium units connected by an unsupported Os—Os bond (303.8 pm), with the two units rotated by 90°. It reacts with all solvents in which it dissolves, and apparently is isolable due to its rapid precipitation. The dimerization is an interesting example of how a 17-electron species (Cp_2Os^+) is stabilized by formation of a metal-metal bond, even though two positively-charged metals are involved in the reaction.

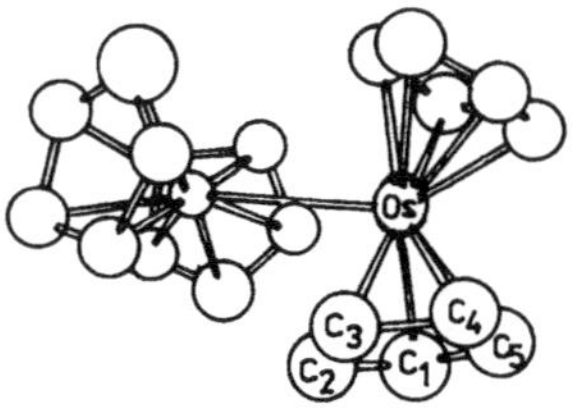

9

Once again, permethylation of the cyclopentadienyl rings simplifies the electron-transfer reaction. Cp^*_2Os undergoes a highly reversible oxidation to $Cp^*_2Os^+$

with an $E^\circ = +0.46$ V *vs.* SCE in CH_2Cl_2 [85]. The monocation was isolated after treatment of the neutral complex with Ag^+ [85, 87], and shown by ESR to have the expected ${}^2E_{2g}$ ground state. It was shown to form 1 : 1 charge-transfer salts with a series of organic acceptor ions [85].

The three complexes $Cp^*{}_2M$, where M = Fe, Ru, Os, provide the only known comparison of the same metallocene redox couple for the first to third row metals in the same group. The iron complex is seen to be considerably easier to oxidize ($E^\circ = -0.11$ V *vs.* SCE) than is the Ru complex (+0.55 V) or the Os complex (+0.46 V).

4.6 Cobalt Group

The d^6–d^7 couple, Cp_2Co^+/Cp_2Co, has been widely studied [88–91], and in many cases could act as a reversible one-electron standard. Its heterogeneous charge-transfer reaction is very fast at Hg electrodes in DMF or acetonitrile [47, 48, 92]. The E° of the permethylated analogue, $Cp^*{}_2Co^+/Cp^*{}_2Co$, is, predictably, about one-half volt negative of the cyclopentadienyl complex (−1.48 V *vs.* SCE compared to −0.94 V, in CH_3CN) [67]. However, the neutral complex can be isolated after reduction of the cation by alkali metals [67, 91].

Cobaltocene can also be reduced by one electron to a d^8, 20 valence electron, monoanion. The E° for the couple Cp_2Co/Cp_2Co^- is −1.88 V *vs.* SCE in CH_3CN. This was the first metallocene system to give evidence of a persistent anion [90, 93]. The reactions of this and other metallocene anions are covered in Sec. 5.1.

The thiadiborol Co(III) complex $CpCo(C_2Et_2B_2Me_2S)$ is reduced by one-electron with potassium in THF. The ESR spectrum of the monoanion is consistent with a d^9 complex [95].

The rhodocenium ion, like the cobaltocenium ion, is subject to two consecutive one-electron reductions. However, the reduced Rh complexes are much less stable than their Co analogues. The E° potentials for the series Cp_2Rh^+/Cp_2Rh and Cp_2Rh/Cp_2Rh^- are −1.41 V and −2.18 V *vs.* SCE in CH_3CN [94]. Neutral rhodocene has a lifetime in solution of only a few seconds at room temperature,

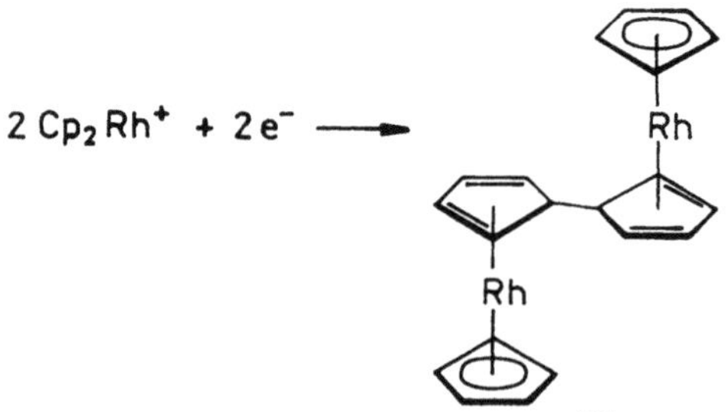

10

giving a high yield of the ligand-coupled dimer **10** in which each metal is bonded to one pentahapto and one tetrahapto ring. This mode of coupling of two 19-electron complexes reduces the electron count to 18 at each metal. The anion, Cp_2Rh^-, has a lifetime of only a few seconds at −50 °C, and the reduction products are unknown [94].

To our knowledge, no electrochemical waves have been reported for the Cp_2Ir^+ ion.

4.7 Nickel Group

The 20-electron, d^8, complex nickelocene is, not surprisingly, oxidizable in two one-electron steps to the nickelocenium dication. A one-electron reduction is also possible, so that the bis(cyclopentadienyl)nickel system displays the well-established four-membered electron-transfer series (Eq. 4):

$$Cp_2Ni^{2+} + e^- \rightleftharpoons Cp_2Ni^+ + e^- \rightleftharpoons Cp_2Ni + e^- \rightleftharpoons Cp_2Ni^- \qquad (4)$$

Oxidation of Cp_2Ni (−0.09 V *vs.* SCE) by one electron gives an isolable monocation, which is isoelectronic with cobaltocene and has the expected ${}^2E_{1g}$ ground state. The half-occupied orbital of the 19-electron Cp_2Ni^+ is significantly more ligand-based than is the corresponding Cp_2Co orbital (*ca.* 40% Ni *vs.* 60% Co) [96]. This greater delocalization may account for the extensive electron-transfer series observed for nickelocene.

The unsymmetrical sandwich complex Cp(pentaphenylcyclopentadienyl)Ni has an $E°$ slightly positive (*ca.* 100 mV) of unsubstituted nickelocene, but the monocation displays increased kinetic stability [97]. The nickel complex with two tetraphenylcyclopentadienyl ligands, $(C_5Ph_4H)_2Ni$, has a reversible one-electron wave at 0 V *vs.* SCE in THF, but attempts at isolation of the monocation were not reported [68]. A monocation was isolated from the oxidation of bis(pentamethylcyclopentadienyl)nickel, $Cp^*{}_2Ni$, with Cp_2Fe^+ [67] or with Fe^{3+} [91]. Consistent with all electrochemical measurements on permethylated metallocenes, the $E°$ value for the +/0 couple was shifted about 0.5 V negative of the bis-cyclopentadienyl complex (−0.65 V *vs.* SCE) [67, 91].

Nickelocene and its derivatives are the only metallocenes showing relatively facile oxidation to a dication, a feature ascribed to the fact that Cp_2Ni^{2+} is isoelectronic with Cp_2Fe. The unsubstituted nickelocenium dication has been observed only under electrochemical conditions, either at low temperatures [98], in rigorously purified solvents [99], or in molten salts [81]. The 2+/+ waves for Cp_2Ni, Cp(pentaphenylcyclopentadienyl)Ni [97], and $Cp^*{}_2Ni$ [67, 91] are all about 0.9 V positive of the 0/+ waves. The second oxidation of the bis-tetraphenylcyclopentadienyl complex is only 0.6 V positive of the first oxidation [68].

All of the dications of the phenyl- or methyl-substituted bis(cyclopentadienyl)-nickel complexes appear to be more stable than Cp_2Ni^{2+}, but the only one which has been isolated is $Cp^*{}_2Ni^{2+}$, which has a relatively mild $E°$ (2+/+) of +0.31 V *vs.* SCE. Oxidation of the neutral complex with either O_2, Br_2, [100] or Hg^{2+} [67] gives the desired dication. It has limited stability, even as a solid protected from air, but NMR measurements suggest that it has the symmetrical sandwich structure [67]. The dication reacts with electrophiles such as BH_4^- and CN^- to give

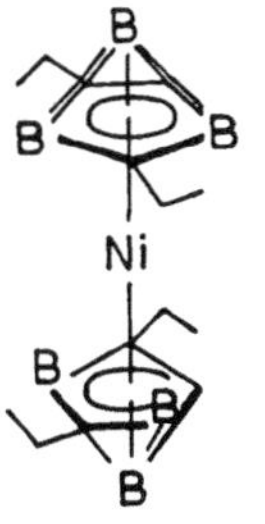

11

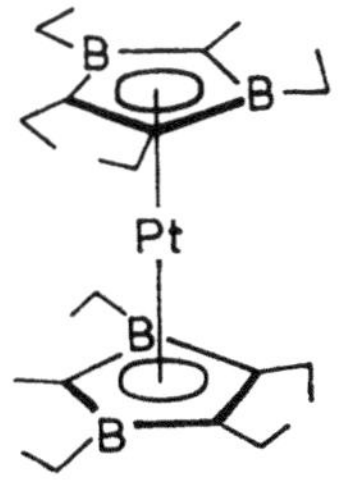

12

the cyclopentadiene complexes $[Cp^*Ni(\eta^4\text{-}C_5Me_5R)]^+$. Oxidation of the latter (R = H) re-forms $Cp^*_2Ni^{2+}$ with release of dihydrogen [100].

The 20-electron bis(tricarbahexaboranyl)Ni complex **11** is isoelectronic with Cp_2Ni, and undergoes oxidation to a stable monocation ($E^\circ = -0.11$ V *vs.* SCE in CH_2Cl_2). Further oxidation is a multi-electron, irreversible process. A reduction (−2.0 V) is also irreversible [101].

The E° of the 0/− couple for Cp_2Ni is the most positive of any unsubstituted metallocene (−1.66 V *vs.* SCE in CH_3CN) [65]. It is ironic that the most electron-rich metallocene should also be the most willing to receive another electron. The nickelocene anion is, however, short-lived. A temperature of −60 °C is necessary to observe a chemically reversible cyclic voltammogram in DMF. The products resulting from the decomposition of Cp_2Ni^- are discussed in Section 5.1.

The heterogeneous charge-transfer rate for the Cp_2Ni/Cp_2Ni^- couple is quite low at a Pt electrode (k_s *ca.* 3×10^{-3} cm/s). It is possible that a structural change such as cyclopentadienyl ring slippage occurs during formation of the 21-electron monoanion [65], but Cp_2Ni^- is too unstable to allow confirmation of this hypothesis.

The one-electron reduction of $(C_5Ph_4H)_2Ni$ is reported to be reversible ($E^\circ = -1.46$ V *vs.* SCE in THF) [68].

Neither palladocene nor platinocene has been reported. An analogous platinocarborane **12** has been reported. This complex is formally Pt(IV), containing 16 valence electrons. Although no electrochemistry has been reported on it, **12** may be reduced to either a monoanion or dianion by potassium, and the anions were characterized by ESR (−) or NMR (2−) spectroscopy [102].

5. Metallocene Anions

References to metallocene anions are sprinkled throughout the metal-by-metal coverage above. It is noteworthy that the first evidence for a metallocene anion [90, 93] was uncovered over 20 years after the preparation of a metallocene cation. There is now confirmation of Cp_2M^- for M = V, Cr, Mn, Fe, Co, Rh, and Ni. The V, Mn (as the Cp* complex), and Co anions have been isolated, although their E° potentials are so negative that this must be viewed as a remarkable achievement. In every case, the first published evidence for a particular metallocene anion has come from cyclic voltammetry data.

It has been recognized that since the pi-cyclopentadienyl ligand is not a strong pi-acceptor, metallocene anions, if isolable, would be extremely reactive [103]. INDO-SCF calculations predicted that the ferrocene anion would have a d^6 lig^1 electronic configuration, in contrast to d^7 for isoelectronic cobaltocene [104]. However, this has not been confirmed by electron transmission spectroscopy (ETS) studies. This technique has the ability to characterize the gas phase negative ion states of a molecule arising from electron capture into an unoccupied orbital. Two such studies of the metallocenes agree on the electronic structure of the ferrocene anion. The anion state is assigned as $^2E_{1g}$, $(e_{2g})^4(a_{1g})^2(e_{1g})^1$, leading to the conclusion that the LUMO of Cp_2Fe is quite like the half-filled orbital of Cp_2Co [105,106].

5.1 Reactions of Metallocene Anions

There appear to be three known types of reactions of metallocene anions: (1) one-electron transfer; (2) reductive elimination of a cyclopentadienyl ligand; (3) attack by an electrophile. As for one-electron transfer reactions, it is not surprising that the anions should be very strong reducing agents, since their E° potentials are 1.4 V (Ni) to 2.2 V (V) lower than the normal hydrogen electrode. However, no studies seem to have appeared taking advantage of these powerful reductants as simple one-electron transfer agents.

The most commonly reported reaction of metallocene anions to date is that of cyclopentadienyl elimination. This reaction was first observed in 1957 [107], but a lengthy hiatus followed in which little advantage was taken of the cyclopentadienyl-stripping reaction. More recently, Jonas *et al.* have reacted metallocene anions with two electron ligands to produce a variety of interesting complexes.

When Cp_2Co is treated with Li in THF at $-30\,°C$, a salt of composition $[Li(THF)_y][Cp_2Co]$ is isolated [108]. In the presence of olefins, the anion goes on to form $CpCo(olef)_2$. What makes this method particularly useful is the mild set of conditions used in the reaction, allowing high-yield preparations of thermally labile complexes. In the presence of excess olefin, for example, the final product is $[M_A][Co(olef)_4]$ where M_A represents an alkali metal.

The interesting 18-electron anion $[CpFe(\eta^4\text{-cyclooctadiene})]^-$ was prepared as the Li salt by the action of Li on ferrocene in the presence of C_8H_{12}, and $[Ni(\eta^4\text{-cyclooctadiene})_2]^{2-}$ was isolated from the reaction of Cp_2Ni with Li and C_8H_{12} [108–110].

A series of papers has also appeared detailing the reductive elimination of the cyclopentadienyl ligand from Cp_2V through the use of alkali metal reducing agents or sodium naphthalenide. The vanadocene anion has been isolated as $[K(THF)(Et_2O)][Cp_2V]$ [111], which reacts with other pi-ligands to form a variety of mixed-sandwich and other complexes [111, 112]. This work has been recently reviewed [113, 114]. In at least one case, "reductive substitution" has been accomplished without the aid of reducing agents, by heating Cp_2Co in the presence of diolefins such as duroquinone and cyclooctadiene. Half-sandwich CpCo(diolef) compounds were produced, but in no higher than 12% yield [115].

Other examples of reductive elimination of Cp ligands from metallocenes involve Cr, Fe, and Ni. The chromocene anion, prepared electrochemically under argon, forms the half-sandwich complex $CpCr(CO)_3^-$ when the electrolysis is performed under CO [72]. El Murr and co-workers have made use of reductive Cp elimination to prepare solutions of substituted cyclopentadienyl anions which are then used for coordination to other metals ions [31, 32]. Finally, the major product of the electrochemical reduction of Cp_2Ni in THF was identified as a probable multinuclear complex, but it was not isolated [65]. When the reduction is carried out using [Na][naphthalene] in THF, a series of cluster compounds is isolated which contain the CpNi unit. One of these clusters, Cp_6Ni_6, has been structurally characterized, and consists of an octahedron of nickel atoms [116]. The reaction appears to be a further example of how unusual complexes can be prepared from coordinatively unsaturated CpM moieties.

Only a few examples have appeared in which electrophilic attack occurs on a metallocene anion. This is probably because such an encounter would tend to be dominated by simple electron transfer to the electrophile, with regeneration of the neutral metallocene. Protonation of the cobaltocene anion by weak acids, such as water or phenol, occurs at the cyclopentadienyl ligand [117], in contrast to the reaction of acids with neutral Cp_2Co, wherein attack occurs exclusively at the metal [118]. When Cp_2Co^- is produced in DMF, other electrophiles such as CO_2 may be added to the cyclopentadienyl ring to produce the substituted cyclopentadiene complexes $CpCo(C_5H_5R)$ [119].

6. Bis-Arene Complexes

In contrast to the metallocenes, the synthetic redox chemistry of bis-arene metal complexes preceeded the widespread use of electrochemistry to study the electron-transfer reactions. A large number of arene sandwich complexes were prepared during the mid-1950's through the 1960's, principally by the E. O. Fischer group, many of the complexes being isolated in more than one oxidation state. This earlier work has been reviewed by Silverthorn [120]. In many cases, the most stable complexes were those of the hexamethylbenzene, C_6Me_6, ligand. Table 3 gives a compilation of the molecules and ions known for bis-arene complexes, arranged by *d*-orbital occupancy. The significance of the entries in the Table is complicated by the tendency of the coordinated arene to distort to a tetrahapto bonding mode, especially in electron-rich complexes of second and third row metals [121], and the text should be consulted for details. In the discussion below, all arene rings are assumed to be hexahapto-coordinated, unless otherwise noted.

6.1 Titanium Group

Bis(benzene)titanium(0) has a very low first ionization potential, less than 6 eV, consistent with the relatively strong electron-releasing properties of the arene ligand [122]. However, the oxidation of this d^4 complex has apparently not been reported. Rather, the reduction of $(benzene)_2Ti$ by potassium has been shown to give deep-green solutions of the anion, $(Bz)_2Ti^-$ [123]. Blue crystals of composition $[K][(benzene)_2Ti]$ were isolated from THF, and appear to be the first example of an isolated $(arene)_2M$ anion. The ESR spectrum of the d^5 anion showed coupling to 12 equivalent hydrogens, with an isotropic hyperfine splitting, a_H, of 4.5 G. Recall that the d^5 metallocenes generally have a $^2E_{2g}$ ground state, and are subject to extensive spin-orbit coupling which makes their ESR spectra impossible to observe in fluid media [96, 124]. Such is not the case with d^5 bis-arenes, in which the $^2A_{1g}$ ground state quenches most of the spin-orbit coupling, allowing long relaxation times and well-resolved spectra in fluid solutions. The hydrogen hyperfine splitting, hfs, is slightly larger in $(benzene)_2Ti^-$ than in isolectronic $(benzene)_2V$ (4.1 G) or $(benzene)_2Cr^+$ (3.56 G).

6.2 Vanadium Group

There appear to be no electrochemical data reported on $(arene)_2V$. However, there is ample evidence for the three-member electron-transfer series of Eq. 5, encompassing d^4–d^6 (16–18 e⁻)

Eq. 5 $(benzene)_2V^+ + e^- \rightleftharpoons (benzene)_2V + e^- \rightleftharpoons (benzene)_2V^-$

complexes. The monocation is the oxidation state formed under the conditions of the Friedel-Crafts synthesis [125]. The d^5, 17 e^-, neutral complexes of V, Nb, and Ta are quire stable, exhibiting no tendency to undergo the dimerization reactions found for neutral d^5 metallocenes (e.g., Cp_2Re) [122].

The vanadium complex has been reduced by potassium to a monoanion, which was characterized by NMR spectroscopy [126]. The results of competitive *K* reduction experiments allowed estimation of the $E°$ for the couple $(benzene)_2V/(benzene)_2V^-$ as approximately that of biphenyl [127]. Oxidation of the anion by O_2 or by benzophenone regenerated the neutral complex. A linear correlation exists between the ^1H chemical shift of a d^6, 18 e^-, sandwich complex, and the overall charge on the complex, as shown in Fig. 2. An upfield shift of 1.5 ppm is observed with an increase of (−1) in the total charge, from dications through monoanions [126].

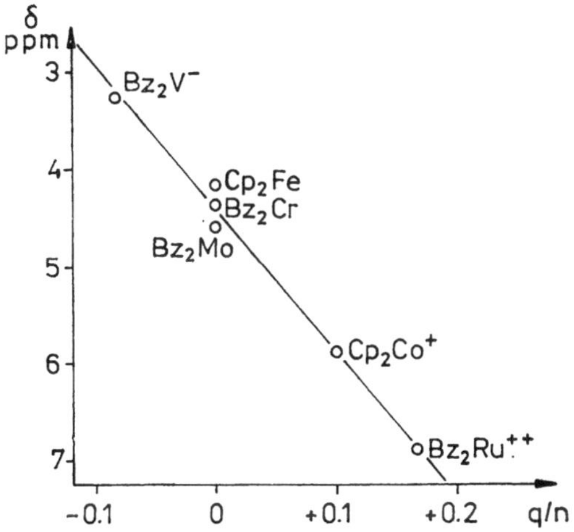

Fig. 2. Plot of ^1H chemical shifts as a function of normalized formal changes on the carbons in the rings, for a series of 18-electron sandwich complexes. Redrawn from Ch. Elschenbroich and F. Gerson, J. Amer. Chem. Soc. **97** (1975) 3557.

Electrochemistry has been reported for $(arene)_2Nb$ compounds in which arene = toluene or mesitylene. The toluene complex undergoes oxidation to a cation at $E° = -0.90$ V *vs.* SCE in THF, and one-electron reduction occurs at $E° = -2.46$ V [128]. Oxidation of the neutral complex by Ag(I) gave no tractable product, but repeating this procedure in the presence of a phosphine gave the monocation $[(arene)_2Nb(PR_3)][BF_4]$. Consistent with earlier unsuccessful attempts to reduce $(arene)_2Nb$ [126], reduction of the bis(mesitylene) complex with K/THF gave no isolable product, but when the reduction was performed in the presence of the crown ether 18-crown-6, maroon pyrophoric crystals of composition $[K(crown)][(arene)_2Nb]$ were isolated. With CO, the Nb anion gave $Nb(CO)_6^-$, demonstrating that, like the metallocene anions, bis-arene sandwich anions are also subtitutionally labile [128].

6.3 Chromium Group

6.3.1 Oxidations

The d^5–d^6 couple for $(arene)_2Cr^+/(arene)_2Cr$ is analogous to the ferrocenium/ferrocene couple, and as such has been widely studied. As mentioned in Sec. 3.1.3, it is an acceptable reference redox couple [6(a)], having been studied by a number of electrochemical methods [129–132]. The E° of the benzene complex **13** is generally reported to be -0.7 V $\pm$ 0.1 V *vs.* SCE. The effect of arene substituents on the E° potential has been the subject of several investigations [132–139], and correlations with both inductive- [130] and resonance- [136] sensitive parameters have been noted. The recent study of 20 $(arene)_2Cr$ derivatives, which favors an inductive mechanism for substituent effect transmission, found a good correlation between the E° potential and the gas-phase ionization potential of the complex [132]. A spread of over 1 volt in E° is possible by varying the substituent on the benzene ring $\{E^\circ = -0.98$ V *vs.* SCE for (mesitylene)$_2$-$Cr^{0/+}$; $E^\circ = +0.34$ V for $[1,4\text{-}(CF_3)_2C_6H_4]_2Cr^{0/+}\}$.

When the benzene ring in **13** is replaced by a larger polycyclic hydrocarbon, the E° value remains essentially unchanged. This was confirmed for $(arene)_2Cr$ complexes in which arene = naphthalene, phenanthrene, and pyrene, in spite of the fact that the free ligands show very large changes in E^0 values. The HOMO of the $(arene)_2Cr$ complex, which is composed mostly of the $3d_{z^2}$ Cr orbital, is apparently σ-bonding with respect to the arene ring, and the π-levels of the arene have little effect on the energy of the HOMO [131].

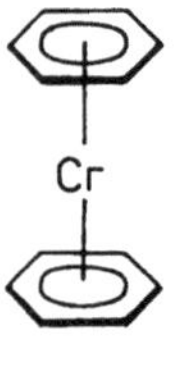

13

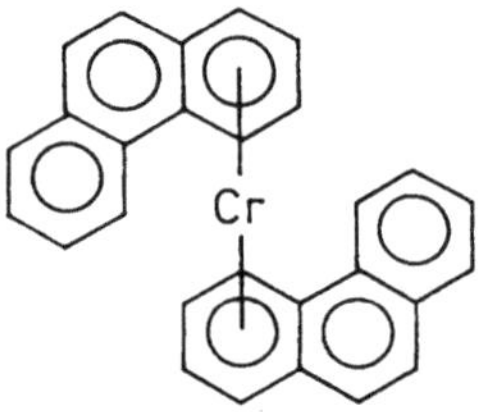

14

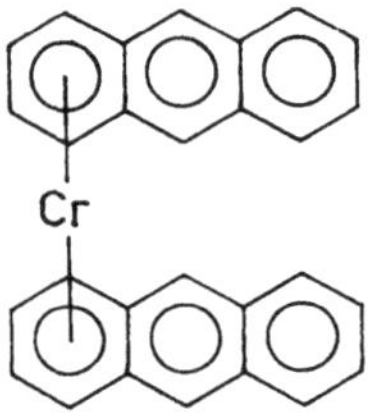

15

The ESR spectra of $(arene)_2Cr^+$ establish that the electronic ground state of the 17-electron, d^5, cation is $^2A_{1g}$, predominantly metal based [140], derived from oxidation of the fully occupied d_{z^2} HOMO of the neutral complex [121]. Several $(arene)_2Cr^+$ cations with polycyclic hydrocarbons as the arene have also been studied by ESR. Usually, the d^5 complex is prepared by controlled air oxidation of the d^6 precursor. The bis-phenanthrene complex **14** is oxidized at -0.65 V *vs.* SCE in DMF to an ESR-active monocation [141]. It and the cation derived from the bis-anthracene complex **15** both have the predicted $^2A_{1g}$ ground state, but are more prone to arene displacement than are analogous monocyclic ring complexes [141, 142]. This may be due to the more facile formation of a tetrahapto intermediate in the anthracene and phenanthrene cases, allowing solvent attack on the coordinatively-unsaturated slipped arene complex. The cation from **15** is also easily hydrogenated at the 9, 10 positions of the anthracene ring. In contrast, the bis-biphenylene complex **16** is solvolytically inert in the oxidized form. In this case, there is no driving force for formation of the slipped, tetrahapto, arene [143]. The subject of arene slippage will be treated in greater detail below.

Elschenbroich, *et al.*, have also studied the monocation derived from octamethyltetrasila-η^{12}-[2.2]paracyclophane chromium, **17**. Its E° was essentially unchanged from that of $(benzene)_2Cr^{0/+}$ [144]. The ESR spectra indicated that the metal/ligand mixing of the HOMO was only slightly greater in **17**$^+$ than in the analogous cation without annelated rings, $[(SiMe_3)C_6H_5]_2Cr^+$ [145].

Studies of two heterocyclic analogues of $(arene)_2Cr$ complexes have recently appeared. Bis (arsabenzene)Cr, **18**, may be prepared by co-condensation of Cr

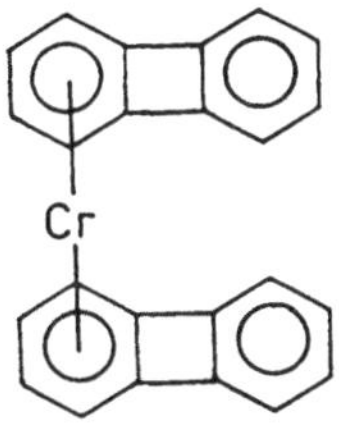

16

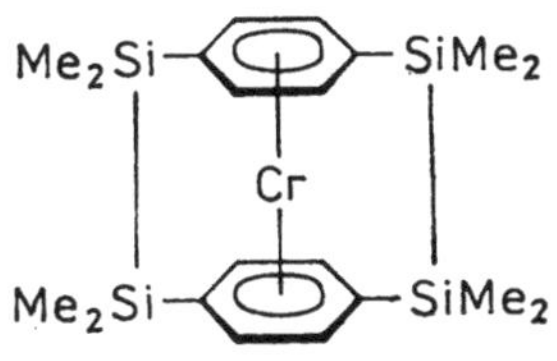

17

with arsabenzene. It was shown by cyclic voltammetry to undergo a reversible one-electron oxidation at $E° = -0.51$ V *vs.* SCE in dimethoxyethane, shifted about 170 mV positive from $(benzene)_2Cr^{0/+}$. The ESR spectrum of the cation showed considerably more ligand character in the half-filled orbital than does the corresponding benzene complex [146]. The analogous bis-pyridine complex **19** and the mixed-sandwich complex **20** have also been prepared, and studied by electrochemistry and ESR spectroscopy [147]. The $E°$ of **19** is -0.45 V *vs.* SCE.

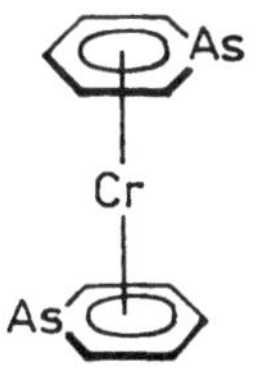

18

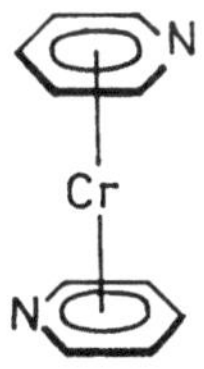

19

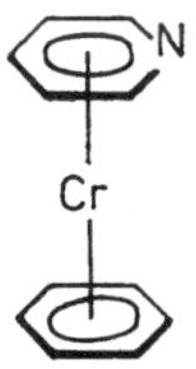

20

Chromium complexes pi-bonded to two $(\eta^6\text{-}C_5H_5BR)^-$ ligands undergo reversible one-electron reduction to their Cr(I) monoanions at $E° = -1.26$ V *vs.* SCE (R = Me) or -1.01 V (R = Ph) [66]. Oxidations to the Cr(III) complexes at *ca.* +0.5 V were irreversible.

The heavier metal analogues $(arene)_2M$, M = Mo or W, remain relatively unexplored. Both neutral and cationic complexes of this type are known [122, 148], and the cations have the expected $^2A_1$ ground state. The ionization potentials are very low (<6 eV) [122]. However, the only electrochemistry of which we are aware, involving the Mo or W complexes is unpublished work by W. J. Bowyer, who found $E° = -0.76$ V *vs.* SCE for the oxidation of $(mesitylene)_2W$ in CH_2Cl_2 [149]. The stability of the monocation was confirmed by controlled-potential coulometry and by ESR spectroscopy, for which the latter showed $\langle g \rangle = 1.968$, consistent with a $d^5\ ^2A_1$ formulation.

There has been brief mention of irreversible oxidations of the $(arene)_2M$ monocations to dications [141, 144, 149] *ca.* 1.7 V positive of the 0/+ couple.

6.3.2 *Reductions*

The $E°$ of $(benzene)_2Cr/(benzene)_2Cr^-$ is apparently too negative to be measured by voltammetry in conventional solvents [131], but the 18-electron complex can be reduced by potassium in dimethoxyethane containing a crown ether. The resulting anion displays a sharp-line ESR spectrum in fluid solution at $-55\,°C$ [150], revealing coupling of the unpaired electron to four types of hydrogens. One is a set of 6 equivalent nuclei from a planar benzene ring, $a_H = 4.65$ G. The other arene ring shows coupling to three different sets of 2 H, with a_H values of 1.00 G, 1.57 G, and 7.75 G. The half-filled orbital appears to contain predominantly ligand pi character. This was surprising, since the LUMO of the neutral complex was thought to be of e_{1g} character, approximately 75% metal d_{xz}, d_{yz} [151]. The assigned $^2E_{2u}$ ground state for the monoanion, $(a_{1g})^2(e_{2u})^1(e_{1g})^0$, is subject to a static Jahn-Teller distortion [150]. The most likely structural distortion of the anion is a slippage of the arene from a planar, η^6, bonding mode, to one **(21)** in which the arene donates only four electrons to the metal. Arene bending is a well-established phenomenon for electron rich metals (*vide infra*), but the (benzene)$_2$Cr$^-$ anion is the only reported example of such a distortion taking place in a formal 19-electron complex of a first-row metal. Electron transmission spectroscopy measurements on (benzene)$_2$Cr assign the LUMO as the e_{1g} orbital [152], leading one to conclude that there is disagreement about the electronic ground state of (benzene)$_2$Cr$^-$, unless the e_{2u} and e_{1g} orbitals invert position in going from the gas-phase to solution-phase ions.

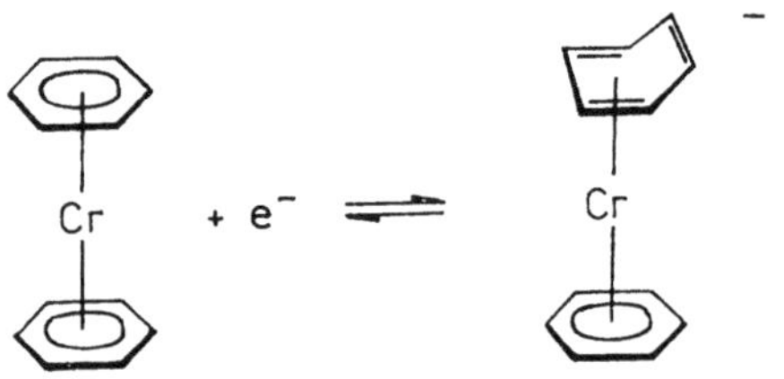

21

The reduction of other (arene)$_2$Cr complexes has been reported. In some cases, such as those in which a nitrophenyl substituent is on one benzene, the anion should clearly be classified as a metal-stabilized organic anion radical [153]. In other cases, the issue is not so obvious. The bis-phenanthrene complex **14** is reduced to its anion at -2.24 V *vs.* SCE, the half-filled orbital being assigned to e_1-type in this case, a mixture of Cr d_{xz}, d_{yz} and ligand π^*. The situation for the bis(naphthalene)chromium anion [154] is ambiguous [131], but calculations show that the LUMO of (naphthalene)$_2$Cr is essentially a naphthalene π^* orbital [155].

The bis(arsabenzene) complex **18** has a relatively positive $E°$ for the 0/− couple (-1.36 V *vs.* SCE), although lower temperatures are needed to stabilize the anion [146]. Finally, the monanions and ESR spectra of trimethylsilyl-derivatives of (benzene)$_2$Cr have been reported [156].

Two recent papers have implicated 19-electron anions of (arene)$_2$M in synthetic transformations. Arene labilization occurs when bis(toluene)tungsten is treated with K/THF. In the presence of PMe$_3$, the product is (toluene)W(H)$_2$(PMe$_3$)$_2$ [157]. In another study, (benzene)$_2$M, M = Cr or Mo, with one equivalent of K in glyme catalyzes the dehydrogenation of 1,4 cyclohexadiene to benzene [158].

6.4 Manganese Group

The stable form of (arene)$_2$Mn complexes is the 18 e$^-$ monocation [159, 160]. These complexes are not available in high yield and no redox chemistry has been reported. The cations can also be prepared by metal atom vapor deposition in

1–2% yield when I_2 is present (arene = toluene, xylene, mesitylene). Presumably the 19-electron neutral complexes are initially formed, followed by oxidation by iodine to give $(arene)_2Mn^+$ [161].

6.5 Iron Group

Complexes of $(arene)_2Fe^{n+}$, where n = 0, 1, 2, have been prepared [148]. An improved method of preparation has facilitated studies of the redox reactions of the 20-electron neutral complex $(C_6Me_6)_2Fe$. By controlling the conditions of the Na/Hg reduction of $(C_6Me_6)_2Fe^{2+}$, either $(C_6Me_6)_2Fe^+$ or $(C_6Me_6)_2Fe$ may be prepared in over 80% yield. The neutral complex is paramagnetic, with two unpaired electrons, so it most likely has two planar hexamethylbenzene rings, and is a true 20-electron species [162]. We will see below that the analogous Ru complex has one bent hexamethylbenzene ligand. Mossbauer spectroscopy on both $(C_6Me_6)_2Fe^+$ and $(C_6Me_6)_2Fe$ indicate that the HOMO in the 19- and 20-electron systems is the expected e_{1g}, and a high degree of metal character (>80% d_{xz}, d_{yz}) was indicated, on the basis of the Mossbauer quadrupole splittings. An ESR study of the Fe(I) monocation has confirmed this formulation of the half-filled orbital [163].

The $E°$'s of the couples $(C_6Me_6)_2Fe^{2+}/(C_6Me_6)_2Fe^+$ and $(C_6Me_6)_2Fe^+/(C_6Me_6)_2Fe$ are, respectively, −0.48 V and *ca.* −1.46 V *vs.* SCE in aqueous ethanol [164]. The more negative reduction was irreversible, ubdoubtedly due to the reaction of the neutral complex with the solvent. Given the value of the Fe(I)/Fe(0) wave, it is not surprising that the neutral complex is a strong reducing agent, about as strong as decamethylcobaltocene. Astruc has investigated the reactions of $(C_6Me_6)_2Fe$ with reducible substrates. Aryl or alkyl halides add an R group to a hexamethylbenzene ring, giving the cyclohexadienyl compounds **22**. It is likely

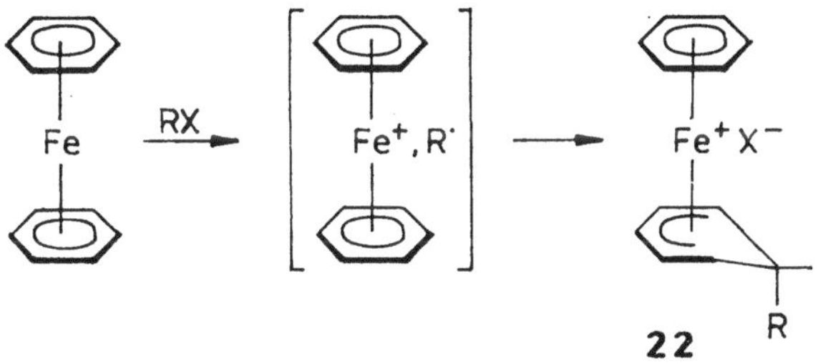

22

that the mechanism of the reaction involves initial electron transfer from $(C_6Me_6)_2Fe$ to RX and formation of the [Fe(I), R] radical pair shown above [165]. When $(C_6Me_6)_2Fe$ is treated with dry O_2 at room temperature, the monocation $(C_6Me_6)_2Fe^+$ is produced in a simple electron-transfer reaction. However, when the oxidation is carried out at −40 °C, the *o*-xylylene complex **23** is produced. This represents an example of C—H bond activation, and probably arises through an electron-transfer reaction producing the radical pair [Fe(I), O_2^-], followed by H^+ abstraction by the superoxide. Complex **23** can be derivatized by addition to the exo double bond [165].

Bis-arene complexes of ruthenium make a fascinating story in relating metal electron counts to the structures of coordinated aromatic ligands. The bis (hexamethylbenzene)Ru dication, **24**, is a d^6, 18-electron complex with two planar hexamethylbenzene rings bonded to the Ru in a symmetrical fashion. However, the neutral complex **25**, isolated from the reduction of **24** by sodium in liquid ammonia [166], has one partially de-coordinated arene ring and must be formulated as $[(C_6Me_6)Ru(\eta^4\text{-}C_6Me_6)]$ [167]. The latter is an active catalyst for the hydrogenation of arenes [168].

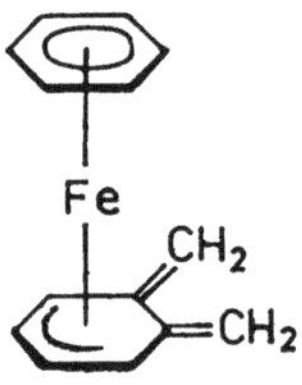

23

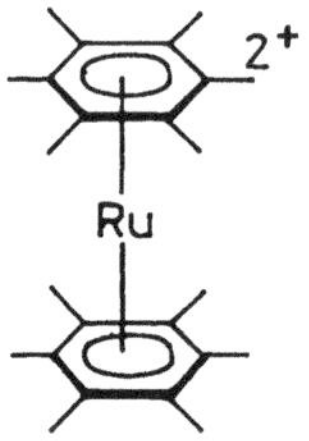

24

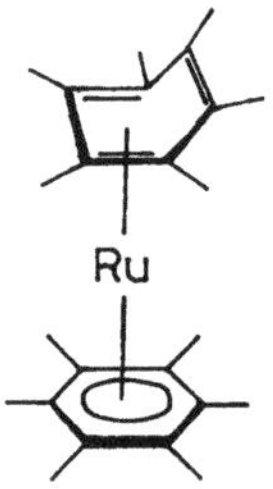

25

It has been shown that **24** and **25** are related by a two-electron couple with a formal potential of $E^\circ = -1.02$ V *vs.* SCE in acetonitrile [169]. The unsymmetrical complexes comprised of one hexamethylbenzene ring and one durene or p-xylene ring show similar behavior [169]. When the solvent is dichloromethane [170], splitting of the 2-electron cyclic voltammetric wave occurs, and the two-one-electron processes of Eq. 6 are observed directly:

$$(C_6Me_6)_2Ru^{2+} + e^- \rightleftharpoons (C_6Me_6)_2Ru^{+} + e^- \rightleftharpoons (C_6Me_6)_2Ru \qquad (6)$$

Since the overall two-electron reduction of the $(arene)_2Ru$ dications requires bending of one arene, it was reasoned that the process should be facilitated by an arene that is predisposed to the bent configuration. Cyclophanes are just such a class of arenes. Thus, the authors of Refs. [169] and [170] replaced one hexamethylbenzene ring with any one of a number of cyclophanes and investigated the redox chemistry of the unsymmetrical complexes. Taking the data on complex **26**, which employs a $[2_4](1,2,4,5)$cyclophane ligand, as an example, the E° (−0.50 V *vs.* SCE in CH_3CN) of all cyclophane complexes were shifted positive compared to $(C_6Me_6)_2Ru^{2+/0}$, indicative of the stabilization of the tetrahapto arene bonding in the fully reduced, Ru(0), product [170].

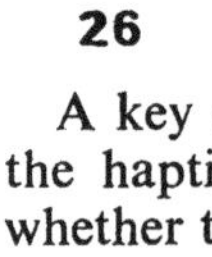

26

A key question in the consecutive electron-transfer reactions of Eq. 6 involves the hapticity of the bendable arene in the Ru(I) intermediate. Depending on whether the arene is hexa- or tetra-hapto-bound, the Ru(I) monocation has either the 19-electron or 17-electron configuration, respectively. This question cannot be answered by the available electrochemical data [170]. However, an analogous system involving Cp*Rh in place of $(C_6Me_6)Ru$ has been shown to have a planar arene intermediate with a 19-e$^-$ metal in an identical electron-transfer series involving the sequence $Cp^*Rh(\eta^6\text{-}C_6Me_6)^{2+}/Cp^*Rh(\eta^6\text{-}C_6Me_6)^+/Cp^*Rh(\eta^4\text{-}C_6Me_6)$ [171–174].

It is interesting to compare the separation between the $E°$ values of two one-electron couples for the Fe(II)/Fe(I)/Fe(0) processes with that found for the analogous Ru complexes. The two processes are separated by about one volt in the Fe case, but essentially 0 volt in the Ru case. There are probably two reasons why the M(0) complex is so greatly stabilized with respect to the M(I) complex in the Ru case. Certainly, the bending of the arene ring, which occurs in the Ru(0) complex but not in the Fe(0) complex, is one reason. However, it should be recalled that the Ru(I) oxidation state is thermodynamically disfavored with respect to disproportionation to Ru(II) and Ru(0), irrespective of the type of ligands involved, as the dearth of known examples of d^7 Ru will attest.

6.6 Cobalt Group

The cobalt hexamethylbenzene complexes $(C_6Me_6)_2Co^{n+}$ have been reported as isolated in three different oxidation states with n = 0, 1, 2 [175]. The monocation is a paramagnetic, 20-electron, system. The crystal structure of the Co(I) complex shows that both hexamethylbenzene rings are planar and hexahapto-coordinated to the metal [176]. Thus, the 20-electron Fe(0) and Co(I) complexes with two hexamethylbenzene ligands are isostructural, and do not undergo arene bending in order to relieve the high electron density at the metal.

The $E°$ potential of the couple $(C_6Me_6)_2Co^{2+}/(C_6Me_6)_2Co^+$ is +0.52 V *vs.* SCE in propylene carbonate [177, 178]. The ESR spectrum of the 19-electron Co(II) species was investigated at low temperatures, and the half-filled e_{1g} orbital was concluded to be about 2/3 on the metal [177]. When the Co(I) monocation is produced electrochemically in the presence of diolefins, one of the hexamethylbenzene ligands is displaced, leading to high yields of $(C_6Me_6)Co(diolef)^+$ [178].

A second reduction wave is seen in cyclic voltammetry scans of $(C_6Me_6)_2Co^{2+}$ [177, 178]. The wave is chemically reversible, $E° = -1.73$ V *vs.* SCE, arguing for the existence of the 21-electron system $(C_6Me_6)_2Co$, prepared earlier by Fischer under very different conditions [175].

The analogous 19 and 20-electron Rh complexes, $(C_6Me_6)_2Rh^{n+}$ (n = 2, 1) have been reported [175], but apparently the report of the Rh(I) complex is incorrect [176].

7. Conclusions

Electron-transfer studies of symmetrical metal sandwich compounds have provided a fertile ground for growth in our knowledge of structure and bonding of organometallic pi-complexes. However, there are relatively few studies of the reactions of the complexes generated in reactive oxidation states. Some of the results reviewed above, most notably those on the reductive elimination of the cyclopentadienyl ligand from metallocenes, suggest that expansion of reactivity studies could yield synthetic dividends. The work of Astruc and co-workers, not part of the subject of this review because it relates mainly to mixed-sandwich Fe complexes containing one cyclopentadienyl and one arene ligand, strengthens this viewpoint [179].

Finally, a survey of the electrochemical literature shows unacceptably large variations in the reported heterogeneous electron-transfer rates for the ferrocene/ferrocenium couple. Until the reasons for these variations are understood, it cannot be assumed that this couple will behave in a Nernstian (electrochemically reversible) fashion under all electrochemical conditions.

8. Acknowledgment

The author is grateful to the National Science Foundation for support during the writing of this review, and to Jon Merkert for his help with the drawings.

9. References

1. J. A. Page and G. Wilkinson, J. Amer. Chem. Soc. **74** (1952) 6149.
2. (a) S. P. Gubin, Pure Appl. Chem. **23** (1970) 463, (b) C. K. Mann and K. Barnes, Non-aqueous Electrochemistry, Dekker, New York, 1970, p. 418, (c) M. D. Morris in A. J. Bard (Ed.), Electroanalytical Chemistry, Dekker, New York, 1974, vol. **7**, p. 79, (d) L. I. Denisovich and S. P. Gubin, Russ. Chem. Reviews **46** (1977) 27.
3. J. C. Kotz in A. J. Fry and W. E. Britton (Eds.) Topics in Organic Electrochemistry, Plenum Press, New York, 1986, Chapt. 3.
4. J. E. Sheats, J. Organomet. Chem. Libr. **7** (1979) 461; CA 91: 74661s.
5. The yearly reviews of ferrocene literature are a fertile source of information on applications of the ferrocene/ferrocenium couple. See, for example, B. W. Rockett and G. Marr, J. Organomet. Chem. **318** (1987) 231, and other papers in that series. A recent paper providing leading references to the literature of metallocene polymer electrodes is R. A. Saraceno. G. H. Riding, H. R. Allcock, and A. G. Ewing, J. Amer. Chem. Soc. **110** (1988) 7254.
6. (a) G. Gritzner and J. Kuta, Pure Appl. Chem. **56** (1984) 461 (b) R. R. Gagne', C. A. Koval, G. C. Lisensky, Inorg. Chem. **19** (1980) 2854.
7. D. M. P. Mingos in G. Wilkinson and F. G. A. Stone (Eds.) Comprehensive Organometallic Chemistry, Pergamon Press, Oxford, 1982, Vol. **3**, p. 28.
8. F. A. Cotton and G. Wilkinson, Advanced Inorganic Chemistry, J. Wiley and Sons, New York, 1980, 4th ed., p. 101.
9. T. A. Albright, J. K. Burdett, and M.-H. Whangbo, Orbital Interactions in Chemistry, J. Wiley and Sons, New York, 1985, p. 393.
10. A. Veillard and J. Demuynck in H. F. Schaefer (III) (Ed.) Applications of Electronic Structure Theory, Plenum, New York, 1977, p. 203.
11. (a) A. H. Cowley, Progr. Inorg. Chem. **26** (1979) 45, (b) J. C. Green, Struct. Bonding **43** (1981) 37.
12. A. Stebler, A. Furrer, and J. H. Ammeter, Inorg. Chem. **23** (1984) 3493.
13. S. Evans, J. C. Green, and S. E. Jackson, J. Chem. Soc., Faraday II, **68** (1972) 249.
14. T. J. Kealy and P. L. Pauson, Nature **168** (1951) 1039.
15. G. Wilkinson, M. Rosenblum, M. C. Whiting, and R. B. Woodward, J. Amer. Chem. Soc. **74** (1952) 2125.
16. (a) J. G. Mason and M. Rosenblum, J. Amer. Chem. Soc. **82** (1960) 4206, (b) W. F. Little, C. N. Reilley, J. D. Johnson, K. N. Lynn, and A. P. Sanders, J. Amer. Chem. Soc. **86** (1964) 1376.
17. (a) S. Toma, A. Gaplovsky, M. Hudecek, and Z. Langfelderova, Monats. Chem. **116** (1985) 357, (b) A. G. Nagy and S. Toma, J. Organomet. Chem. **282** (1985) 267, (c) *Ibid.* **266** (1984) 257.
18. (a) G. L. K. Hoh, W. E. McEwen, and J. Kleinberg, J. Amer. Chem. Soc. **83** (1961) 3949,

(b) W. F. Little, C. N. Reilley, J. D. Johnson, and A. P. Sanders, J. Amer. Chem. Soc. **86** (1964) 1382, (c) D. W. Hall and C. D. Russell, J. Amer. Chem. Soc. **89** (1967) 2316.

19. A. M. Al-Saeed, E. A. Seddon, K. R. Seddon, A. A. Shimran, S. Tompkins, M. C. Grossel, and J. P. Knychala, J. Organomet. Chem. **347** (1988) C25.
20. R. M. G. Roberts and J. Silver, J. Organomet. Chem. **263** (1984) 235.
21. K.-P. Stahl, G. Boche, and W. Massa, J. Organomet. Chem. **277** (1984) 113.
22. W. E. Britton, R. Kashyap, M. El-Hashash, M. El-Kady, and M. Herberhold, Organometallics **5** (1986) 1029.
23. R. A. Saraceno, G. H. Riding, H. R. Allcock, A. G. Ewing, J. Amer. Chem. Soc. **110** (1988) 980.
24. J. Georges and S. Desmettre, Electrochim. Acta **29** (1984) 521.
25. A. M. Bond, M. Fleischman, and J. Robinson, J. Electroanal. Chem. **180** (1984) 257.
26. A. M. Bond and T. F. Mann, Electrochim. Acta **32** (1987) 863.
27. E. Fujita, B. Gordon, M. Hillman, and A. G. Nagy, J. Organomet. Chem. **218** (1981) 105.
28. M. Hillman and A. C. Larson, J. Organomet. Chem. **280** (1985) 389.
29. R. J. Gale, P. Singh, and R. Job, J. Organomet. Chem. **199** (1980) C44.
30. T. Drewello, C. B. Lebrilla, H. Schwarz, and T. Ast, J. Organomet. Chem. **339** (1988) 333.
31. N. El Murr and A. Chaloyard, J. Organomet. Chem. **193** (1980) C60.
32. N. El Murr, A. Chaloyard, and E. Laviron, Nouv. J. Chim. **2** (1978) 15.
33. Y. Mugnier, C. Moise, J. Tirouflet, and E. Laviron, J. Organomet. Chem. **186** (1980) C49.
34. N. Ito, T. Saji, and S. Aoyagui, J. Organomet. Chem. **247** (1983) 301.
35. Z. M. Koepp, H. Wendt, and H. Strehlow, Z. Elektrochem. **64** (1960) 483.
36. S. I. Bailey, W.-P. Leung, and I. M. Ritchie, Electrochim. Acta **30** (1985) 861.
37. A. M. Bond, E. A. McLennan, R. S. Stojanovic, and F. G. Thomas, Anal. Chem. **59** (1987) 2853.
38. W. E. Geiger, Prog. Inorg. Chem. **33** (1985) 275.
39. K. M. Kadish, J. Q. Ding, and T. Malinski, Anal. Chem. **56** (1984) 1741.
40. M. Sharp, Electrochim. Acta **28** (1983) 301.
41. T. Saji, Y. Maruyama, and S. Aoyagui, J. Electroanal. Chem. **86** (1978) 219.
42. J. Leddy and A. J. Bard, J. Electroanal. Chem. **189** (1985) 203.
43. R. M. Crooks and A. J. Bard, J. Electroanal. Chem. **243** (1988) 117.
44. N. R. Armstrong, R. K. Quinn, and R. Vanderburgh, J. Electrochem. Soc. **123** (1976) 646.
45. A. J. Zara, S. S. Machado, L. O. S. Bulhoes, A. V. Benedetti, and T. Rabockai, J. Electroanal. Chem. **221** (1987) 165.
46. M. Peover in A. J. Bard (Ed.), Electroanal. Chem. **2** (1967) 1.
47. T. Gennett, D. F. Milner, and M. J. Weaver, J. Phys. Chem. **89** (1985) 2787.
48. T. Gennett and M. J. Weaver, J. Electroanal. Chem. **186** (1985) 179.
49. M. I. Montenegro and D. Pletcher, J. Electroanal. Chem. **200** (1986) 371.
50. D. O. Wipf, E. W. Kristensen, M. R. Deakin, and R. M. Wightman, Anal. Chem. **60** (1988) 306.
51. J. Daschbach, D. Blackwood, J. W. Pons, and S. Pons, J. Electroanal. Chem. **237** (1987) 269.
52. J. R. Eyler and D. E. Richardson, J. Amer. Chem. Soc. **107** (1985) 6130.
53. R. M. Nielson, M. N. Golovin, G. E. McManis, and M. J. Weaver, J. Amer. Chem. Soc. **110** (1988) 1745.
54. Z. J. Karpinski, C. Nanjundiah, and R. A. Osteryoung, Inorg. Chem. **23** (1984) 3358.
55. R. Prins, A. R. Korswagen, and A. G. T. G. Kortbeek, J. Organomet. Chem. **39** (1972) 335.
56. A. V. Benedetti, Ecletica Quim. **9** (1984) 13; Chem. Abstr. no. 104: 118340e.
57. P. Lemoine, M. Gross, P. Braunstein, F. Mathey, B. Deschamps, and J. H. Nelson, Organometallics **3** (1984) 1303.
58. A. J. Ashe III, S. Mahmoud, Ch. Elschenbroich, and M. Wunsch, Angew. Chem., Int. Ed. **26** (1987) 229.
59. E. Roman, A. M. Leiva, M. A. Casasempere, C. Charrier, F. Mathey, M. T. Garland, and J.-Y. Le Marouille, J. Organomet. Chem. **309** (1986) 323.

60. Ch. Elschenbroich, E. Bilger, R. D. Ernst, D. R. Wilson, and M. S. Kralik, Organometallics **4** (1985) 2068.
61. L. Stahl, H. Ma, R. D. Ernst, I. Hyla-Kryspin, R. Gleiter, and M. L. Ziegler, J. Organomet. Chem. **326** (1987) 257.
62. J. Blumel, F. H. Kohler, and G. Muller, J. Amer. Chem. Soc. **110** (1988) 4846.
63. W. E. Geiger and D. E. Brennan, Inorg. Chem. **21** (1982) 1963.
64. D. B. Jacobson, G. D. Byrd, and B. S. Freiser, J. Amer. Chem. Soc. **104** (1982) 2320.
65. J. D. L. Holloway and W. E. Geiger, J. Amer. Chem. Soc. **101** (1979) 2038.
66. U. Koelle, J. Organomet. Chem. **157** (1978) 327.
67. J. L. Robbins, N. Edelstein, B. Spencer, J. C. Smart, J. Amer. Chem. Soc. **104** (1982) 1882.
68. M. P. Castellani, S. J. Geib, A. L. Rheingold, and W. C. Trogler, Organometallics **6** (1987) 1703.
69. E. O. Fischer, W. Hafner, and H. O. Stahl, Z. Anorg. Chem. **282** (1955) 47.
70. J. D. L. Holloway, W. L. Bowden, and W. E. Geiger, J. Amer. Chem. Soc. **99** (1977) 7089.
71. J. D. L. Holloway, F. C. Senftleber, and W. E. Geiger, Anal. Chem. **50** (1978) 1010.
72. E. U. van Raaij, S. Monkeberg, H. Kiesele, and H. H. Brintzinger, J. Organomet. Chem. **356** (1988) 307.
73. J. C. Smart and J. L. Robbins, J. Amer. Chem. Soc. **100** (1978) 3936.
74. J. L. Robbins, N. M. Edelstein, S. R. Cooper, and J. C. Smart, J. Amer. Chem. Soc. **101** (1979) 3853.
75. (a) D. P. Freyberg, J. L. Robbins, K. N. Raymond, and J. C. Smart, J. Amer. Chem. Soc. **101** (1979) 892; (b) L. Fernholt, A. Haaland, R. Seip, J. L. Robbins, and J. C. Smart, J. Organomet. Chem. **194** (1980) 351.
76. J. A. Bandy, F. G. N. Cloke, G. Cooper, J. P. Day, R. B. Girling, R. G. Graham, J. C. Green, R. Grinter, and R. N. Perutz, J. Amer. Chem. Soc. **110** (1988) 5039.
77. T. Kuwana, D. E. Bublitz, and G. Hoh, J. Amer. Chem. Soc. **82** (1960) 5811.
78. L. I. Denisovich, N. V. Zakurin, A. A. Bezrukova, and S. P. Gubin, J. Organomet. Chem. **81** (1974) 207.
79. D. E. Bublitz, G. Hoh, and T. Kuwana, Chem. Ind. (London) **78** (1959) 635.
80. S. P. Gubin, S. A. Smirnova, L. I. Denisovich, and A. A. Lubovich, J. Organomet. Chem. **30** (1971) 243.
81. R. J. Gale and R. Job, Inorg. Chem. **20** (1981) 42.
82. P. G. Gassman and C. H. Winter, J. Amer. Chem. Soc. **110** (1988) 6130.
83. U. Koelle and A. Salzer, J. Organomet. Chem. **243** (1983) C27.
84. U. Koelle and J. Grub, J. Organomet. Chem. **289** (1985) 133.
85. D. O'Hare, J. C. Green, T. P. Chadwick, and J. S. Miller, Organometallics **7** (1988) 1335.
86. M. W. Droege, W. D. Harman, and H. Taube, Inorg. Chem. **26** (1987) 1309.
87. D. C. Liles, A. Shaver, E. Singleton, and M. B. Wiege, J. Organomet. Chem. **288** (1985) C33.
88. A. A. Vlcek, Collect. Czech. Chem. Commun. **30** (1965) 952.
89. S. P. Gubin, S. A. Smirnova, and L. I. Denisovich, J. Organomet. Chem. **30** (1971) 257.
90. W. E. Geiger, J. Amer. Chem. Soc. **96** (1974) 2632.
91. U. Koelle and F. Khouzami, Angew. Chem., Int. Ed. **19** (1980) 640.
92. W. E. Geiger and D. E. Smith, J. Electroanal. Chem. **50** (1974) 31.
93. N. El Murr, R. Dabard, and E. Laviron, J. Organomet. Chem. **47** (1973) C13.
94. N. El Murr, J. E. Sheats, W. E. Geiger, and J. D. L. Holloway, Inorg. Chem. **18** (1979) 1443.
95. W. Siebert, El-Essawi, R. Full, and J. Heck, Z. Naturforsch. **40b** (1985) 458.
96. M. V. Rajasekharan, R. Bucher, E. Deiss, L. Zoller, A. Salzer, E. Moser, J. Weber, and J. H. Ammeter, J. Amer. Chem. Soc. **105** (1983) 7516.
97. W. Klaui and L. Ramacher, Angew. Chem. Int. Ed. **25** (1986) 97.
98. R. J. Wilson, L. F. Warren, and M. F. Hawthorne, J. Amer. Chem. Soc. **91** (1969) 758.
99. R. P. Van Duyne and C. N. Reilley, Anal. Chem. **44** (1972) 158.
100. U. Kolle, F. Khouzami, and H. Lueken, Chem. Ber. **115** (1982) 1178.
101. J. Zwecker, H. Pritzkow, U. Zenneck, and W. Siebert, Angew. Chem. Int. Ed. **25** (1986) 1099.

102. H. Wadepohl, and W. Siebert, Z. Naturforsch. **39b** (1984) 50.
103. B. R. Francis, M. L. H. Green, T. Luong-thi, and G. A. Moser, J. Chem. Soc., Dalton Trans. (1976) 1339.
104. D. W. Clack and K. D. Warren, J. Organomet. Chem. **152** (1978) C60.
105. A. Modelli, A. Foffani, M. Guerra, D. Jones, and G. Distefano, Chem. Phys. Lett. **99** (1983) 58.
106. J. C. Giordan, J. H. Moore, J. A. Tossell, and J. Weber, J. Amer. Chem. Soc. **105** (1983) 3431.
107. D. Trifan and L. Nicholas, J. Amer. Chem. Soc. **79** (1957) 2746.
108. K. Jonas and C. Kruger, Angew. Chem., Int. Ed. **19** (1980) 520.
109. K. Jonas, R. Mynott, C. Kruger, J. C. Sekutowski, and Y.-H. Tsay, Angew. Chem., Int. Ed. **15** (1976) 767.
110. K. Jonas, L. Schieferstein, C. Kruger, Y.-H. Tsay, *Ibid.* **18** (1979) 550.
111. K. Jonas, and V. Wiskamp, Z. Naturforsch. **38b** (1983) 1113.
112. K. Jonas, W. Russeler, C. Kruger, and E. Raabe, Angew. Chem., Int. Ed. **25** (1986) 928.
113. K. Jonas, Pure Appl. Chem. **56** (1984) 63.
114. K. Jonas, Angew. Chem., Int. Ed. **24** (1985) 295.
115. D. W. Slocum, T. R. Engelmann, R. L. Fellows, M. Moronski, and S. Duraj, J. Organomet. Chem. **260** (1984) C21.
116. M. S. Paquette and L. F. Dahl, J. Amer. Chem. Soc. **102** (1980) 6621.
117. W. E. Geiger, W. L. Bowden, and N. El Murr, Inorg. Chem. **18** (1979) 2358.
118. U. Koelle, P. P. Infelta, and M. Grätzel, Inorg. Chem. **27** (1988) 879.
119. N. El Murr and E. Laviron, Can. J. Chem. **54** (1976) 3350.
120. W. E. Silverthorn, Adv. Organomet. Chem. **13** (1975) 47.
121. E. L. Muetterties, J. R. Bleeke, E. J. Wucherer, and T. A. Albright, Chem. Rev. **82** (1982) 499.
122. F. G. N. Cloke, A. N. Dix, J. C. Green, R. N. Perutz, and E. A. Seddon, Organometallics **2** (1983) 1150.
123. J. A. Bandy, A. Berry, M. L. H. Green, R. N. Perutz, K. Prout, and J.-N. Verpeaux, J. Chem. Soc., Chem. Commun. (1984) 729.
124. A. H. Maki and T. E. Berry, J. Amer. Chem. Soc. **87** (1965) 4437.
125. F. Calderazzo, Inorg. Chem. **3** (1964) 810.
126. Ch. Elschenbroich and F. Gerson, J. Amer. Chem. Soc. **97** (1975) 3556.
127. Elschenbroich reported in a footnote in J. Amer. Chem. Soc. **106** (1984) 4297 that (benzene)$_2$V/(benzene)$_2$V$^-$ is accessible by cyclic voltammetry, but no potential was given.
128. J. A. Bandy, K. Prout, F. G. N. Cloke, H. C. deLemos, and J. M. Wallis, J. Chem. Soc., Dalton Trans. (1988) 1475.
129. H.-S. Hsiung and G. H. Brown, J. Electrochem. Soc. **110** (1963) 1085.
130. L. P. Yureva, S. M. Peregudova, L. N. Nekrasov, A. P. Korotkov, N. N. Zaitseva, N. V. Zakurin, and A. Yu. Vasilkov, J. Organomet. Chem. **219** (1981) 43.
131. N. Ito, T. Saji, K. Suga, and S. Aoyagui, J. Organomet. Chem. **229** (1982) 43.
132. L. P. Yureva, S. M. Peregudova, D. N. Kratsov, A. Yu. Vasilkov, L. N. Nekrasov, N. L. Asfandiarov, M. M. Timoshenko, and Yu. V. Chizhov, J. Organomet. Chem. **336** (1987) 371.
133. A. Rusina and H.-P. Schroer, Collect. Czech. Chem. Commun. **31** (1966) 2600.
134. H.-P. Schroer and A. A. Vlcek, Z. Anorg. Allg. Chem. **334** (1964) 205.
135. (a) C. Furliani, Ric. Sci. **36** (1966) 989; (b) C. Furliani and G. Sartori, *Ibid.* **28** (1958) 973.
136. P. M. Treichel, G. P. Essenmacher, H. F. Efner, and K. J. Klabunde, Inorg. Chem. Acta **48** (1981) 41.
137. H. Brunner and H. Koch, Chem. Ber. **115** (1982) 65.
138. S. E. Anderson and R. S. Drago, Inorg. Chem. **11** (1972) 1564.
139. T.-T. T. Li and C. H. Brubaker, J. Organomet. Chem. **216** (1981) 223.
140. Ch. Elschenbroich, R. Mockel, U. Zenneck, D. W. Clack, Ber. Bunsenges. Phys. Chem. **83** (1979) 1008.
141. Ch. Elschenbroich, E. Bilger, R. Mockel, Z. Naturforsch. **38b** (1983) 1357.

142. Ch. Elschenbroich, R. Mockel, E. Bilger, *Ibid.* **39b** (1984) 375.
143. Ch. Elschenbroich, J. Schneider, W. Massa, G. Baum, and H. Mellinghoff, J. Organomet. Chem. **355** (1988) 163.
144. Ch. Elschenbroich, J. Hurley, W. Massa, and G. Baum, Angew. Chem., Int. Ed. **27** (1988) 684.
145. Ch. Elschenbroich and J. Koch, J. Organomet. Chem. **229** (1982) 139.
146. Ch. Elschenbroich, J. Kroker, W. Massa, M. Wunsch, and A. J. Ashe III, Angew. Chem., Int. Ed. **25** (1986) 571.
147. Ch. Elschenbroich, J. Koch, J. Kroker, M. Wunsch, W. Massa, G. Baum, and N. Sutin, submitted to Angew. Chem.
148. See references in Table III of Ref. 120.
149. W. J. Bowyer and W. E. Geiger, unpublished work at the University of Vermont, 1983. The sample was provided by M. L. H. Green.
150. Ch. Elschenbroich, E. Bilger, J. Koch, and J. Weber, J. Amer. Chem. Soc. **106** (1984) 4297.
151. J. Weber, M. Geoffroy, A. Goursot, and E. Penigault, J. Amer. Chem. Soc. **100** (1978) 3995.
152. P. D. Burrow, A. Modelli, M. Guerra, and K. D. Jordan, Chem. Phys. Lett. **118** (1985) 328.
153. Ch. Elschenbroich, E. Bilger, J. Heck, F. Stohler, J. Heinze, Chem. Ber. **117** (1984) 23.
154. G. Henrici-Olive' and S. Olive', J. Amer. Chem. Soc. **92** (1970) 4831.
155. J. H. Osborne, W. C. Trogler, P. D. Morand, and C. G. Francis, Organometallics **6** (1987) 94.
156. Ch. Elschenbroich and J. Koch, J. Organomet. Chem. **229** (1982) 139.
157. M. L. H. Green and G. Parkin, J. Chem. Soc., Chem. Commun. (1985) 771.
158. G. Fochi, Organometallics **7** (1988) 2255.
159. E. O. Fischer and M. W. Schmidt, Chem. Ber. **100** (1967) 3782.
160. M. Tsutsui and H. Zeiss, J. Amer. Chem. Soc. **83** (1961) 825.
162. W. E. Billups, A. W. Moorehead, P.-J. Ko, J. L. Margrave, and J. P. Bell, Organometallics **7** (1988) 2230.
161. P. Michaud, J.-P. Mariot, F. Varret, and D. Astruc, J. Chem. Soc., Chem. Commun. (1982) 1383.
163. M. V. Rajasekharan, S. Giezynski, J. H. Ammeter, N. Oswald, P. Michaud, J. R. Hamon, and D. Astruc, J. Amer. Chem. Soc. **104** (1982) 2400.
164. D. M. Braitsch and R. Kumarappan, J. Organomet. Chem. **84** (1975) C37.
165. A. M. Madonik and D. Astruc, J. Amer. Chem. Soc. **106** (1984) 2437.
166. E. O. Fischer and Ch. Elschenbroich, Chem. Ber. **103** (1970) 162.
167. G. Huttner and S. Lange, Acta Crystallogr. **B28** (1972) 2049.
168. M. Y. Darensbourg and E. L. Muetterties, J. Amer. Chem. Soc. **100** (1978) 7425.
169. E. D. Laganis, R. H. Voegeli, R. T. Swann, R. G. Finke, H. Hopf, and V. Boekelheide, Organometallics **1** (1982) 1415.
170. R. G. Finke, R. H. Voegeli, E. D. Laganis, and V. Boekelheide, Organometallics **2** (1983) 347.
171. W. J. Bowyer and W. E. Geiger, J. Amer. Chem. Soc. **107** (1985) 5657.
172. W. J. Bowyer and W. E. Geiger, J. Electroanal. Chem. **239** (1988) 253.
173. W. J. Bowyer, J. W. Merkert, W. E. Geiger, and A. L. Rheingold, Organometallics **8** (1989) 191.
174. J. Merkert, R. M. Nielson, M. J. Weaver, and W. E. Geiger, J. Amer. Chem. Soc. **111** (1989) 7084.
175. E. O. Fischer and H. H. Lindner, J. Organomet. Chem. **1** (1964) 307; *Ibid.* **2** (1964) 222.
176. M. R. Thompson, C. S. Day, V. W. Day, R. I. Mink, and E. L. Muetterties, J. Amer. Chem. Soc. **102** (1980) 2979.
177. U. Koelle, B. Fuss, M. V. Rajasekharan, B. L. Ramakrishna, J. H. Ammeter, and M. C. Bohm, J. Amer. Chem. Soc. **106** (1984) 4152.
178. J. Edwin and W. E. Geiger, Organometallics **3** (1984) 1910.
179. (a) D. Astruc, Accounts Chem. Res. **19** (1986) 377, (b) J.-P. Hamon and D. Astruc, Organometallics **7** (1988) 1036.

CHAPTER 6

Paramagnetic Main Group and Transition Metal Complexes Containing Unsaturated Nitrogen Ligands

Wolfgang Kaim and Barbara Olbrich-Deussner

Institut für Anorganische Chemie, Universität Stuttgart, Pfaffenwaldring 55, D-7000 Stuttgart 80, West Germany

Table of Contents

1. Introduction

Organometallics are generally considered to form radical species more easily than "normal" organic compounds. In some cases, the weakness of bonds between carbon ligands and transition or main group metals is believed to facilitate free radical formation by homolytic dissociation; in other instances, the highly electropositive character of many metals implicates the presence of electron transfer active "carbanionic" ligands, which may be responsible for radical ion formation by electron transfer to an acceptor *A* (1) [1, 2].

$$L_nM{-}R \nearrow L_nM^{\cdot} + R^{\cdot} \qquad [L_nM^{+}(A^{-\cdot})] + R\cdot \swarrow^{+A} [L_nM^{\delta+}{-}R^{\delta-}] \qquad \text{A: acceptor} \tag{1}$$

Both mechanistic alternatives (1) for the propensity of organometallics to engage in radical processes may contribute to a varying extent in actual examples, in this chapter we shall review and discuss radical (or more general: paramagnetic) organometallics, which owe their stability largely to the presence of unsaturated nitrogen ligands.

All major organometallic fragments will be considered, including alkyl or aryl compounds of the main group metals and transition metal fragments with σ (CO, CNR) and π ligands (cyclo-C_nR_n, olefins). Although many of the radical species described may be viewed as intermediates, e.g. in multi-step processes [3, 4], the emphasis will be on relatively long lived species. Electron spin resonance (ESR) continues to be the primary technique to characterize stable as well as short-lived radicals, nitrogen ligands being particularly informative because of the ^{14}N isotope (99.6% natural abundance, I = 1). However, this article will illustrate that recent progress in the stabilization of organometallic radicals has allowed one to study such compounds by all other methods, including "conventional" spectroscopy, electrochemistry and structure determination. As concerns ESR spectroscopy, recent reviews [5, 6] and exhaustive data collections [7, 8] have appeared on various kinds of radical complexes, we shall therefore emphasize those other methods in our presentation and use the results for a discussion of reactivity.

Unsaturated nitrogen compounds comprise a vast class of important ligands [9], ranging from simple diatomic N_2 to molecules of intermediate size such as pyrazine **2** or 2,2′-bipyridine **3** to the biologically important conjugated macrocycles of the porphyrine type (2).

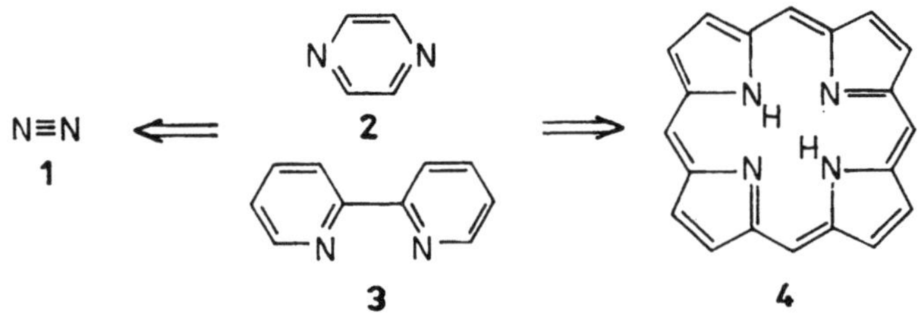

(2)

The excellent capability of many unsaturated nitrogen ligands to stabilize paramagnetic organometallics may be attributed to their potential to accomodate spin in their π system, while the nucleophilic (σ-)lone pair available at most nitrogen centers guarantees a sufficiently strong coordinative bond to the generally electrophilic organometal center (3).

(3)

In fact, most of the compounds falling under the scope of this chapter may be formulated as complexes in which the unsaturated nitrogen ligand has acquired some amount of negative charge. Addition of an electron [10, 11] or electron density (e.g. via back bonding) [12, 13] to a "normal" neutral ligand has been shown to result in a highly increased basicity of such (partially) reduced ligands. In the case of monodentate ligands, this effect amounts to about 5 orders of magnitude of K_{BH^+} [10, 12, 13]; however, chelate ligands such as the ubiquitous 2,2′-bipyridines (**3**) and 1,10-phenanthrolines show an increase of 20 orders of magnitude (4) [11]!

$$\begin{array}{ccc} \mathrm{L} & \xrightarrow[+e^-]{\sim -2.0\,\mathrm{V}} & \mathrm{L}^{-\cdot} \\ pK_{\mathrm{BH}^+}\sim 4.5 \downarrow +\mathrm{H}^+ & & +\mathrm{H}^+ \downarrow pK_{\mathrm{BH}^+}\sim 24 \\ \mathrm{LH}^+ & \xrightarrow[\sim -0.9\,\mathrm{V}]{+e^-} & \mathrm{LH}^{\cdot} \end{array} \qquad \mathrm{L} = \mathbf{3} \tag{4}$$

The increase of charge in the unsaturated ligand may not only be effected by direct chemical or electrochemical addition of electrons [10, 11], but also by the phenomenon of π back donation for which the mostly electron rich organometallic fragments are notorious. As we shall show later (Section 4.4), the consequences of such an increase of electron density within the nitrogen ligand may include (catalytic) activation of the complex [3, 4] and a tendency towards coordinative saturation, i.e. the formation of polynuclear organometallic complexes [14–16].

An important clarification to be made at this point concerns the sometimes imprecise definition of a transition metal containing "radical" [17]. While organometallic chemists often use this term even in those cases where the unpaired electron clearly resides in a *d* orbital at the metal center [18, 19], classical complexes such as $[CuCl_4]^{2+}$ (d^9) [20] or $[IrCl_6]^{3+}$ (l.s. d^5) [21] with substantial delocalization of spin to the ligands are commonly called paramagnetic and not a "radical" species. Another frequent distinction is based on charge, neutral paramagnetic species such as $Mn(CO)_5$ [22] are more often referred to as "radicals" than charged complexes. In this article we shall refer to all paramagnetic organometallics as radicals unless the amount of spin residing on the ligands is very small or uncertain. Those cases where there are comparable contributions from the metal(s) *and* the ligands to the accomodation of unpaired electron density have become of particular interest [23–25] and will be discussed later.

2. Classification of Unsaturated Nitrogen Ligands

Since the π system of the unsaturated ligands is frequently needed to accomodate the unpaired electron in a radical state, such ligands have to bind organometallic fragments primarily through electrostatic forces, especially by acquiring negative charge. The first alternative (class I radical ligands) is that a neutral unsaturated ligand adds an electron so that it becomes a spin *and* charge bearing anion radical [6]. Coordination of such species can then occur with one or more neutral or positively charged electrophilic organometallic fragments, giving rise to a number of possible combinations (5). The use of metal fragments with rather low lying unoccupied orbitals may result in some form of electron spin and charge delocalization from the radical ligand to that fragment with the opportunity for a controlled activation, e.g. in the sense of an (18 + δ) valence electron configuration [3, 4, 14] (Section 4.4).

$$\begin{array}{lll} \mathrm{A} \xrightarrow{+e^-} \mathrm{A}^{-\cdot} \xrightarrow{\mathrm{MR}_n} & [(\mathrm{A}^{-\cdot})(\mathrm{MR}_n)]^{-\cdot} & [54, 97] \\ \qquad\qquad \xrightarrow{n^+\mathrm{MR}_n} & [(\mathrm{A}^{-\cdot})(^{n+}\mathrm{MR}_n)]^{(n-1)+\cdot} & [71] \\ \qquad\qquad \xrightarrow{2\mathrm{MR}_n} & [(\mathrm{A}^{-\cdot})(\mathrm{MR}_n)_2]^{-\cdot} & [52, 101] \\ \qquad\qquad \xrightarrow{2n^+\mathrm{MR}_n} & [(\mathrm{A}^{-\cdot})(^{n+}\mathrm{MR}_n)_2]^{(2n-1)+\cdot} & [46, 102] \\ \qquad\qquad \xrightarrow{\mathrm{MR}_n,\,^+\mathrm{MR}_{n-1}} & [(\mathrm{MR}_n)(\mathrm{A}^{-\cdot})(^+\mathrm{MR}_{n-1})]^{\cdot} & [2, 41, 51] \end{array} \tag{5}$$

Another possibility (6) to arrive at paramagnetic organometallics with unsaturated nitrogen ligands arises in the use of diamagnetic *σ and π* electron rich anions [23, 26] or dianions [25, 27], which can stabilize electron deficient paramagnetic metal centers (6A) by some degree of electron delocalization.

$$R_nM + D{-}H \xrightarrow[-e^-]{-H^+} [(R_nM^{+\cdot})(D^-)]^{\cdot} \leftrightarrow [(R_nM)(D^{\cdot})]^{\cdot} \qquad (6)$$

A B D—H: donor

e.g. $Cp(CO)_2MnD$, $D^- =$ **56–72** (29) [26]

In the alternative resonance structure (6B) of the former type of paramagnetic compounds, the unsaturated nitrogen system may act as a neutral (class II) radical ligand, which then has to be very electron rich in order to bind organometallic fragments. Species with substantial free radical character (6B versus 6A) have thus only been detected in the case of the extremely electron rich p-phenylenediamide systems (**69–72**) [23, 26]. Bifunctional ligands are also possible, the metal centered formulation (7A) then corresponds to a mixed valence dimer situation [24, 28].

$$2R_nM + D{-}H \xrightarrow[-e^-]{-H^+} [(R_nN^{+\cdot})(D^-)(MR_n)]^{\cdot} \leftrightarrow [(R_nM)_2(D^{\cdot})]^{\cdot} \qquad (7)$$

A B

e.g. $[Cp(CO)_2Mn]_2(\mu\text{-}NCC(R)CN)$ [28]

The binding of an organometallic fragment can be enhanced by employing formally dianionic unsaturated nitrogen ligands (8B), either chelating [27] or non-chelating [25]. A resonance situation like (8) may contribute to the stabilization of such odd-electron organometallic fragments as demonstrated in the two examples (**32, 73**) by Trogler and coworkers [25, 27]; while the third resonance formulation (9) with the ligand-centered unpaired electron can be assessed using ESR, the distinction between contributions from (8A) and (8B) is more difficult and may require help from quantum mechanical calculations [25, 27, 86, 87].

$$[(R_nM^{\cdot})(L)]^{\cdot} \leftrightarrow [(R_nM^{2+\cdot})(L^{2-})]^{\cdot} \qquad (8)$$

A B

e.g. $Cp_2^*V(NAr)$ [25]

$$[(R_nM^+)(L^{-\cdot})]^{\cdot} \qquad (9)$$

Paramagnetic ground or very low lying (triplet) excited states may also occur in even electron systems according to (10A, B). An example involving the 2,2′-bipyridine ligand has been reported [29]. There are even 16- [30] and 18-valence electron organometallics with nitrogen ligands [15, 150], which owe their thermally populated triplet states to rather small ligand field splittings (cf. 45).

$$(R_nM)(L) \xrightarrow{700\,cm^{-1}} {}^3[(R_nM^{+\cdot})(L^{-\cdot})] \qquad (10)$$

A B

e.g. $Cp_2Ti(bpy)$ [29]

The following survey of typical ligands serves to illustrate the variety of existing species and to indicate opportunities for further work.

2.1 Class I radical ligands

Beginning with the smallest conceivable system, dinitrogen (N_2) is a common ligand for organometallics [31] and does form an anion radical $\mathbf{1}^{-\cdot}$ at low temperatures [32]. However, there are no true "radical" complexes known; the compound (11) reported by Lappert and coworkers [33] is clearly a metal centered (Zr^{III}, d^1) species with side-on (π) bonded $^{14(15)}N_2$.

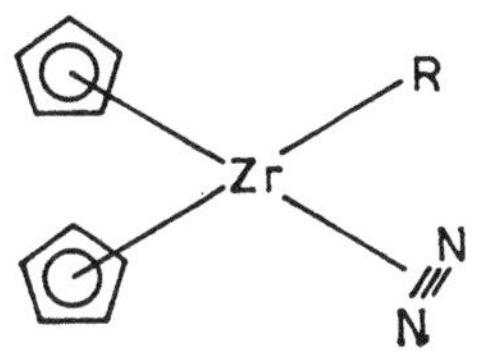

(11)

Another small unsaturated nitrogen ligand is cyanogen N≡C—C≡N which, like N_2, forms binuclear complexes with back-bonding metal fragments [34]. However, no radical species have been reported here either, the tendency of NCCN to undergo reductive cleavage and yield two cyanide ions may require a careful choice of organometallic fragments and reaction conditions to form radical species.

Nitrile substituted π ligands (12) such as fumarodinitrile, tetracyanoethene (TCNE) [35–37], terephthalodinitriles [38], cyanopyridines [38], N, N′-dicyano-p-quinonediimines (DCNQI) [39], and tetracyanoquinodimethane (TCNQ) [40] may be viewed in part as "vinylogous" or "phenylogous" cyanogen compounds; radical complexes with transition metal fragments have been reported [15, 35–40].

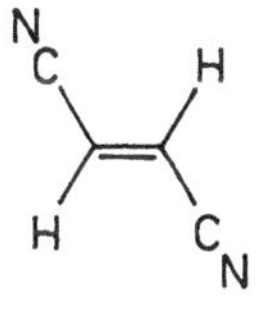

5

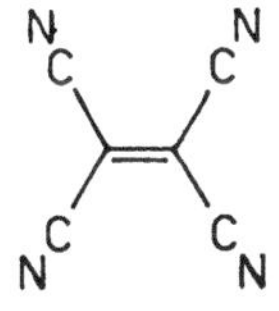

6

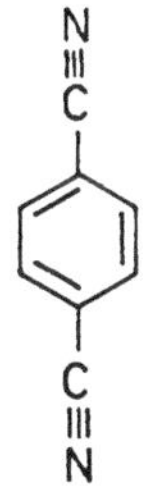

7

(12)

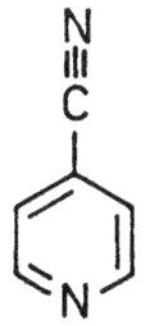

8

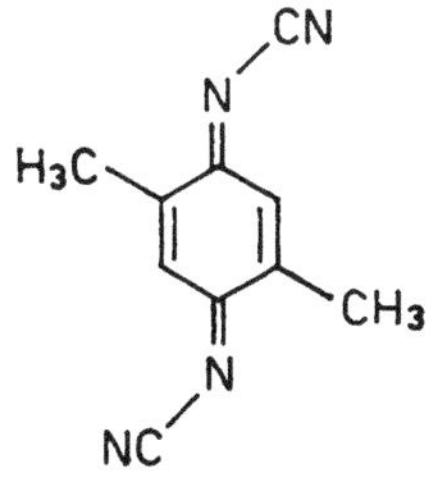

9

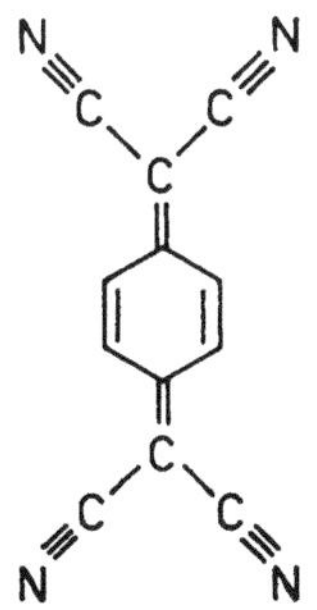

10

Some of these radical complexes show the phenomenon of charge transfer-assisted polynucleation [14], i.e. coordinative saturation of all available nitrile sites occurs only after increase of electron density within the ligand through external (13A) [38] or internal electron transfer (13B, back donation [16, 38] or "one-electron oxidative addition" [15]). A special case is the extremely π electron deficient, yet relatively small TCNE molecule, the "E. coli" of electron transfer chemistry, which engages in *inner* and *outer sphere* radical reactions with main group and transition metal species [1, 35, 37, 41] as will be shown in more detail in Section 4.3.

$$2\,R_nM + L \rightarrow (R_nM)(L) + R_nM \xrightarrow{+e^-} [(R_nM)_2(L^{-\cdot})]^{-\cdot} \qquad (13A)$$

e.g. $Cp(CO)_2Mn$/**7** [38]

$$4\,R_nM + L \rightarrow [(R_nM^{+/4})_4(L^{-\cdot})] \qquad (13B)$$

e.g. $Cp(CO)_2Mn$/**6** [15]

Coordination polymers of Cu(I) with DCNQI "radical" ligands as prepared by Hünig and coworkers have recently received much attention because electron propagation through the copper chain as well as through the ligand radical stacks seems to cause an unprecedentedly high conductivity of this "molecular metal" [39].

$$4\ \mathbf{9} + 2\,CuI \rightarrow I_2 + 2[(\mathbf{9}_2)^{-\cdot}(Cu)^+] \quad [39] \qquad (14)$$

Aryl substituted nitrogen compounds such as nitroso and diazomethane systems [42] can also form anion radicals with potentially σ coordinating N sites, neutral complexes of nitrosobenzene (11) anion radical with cationic d^4, d^6 and d^8 transition metal fragments have been described (15) [43].

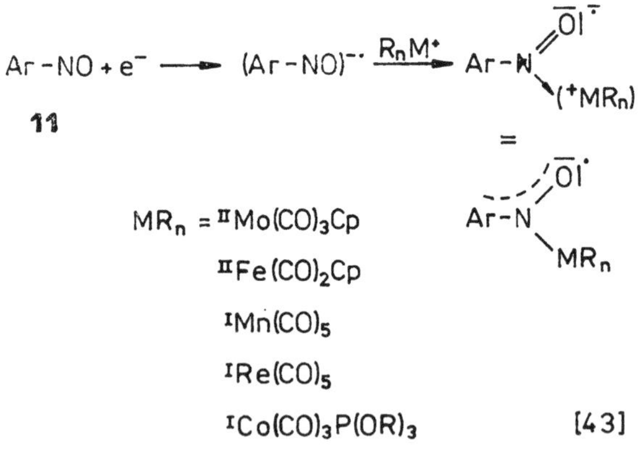

(15)

A major class of unsaturated nitrogen radical ligands is made up by reduced [44] electron deficient [45] N-heterocycles of the pyridine type. Six- and five-membered ring systems (16) have been employed as single components, in substituted, benzannelated, fused, or connected form (17) as non-chelating, often potentially bridging anion radicals with neutral and cationic main group (groups 2–4 and 12 of the periodic table) as well as transition (groups 6 and 7) organometallic fragments [38, 46–60, 120].

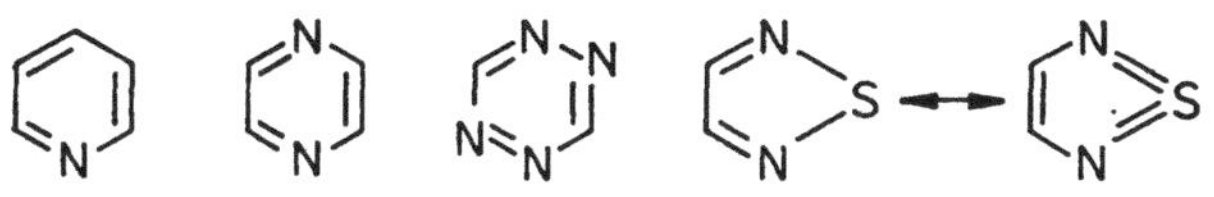

12 2 13 14 (16)

Among the systems (17), the 4,4′-bis(2-pyridyl)tetrazine anion radical ligand **24**$^{-\cdot}$ is unique in that it exhibits a nodal plane at the basic pyridine coordination centers in the singly occupied π MO [58, 60a], the effect of coordination on the spin distribution in the radical is therefore extremely small. N-Methylpyrazinium cation **16** gets reduced to a neutral radical very easily [61], while still offering a coordination site for efficient back-bonding [62, 63]; inorganic [64] and organometallic fragments [60b] can thus be coordinated to this "spin label" system.

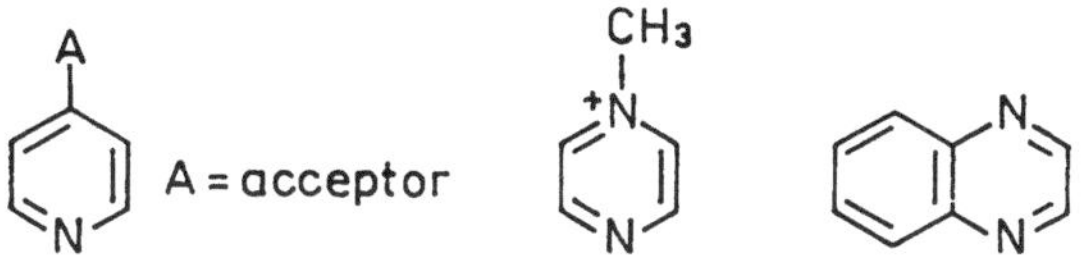

15 16 17

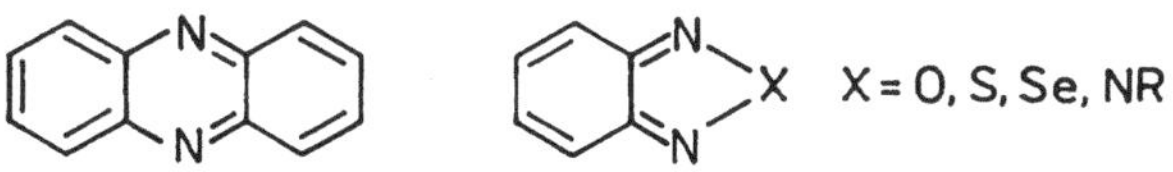

18 19

(17)

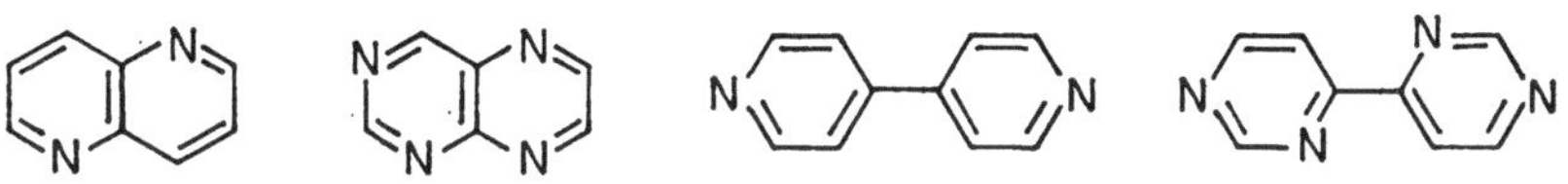

20 21 22 23

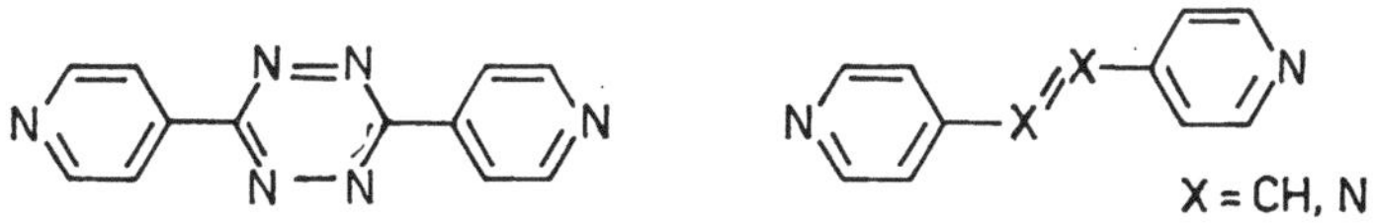

24 25

The coordination-affected equilibrium (18) of organometallic compounds of groups 2 and 3 [65] has left some ambiguity [2] as to what electrophiles coordinate to reduced ligands of the type (17); ESR data for corresponding metal hydride radicals [55] and ENDOR results [52] for escape products from SET processes (cf. Section 4.1) suggest a preference of the anion radicals for the neutral Lewis acidic species of group 3 elements [41]. The highly polar organolithium reagents, on the other hand, were found to form cationic triple ions on SET reactions of RLi with pyrazine **2** or 4,4′-bipyridine **22** [2, 66]. Organomagnesium and bulky organoaluminium compounds take an intermediate position by coordinating as molecules R_nM *and* as cations $R_{n-1}M^+$ (5) [2, 41, 51]; coordinatively unsaturated organosilicon species such as $^{+}SiR_3$ can only act as cations in mono- [67] or binuclear "complexes" [46, 51] with anion radical ligands.

$$2R_nM + \text{solvent} \rightleftharpoons R_{n+1}M^- + R_{n-1}M^+(\text{solvate}) \qquad (18)$$

Among potentially *chelating* paramagnetic nitrogen ligands are the anion radicals of (deprotonated) porphyrins (**4**) [68] and of α-diimines (19).

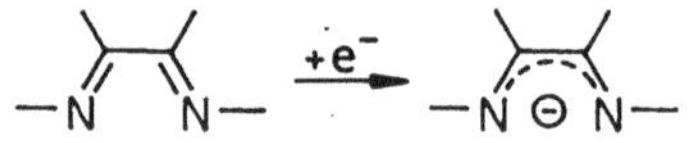

(19)

With respect to cyanogen N≡C—C≡N, the 1,4-diaza-1,3-dienes (dad) [69] are tetraorgano derivatives RN=C(R')—C(R')=NR (**26**). N,N'-Alkyl and -aryl substituted dad anions [70] have been used to form complexes with main group organometallic cations R_nM^+, M = Zn [71], Al [72], Si [73], and with η^2-transition metal carbonyl fragments $M(CO)_{4-n}(PR_3)_n$ (M = Cr, Mo, W) [70, 74, 75], $^{+}M(CO)_4$ (M = Mn, Re) [76a] or $Re(CO)_3Re(CO)_5$ [76b]. Recent work by the groups of van Koten [77, 78] and tom Dieck [79] has shown that the reactions leading to and from the main group radicals can be of considerable synthetic value in organic chemistry. Using modified ligands of the α-iminoketone type (cf. 27, 28), these efforts have recently culminated in simple stereospecific syntheses (20) of β-lactams via electron transfer from organozinc compounds [78, 80]. Organometallic radicals with the spin bearing unsaturated N—C—C—N backbone also include some open chain species such as (21) [81].

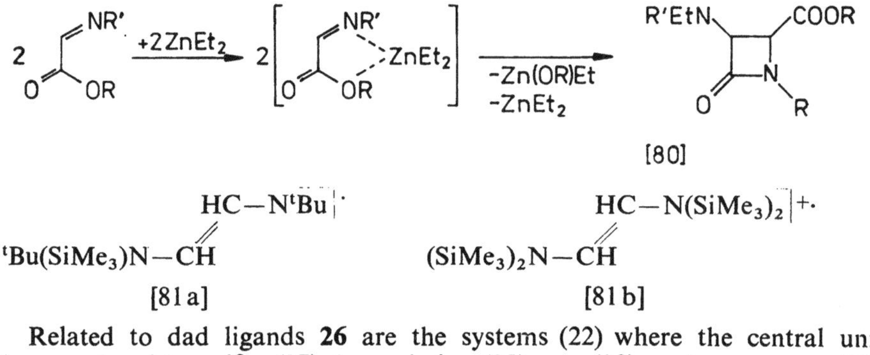

(20)

[80]

$$^{t}Bu(SiMe_3)N{-}CH{=}HC{-}N^{t}Bu\,]^{\cdot} \qquad (SiMe_3)_2N{-}CH{=}HC{-}N(SiMe_3)_2\,]^{+\cdot} \qquad (21)$$

[81a] [81b]

Related to dad ligands **26** are the systems (22) where the central unit has been replaced by sulfur (**27**), by an imino (**28**), azo (**29**), or benzo group (**30**).

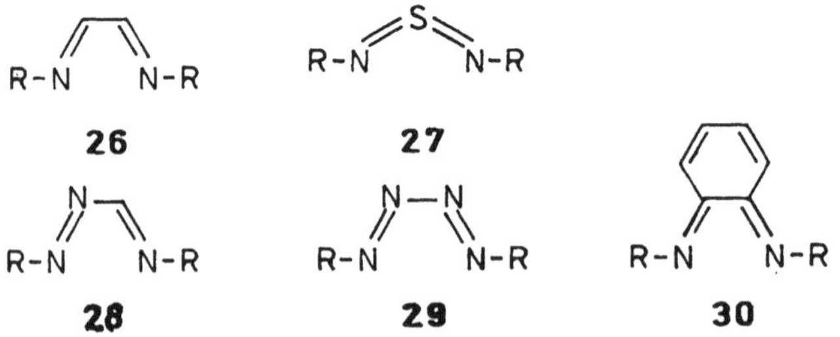

(22)

Main group organometallic [82] and transition metal carbonyl complexes [83] of chelating sulfurdiimine anion radicals are known (cf. the non-chelating analogues **14** and **19** [50]), with the strongly polarizing [84a] $Re(CO)_3Cl$ fragment it is not the reduction but the oxidation of the neutral complex that becomes electrochemically reversible [84b]. This result points to an alternative like (8A, B) for the distribution of oxidation states between metal and ligand, which has been discussed for complexes of triaza- [85] and tetraazadiene (tad) ligands [27, 86, 87]: Mixing of the very low lying unoccupied tad ligand orbitals (π^*) [88] with high lying occupied metal *d* orbitals may result in a highly covalent character of the bonding. Structural studies for **31** have led van Koten and coworkers to favour a formulation **31B**, corresponding to (8B) [87]; ESR data for **32**, which also indicate metal character of the singly occupied MO, were interpreted by Maroney and Trogler in terms of a Co(0) species **32A** with the neutral tad ligand [27], although the tad π system should be highly susceptible to accept additional electrons.

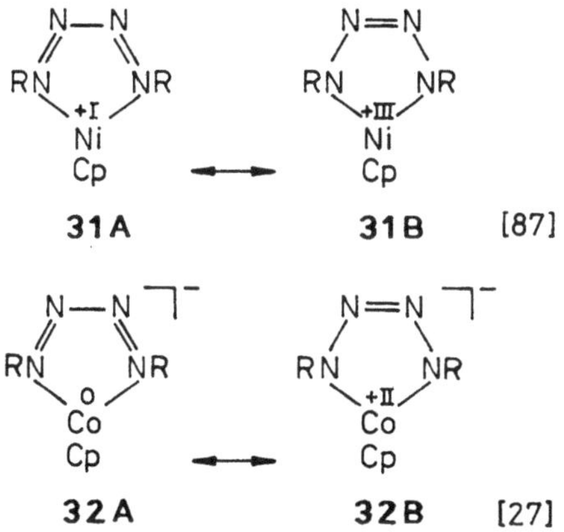

(23)

An important group among the α-diimine radical ligands are the o-semiquinonediimines **30**$^{-}$. [89]. Very facile formation of stable triorganotin(IV) [90] and diorganothallium(III) complexes of such anion radical ligands has been reported by Stegmann and coworkers [91, 92].

As in the series of non-chelating ligands (17), anion radical molecules of the α-diimine type may be "constructed" by combining N-heterocyclic components (16) of the pyridine type [89, 93, 94]. The most famous among these systems is the century-old [95] 2,2'-bipyridine **3**, the "E. coli of coordination chemistry" [96] which, like many related systems (24), forms radical complexes with organometallic main group (RBe^+, RMg^+, R_2B^+, R_2Al^+, R_2Ga^+, RZn^+ [56, 98, 100]) and transition metal fragments ($M(CO)_4$, M = Cr, Mo, W [75, 97]; $Mo(CO)_n(PR_3)_{4-n}$, n = 2–4; $ReCl(CO)_3$ [84]; $^+Rh(nbd)$ [99]). Not only heterocycles (16) but also the imino (C=NR), azo (N=N), and carbonyl (C=O) groups may be combined to form an enormous variety of possible arrangements of potentially radical anionic mono and bis chelate ligands (24) [58, 59, 101–105].

Some of the polynucleating systems have proven so stable that low valent platinum metal and even gold(I) radical complexes (**44**) could be observed for the first time [105]; furthermore, structural (25) and other spectroscopic studies were obtained for completely stable complexes such as **45** [102, 106].

Interest in the electronic structure of these organometallic radicals has come from efforts to understand and modify photo- and electrocatalysts (26), which often contain heterocyclic α-diimine ligands and go through essential "radical" states (26) during the catalytic cycle [107].

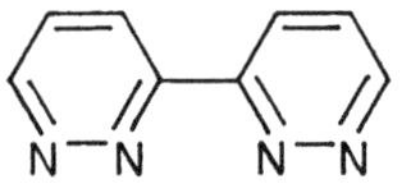

33

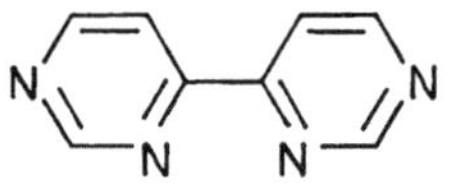

34

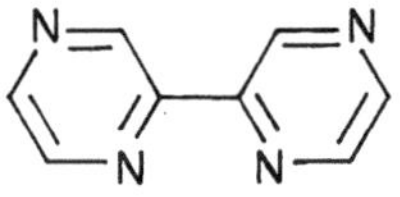

35

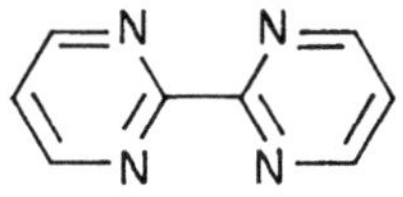

36

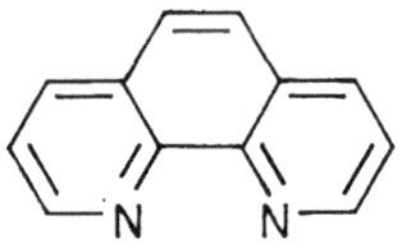

37

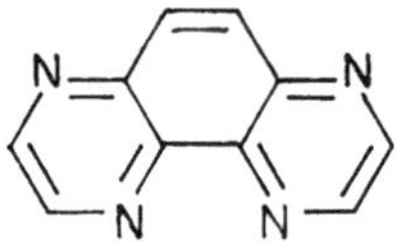

38

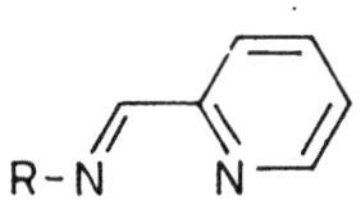

39

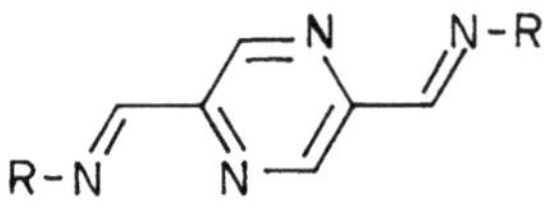

40

(24)

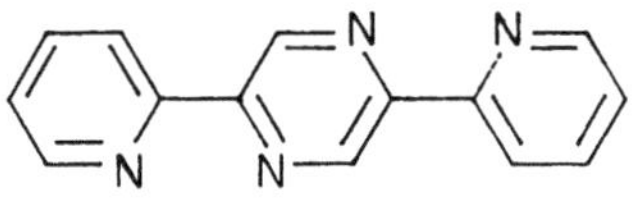

41

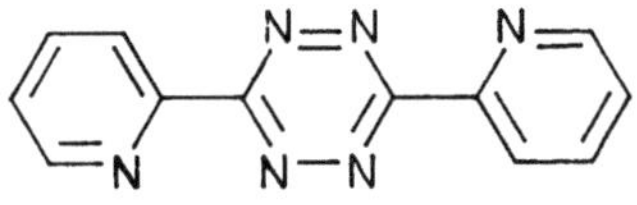

42

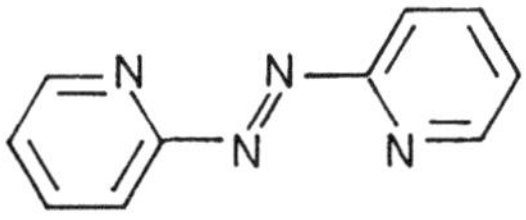

43

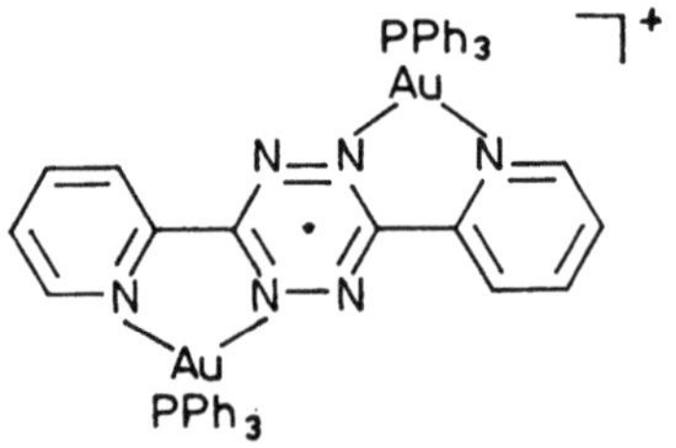

44

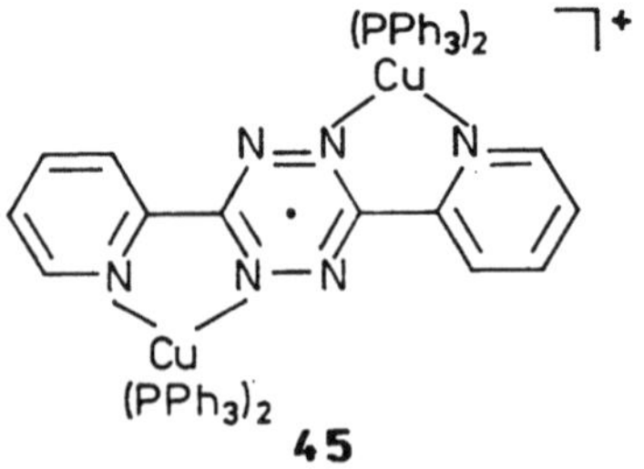

45

(25)

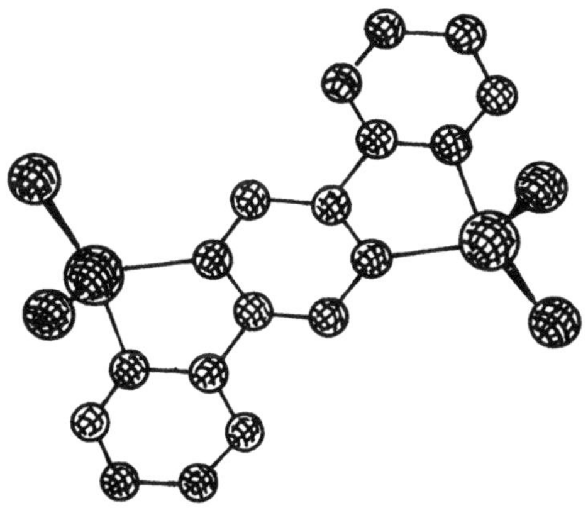

(25)

structure of dicopper radical cation **45**
phenyl groups omitted for clarity

$$CO_2 + 2H^+ \xrightarrow[+h\nu,\ \substack{+2e^-,\ \text{or} \\ +2D,\ -2D^{+\cdot}}]{(CO)_3ClRe(bpy)} CO + H_2O \tag{26}$$

$$(CO)_3ClRe(bpy) + e^- \rightarrow (CO)_3ClRe(bpy^{-\cdot})$$

$$(CO)_3ClRe(bpy) \xrightarrow[MLCT]{h\nu} {}^*[(CO)_3ClRe^+(bpy^{-\cdot})]$$

A remarkable ambiguity concerns the electronic ground state of 1,10-phenanthroline (**37**) anion radical [100] and its diaza derivatives (e.g. **38**) [88, 108]: Two close lying molecular orbitals, a_2 and b_1, are available to accomodate the unpaired electron and only careful ESR measurements have allowed an unambiguous determination of the spin distribution in each particular case [100, 108].

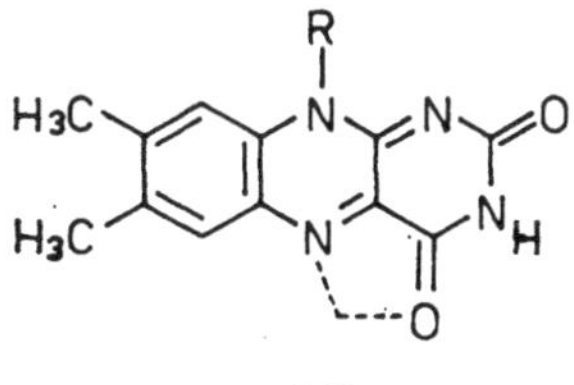

46

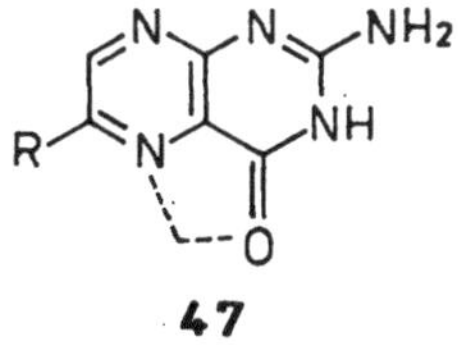

47

(27)

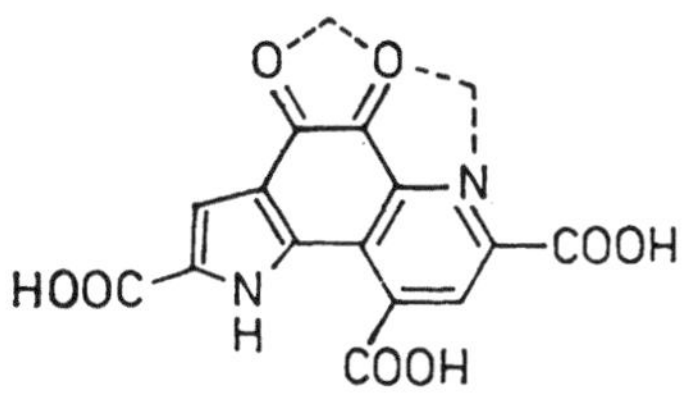

48

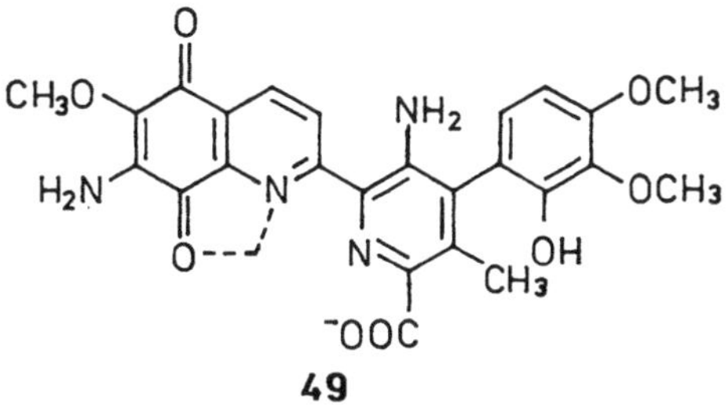

49

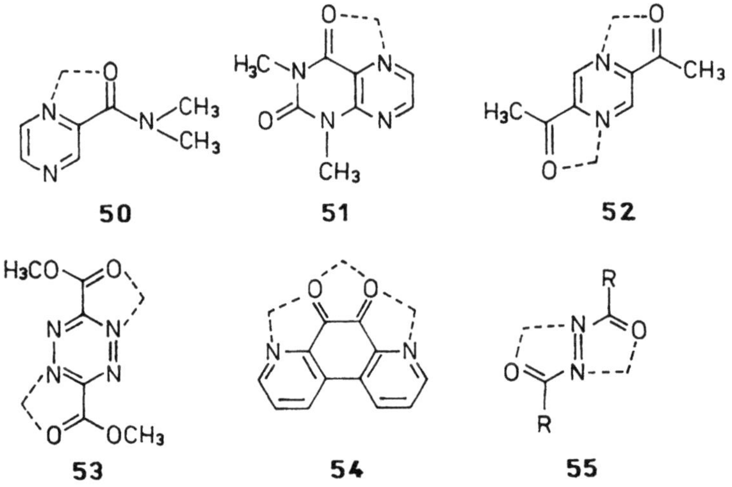

(28)

Radical systems containing α-iminoketone, i.e. mixed N,O-coordinating chelate ligand anions [109] are interesting in context with biochemical electron transfer processes [110]; many naturally occurring electroactive molecules such as dehydrogenases [110–112] and cytostatika [113] (27) show this kind of reactivity. A few organometallic [114] and other complexes [111, 116] of corresponding model ligands (28) have been studied [102, 117–120].

2.2 Class II radical ligands

Unsaturated nitrogen ligands of the second category (class II) stabilize electron deficient organometallic "radicals" through their σ and π donation capability. An extensive investigation of such a system involving paramagnetic $Cp(CO)_2Mn^+$ (low spin d^5) [23, 26] has shown that appropriate nitrogen ligands can accomodate more than one third of the total spin density according to formulation (6B); the attention given recently to organometallics with higher metal oxidation states [121] should lead to the discovery of more such examples.

Among the σ and π electron rich mono-anionic nitrogen ligands available for such purpose are deprotonated electron rich heterocycles, malonodinitrile anions, hydrazides(1-), and anilides [17, 18, 26], including the extremely strong p-phenylenediamide(1-) donors (29) [23]. Some of the bifunctional anions such as **56** or **63** can coordinate a second (diamagnetic) organometallic fragment, which results in the formation of mixed valence organometallic dimers with effective spin delocalization via the ligand bridge (7) [28].

Even more strongly bonding than monoanionic ligands are unsaturated nitrogen ligands with a double negative charge. Frequently these systems combine with one or two organometallic fragments in such a way that even electron species result (30) [23, 24]; odd electron systems containing formally doubly deprotonated anilines have recently been presented by Trogler and coworkers (8, 31) [25], strong metal/ligand interaction apparently causing a high degree of covalent bonding within those systems.

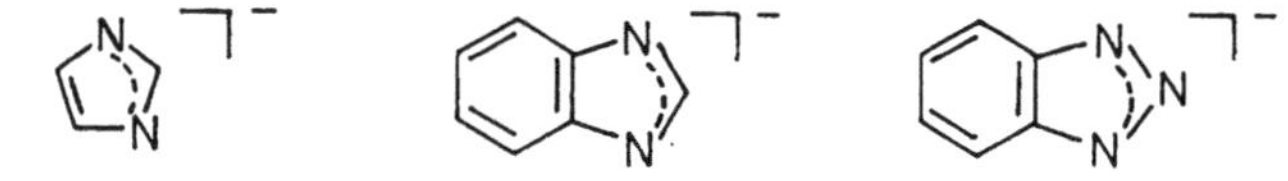

56 **57** **58**

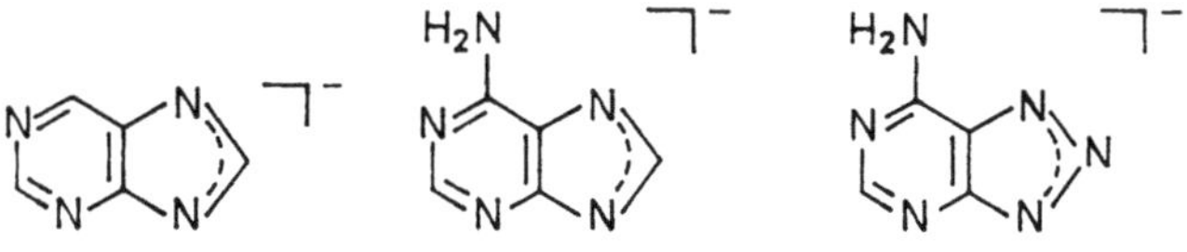

59 **60** **61**

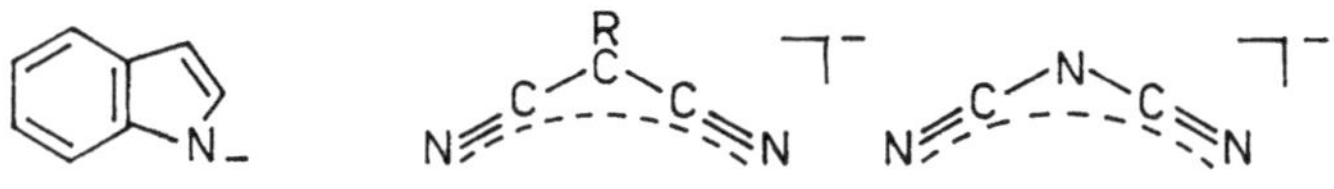

62 **63** **64**

(29)

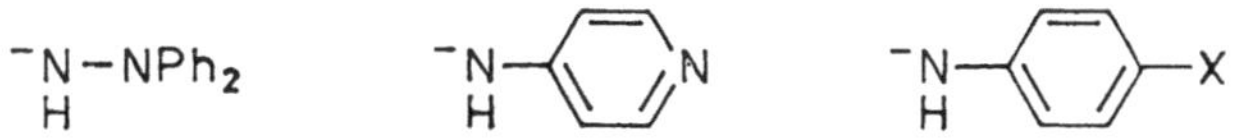

65 **66**

X = H : **67**
X = Me : **68**
X = NH_2 : **69**
X = NMe_2 : **70**

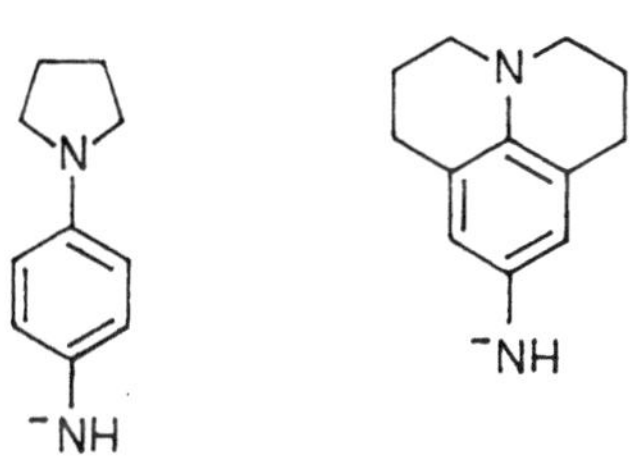

71 **72**

$$
\begin{array}{c}
L + 2e^- \\
\downarrow \\
2R_nM^{+\cdot} + L^{2-} \rightarrow [(R_nM)(L)(MR_n)] \\
\uparrow \\
L{-}H - 2H^+
\end{array}
\qquad (30)
$$

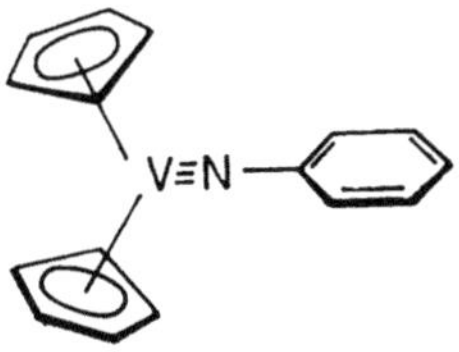

73

(31)

3. Formation and Reactivity of Organometallic Radicals

It is of foremost interest to the non-expert to estimate when organometallic radicals may be formed, either as intermediates (altering the reaction pathway) or even as stable products.

One rather obvious route to organometallic radicals involves the use of *external* reductants or oxidants. In addition, however, many organometallics can be viewed as powerful reductants by themselves [1, 2, 41] and a considerable portion of the driving force necessary to transfer electrons and cleave bonds may come from the formation of coordinate bonds between metals and basic nitrogen centers in the ligands discussed here. Even in those cases where external redox agents are applied, their action may often be supported by non-electron transfer interactions as illustrated in the following.

$$\begin{array}{ccc} \mathbf{2} + 2\,Al^{i}Pr_3 & \rightarrow & (\mathbf{2})(Al^{i}Pr_3)_2 \\ \downarrow {\scriptstyle +e^-,\ -2.15\,V\ vs.\,SCE} & & \downarrow {\scriptstyle +e^-,\ -0.77\,V} \\ \mathbf{2}^{-\cdot} & & [(\mathbf{2}^{-\cdot})(Al^{i}Pr_3)_2]^{-\cdot} \end{array} \tag{32}$$

Reduction of unsaturated (π) systems is often facilitated to a large extent by coordination of neutral or positively charged (organometallic) electrophiles [6]. A striking example is the bifunctional small ligand pyrazine **2**, which experiences an increase of the reduction potential by about 1.4 V through coordination of two neutral AlR_3 acceptor molecules at the nitrogen centers [41], the complex thus becomes as reducible as a quinone (32) [7]. Coordination alone may thus favour susceptibility towards electron transfer by a vast amount, the formation of the coordinative bond providing the necessary energy via polarization of the π system. Complementary to this facilitation of electron transfer by complexation, the coordination of electrophiles is generally promoted through electron transfer since reduction causes a significant increase in ligand basicity (4) [6, 10, 11]. Coordination and electron transfer may thus combine in a synergistic, cooperative fashion (33) to result in organometallic reactions, which often require very little activation (→ oxidative addition, catalysis [1]).

$$L \xrightarrow[+e^-]{+R_nM^{\delta+}} [(L^{-\cdot})(R_nM^{\delta+})] \tag{33}$$

synergistic interaction:
better organometal coordination after ligand reduction,
better ligand reduction after organometal coordination

On the other hand, oxidation of a complex is facilitated if the ligand donor strength is increased. Such an increase can be brought about through the charge effect alone, e.g. by deprotonation of an already basic ligand (6, 30). Oxidants that can simultaneously function as bases and are thus more effective than simple *outer sphere* [1, 122, 123] electron transferring systems include O_2, H_2O_2 and PbO_2; the oxidative deprotonation of inorganic complexes [124] or of organometallics [23, 26], which leads to stable paramagnetic species, may again be regarded as a synergistic process (34).

$$D{-}H \xrightarrow[-e^-,\ -H^+]{+R_nM} [(D^-)(R_nM^{+\cdot})] \tag{34}$$

synergistic interaction:
better coordination after (reversible) organometal oxidation,
better organometal oxidation after coordination of deprotonated donor

If the above steps are not sufficient to bring about radical formation, the additional energy required may come from the absorption of photons, i.e. via photochemical processes [1]. In a single electron transfer (SET) mechanistic scheme [2] (35), the charge transfer complex acts as a precursor of electron transfer products. This complex may be induced to form the primary radical pair by irradiation into the charge transfer absorption band which leads to an excited state with charges separated between complex components (35) [1, 125, 130]. Another pathway leading to organometallic radicals is often referred to as photochemical "spin trapping" (36) [5, 43, 126, 127]: Irradiation of organometallic molecules produces reactive free radical intermediates e.g. by metal-metal bond dissociation [22], which can then be stabilized by binding and electron spin delocalization to unsaturated nitrogen ligands [5, 43, 76, 126, 127].

(35)

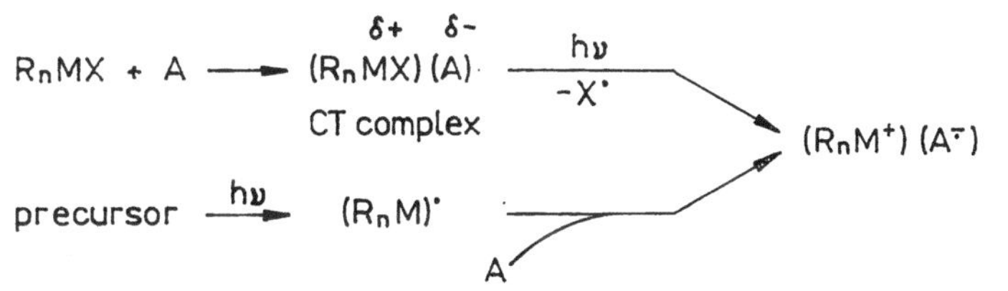

(36)

Thermal and photoinduced electron transfer between "normal" i.e. diamagnetic species has received particular attention in recent years [1, 2, 123, 128–130]. It can be safely anticipated that this kind of reactivity will be recognized as extremely important in all areas of organometallic chemistry [129] for the reasons outlined in the beginning. Furthermore, the hydrophobic environment often present in organometallic chemistry favours the efficiency of electron transfer because of the small reorganization energies involved (absence of charge trapping) [131]; electron transfer mechanisms of high efficiency are required in the area of homogeneous catalysis, for highly specific organic reactions and within energy transforming systems.

4. Selected Examples of Organometallic Radical Reactivity

4.1 Single electron transfer between diorganozinc compounds and α-diimines – the charge transfer precursor complex, escape and cage products

The mechanistic steps in radical producing electron transfer reactions (35) [2] will be introduced here by example of the well documented reaction of diorganozinc species with aromatic and non-aromatic α-diimines [71, 78]. Reaction between the organometallic electrophiles ZnR_2 and the (σ-)chelating but π accepting α-diimines (22, 24) results in the formation of precursor complexes with tetrahedrally coordinated zinc and varying degrees of metal-to-ligand charge transfer $\sigma_{Zn-C} \rightarrow \pi^*$ in the excited state [78, 132].

(37)

Table 1. Spectroscopic and reactivity data of ZnR_2/α-diimine complexes

R	α-diimine	E_{op}^{CT} (eV), color	conversion temp. (°C) cage/escape reaction
Ph	**3**[a]	3.54 (yellow)[b]	>70 [132]
Mes	**3**[a]	3.87 (yellow)[c]	>50 [132]
o-Tol	**26**[d]	colorless	>110 [78]
Me	**26**[d]	3.10 (yellow)	25 [78]
Et	**26**[d]	2.53 (orange/red)	−50 [78]
i-Pr	**26**[d]	2.10 (blue)	−65 [78]
t-Bu	**26**[d]	1.91 (purple)	−85 [78]

[a] Reduction potential $E^0_{red} = -2.10$ V vs. SCE.
[b] $E^0_{red} = -1.60$ V, $E^{pa}_{0x} = +0.87$ V vs. SCE [132].
[c] $E^0_{red} = -1.43$ V, $E^{pa}_{0x} = +0.85$ V vs. SCE [132].
[d] R' = t-Bu (37).

The numbers given in Table I illustrate the strong dependence of this HOMO-LUMO gap on the donor capability of the alkyl or aryl group and on the nature of the α-diimine. Most useful is the correlation between the charge transfer absorption energy and the conversion temperature, i.e. the rate of intra-complex electron transfer, leading to cage- and escape-products (35) [71]. The charge-transfer energy also correlates with the difference in redox potentials according to Eq. (38), the necessary decrease of the HOMO-LUMO gap for intramolecular charge transfer is favoured i) by the lowering of the oxidation potential of the electron rich metal carbon bond (inductive effect), and ii) by an increase of the reduction potential of the unsaturated ligand, following metal coordination to nitrogen centers of high electron density in the acceptor orbital [132].

$$E_{op}^{CT}[eV] = (E^0_{ox} - E^0_{red})\,[V] + \chi \qquad (38)$$

χ: contributions to Franck-Condon energy from intra- and intermolecular reorganisation

In a step which may be described as "inverse" inner sphere electron transfer [2, 41], an electron may be exchanged between the carbanionic alkyl or aryl ligands via the metal "bridge" to the unsaturated acceptor ligand. Such an electron transfer is favored in the rigid chelate situation shown here (37) because of maximal possible orbital overlap between the ligand π system and the metal carbon σ bonds.

The pair of oppositely charged ion radicals as formed after electron exchange in a solvent cage (35) [2] is generally unstable, only one example involving an unusually persistent organometallic cation radical has been reported (cf. 4.3.4) [133]. In the present case, the escape reaction leads to a free alkyl radical and a fairly stable neutral radical complex (RZn^+)(α-diimine$^{-\cdot}$) which may dimerize in the solid state [71]; the cage pathway, on the other hand, produces synthetically interesting complexes of C- and N-alkylated ligands (39) from which the metal fragment can be removed by hydrolysis [77, 78].

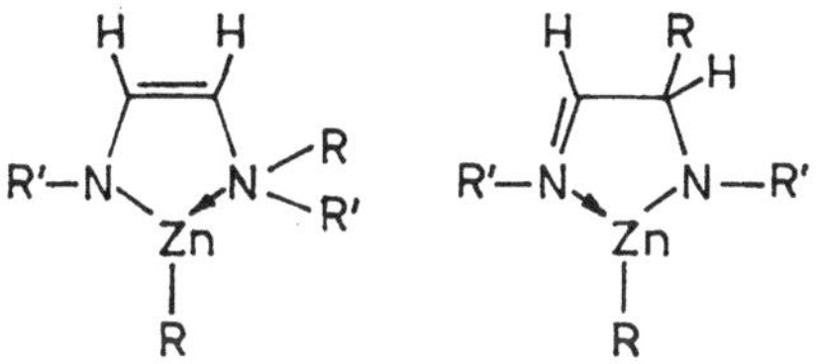

(39)

4.2 A solvent induced single electron transfer reaction

Although organometallic compounds generally act as electron donors in electron transfer processes, cationic systems such as $^+SiR_3$ or $^+MnCp(CO)_2$ may also function as acceptors. For instance, the pentacarbonyltungsten(0) cation **74** (40) with an isonitrile ligand is stable with the hardly nucleophilic $AlBr_4$-anion in dichloromethane solution, however, addition of the basic solvent tetrahydrofuran (THF) triggers a truly elucidating reaction sequence [134]: THF forms a very stable complex with the strong Lewis acid $AlBr_3$, leaving bromide instead of tetrabromoaluminate as counterion for the organometallic cation **74**. Yet, the bromide ion is a sufficiently strong reductant in dichloromethane solution to bring about electron exchange with the organometallic cation, resulting in an intermediate pair of *neutral* radicals within the solvent cage. In the cage (or "collapse") process the bromine radical substitutes the (trans) carbon monoxide ligand, while the "escaped" organometallic free radical can either dimerize or, if sterically protected, crystallize as a stable species out of the solution [134].

Solvent-Induced Electron Transfer:

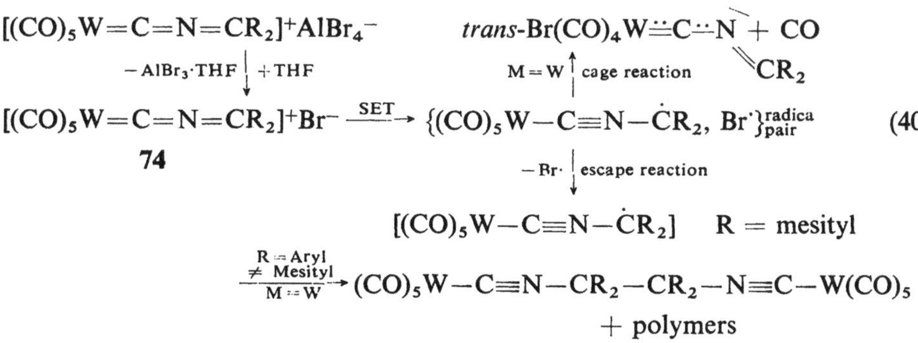

(40)

4.3 One more time: reaction of TCNE with organometallics

Tetracyanoethene (TCNE, **6**) is an unsaturated nitrogen ligand, which can safely be described as the "E. coli" of electron transfer chemistry [1, 35, 37, 41]. It is a small, symmetric and stable compound, which yet is among the most electron deficient organic molecules [135]; it gets reduced almost as easily as ferricinium cation. Paradoxically as it may seem, the extremely facile formation of the TCNE anion radical and of the diamagnetic dianion does not favour the stability of true "radical" complexes [36, 37], yet the remarkable variety of charge or electron transfer reactions and coordination arrangements require an update to former treatments of this topic [1, 136], including the excellent recent review by Fatiadi [35].

Main group organometallics such as organosilicon compounds may react with TCNE as illustrated by the examples in (41):

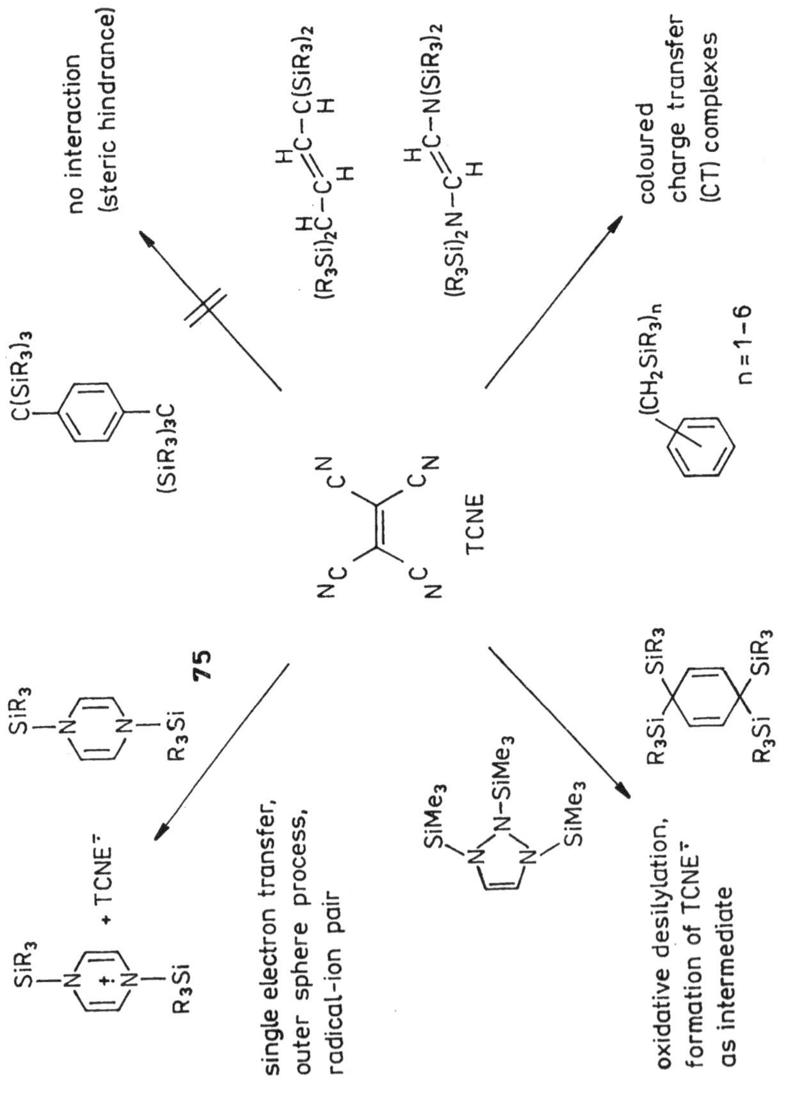

(41)

4.3.1 Very bulky organosilicon donors such as $\alpha,\alpha,\alpha,\alpha',\alpha',\alpha'$-hexakis(trimethylsilyl)-p-xylene [137] or 1,1,4,4-tetrakis(trimethylsilyl)-2-butene [81b] (41) do not interact at all with TCNE despite low enough oxidation and ionization potentials for some form of charge transfer [138]. The distance between donor and acceptor must fall below a critical value for frontier orbital interaction to be significant [139].

4.3.2 If sufficient overlap is possible, the interaction between an acceptor molecule and the organometallic donor may lead to a σ- or π-coordinated charge transfer complex, which exhibits electronic transitions due to electron exchange in the excited state (35) [139]. Many organosilicon compounds [138] including silylamines as shown in (41) [81b] and polysilanes [140] exhibit this behaviour.

4.3.3 Electron exchange may occur in the ground state if energy differences and (*inner sphere*) donor/acceptor interactions favor such a situation [123]. The primary pair of oppositely charged radical ions may collapse to form the diamagnetic cage products (35) or dissociate to be detected as "escape" species [2]. In the case of main group organometallic donors, the cation radical escape products have never been obtained as stable species before the studies described with the very special 1,4-dihydropyrazines (41) [2, 41, 133]; metal-element bond cleavage was found to occur as the dominant follow-up process after single electron transfer (41) [81b, 141].

4.3.4 Protection of the sensitive bonds liable to dissociate after SET is possible and was shown recently to lead to the first observation of the pair of oppositely charged radical ions as solvent separated species by ESR (41) [2, 41, 133]. The steric shielding needed in such cases implicates the involvement of an *outer sphere* electron transfer mechanism and the requirement for a rather extreme redox potential in order to make such a probe useful for SET studies; all these requirements are met by compound **75**, R = ^{i}p$_r$ [133].

4.3.5 Alternative pathways of single electron transfer between organometallic compounds and acceptors, such as TCNE, are possible in the transition metal series where the metal center itself can be 1e-oxidized *reversibly* and then bind more or less covalently to the anion radical of the acceptor; such an alternative is not possible for the silicon(IV) compounds (41) or for other main group organometallic systems. In addition, there are also many organotransition metal compounds which show bond-cleavage after electron transfer with TCNE [1, 35].

Before discussing the extent of electron transfer between transition metal systems and TCNE, the coordination properties of TCNE should be reviewed.

TCNE and related molecules have been shown to form "charge transfer complexes" with coordinatively saturated organometallics, such as certain metallocenes [142–145]. The correlation between crystal structure (e.g. stacking) and the electrical or magnetic properties of these solids continues to be of considerable interest [144, 145]. If electron transfer occurs between TCNE and such coordinatively saturated molecules, it may be described as an *outer sphere* process in the absence of close contacts. Secondly, TCNE may form tightly bonded π complexes with organometallic fragments just like other olefins, however, intra-complex electron transfer [37] may favour species with predominant metallacyclopropane character [136]. Finally, the nitrile groups of TCNE and related molecules contain lone pairs, which may be used for regular σ bonding of metal fragments [146], as in e.g. complexes of acetonitrile. Since there are four available nitrile groups in TCNE, all these may coordinate even to a neutral organometallic fragment after activation by electron transfer as was shown recently [15].

For the first class of organotransition metal compounds with TCNE, structural, magnetic, and spectroscopic studies have revealed considerably different extents of electron transfer, e.g. on going from ferrocene to the stronger decamethylferrocene donor [142–145]. According to the extent of charge separation and to the size ratio between donor and acceptor, radical salts with remarkably different magnetic behaviour are observed [145].

The π complexes of TCNE are primarily formed with electron rich organometallic fragments containing "soft", mostly late transition metals [136]. Structural and spectroscopic studies have revealed much $TCNE^{2-}$ character of the coordinated ligand; however, the situation is more ambiguous with the $M(CO)_5$ complexes (42, M = Cr, W). According to electrochemical and absorption spectroscopic data [37, 147], these diamagnetic species show considerable electron delocalization between metal and ligand corresponding to a spin-paired formulation M^I (l.s. d^5)/$TCNE^{-\cdot}$; the ESR of the radical anion of the tungsten system suggests an unsymmetrical arrangement (σ-bonding), although the neutral complex shows clearly π coordination [37].

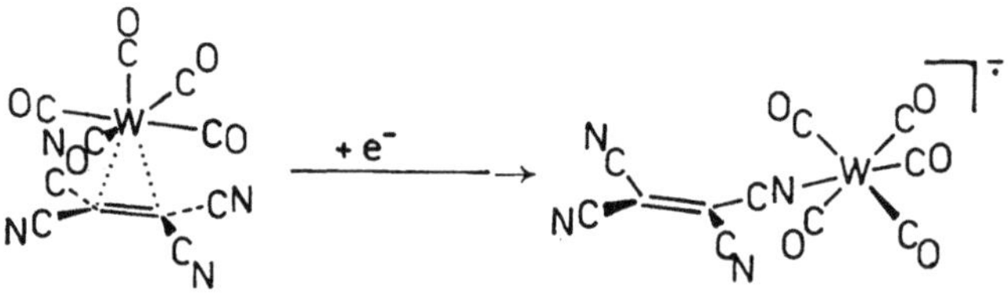

(42)

Exchange of three carbonyls in $Cr(CO)_5$ by the stronger electron donating η^6-bonded hexamethylbenzene (HMB) ligand results in a change of the coordination mode: The $Cr(CO)_2$ (HMB) fragment is σ bonded to a TCNE nitrogen center [37, 148], just like the related $Mn(CO)_2Cp$ [37] and the $VCp_2(Hal)$ [146] fragments. A crystal structure determination, albeit of poor quality, has established this coordination in the latter case (Hal = Br) [146]. All three mononuclear σ complexes with TCNE show indications of considerable $d_\pi \rightarrow \pi^*$(TCNE) electron transfer from the organometallic fragment to the ligand (43) in the ground state [37, 146]. This amount of metal-to-ligand electron transfer and the ionic character of the bond increases in the series (43), as indicated by chemical, electrochemical, and spectroscopic results [37, 147]. As a consequence, solvents of high dielectric constant induce dissociation of the TCNE and of related TCNQ (**10**) complexes into simultaneously ESR detectable paramagnetic species; the TCNE or TCNQ anion radicals and the d^5 metal fragments (44) [15, 147].

$$(TCNE)(MR_n) = (TCNE^{n-})(^{n+}MR_n) \quad 0 < n < 2 \tag{43}$$

MR_n: $W(CO)_5$, $Cr(CO)_5$, $Mn(C_5Me_5)(CO)_2$, $V(C_5H_5)_2Br$, $Cr(C_6Me_6)(CO)_2$

$$[(R_nM^{\delta+})(L^{\delta-})] \xrightarrow{\text{Solvent } S} [R_nM(S)]^{+\cdot} + [L(S)]^{-\cdot} \tag{44}$$

With the more covalently σ bonded fragments $Mn(CO)_2Cp$ and VCp_2Br it has been possible to obtain neutral polynuclear complexes $(TCNE)(ML_n)_m$ (m = 1–4) with singular spectroscopic properties [15, 149]. This result can only be rationalized assuming high electron density in the TCNE ligand as a consequence of metal-to-ligand electron transfer and considerable electron delocalization, as exemplified by the apparent D_{2h} symmetry of the tetranuclear system [15, 24].

The $MnCp(CO)_2$ complexes exhibit a further unexpected behaviour, which is pertinent to the topic of this volume: Although these species contain an even number of electrons and despite the 18 valence electron configuration of all carbonyl-metal centers, the complexes possess easily accessible $S \neq 0$ magnetic states ($S = 1$ for the mononuclear TCNE complex) [150]. Studies on other complexes [151] of this organometallic fragment have indicated that the paramagnetism is due to a small ligand field splitting of the lowest four orbitals of the $3d$ set so that an intermediate spin configuration (45) with two unpaired electrons becomes available [150].

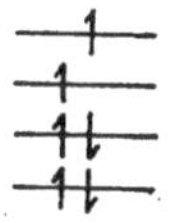

(45)

This result shows that some "typical" organotransition metal species may have much in common with classical transition metal complexes [24] for which higher spin states are not exceptional; this rather new discovery [150, 151] explains, among other things, the special photoreactivity of such compounds [152, 153].

5. Electron Transfer Catalysis Involving Organometallics and Nitrogen Ligands

The synthesis of the TCNE and TCNQ complexes e.g. with the $Mn(CO)_2Cp$ fragments as described above involves a substitution of THF in the photogenerated precursor complexes $(THF)Mn(CO)_2Cp$ [15]. It was recognized that this reaction proceeds several orders of magnitude faster with the electron transfer active TCNE and TCNQ ligands than with other "normal" unsaturated nitrogen ligands [37]. The reason for this apparently electron-transfer induced acceleration of substitution may be rationalized [37] using the electron transfer catalysis mechanism [130, 154, 155] presented for $Mn(CO)_2Cp$ complexes by Kochi and Amatore [156–158]:

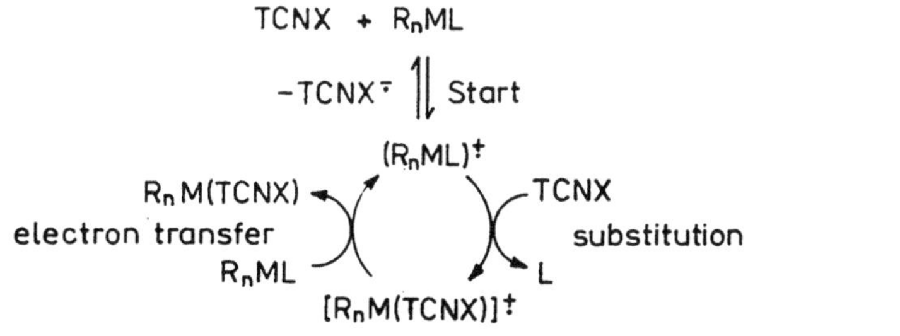

(46)

TCNX: TCNE (6) or TCNQ (9)

L: oxidatively labilized ligand (THF or PR_3)

In contrast to the examples described by Kochi and coworkers [156–158], the TCNE and TCNQ ligands are sufficiently electron accepting to cause one electron oxidation of the THF complex *in the absence* of external oxidation (e.g. at an anode). The oxidized, i.e. 17 valence electron (Mn^{II}) species is much more labile towards substitution than the 18 valence electron precursor [154–159] and an electron transfer *auto-catalytic* cycle (46) is closed because the substituted complex cation can oxidize the neutral THF precursor [37].

The electron transfer catalytic cycle (46) involving oxidative activation and reactive 17 valence electron species as very short lived intermediates has its precedent [155] in mechanisms based on substitutionally labile 19 valence electron species, which are formed by reductive activation. The synthetic scope of these processes is described in chapters 4, 9 and 10 of this volume. In keeping with the title of this contribution we would like to point out how the involvement of electron delocalizing unsaturated nitrogen ligands can moderate this kind of reactivity and stabilize the radical intermediates.

Group 6 metal carbonyl complexes of chelating and non chelating unsaturated nitrogen ligands (17, 24) may be reduced chemically or electrochemically to anion radical complexes. These species are stable enough to be characterized by IR or ESR spectroscopy, yet, the metal carbonyl fragment is activated to a small extent ("18 + δ electron configuration") via spin delocalization from the nitrogen ligand to the cis carbonyls (47) [47]. In the presence of e.g. phosphanes, this activation results in a rather slow substitution of one and only one CO group in cis position at each metal center [53, 101, 160].

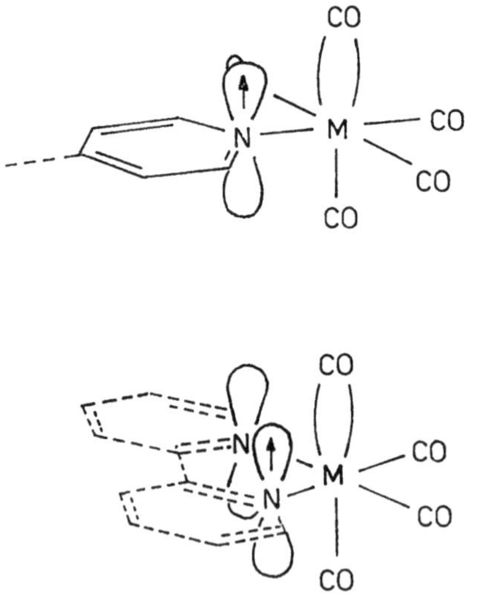

(47)

It is believed that this selectivity is due to the π/σ-hyperconjugative nature of the spin delocalization from the ligand to the metal (47); the substituted radical complex may be analyzed separately by ESR, showing a predominant spin delocalization to the metal/phosphorus bond, which disfavours a second CO substitution [3, 4, 53, 101]. During actual catalysis the substituted radical complex is rapidly consumed in the reduction of the precursor carbonyl complex, thus completing the catalytic cycle (48) [3, 4, 14].

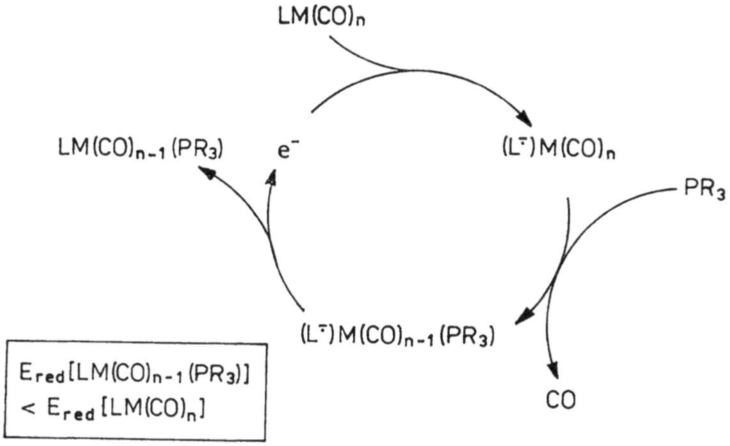

(48)

Although the small extent of activation in the 18 + δ electron species leaves the substitution reaction relatively slow and the necessary amount of chemical [3, 4] or electrochemical [160] redox equivalents rather high, there are some synthetic advantages for this kind of carbonyl substitution [3, 4]. Apart from the substitutional selectivity and the possibility to characterize the rather stable intermediates, this process (48) allows the facile preparation of metal carbonyl complexes containing both donor (PR_3) *and* weakly basic π acceptor ligands. "Redox-umpolung" (acceptor $\rightarrow$ donor) of the unsaturated nitrogen system by electron transfer is responsible for this opportunity, as well as for the potential formation of polynuclear complexes ("electron transfer assisted polynucleation"); the mixed donor/acceptor-substituted carbonyl complexes are distinguished by long wavelength metal-to-ligand charge transfer (MLCT) transitions [3, 4, 14].

6. Summary and Outlook

The previous examples have demonstrated that many different unsaturated nitrogen ligands are suitable to stabilize paramagnetic organometallic compounds or to facilitate their formation in the first place. In most instances, the high basicity of lone pair containing N atoms relative to e.g. O centers is responsible for initial coordinative stability, while the π system of the unsaturated N ligand can moderate the reactivity of the organometallic radical via spin delocalization.

The kind of π type interaction of unsaturated nitrogen ligands with organometallics may vary from p_π/d_π overlap (49 A, σ and π bonded systems included) in transition metal complexes [37] via synthetically useful metal carbon bond activating $p_\pi/\sigma_{(M-C)}$ interactions (49B) [2, 78] to $p_\pi/\sigma_{(M-M)}$ (49C) and $p_\pi/n_{(M)}$ interactions (49D). The latter two arrangements have received little attention so far; while (49C) [76a, 161] represents an interesting new approach toward the activation of metal-metal bonded systems ($\sigma^2 \rightarrow \sigma^1$), situation (49D) may lead to an electron transfer induced "deactivation" of a stereoactive lone pair and thus to the possibility of element-element bond formation ($\sigma^4 \rightarrow \sigma^{3,2}$) [162]. Interactions of the latter kind, including "inorganic" $n \rightarrow \pi^*$ transitions, have been observed in Sn(II) complexes of α-diimine ligands [162–164].

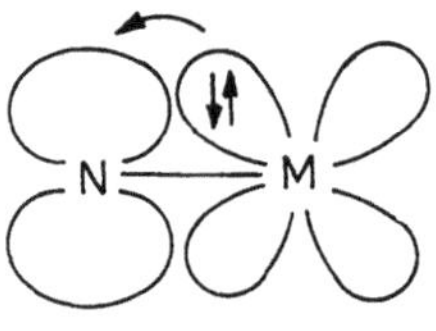

A

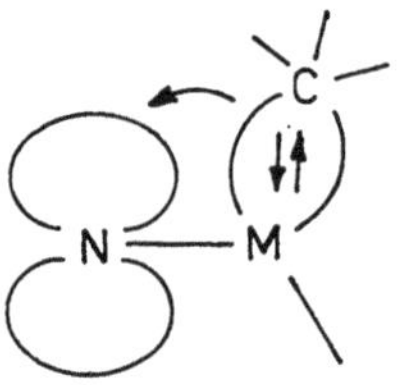

B

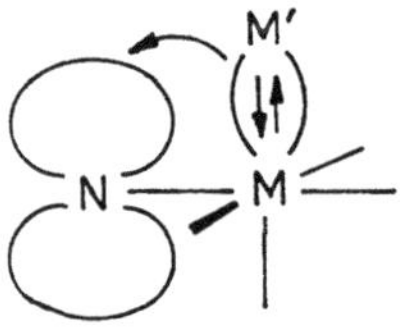

C

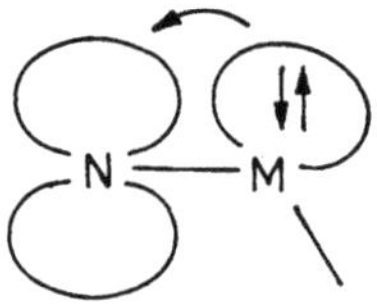

D

Finally, one of the most fascinating aspects of organometallic radicals is their propensity to participate as essential intermediates in electron transfer catalytic processes [130, 154, 155]; the detection and development of such mechanisms in unusually fast reactions, in conventional organometallic catalysis and in main group or transition metal organometallic reactions directed towards organic synthesis holds a promising future.

7. Acknowledgment

Our own contributions to this field have been made possible through generous support by Deutsche Forschungsgemeinschaft, Stiftung Volkswagenwerk, Flughafen Frankfurt/Main AG, and BASF AG and through the dedicated efforts by the coworkers whose names appear on the publications cited. A Karl Winnacker-Fellowship by Hoechst AG for W. Kaim is also gratefully acknowledged.

8. References

1. J. K. Kochi, Organometallic Mechanisms and Catalysis, Academic Press, New York, 1978.
2. W. Kaim, Acc. Chem. Res. **18** (1985) 160.
3. B. Olbrich-Deussner and W. Kaim, J. Organomet. Chem. **340** (1988) 71.
4. B. Olbrich-Deussner and W. Kaim, J. Organomet. Chem., **361** (1989) 335.
5. K. A. M. Creber, T. I. Ho, M. C. Depew, D. Weir and J. K. S. Wan, Can. J. Chem. **60** (1982) 1504.
6. W. Kaim, Coord. Chem. Rev. **76** (1987) 187.
7. J. A. Pedersen, Handbook of EPR Spectra from Quinones and Quinols, CRC Press, Boca Raton, FL, 1985.
8. A. von Zelewsky, C. Daul, C. W. Schläpfer, in Landolt-Börnstein, Physical and Chemical Tables, New Series, Vol. II/17a, Springer-Verlag, West Berlin, 1987, pp. 199–505.
9. F. A. Cotton, G. Wilkinson, Advanced Inorganic Chemistry, 5th Edn., Wiley, New York, 1988, pp. 334–381.

10. K. Wieghardt, H. Cohen and D. Meyerstein, Angew. Chem., **90** (1978) 632; Angew. Chem. Int. Ed. Engl. **17** (1978) 608.
11. C. V. Krishnan, C. Creutz, H. A. Schwarz and N. Sutin, J. Am. Chem. Soc. **105** (1983) 5617.
12. P. Ford, D. F. P. Rudd, R. Gaunder and H. Taube, J. Am. Chem. Soc. **90** (1968) 1187. J. Sen and H. Taube, Acta Chem. Scand. Ser. **A, 33** (1979) 125.
13. M. L. Bento and E. Tfouni, Inorg. Chem. **27** (1988) 3410.
14. W. Kaim, B. Olbrich-Deussner, R. Groß, S. Ernst, S. Kohlmann and C. Bessenbacher, in M. Chanon (Ed.), Importance of Paramagnetic Organometallic Species in Activtion, Selectivity, and Catalysis, D. Reidel, Dordrecht, 1989, p. 283.
15. R. Groß and W. Kaim, Angew. Chem. **99** (1987) 257; Angew. Chem., Int. Ed. Engl. **26** (1987) 251.
16. S. Ernst and W. Kaim, Inorg. Chim. Acta **144** (1988) 223.
17. R. Groß and W. Kaim, Angew. Chem. **97** (1985) 869; Angew. Chem., Int. Ed. Engl. **24** (1985) 856.
18. D. Sellmann, J. Müller and P. Hofmann, Angew. Chem., Int. Ed. Engl. **21** (1982) 691; A. Winter, G. Huttner, L. Zsolnai, P. Kroneck, and M. Gottlieb, Ibid. **23** (1984) 975.
19. N. C. Schroeder and R. J. Angelici, J. Am. Chem. Soc. **108** (1986) 3688; D. Ballivet-Tkatchenko, B. Nickel, A. Rassat and J. Vincent-Vaucquelin, Inorg. Chem. **25** (1986) 3497; H. H. Karsch, B. Milewski-Mahrla, J. O. Besenhard, P. Hofmann, P. Stauffert and T. A. Albright, Ibid. **25** (1986) 3811.
20. S. R. Desjardins, K. W. Pebfield, S. L. Cohen, R. L. Musselman and E. I. Solomon, J. Am. Chem. Soc. **105** (1983) 4590.
21. J. Owen and K. W. H. Stevens, Nature **171** (1953) 836.
22. Cf. T. R. Herrinton and T. L. Brown, J. Am. Chem. Soc. **107** (1985) 5700.
23. R. Groß and W. Kaim, Inorg. Chem. **26** (1987) 3596.
24. W. Kaim and R. Groß, Comments Inorg. Chem. **7** (1988) 269.
25. J. H. Osborne, A. L. Rheingold and W. C. Trogler, J. Am. Chem. Soc. **107** (1985) 7945.
26. R. Groß and W. Kaim, J. Chem. Soc., Faraday Trans. 1, **83** (1987) 3549.
27. M. J. Maroney and W. C. Trogler, J. Am. Chem. Soc. **106** (1984) 4144.
28. R. Groß and W. Kaim, Inorg. Chem. **25** (1986) 4865.
29. A. M. McPherson, B. F. Fieselmann, D. L. Lichtenberger, G. L. McPherson and G. D. Stucky, J. Am. Chem. Soc. **101** (1979) 3425.
30. M. H. Chisholm, E. M. Kober, D. J. Ironmonger and P. Thornton, Polyhedron **4** (1985) 1869.
31. R. A. Henderson, G. J. Leigh and C. J. Pickett, Adv. Inorg. Chem. Radiochem. **27** (1983) 197. J. Chatt, J. R. Dilworth and R. L. Richards, Chem. Rev. **78** (1978) 589.
32. Cf. P. W. Atkins and M. C. R. Symons, The Structure of Inorganic Radicals, Elsevier, Amsterdam, 1967.
33. M. J. S. Gynane, J. Jeffery and M. F. Lappert, J. Chem. Soc., Chem. Commun. (1978) 34.
34. G. M. Tom, C. Creutz and H. Taube, J. Am. Chem. Soc. **96** (1974) 7827.
35. A. J. Fatiadi, Synthesis (1987) 959.
36. P. J. Krusic, H. Stoklasa, L. E. Manzer and P. Meakin, J. Am. Chem. Soc. **97** (1975) 667.
37. B. Olbrich-Deussner, R. Gross and W. Kaim, J. Organomet. Chem., **366** (1989) 155.
38. R. Groß and W. Kaim, J. Organomet. Chem. **333** (1987) 347.
39. A. Aumüller, P. Erk, G. Klebe, S. Hünig, J. U. von Schütz and H. P. Werner, Angew. Chem. **98** (1986) 759; Angew. Chem., Int. Ed. Engl. **25** (1986) 740.
40. H. Endres in J. S. Miller (Ed.), Extended Linear Chain Compounds, Plenum Press, New York, 1983, p. 263.
41. J. Baumgarten, C. Bessenbacher, W. Kaim and T. Stahl, J. Am. Chem. Soc. **111** (1989) 2126 and 5017.
42. Y. Mizobe, R. Ono, Y. Uchida, M. Hidai, M. Tezuka, S. Moue and A. Tsuchiya, J. Organomet. Chem. **204** (1981) 377.
43. A. Hudson, M. F. Lappert and B. K. Nicholson, J. Chem. Soc., Dalton Trans., (1977) 551.
44. P. Hanson, Adv. Heterocycl. Chem. **25** (1979) 205 and **27** (1980) 31.
45. G. R. Newkome and W. W. Paudler, Contemporary Heterocyclic Chemistry, Wiley, New York, 1982.

46. W. Kaim, Angew. Chem. **92** (1980) 940; Angew. Chem., Int. Ed. Engl. **19** (1980) 911.
47. W. Kaim, Inorg. Chim. Acta **53** (1981) L151; W. Kaim, Chem. Ber. **115** (1982) 910.
48. W. Kaim, Z. Naturforsch. **36b** (1981) 677.
49. W. Kaim, Chem. Ber. **114** (1981) 3789.
50. W. Kaim and V. Kasack, Angew. Chem. **94** (1982) 712; Angew. Chem. Int. Ed. Engl. **21** (1982) 700; W. Kaim, J. Organomet. Chem. **264** (1984) 317.
51. W. Kaim, J. Organomet. Chem. **241** (1983) 157.
52. W. Kaim and W. Lubitz, Angew. Chem. **95** (1983) 915; Angew. Chem., Int. Ed. Engl. **22** (1983) 892; Angew. Chem. Suppl. (1983) 1209.
53. W. Kaim, J. Organomet. Chem. **262** (1984) 171.
54. W. Kaim, Inorg. Chem. **23** (1984) 504.
55. W. Kaim, Z. Naturforsch. **39b** (1984) 801.
56. W. Kaim, J. Am. Chem. Soc. **106** (1984) 1712 and Z. Naturforsch. **37b** (1982) 7843.
57. R. Groß and W. Kaim, Inorg. Chem. **25** (1986) 498.
58. W. Kaim and S. Kohlmann, Inorg. Chem. **25** (1986) 3442.
59. W. Kaim and S. Ernst, J. Phys. Chem. **90** (1986) 5010.
60. (a) S. Kohlmann, Ph.D. Thesis, Universität Frankfurt, 1988. (b) A. Schulz, Ph.D. Thesis, Universität Stuttgart, 1988. Cf. also W. Kaim, S. Kohlmann, S. Ernst, B. Olbrich-Deussner, C. Bessenbacher and A. Schulz, J. Organomet. Chem. **321** (1987) 215.
61. K. B. Wiberg and T. P. Lewis, J. Am. Chem. Soc. **92** (1970) 7154. W. Kaim, J. Chem. Soc., Perkin Trans. **2** (1984) 1357.
62. W. L. Darby and L. M. Vallarino, Inorg. Chim. Acta **75** (1983) 65.
63. J. F. Wishart, A. Bino and H. Taube, Inorg. Chem. **25** (1986) 3318.
64. D. R. Eaton, J. M. Watkins and R. J. Buist, J. Am. Chem. Soc. **107** (1985) 5604.
65. W. Strohmeier, Z. Elektrochem., Ber. Bunsenges. Phys. Chemie **60** (1956) 396 and **61** (1957) 1010.
66. W. Kaim, Angew. Chem. **95** (1983) 201; Angew. Chem., Int. Ed. Engl. **22** (1983) 171.
67. A. Alberti and G. F. Pedulli, J. Organomet. Chem. **248** (1983) 261.
68. J. Fajer and M. S. Davis in D. Dolphin (Ed.), The Porphyrins, Vol. **IV**, Academic Press, New York, 1979, pp. 197–256.
69. G. van Koten and K. Vrieze, Recl. Trav. Chim. Pays-Bas, **100** (1981) 129, and Adv. Organomet. Chem. **21** (1982) 152.
70. P. Clopath and A. von Zelewsky, Helv. Chim. Acta **55** (1972) 52.
71. G. van Koten, J. T. B. H. Jastrzebski and K. Vrieze, J. Organomet. Chem. **250** (1983) 49.
72. J. N. Louwen, D. J. Stufkens and A. Oskam, J. Chem. Soc., Dalton Trans. (1984) 2683.
73. H. tom Dieck, B. Bruder and K. D. Franz, Chem. Ber. **116** (1983) 136.
74. K. D. Franz, H. tom Dieck, U. Krynitz and I. W. Renk, J. Organomet. Chem. **64** (1974) 361; K. D. Franz, H. tom Dieck, K. A. Ostoja-Starzewski and F. Hohmann, Tetrahedron **31** (1975) 1465; H. tom Dieck and E. Kühl, Z. Naturforsch. **37b** (1982) 324.
75. H. tom Dieck, K. D. Franz and F. Hohmann, Chem. Ber. **108** (1975) 163.
76. (a) M. W. Kokkes, D. J. Stufkens and A. Oskam, Inorg. Chem. **24** (1985) 2934; R. R. Andrea, W. G. J. de Lange, T. van der Graaf, M. Rijkhoff, D. J. Stufkens and A. Oskam, Organometallics **7** (1988) 1100. (b) A. Alberti and A. Hudson, J. Organomet. Chem. **248** (1983) 199.
77. J. M. Klerks, J. T. B. H. Jastrzebski, G. van Koten and K. Vrieze, J. Organomet. Chem. **224** (1982) 107.
78. G. van Koten in A. de Meijere, H. tom Dieck (Eds.), Organometallics in Organic Synthesis, Springer-Verlag, West Berlin, 1988, pp. 277–295.
79. L. Stamp and H. tom Dieck, J. Organomet. Chem. **277** (1984) 297. M. Zettlitzer, H. tom Dieck, E. T. K. Haupt and L. Stamp, Chem. Ber. **119** (1986) 1868.
80. M. R. P. van Vliet, J. T. B. H. Jastrzebski, W. J. Klaver, K. Goubitz and G. van Koten, Recl. Trav. Chim. Pays-Bas **106** (1987) 132; J. T. B. H. Jastrzebski, F. H. van der Steen and G. van Koten, Ibid. **106** (1988) 516.
81. (a) A. Alberti and A. Hudson, J. Organomet. Chem. **241** (1983) 313. (b) C. Bessenbacher and W. Kaim, Z. Naturforsch., **446** (1989) 511.

82. C. Bessenbacher and W. Kaim, J. Chem. Soc., Chem. Commun., (1989) 469 and 888.
83. J. A. Hunter, B. King, W. E. Lindsell and M. A. Neish, J. Chem. Soc., Dalton Trans. (1980) 880.
84. (a) W. Kaim and S. Kohlmann, Chem. Phys. Lett. **139** (1987) 365; (b) C. Bessenbacher, W. Kaim and C. Vogler, unpublished results.
85. R. Mason, K. M. Thomas, A. R. Galbraith, B. L. Shaw and C. M. Elson, J. Chem. Soc., Chem. Commun. (1973) 297.
86. M. E. Gross, W. C. Trogler and J. A. Ibers, J. Am. Chem. Soc. **103** (1981) 192 and Organometallics **1** (1982) 732.
87. P. Overbosch, G. van Koten, A. L. Spek, G. Roelofsen and A. J. M. Duisenberg, Inorg. Chem. **21** (1982) 3908.
88. S. Ernst, Ph.D. Thesis, University of Frankfurt, 1987.
89. Cf. M. E. Gross, J. A. Ibers and W. C. Trogler, Organometallics **1** (1982) 530.
90. H. B. Stegmann, W. Uber and K. Scheffler, Tetrahedron Lett., (1977) 2697.
91. H. B. Stegmann, K. B. Ulmschneider, K. Hieke and K. Scheffler, J. Organomet. Chem. **118** (1976) 259.
92. H. B. Stegmann, M. Schnabel, and K. Scheffler, Angew. Chem. **91** (1979) 1007; Angew. Chem., Int. Ed. Engl. **18** (1979) 943.
93. S. Ernst and W. Kaim, J. Am. Chem. Soc. **108** (1986) 3578.
94. S. D. Ernst and W. Kaim, Inorg. Chem., **28** (1989) 1520.
95. F. Blau, Ber. Dtsch. Chem. Ges. **21** (1988) 1077.
96. J. M. Lehn, lecture at the 1st European Inorganic Chemistry Seminar, Dourdan (France) 1988.
97. Y. Kaizu and H. Kobayashi, Bull. Chem. Soc. Jpn. **45** (1972) 470.
98. W. Kaim, Chem. Ber. **114** (1981) 3789.
99. W. A. Fordyce and G. A. Crosby, Inorg. Chem. **21** (1982) 1023.
100. W. Kaim, J. Am. Chem. Soc. **104** (1982) 3833.
101. W. Kaim, Inorg. Chem. **23** (1984) 3365.
102. W. Kaim and S. Kohlmann, Inorg. Chem. **26** (1970) 1469.
103. S. Ernst, S. Kohlmann and W. Kaim, J. Organomet. Chem. **354** (1988) 177.
104. C. Bessenbacher, S. Ernst, W. Kaim, V. Kasack, S. Kohlmann, E. Roth and J. Jordanov, J. Chem. Soc., Faraday Trans. 1, in print.
105. S. Kohlmann, V. Kasack and W. Kaim, J. Chem. Soc., Faraday Trans. 1, in print.
106. D. Fenske, W. Kaim and S. Kohlmann, to be submitted.
107. J. M. Lehn, J. Hawecker and R. Ziessel, Helv. Chim. Acta **69** (1986) 1990.
108. W. Kaim, C. Bessenbacher, S. Ernst and C. Vogler, to be submitted.
109. A. Alberti and C. M. Camaggi, J. Organomet. Chem. **181** (1979) 355.
110. P. Hemmerich, V. Massey, H. Michel and C. Schug, Struct. Bonding **48** (1982) 93.
111. P. Hemmerich and J. Lauterwein in G. L. Eichhorn, Inorganic Biochemistry, Elsevier, Amsterdam, 1975, pp. 1168–1190.
112. J. B. Noar, E. J. Rodriguez and T. C. Bruice, J. Am. Chem. Soc. **107** (1985) 7198.
113. J. Hajdu and E. C. Armstrong, J. Am. Chem. Soc. **103** (1981) 232. J. W. Lown, S. K. Kim and H. H. Chen, Can. J. Biochem. **56** (1978) 1042.
114. C. Bessenbacher and W. Kaim, Z. Anorg. Allg. Chem., in print.
115. (a) F. Müller, L. E. G. Eriksson and A. Ehrenberg, Eur. J. Biochem. **12** (1970) 93. (b) M. J. Clarke, Comments Inorg. Chem. 3 (1984) 133 and literature cited.
116. B. Schwederski, V. Kasack, W. Kaim, E. Roth and J. Jordanov, Angew. Chem., in print.
117. S. Ernst, V. Kasack, C. Bessenbacher and W. Kaim, Z. Naturforsch. **42b** (1987) 425.
118. S. Ernst, P. Hänel, J. Jordanov, W. Kaim, V. Kasack and E. Roth, J. Am. Chem. Soc. **111** (1989) 1733.
119. W. Kaim, V. Kasack, H. Binder, E. Roth and J. Jordanov, Angew. Chem. **100** (1988) 1229; Angew. Chem., Int. Ed. Engl. **27** (1988) 1174.
120. C. Bessenbacher, Ph.D. Thesis, Universität Frankfurt, 1988.
121. W. A. Herrmann, Angew. Chem. **100** (1988) 1269; Angew. Chem., Int. Ed. Engl. **27** (1988) 1297.
122. H. Taube, Angew. Chem. **96** (1984) 315; Angew. Chem., Int. Ed. Engl. **23** (1984) 329.

123. L. Eberson, Electron Transfer Reactions in Organic Chemistry, Springer-Verlag, West Berlin, 1987.
124. K. Pohl, K. Wieghardt, W. Kaim and S. Steenken, Inorg. Chem. **27** (1988) 440.
125. S. Emori, D. Weir and J. K. S. Wan, Chem. Phys. Lett. **84** (1981) 512.
126. K. A. M. Creber, K. Chen and J. K. S. Wan, Rev. Chem. Intermed. **5** (1984) 37.
127. Cf. J. Basters, Recl. Trav. Chim. Pays-Bas **91** (1972) 50; D. Mulvey and W. A. Waters, J. Chem. Soc., Perkin Trans. **2** (1974) 666; P. J. Krusic, P. J. Fagan and J. San Filippo, J. Am. Chem. Soc. **99** (1977) 250.
128. A. Pross, Acc. Chem. Res. **18** (1985) 212.
129. M. Channon (Ed.), Importance of Paramagnetic Organometallic Species in Activation, Selectivity and Catalysis, D. Reidel, Dordrecht, 1989.
130. J. K. Kochi, Angew. Chem. **100** (1988) 1331; Angew. Chem., Int. Ed. Engl. **27** (1988) 1227.
131. I. Rips and J. Jortner, J. Chem. Phys. **87** (1987) 2090.
132. W. Kaim and T. Stahl, unpublished results.
133. W. Kaim, Angew. Chem. **96** (1984) 609; Angew. Chem., Int. Ed. Engl. **23** (1984) 614.
134. F. Seitz, H. Fischer, J. Riede, T. Schöttle and W. Kaim, Angew. Chem. **98** (1986) 753; Angew. Chem., Int. Ed. Engl. **25** (1986) 744.
135. A. J. Fatiadi, Synthesis (1986) 249.
136. W. H. Baddley, Inorg. Chim. Acta **2** (1968) 7.
137. H. Bock. W. Kaim and H. E. Rohwer, Chem. Ber. **111** (1978) 3573.
138. H. Bock and W. Kaim, Chem. Ber. **111** (1978) 3552.
139. R. Foster, Organic Charge Transfer Complexes, Academic Press, London, 1969.
140. V. F. Traven and R. West, J. Am. Chem. Soc. **95** (1973) 6834. H. Sakurai, M. Kira and T. Uchida, Ibid. **95** (1973) 6826.
141. H.-D. Hausen, C. Bessenbacher and W. Kaim, Z. Naturforsch. **43b** (1988) 1087.
142. J. W. Fitch and J. J. Lagowski, Inorg. Chem. **4** (1965) 864.
143. E. Adman, M. Rosenblum, S. Sullivan and T. N. Margulis, J. Am. Chem. Soc. **89** (1967) 4540.
144. J. S. Miller, J. C. Calabrese, H. Rommelmann, S. R. Chittipeddi, J. H. Zhang, W. M. Reiff and A. J. Epstein, J. Am. Chem. Soc. **109** (1987) 769.
145. J. S. Miller, A. J. Epstein and W. M. Reiff, Acc. Chem. Res. **21** (1988) 114.
146. M. F. Rettig and R. M. Wing, Inorg. Chem. **8** (1969) 2685.
147. B. Olbrich-Deussner, W. Kaim and R. Gross-Lannert, Inorg. Chem., **28** (1989) 3113; H. Braunwarth, G. Huttner and L. Zsolnai, J. Organomet. Chem. **372** (1989) C23.
148. K. L. Amos and N. G. Connelly, J. Organomet. Chem. **194** (1980) C57.
149. W. Kaim and A. Schulz, unpublished results.
150. R. Gross, J. Jordanov, W. Kain, B. Olbrich-Deussner and E. Roth, to be submitted.
151. M. L. Ziegler, R. Balles, B. Nuber, G. Becker, H. Schmidt, B. Kanellakopoulos and B. Powietzka, Angew. Chem. **99** (1987) 929; Angew. Chem., Int. Ed. Engl. **26** (1987) 900.
152. P. J. Giordano and M. S. Wrighton, Inorg. Chem. **16** (1977) 160.
153. G. L. Geoffroy and M. S. Wrighton, Organometallic Photochemistry. Academic Press, New York, 1979.
154. M. Chanon and M. L. Tobe, Angew. Chem. **94** (1982) 27; Angew. Chem., Int. Ed. Engl. **21** (1982) 1.
155. D. Astruc, Angew. Chem. **100** (1988) 662; Angew. Chem., Int. Ed. Engl. **27** (1988) 643.
156. J. W. Hershberger, R. J. Klingler and J. K. Kochi, J. Am. Chem. Soc. **105** (1983) 61.
157. P. M. Zizelman, C. Amatore and J. K. Kochi, J. Am. Chem. Soc. **106** (1984) 3771.
158. J. K. Kochi, J. Organomet. Chem. **300** (1986) 139.
159. T. L. Brown, N.Y. Acad. Sci. **333** (1980) 80.
160. D. Miholova and A. A. Vlcek, J. Organomet. Chem. **279** (1985) 317.
161. R. R. Andrea, W. G. J. de Lange, D. J. Stufkens and A. Oskam, Inorg. Chim. Acta **149** (1988) 77.
162. W. Kaim, S. Kohlmann and T. Stahl, unpublished results.
163. F. X. Kohl, E. Schlüter, P. Jutzi, C. Krüger, G. Wolmershäuser, P. Hofmann and P. Stauffert, Chem. Ber. **117** (1984) 1178.
164. R. D. Barreto, T. P. Fehlner and N. S. Hosmane, Inorg. Chem. **27** (1988) 453.

CHAPTER 7

Activation of Diamagnetic Metal Carbonyls by Electrochemical, Thermal, and Photochemical Electron Transfer

J. K. Kochi

Department of Chemistry, University of Houston, University Park, Houston, Texas 77204-5641, U.S.A.

Table of Contents

1. Introduction: Organometallic radicals from electron transfer of carbonylmetal cations and anions

Metal carbonyls play a key role in the catalytic sequences of a variety of important processes leading to the fixation of carbon monoxide [1, 2]. As such, the ready incorporation of substrates (as ligands) into the metal carbonyl represents a critical step in the development of efficient catalytic cycles [3]. Indeed, extensive mechanistic studies have been carried out on the ligand substitution of a variety of metal carbonyl systems, and they are commonly considered to proceed via the even numbered, 16- and 18-electron intermediates [4–6]. Thus, the displacements of ligands from a series of metal carbonyls $LM(CO)_5$ by nucleophiles **N** have been reported to involve competitive dissociative (S_N1) and associative (S_N2) pathways, as summarized below.

Scheme I:

$$LM(CO)_5 \xrightarrow{N} NM(CO)_5 + L \quad S_N2$$

$$\upharpoonleft\downharpoonright -L$$

$$M(CO)_5 \xrightarrow{N} NM(CO)_5 \quad S_N1$$

Ligands **L**, which serve as σ donors, are particularly good leaving groups. Consequently, such derivatives are versatile synthetic precursors. A few years ago, three research groups independently discovered a much more facile process for ligand substitution. Thus, Rieger and coworkers found that a rapid ligand substitution of the tricobalt cluster $PhCCo_3(CO)_9$ occurred when a small cathodic current [+e] was passed through the solution [7], e.g.

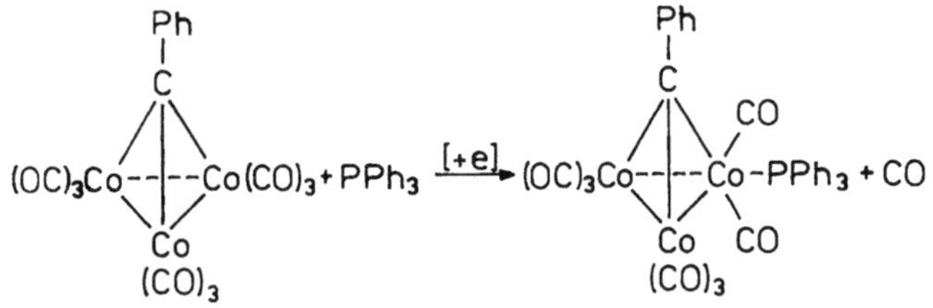

(1)

Robert Klingler working in my laboratory discovered that the tris-acetonitrile complex fac-$(MeCN)_3W(CO)_3$ spontaneously underwent rapid ligand substitution of triphenylphosphine when a small anodic current [—e] was passed through the solution [8], i.e.

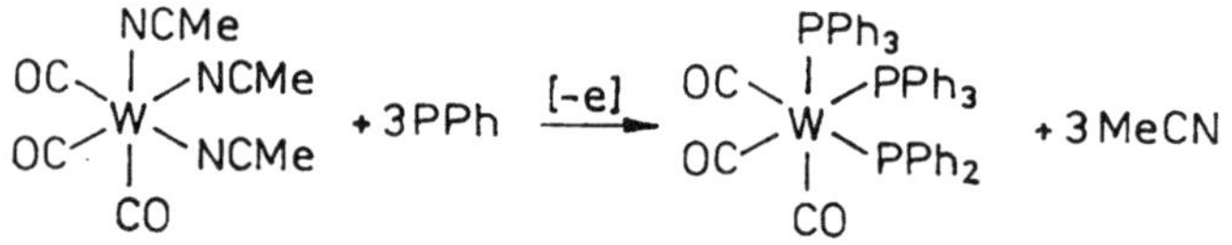

(2)

Wrighton and coworkers reported that a similar process could be induced photochemically [9], i.e.

$$(MeCN)Re(CO)_3phen^+ + PPh_3 \xrightarrow{h\nu} (PPh_3)Re(CO)_3phen^+ + MeCN \quad (3)$$

In each case the conventional thermal processes, the mechanisms of which are presumably depicted in Scheme I, were too slow to compete. The three ligand substitutions illustrated by Eqs. 1, 2, and 3 share in common the corresponding radicals of the metal carbonyls as the key reactive intermediates. The anion- and cation-radicals (designated as $LM^{\overline{\cdot}}$ and $LM^{\dot{+}}$ for simplicity) in Eqs. 1 and 2, respectively, participate in basically the same chain mechanism – the propagation cycle of which consists of two sequential steps, namely (a) ligand substitution and (b) electron transfer as depicted below.

Scheme II:

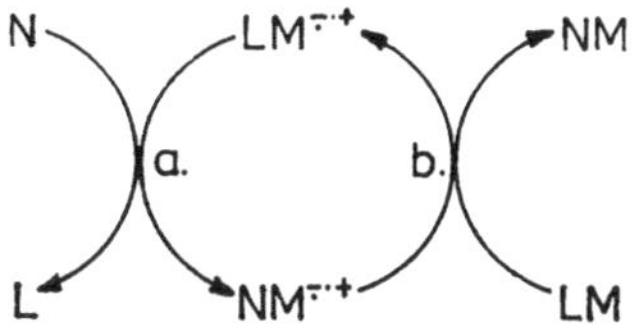

The rapid turnover of this catalytic cycle requires the 19- or 17-electron reactant ion radical ($LM^{-\cdot}$ or $LM^{+\cdot}$) to undergo facile ligand substitution (step a) followed by efficient electron transfer of the product ion radical ($NM^{-\cdot}$ or $NM^{+\cdot}$) to regenerate the reactant ion radical (step b). Similarly the neutral 19-electron Re(0) radical in Eq. 3 undergoes rapid ligand substitution (step a) followed by electron transfer with the cationic Re(I) reactant to regenerate the Re(0) radical (step b). Accordingly, this type of propagation cycle may be referred to as an electron-transfer chain or ETC catalysis. The enhanced rates of ligand substitution of the ion radicals in Scheme II (step a) are directly related to the substitution lability of various uncharged 17-electron carbonyl-metal radicals produced from the homolysis of the corresponding metal-metal bonded dimers, as established by the seminal studies of Brown and coworkers [10, 11].

The mechanistic pathway in Scheme II incorporates two important facets of ETC catalysis of metal carbonyls, namely, the activation by electron transfer (step b) followed by the facile reactions of the paramagnetic carbonylmetal radicals (step a). Thus the initiation of the chain process requires either the oxidation of the metal carbonyl ($LM \xrightarrow{[-e]} LM^{+\cdot}$) or its reduction ($LM \xrightarrow{[+e]} LM^{-\cdot}$). Furthermore, the inhibition of ligand substitution by ETC catalysis occurs when the 19- or 17-electron carbonylmetal radical from either the reactant $LM(CO)_5$ or product $NM(CO)_5$ is diverted, either by self-decomposition or reaction with a contaminant. The latter is often adventitious air (dioxygen) or traces of the by-products of degradation or synthesis. It is thus possible that a number of the substitution reactions which have been considered heretofore as proceeding via 16- or 18-electron intermediates may actually be inhibited forms of chain processes. Since the critical aspects of ETC catalysis according to Scheme II have been recently reviewed, [12] the more recent developments relating to the activation of diamagnetic metal carbonyls will be the principal focus here. Thus, in order to address some particularly relevant questions, our attention is directed to three separate but interrelated methodologies based on the electrochemical, thermal, and photochemical electron transfer of charged diamagnetic metal carbonyls. Before proceeding, it is important to emphasize that carbonylmetal radicals, be they 19-electron (19-e) or 17-electron (17-e) species, are not to be fundamentally distinguished by the ionic charge they may carry. Except for the coulombic forces extant in ion pairs, the neutral, cationic and anionic carbonylmetal radicals behave similarly, all sharing in common their enhanced reactivity as paramagnetic species. The earlier studies demonstrated how various 19-e anion radicals and 17-e cation radicals derive by electron attachment and electron detachment from uncharged diamagnetic metal carbonyls and their polynuclear cluster analogues [13–15]. Electron transfer is likewise pertinent to the charged counterparts. Accordingly, the behavior of 19-e and 17-e carbonylmetal radicals produced via electron attachment and detachment from carbonylmetal cations and anions, respectively, is the principal subject of the following electrochemical, thermal and photochemical treatments.

2. Electrochemical probes for transient 19-e and 17-e carbonylmanganese(0) radicals

Transient electrochemical methods provide invaluable access to metastable and reactive intermediates produced upon the oxidation-reduction of organometallic compounds, among which are an abundant variety of stable, diamagnetic metal carbonyls [16, 17]. Thus, the electron-rich anions are nucleophiles [18, 19] and the electron-poor cations are electrophiles [3, 6] that yield 17-electron and 19-electron carbonylmetal radicals, respectively, upon the electron loss from and the electron accession to their 18-electron precursors [20]. Since such electron-transfer processes are relevant to carbonylmetal chemistry, [21–23] it is important to establish the temporal limits for the detection and characterization of transient 17-e and 19-e radicals for the study of reversible electron transfer. For example, let us consider the carbonylmanganate(-I) anion and the carbonylmanganese(I) cation.

Scheme III:

$$\begin{array}{ccc} Mn(CO)_5^- & & Mn(CO)_6^+ \\ -e \Downarrow\!\upharpoonleft & & \upharpoonleft\!\Downarrow +e \\ Mn(CO)_5\cdot & \underset{}{\overset{CO}{\rightleftharpoons}} & Mn(CO)_6\cdot \\ (17\text{-e}) & & (19\text{-e}) \end{array} \qquad (4)$$

The recent development of microelectrodes offers the potential to carry out cyclic voltammetry of $Mn(CO)_5^-$ and $Mn(CO)_6^+$ at scan rates in excess of 500,000 V s^{-1} to detect the transient $Mn(CO)_5\cdot$ and $Mn(CO)_6\cdot$ with lifetimes that could be as short as 100 ns [24, 25]. In addition, a variety of substituted phosphine (P) derivatives such as the anionic $Mn(CO)_4P^-$, $Mn(CO)_3P_2^-$, etc. and the cationic $Mn(CO)_5P^+$, $Mn(CO)_4P_2^+$, etc. are available to modulate the reversible redox potentials E^0 as well as the stabilities of the corresponding 17-e electron and 19-electron carbonylmanganese(0) radicals [26, 27]. The 17-e radicals, being electron deficient and coordinatively unsaturated, generally undergo rapid ligand substitution primarily by an associative mechanism [28]. Indeed the lifetimes of 17-e radicals are largely limited by second-order kinetics to diffusional processes. On the other hand, the 19-e radicals can be formally considered as electron-excess species and coordinatively saturated. As such, it is not clear whether they actually are discrete intermediates (i.e., exist in bound states) or they represent activated complexes for ligand loss to afford the 17-e radicals in the course of dissociative electron attachment [22, 26]. Thus, there is only limited unambiguous experimental evidence heretofore, either direct or otherwise, that interrelates 17-e and 19-e carbonylmetal radicals – as in the ligand interchange presented in Scheme III [29, 30]. Accordingly, the initial task is to probe for chemical reversibility in the electrooxidation of carbonylmanganate(-I) anions and in the electroreduction of carbonylmanganese(I) cations by microvoltammetric techniques. Furthermore, mechanistic electrochemistry [31] is an alternative probe to examine the equilibrium between 17-e and 19-e carbonylmanganese(0) radicals, as offered by the chelating diphosphine ligand $PPh_2CH_2CH_2PPh_2$ (DPPE) in the reduction of cis-$Mn(CO)_2(DPPE)_2^+$.

2.1 Reversible electron transfer from carbonylmanganate(—I) anions

The chemical oxidation of $Mn(CO)_5^-$ and $Mn(CO)_4PPh_3^-$ by one-electron oxidants such as tropylium ion affords the dimeric $Mn_2(CO)_{10}$ and $Mn_2(CO)_8(PPh_3)_2$ in good yields [32]. Similarly, the anodic oxidation of the same carbonylmanganate (—I) anions at a platinum electrode held at constant potential yields the same results.

$$Mn(CO)_5^- \xrightarrow{0.04V} 1/2\,Mn_2(CO)_{10} \tag{5}$$

$$Mn(CO)_4PPh_3^- \xrightarrow{-0.39V} 1/2\,Mn_2(CO)_8(PPh_3)_2 \tag{6}$$

With both anions, coulometric measurements establish the passage of only 1 faraday of charge for the stoichiometry in Eqs. 5 and 6. However, the oxidation of the analogous bisphosphine substituted carbonylmanganate under the same conditions (i.e., tropylium or electrochemical) affords the corresponding hydridomanganesecarbonyl in almost quantitative yields, i.e.

$$Mn(CO)_3(PPh_3)_2^- \xrightarrow{-0.94V} HMn(CO)_3(PPh_3)_2,\ \text{etc.} \tag{7}$$

The difference between the oxidation stoichiometry in Eqs. 5/6 and Eq. 7 can be reconciled with the rates of dimerization of the carbonylmanganese(0) intermediates. Thus the second-order rate constants [33] for $Mn(CO)_5\cdot$ and $Mn(CO)_4PPh_3\cdot$ of $k_d = 9 \times 10^8$ and $1 \times 10^7\ M^{-1}\ s^{-1}$, respectively, approach the diffusion-controlled limit to favor dimerization, i.e.

Scheme IV:

$$Mn(CO)_5^- \underset{}{\overset{-e}{\rightleftharpoons}} Mn(CO)_5\cdot \tag{8}$$

$$2\,Mn(CO)_5\cdot \xrightarrow{k_d} [Mn(CO)_5]_2 \tag{9}$$

On the other hand, the sterically encumbered bis-substituted radical $Mn(CO)_3$-$(PPh_3)_2\cdot$ is persistent and shows no or limited inclination to dimerize, [34] and its only recourse is hydrogen atom transfer to produce the manganese hydride according to [27]

Scheme V:

$$Mn(CO)_3P_2^- \overset{-e}{\rightleftharpoons} Mn(CO)_3P_2\cdot \tag{10}$$

$$2\,Mn(CO)_3P_2\cdot \rightleftharpoons [Mn(CO)_3P_2]_2 \tag{11}$$

$$Mn(CO)_3P_2\cdot + SH \longrightarrow HMn(CO)_3P_2 + S\cdot,\ \text{etc.} \tag{12}$$

The facility of the followup steps in Eq. 9 and 12 largely determine whether the chemical reversibility of electron transfer in Eqs. 8 and 10, respectively, can be observed by cyclic voltammetric techniques [35].

The reversible cyclic voltammograms in Fig. 1 demonstrate that electron-transfer reversibility for the persistent radical $Mn(CO)_3(PPh_3)_2\cdot$ is attained with a cathodic/anodic peak current ratio of $i_c/i_a = 1.0$ (theoretical) at a scan rate of $v = 0.5\ V\ s^{-1}$, whereas the more transient radical $Mn(CO)_4PPh_3\cdot$ requires a 10^4

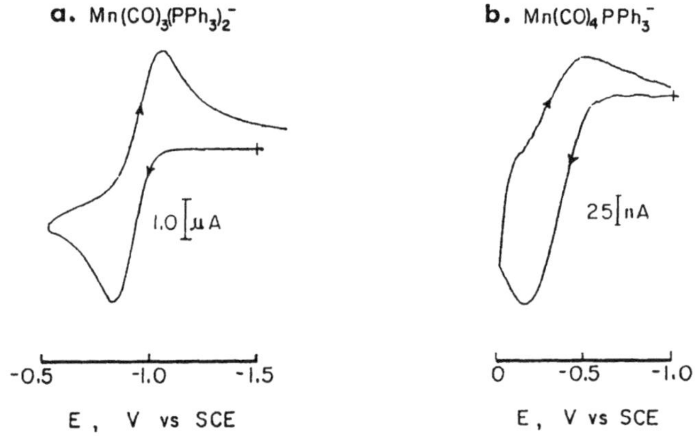

Fig. 1. Initial positive-scan cyclic voltammogram of (left) 1.0×10^{-2} M $Mn(CO)_3(PPh_3)_2^-$ at $v = 0.5$ V s^{-1} with a Pt macroelectrode ($r = 0.5$ mm) and (right) 1.0×10^{-2} M $Mn(CO)_4(PPh_3)^-$ at $v = 4{,}000$ V s^{-1} with a Pt microelectrode ($r = 5$ μm) in MeCN containing 0.3 M $TBA^+PF_6^-$ at 23 °C.

faster scan rate to achieve reversibility. By comparison the parent manganese carbonyl $Mn(CO)_5\cdot$ is the most transient radical, and the initial positive scan cyclic voltammogram in Fig. 2a is totally irreversible at the conventional scan rate of $v = 0.5$ V s^{-1}. Moreover the cyclic voltammogram is still irreversible at $v = 5{,}000$ V s^{-1} with a platinum microelectrode, as shown by the absence of a clearly defined cathodic peak current on the return scan in Fig. 2b. Indeed a scan rate in excess of 40,000 V s^{-1} is required before the cathodic peak current can be simultaneously discerned. The reversible cyclic voltammogram at $v = 175{,}000$ V s^{-1} in Fig. 2c with the peak broadening arising from uncompensated iR drop [36] and the nonfaradaic charging current is shown by comparison with the normal cyclic voltammogram represented by the dashed curve. The arithmetic average of the anodic and cathodic peak potentials $(E_a + E_c)/2$ obtained in this manner repre-

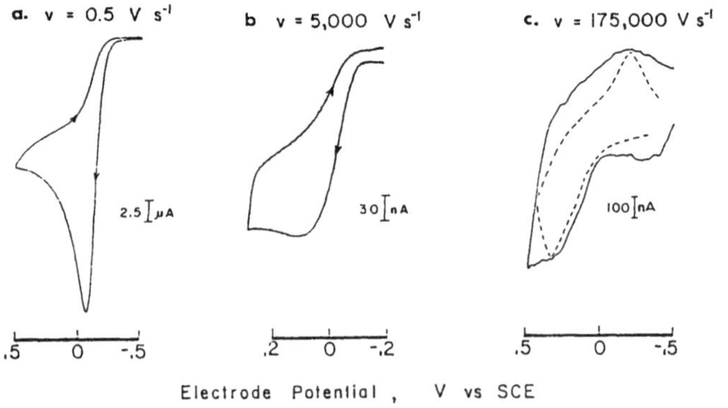

Fig. 2. Initial positive-scan cyclic voltammograms of 1.0×10^{-2} M $Mn(CO)_5^-$ at (a) $v = 0.5$ V s^{-1} with Pt macroelectrode, (b) $v = 5{,}000$ V s^{-1} and (c) $v = 175{,}000$ V s^{-1} with Pt microelectrode in MeCN containing 0.3 M $TBA^+PF_6^-$ at 23 °C.

sents a reliable measure of the reversible potential E^0 for the redox couple [35]. The constancy of E^0 at various scan rates is an indication of the reproducibility of the microvoltammetric method for $Mn(CO)_5^-$ and $Mn(CO)_4PPh_3^-$.

The chemically reversible oxidations of the various carbonylmanganate(−I) anions yield the values of the redox potential E^0 in Table 1. For the EC process depicted in Scheme IV, the second-order dimerization rate constant k_d can be calculated from a knowledge of the ratios of the cathodic and anodic peak currents i_c/i_a at various scan rates according to the method of Nicholson and coworkers [37]. Owing to the well-behaved values of i_c/i_a at CV scan rates between 1,000 and 6,000 V s^{-1} for $Mn(CO)_4PPh_3^-$, the bimolecular rate constant $k_d = 1.4 \times 10^7$ M^{-1} s^{-1} can be successfully extracted for the dimerization of $Mn(CO)_4PPh_3\cdot$ in Eq. 13.

$$2\,Mn(CO)_4PPh_3\cdot \xrightarrow{k_d} Mn_2(CO)_8(PPh_3)_2 \qquad (13)$$

Table 1. Reversible Potentials E^0 for Carbonylmanganate(−I) Anions from Cyclic Voltammetry.[a]

Carbonylmanganate	v,[b] (V s^{-1})	E^0,[c] (V vs SCE)
$Mn(CO)_5^-$	175,000[d]	+0.08
$Mn(CO)_4PPh_3^-$	2,000[d]	−0.40
$Mn(CO)_3[PPh_3]_2^-$	>0.02[e]	−0.94
$Mn(CO)_2(DPPE)_2^-$	>0.02[e]	−1.44

[a] In 1.0×10^{-2} M solutions in acetonitrile containing 0.1 M tetra-n-butylammonium hexafluorophosphate at 23 °C unless indicated otherwise. [b] Scan rate to attain $i_c/i_a \sim 0.5$. [c] $E^0 = (E_a + E_c)/2$ as described in Ref. 35. [d] With Pt microelectrode (r = 5 μm). [e] With conventional Pt electrode (r = 0.5 mm) in THF containing 0.5 M $TBA^+PF_6^-$.

This value is in excellent agreement with the rate constant (vide supra) evaluated from the flash photolysis of $Mn_2(CO)_8(PPh_3)_2$ by Brown and coworkers [33]. For $Mn(CO)_5^-$, the limited observation of well-behaved peak current ratios owing to problems with uncompensated ohmic drop and large charge currents at the requisite higher scan rates necessitates an estimate for $Mn(CO)_5\cdot$ of $k_d = 7 \times 10^8$ M^{-1} s^{-1} based on a current ratio of unity at v = 175,000 V s^{-1}. Nonetheless this value is clearly within experimental uncertainty the same as $k_d = 9 \times 10^8$ M^{-1} s^{-1} from the flash photolysis study of $Mn_2(CO)_{10}$ [33].

The ready chemical reversibility of disubstituted carbonylmanganates $Mn(CO)_3P_2^-$ (where P = alkyl phosphines and phosphites) is shown by the well-behaved cyclic voltammograms with i_c/i_a = 1.0 that are easily attainable even at sweep rates of v = 0.050 V s^{-1} [27], e.g.

$$Mn(CO)_3[P(OPh)_3]_2^- \xrightleftharpoons{-0.55V} Mn(CO)_3[P(OPh)_3]_2\cdot + e \qquad (14)$$

The trend in the values of the reversible oxidation potentials in Table 2 largely follow the σ-donor properties of the phosphine ligand as evaluated by the pK_a of the conjugate acid [38]. However the necessity for including a different correlation for phosphite ligands in Fig. 3 suggests that π-acidity in back bonding is also a contributing factor in E^0. Moreover the effects of additional phosphine ligands in the heavily substituted $Mn(CO)_2(DPPE)_2^-$ are reflected in a highly reversible cyclic voltammogram and the most negative value of $E^0 = -1.45$ V [39].

Table 2. Microvoltammetry of Carbonylmanganese(I) Cations[a]

Carbonylmanganese(I)	v^b (V s^{-1})	E_c^c (V vs SCE)	τ^d (μs)
$Mn(CO)_6^+BF_4^-$	460,000	−1.27	0.06
$Mn(CO)_6^+BF_4^{-e}$	460,000	−1.27	0.06
$Mn(CO)_5PPh_3^+BF_4^-$	200,000	−1.29	0.13
$Mn(CO)_3[NCCH_3]_3^+BF_4^-$	200,000	−1.87	0.13
trans-$Mn(CO)_2[DPPE]_2^+PF_6^-$	225,000	−1.94	0.11
cis-$Mn(CO)_2[CNCH_3]_4^+PF_6^-$	150,000	−2.48	0.17
$Mn(CN-C_6H_4CH_3)_6^+PF_6^-$	170,000	−2.41	0.15
cis-$Mn(CO)_2[P(OMe)_3]_4^+BF_4^+$	225,000	−2.10	0.11

[a] In 1.0×10^{-2} M solutions in acetonitrile containing 0.1 M tetra-n-butylammonium hexafluorophosphate at 23 °C using a Pt microelectrode ($r = 5$ μm). [b] Irreversible cyclic voltammogram at the v indicated. [c] Irreversible cathodic peak potential at $v = 0.5$ V s^{-1}. [d] Estimated upper limit to lifetime of 19-e radical, see text. [e] In the presence of 1 atmosphere of carbon monoxide

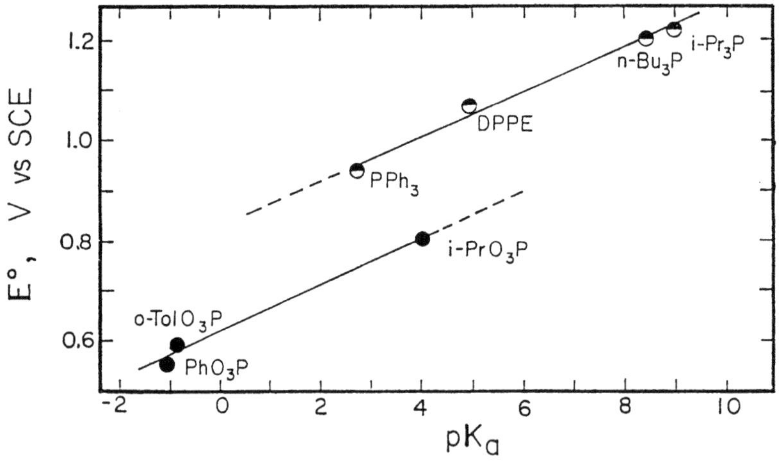

Fig. 3. Correlation of the reversible oxidation potential E^0 for $Mn(CO)_3P_2^-$ and the pK_a of HP^+ for various alkyl-phosphine and phosphite ligands (P) indicated.

2.2 Electroreduction of carbonylmanganese(I) cations

The cathodic reduction of the carbonylmanganese(I) cation $Mn(CO)_6^+$ occurs with the uptake of 1.4 faradays of charge and the evolution of 1.0 mole of carbon monoxide according to the partial stoichiometry in Eq. 15 [26].

$$Mn(CO)_6^+ \xrightarrow{-1.20V} CO + [Mn(CO)_5^- + Mn_2(CO)_{10}] \tag{15}$$

The carbonylmanganese products in Eq. 15 derive most likely via the 17-e radical according to [40]

Scheme VI:

$$Mn(CO)_6^+ \overset{e}{\rightleftharpoons} Mn(CO)_6^{\cdot} \tag{16}$$

$$Mn(CO)_6^{\cdot} \longrightarrow CO + Mn(CO)_5^{\cdot}, \quad \text{etc.} \tag{17}$$

The cyclic voltammetric behavior of $Mn(CO)_6^+$ can be accordingly examined for signs of chemical reversibility. The initial positive scan cyclic voltammogram at $v = 0.5$ V s^{-1} with a conventional platinum electrode is totally irreversible, showing only the cathodic peak for the reduction of $Mn(CO)_6^+$. Calibration of the peak current i_c with a ferrocene standard [41] indicates that the cathodic wave corresponds to an uptake of 1.3 faradays of charge, in agreement with the coulometric results (vide supra). The cyclic voltammogram at $v = 5{,}000$ V s^{-1} with a platinum microelectrode (radius = 5 μm) is fundamentally unchanged relative to that at $v = 0.5$ V s^{-1}. Furthermore even at the fastest practicable limit of $v = 460{,}000$ V s^{-1}, no anodic peak current is detected on the return scan. From results such as these, the first-order rate of decomposition of the 19-electron $Mn(CO)_6^{\cdot}$ can be estimated to be $>10^7$ s^{-1}.

In order to determine whether the presence of different ligands would otherwise stabilize the 19-electron radicals, the various substituted carbonylmanganese(I) cations are included in Table 2. Although large changes in the cathodic peak potentials E_c are induced by structurally different types of ligands, no evidence for chemical reversibility in electron transfer to carbonylmanganese cations is found, even at the enhanced scan rates listed in Table 2. In every case the lifetime of the 19-e carbonylmanganese(0) radicals is estimated to be less than 100 ns when they are limited by the dissociative loss of ligand such as that depicted below.

Scheme VII:

$$Mn(CO)_5P^+ \overset{e}{\rightleftharpoons} Mn(CO)_5P^{\cdot} \tag{18}$$

$$Mn(CO)_5P^{\cdot} \xrightarrow{>10^7 s^{-1}} Mn(CO)_5^{\cdot} + P \tag{19}$$

2.3 Reversible interchange of 19-e and 17-e radicals during the reduction of the bis-chelated $Mn(CO)_2(DPPE)_2^+$

A carbonylmanganese(I) cation that contains a pair of chelating disphosphine ligands such as $PPh_2CH_2CH_2PPh_2$ (DPPE) is stabilized relative to other types of reduced carbonylmetals [42]. In order to optimize the equilibration of 19-e and 17-e organometallic radicals, the bis-chelated carbonylmanganese(0) radical can be examined, since the tethering of the difunctional ligand also favors the reversible interchange, e.g.

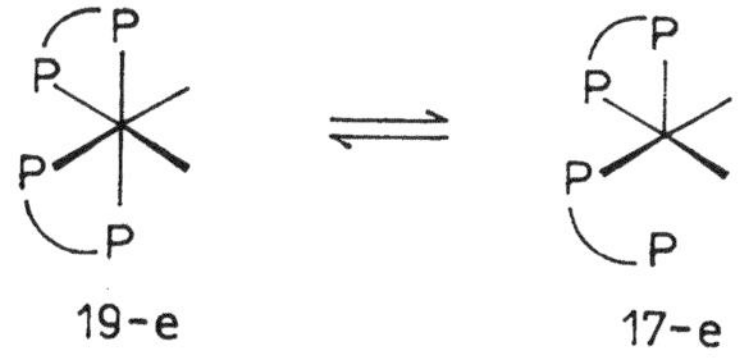

(20)

The equilibrium between such chelated radicals is expected to be especially facile where the reversible interchange in Eq. 20 involves no or minimal structural changes. As such, the recent synthesis [43] of cis-$Mn(CO)_2(DPPE)_2^+$, designated hereafter simply as **I**, provides the opportunity to generate directly the isomeric 19-e radical required for Eq. 20 in the following way. The preparative-scale electroreduction of cis-$Mn(CO)_2(DPPE)_2^+$ **I** at a controlled potential of −2.0 V in acetonitrile containing 0.1 M TBAP as the supporting electrolyte is characterized by the uptake of 2.0 faradays of charge per mole of cation to produce the cis-carbonylmanganate(−I) **II** in Eq. 21.

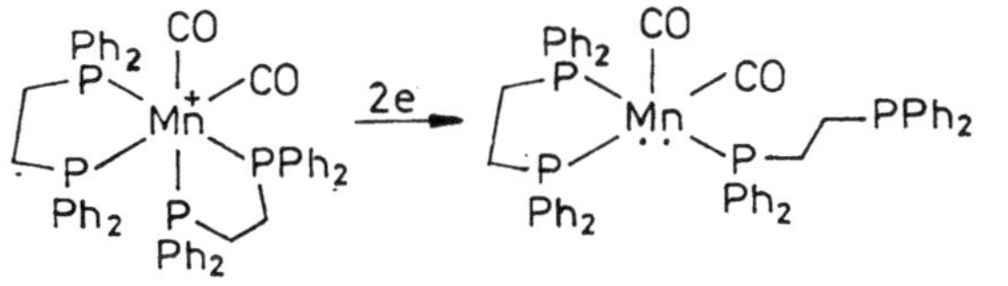

(21)

I II

The 2e reduction of the cation **I** is accompanied by the dramatic color change of the pale yellow solution to the dark red anion **II**, and the concomitant IR change in the carbonyl region from ν_{CO} = 1894 and 1950 cm^{-1} for **I** to ν_{CO} = 1710 and 1766 cm^{-1} for **II** [39].

The transient electrochemical behavior accompanying the 2e reduction of cis-$Mn(CO)_2(DPPE)_2^+$ in Eq. 21 can be examined by cyclic voltammetry. The initial negative-scan cyclic voltammogram of **I** in Fig. 4 shows an irreversible cathodic wave **A** with the peak potential $E_c = -1.71$ V corresponding to a 2e reduction of **I**, as calibrated with a ferrocene standard. The formation of the anion **II** for

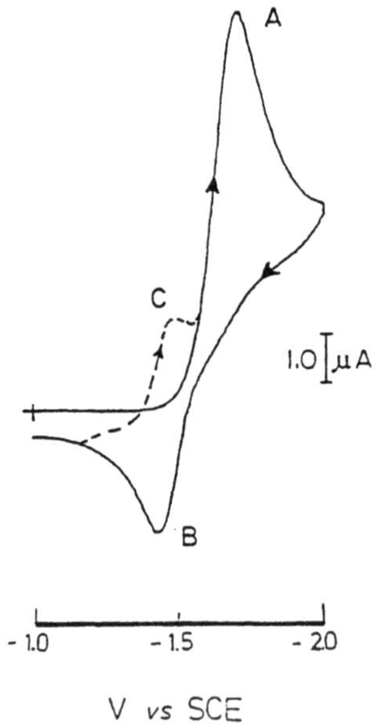

Fig. 4. Initial negative-scan cyclic voltammogram at 23 °C of 5×10^{-3} M cis-$Mn(CO)_2(DPPE)_2^+AsF_6^-$ in THF containing 0.3 M TBAP at $v = 0.5$ v s^{-1} with a Pt macroelectrode. Dashed curve shows cathodic wave on return scan.

Eq. 21 is shown in the cyclic voltammogram (Fig. 4) by the presence of the anodic wave **B** on the return positive-scan. The latter $E_a = -1.39$ V represents the oxidation of the anion $Mn(CO)_2(DPPE)_2^-$ **II** to the corresponding 17-electron radical (hereafter referred to simply as 17-e), i.e.

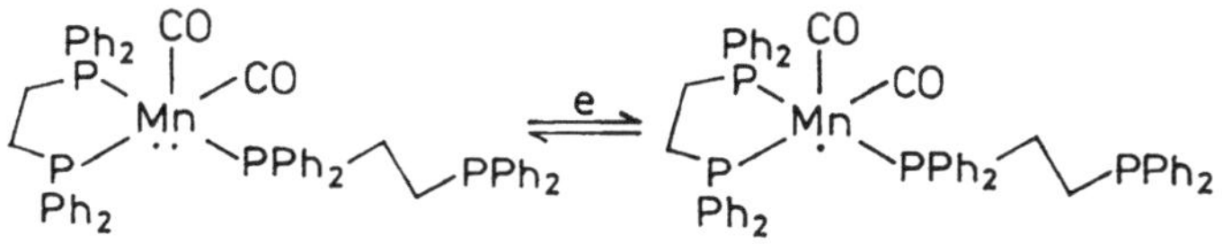

II 17-e (22)

The anodic wave **B** is the same as that in a cyclic voltammogram obtained from the anodic oxidation of authentic anion **II** prepared independently [39]. Furthermore, its reversible character is shown by the presence of the dashed cathodic wave **C** on the reverse negative scan. It is important to note that wave **C** occurs at a potential of $E_c = -1.52$ V, which is substantially more positive than that of wave **A** at $E_c = -1.71$ V for the initial reduction of the cation **I**. As a result, the reduction of the 17-e species is also included in the 2e cathodic wave **A** that remained chemically irreversible even at CV scan rates exceeding 100,000 V s^{-1}, easily achieved with a platinum microelectrode (vide supra). Such an electrochemical behavior is indicative of an ECE process for the 2e reduction of the cation **I** to the anion **II** in Eq. 21, e.g.

Scheme VIII:

$$\text{I} \xrightarrow{e} \text{19-e} \tag{23}$$

$$\text{19-e} \xrightarrow{\circ} \text{17-e} \tag{24}$$

$$\text{17-e} \overset{e}{\rightleftharpoons} \text{II} \tag{25}$$

According to the ECE mechanism in Scheme VIII, the repeated cycling of the potential scans should lead to a steady-state cyclic voltammogram like that shown in Fig. 5a. Thus after only ~5 cycles, the cathodic wave **A** for cis-$Mn(CO)_2$-$(DPPE)_2^+$ is markedly diminished and the **B**/**C** couple is dominant. The isopotential point (IPP being akin to isosbestic point in spectroscopy [44]) at -1.64 V indicates that one electroactive species (**I**) is quantitatively converted to the other (**II**), the sum of the reactant and the product concentrations thus remaining constant throughout the repetitive scans. The essential correctness of the ECE formulation is confirmed by the computer simulation of the cyclic voltammograms in Fig. 5b over a sequence of four repetitive cycles. The CV simulations are carried out with Feldberg's finite difference (digital) method by use of the following optimized electrochemical parameters [45]. The heterogeneous rate constant $k_s = 4.0 \times 10^{-3}$ cm s^{-1} for wave **A** and the standard reduction potential $E_1{}^0 = -1.61$ V are set of yield an $E_c = -1.71$ V with the experimentally determined transfer coefficient $\alpha - 0.40$ [46]. The parameters for the **II**/17-e couple are the experimental values of $k_s = 1.1 \times 10^{-2}$ cm s^{-1}, $E_2{}^0 = -1.45$ V and $\alpha = 0.45$. The diffusion coefficients of all species are set equal to the experimentally determined value of $D = 6 \times 10^{-6}$ cm^2 s^{-1} for **II** [44a]. The first-order rate constant for the 19-e/17-e conversion in Eq. 24 is set as $k_f = 1 \times 10^6$ s^{-1} to accord with the microvoltammetric results [47].

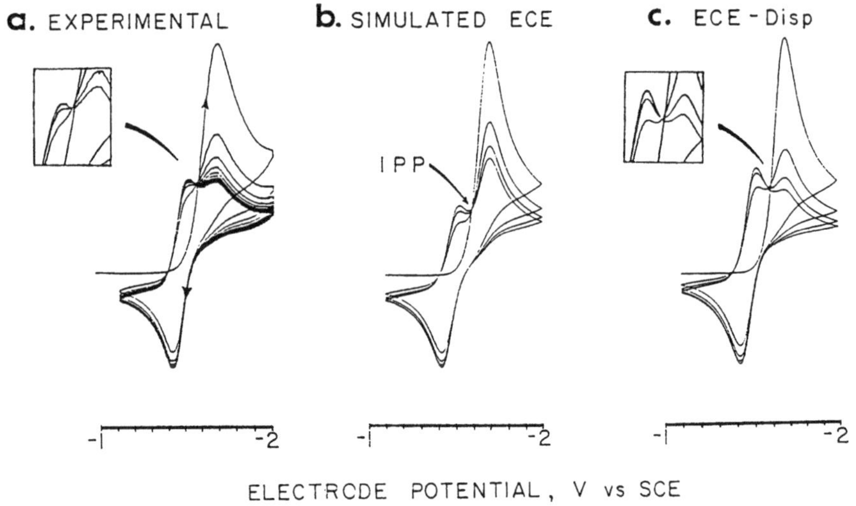

Fig. 5. (a) Initial negative-scan (15-cycle) cyclic voltammograms of 5×10^{-3} M cis-$Mn(CO)_2(DPPE)_2{}^+AsF_6{}^-$ in THF containing 0.3 M TBAP at $v = 0.4\ V\ s^{-1}$. Simulated (4-cycle) cyclic voltammogram based on the (b) ECE and (c) ECE-Disp mechanism, as described in the text.

Although good qualitative agreement exists between the experimental (Fig. 5a) and simulated (Fig. 5b) cyclic voltammograms, there is a major discrepancy that could not be simply resolved by modifying the input parameters [48]. Thus, it is seen that the concentration of **I** (i.e., height of the cathodic wave **A**) after 4 cycles is substantially greater in Fig. 5b than that observed experimentally in Fig. 5a. The unusually rapid dropoff of the concentration of **I** after the first cycle in Fig. 5a indicates the necessity of including in the mechanism an additional step, which would reduce the concentration of **I** faster on successive scans, and simultaneously produce an electroactive species reducible at a more positive potential than **I**. A disproportionation (disp) process between the cation **I** and its reduced anion **II** that gives rise to radicals, i.e.

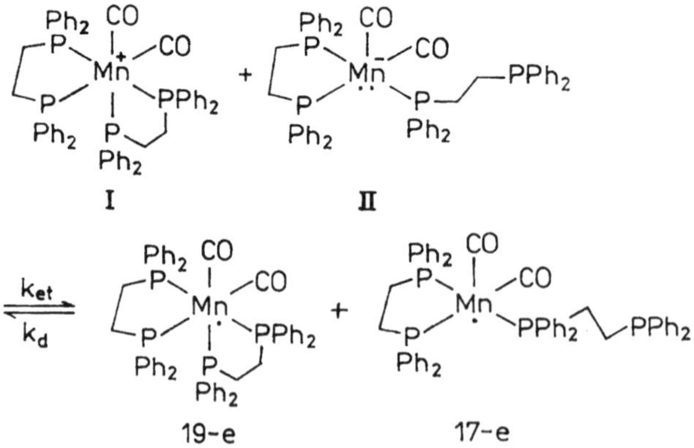

(26)

produces the desired effect, as shown by the CV simulation in Fig. 5c that is generated by the addition of the optimized rate constants $k_{et} = 2.0 \times 10^2$ M^{-1} s^{-1} and $k_d = 20$ M^{-1} s^{-1} for the electron transfer and disproportionation in Eq. 26 to the ECE scheme. Indeed the rapidity of such an ion-pair annihilation can be independently demonstrated by mixing solutions of cis-$Mn(CO)_2(DPPE)_2^+AsF_6^-$ and mer-$Mn(CO)_2(DPPE)_2^-Na^+$: whereupon the ESR spectrum at $\langle g \rangle = 2.025$ taken immediately of the 17-e radical is the same as that obtained from the anodic oxidation of **II** [39].

Although the ECE-Disp mechanism correctly predicts the rapid falloff of the CV wave **A**, a closer scrutiny of Fig. 5c reveals the following three important discrepancies from the experimental CV in Fig. 5a.

(a) The height of the cathodic wave **A** on the fourth cycle is now less than wave **C**.
(b) The height of the cathodic wave **C** is larger on all cycles.
(c) The IPP is positioned at a lower current between peaks **C** and **A**.

The analysis of the behavior (a)–(c) indicates that the 17-e radical is actually undergoing a reaction that reduces its concentration and simultaneously increases the concentration of the cation **I** relative to the ECE-Disp simulation. An obvious remedy to this kinetics situation is to include reversibility (rev) in the formation of the 17-e radical from the cation **I**, viz.

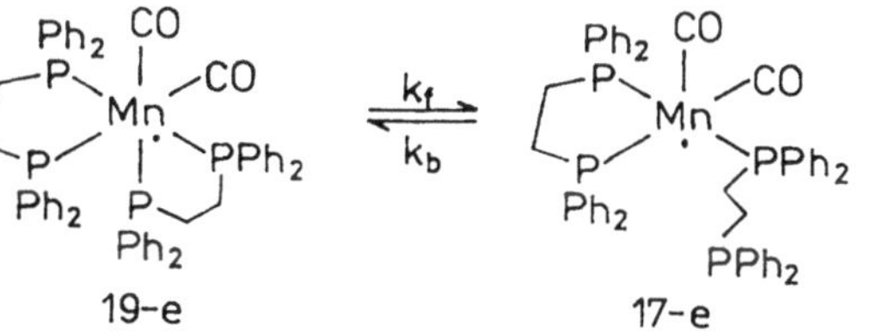

(27)

Inclusion of this reversible interchange to generate the $EC_{rev}E$-Disp mechanism in Scheme IX results in the CV simulation presented in Fig. 6B.

Scheme IX:

$$\text{I} \xrightarrow{e} \text{19-e} \quad (28)$$

$$\text{19-e} \underset{k_b}{\overset{k_f}{\rightleftharpoons}} \text{17-e} \quad (29)$$

$$\text{17-e} \overset{e}{\rightleftharpoons} \textbf{II} \quad (30)$$

$$\textbf{I} + \textbf{II} \rightleftharpoons \text{19-e} + \text{17-e} \quad (31)$$

The excellent agreement of the simulated cyclic voltammogram (Fig. 6b) with the experimental one (Fig. 6a) is carried out with the same optimized rate constants and $k_f = 1 \times 10^6$ s^{-1}, $k_b = 3.5$ s^{-1}. Inspection of the simulated CV (Fig. 6b) in the region of the IPP now reveals its position in both the current and potential scales to be in perfect agreement with the experimental curves (Fig. 6a). Furthermore there are two critical features, namely (a) the rate of depletion of the cation **I** (wave **A**) upon recycling and (b) the passage of the first scan at a point negative of IPP that are both seen to match the experimental trace. At first glance the value of $k_b = 3.5$ s^{-1} may seem to be unusually small to have any drastic effect on the CV traces. However, cognizance must be taken of the fact that such a 17-e/19-e conversion always occurs in a potential region $\leqq -1.45$ V (i.e., E_2^0 for the oxida-

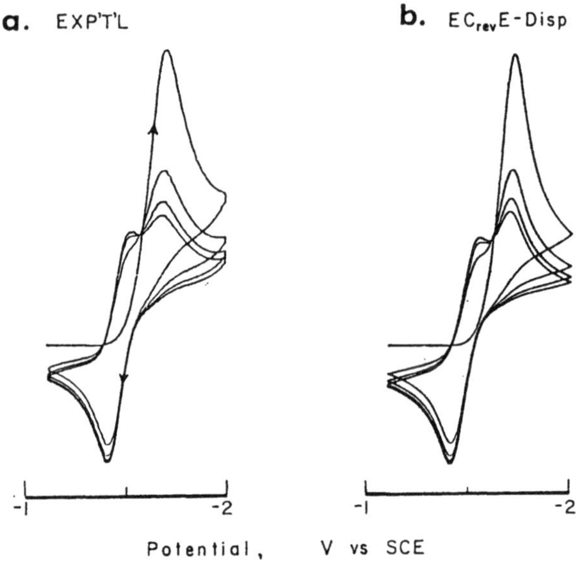

Fig. 6. (a) Initial negative-scan (4-cycle) cyclic voltammograms of 5×10^{-3} M cis-$Mn(CO)_2(DPPE)_2{}^+AsF_6{}^-$ in THF containing 0.3 M TBAP at $v = 0.4$ V s^{-1}. (b) Simulated (4-cycle) cyclic voltammogram based on $EC_{rev}E$-Disp mechanism in Scheme IX.

tion of anion **II**), which is ~150 mV more positive than $E_1{}^0 = -1.61$ V for the reduction of cation **I**! As a result, the 19-e radical is immediately oxidized at the electrode surface; and the reversible interchange in Eq. 29 (however slow), is tantamount to the spontaneous re-generation of the cation **I**. The validity of this analysis is underscored in Fig. 7 by the excellent agreement between experimental

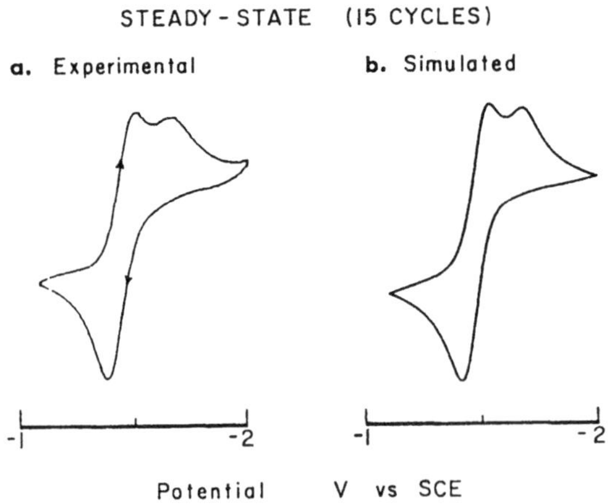

Fig. 7. Comparison of the experimental (left) and simulated (right) steady-state cyclic voltammogram after 15-cycles of cis-$Mn(CO)_2(DPPE)_2{}^+$ based on the $EC_{rev}E$-Disp mechanism (see text).

and simulated steady-state cyclic voltammograms after 15 cycles based on the ECE-Disp mechanism in Scheme IX.

It is singularly noteworthy that the optimized magnitudes of the first-order rate constants of $k_b = 3.5\ s^{-1}$ relative to $k_f = 10^6\ s^{-1}$ in Eq. 29 indicate an "equilibrium" constant $K \simeq 10^6$ that favors the *opened* 17-e radical over the *closed* 19-e counterpart [49]. Therefore, it is not surprising that the reversible interchange between 17-e and 19-e radicals is not possible during the reduction of the non-chelated carbonylmanganese(I) cations listed in Table 2. However short their lifetime, 19-e carbonylmanganese(0) radicals are thus ascertained to be discrete intermediates, as distinguished from activated complexes located at maxima of potential energy surfaces.

2.4 Oxidatively induced cis → trans isomerization of $Mn(CO)_2(DPPE)_2^+$

The lability of cis-$Mn(CO)_2(DPPE)_2^-$ following its reduction (Section 2.3) is also manifested in its dynamic behavior following oxidation. Thus a bright yellow solution of cis-$Mn(CO)_2(DPPE)_2^+$ **I** in acetonitrile on standing slowly darkens to orange, and inspection of the IR spectrum indicates the disappearance of the characteristic carbonyl stretching band at ν_{CO} 1950 cm^{-1} and concomitant growth of the band at $\nu_{CO} = 1897\ cm^{-1}$ of the trans isomer, **III**, i.e. [50]

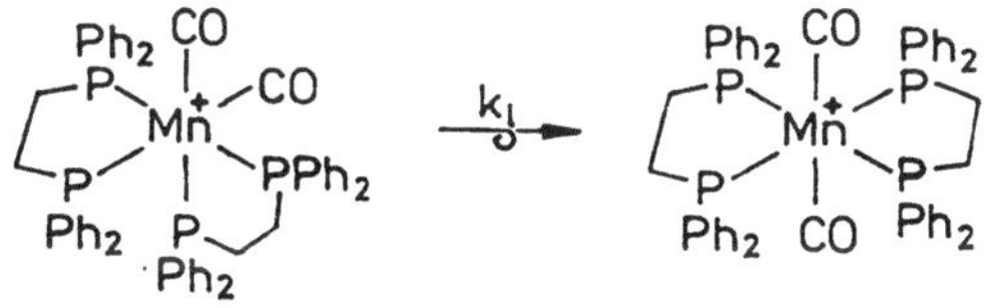

(32)

I III

The rather slow cis → trans isomerization follows first-order kinetics with $k_1 = 1.5 \times 10^{-5}\ s^{-1}$ at 23 °C. However when a yellow solution of 5×10^{-3} M cis-$Mn(CO)_2(DPPE)_2^+$ in acetonitrile containing 0.1 M TBAP is exhaustively oxidized at a controlled potential of $E = 1.50$ V, it turns dark blue green with an electron uptake of 1.0 ± 0.05 coulombs per mole of the *cation I charged.* IR analysis of the anolyte indicates the complete disappearance of **I** with its carbonyl band at ν_{CO} 1950 cm^{-1}, and the simultaneous appearance of a new band at ν_{CO} 1966 cm^{-1}. The latter is associated with the paramagnetic dication trans-$Mn(CO)_2(DPPE)_2^{2+}$ with an isotropic ESR spectrum at $\langle g \rangle = 2.00014$ in Fig. 8 showing a sextet split-

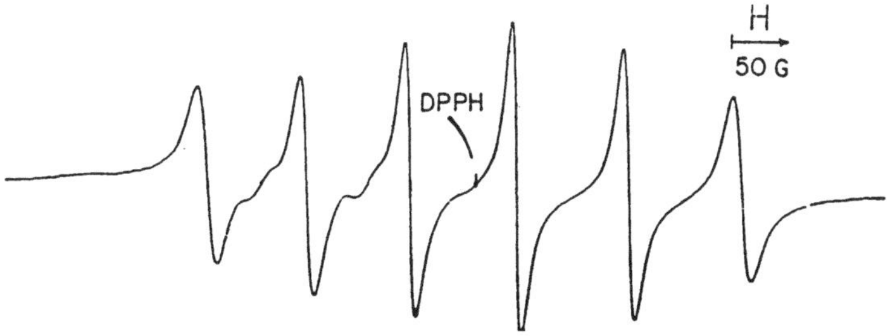

Fig. 8. ESR spectrum of $Mn(CO)_2(DPPE)_2^{++}$ (with DPPH field marker) formed from the anodic oxidation of either cis- or trans-$Mn(CO)_2(DPPE)_2^+$ in acetonitrile.

ting due to ^{55}Mn (I = 5/2) with the hyperfine splitting of $a_{Mn} = 95$ gauss. Thus in the course of anodic oxidation (approximately 30 min), the cis-monocation is completely converted to the trans dication **III**$^+$, i.e.

$$\underset{\mathbf{I}}{\text{cis-Mn(CO)}_2\text{(DPPE)}_2{}^+} \xrightarrow{-e} \underset{\mathbf{III}^+}{\text{trans-Mn(CO)}_2\text{(DPPE)}_2{}^{2+}} \qquad (33)$$

It is important to emphasize that the thermal rearrangement in Eq. 32 occurs at least 10^7 times too slowly to participate in any significant way to the process in Eq. 33. In order to ascertain the source of the rearrangement, the cyclic voltammetry of **I** is examined at 0 °C at several scan rates. The initial positive-scan cyclic voltammograms of **I** in acetonitrile containing 0.1 M TBAP at $v = 2$ and 100 V s^{-1} are shown in Fig. 9a and 9b, respectively. Two features of these sweep-dependent CV's are noteworthy. First, the irreversible anodic wave **A** with $E_a = 1.22$ V in Fig. 9a becomes chemically reversible with $E^0 = 1.19$ V at the higher scan rate shown in Fig. 9b. Such a CV behavior underscores the short lifetime of the cis dication **I**$^+$. Second, the cathodic wave **B** at $E_c = 0.84$ observed on sweep reversal in Fig. 9a is associated with the trans dication **III**$^+$ shown by the reversible CV couple (with the coupled anodic wave **B** indicated as a dashed curve) at $E = 0.88$ V. Since **III**$^+$ appears only after the reversible oxidation of the cis cation **I**, it follows that it is formed by the spontaneous rearrangement of the metastable cis dication **I**$^+$. This facile cis → trans rearrangement of the dication $Mn(CO)_2(DPPE)_2{}^{2+}$ is included in the ECE mechanism below.

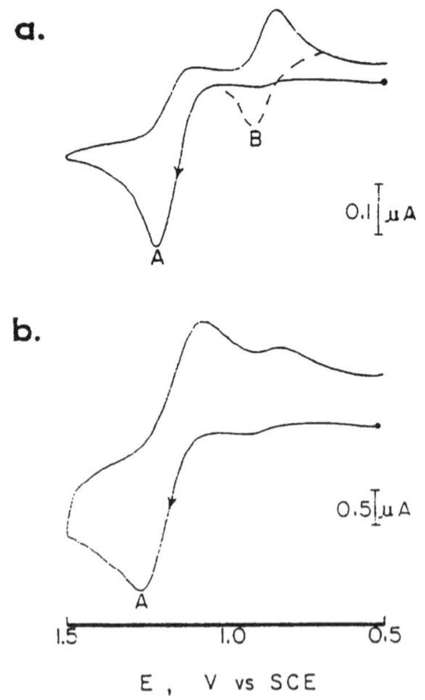

Fig. 9. Initial positive-scan cyclic voltammogram at 0 °C of 5×10^{-3} M cis-Mn(CO)$_2$(DPPE)$_2{}^+$ in acetonitrile containing 0.1 M TBAP with a platinum electrode at (a) $v = 2$ V s^{-1} and (b) $v = 100$ V s^{-1}. Dashed curve shows the reversible anodic wave **B** on the return scan.

Scheme X:

$$\mathrm{I} \underset{}{\overset{-e}{\rightleftharpoons}} \mathrm{I}^+ \tag{34}$$

$$\mathrm{I}^+ \xrightarrow{k_2} \mathrm{III}^+ \tag{35}$$

$$\mathrm{III}^+ \underset{}{\overset{+e}{\rightleftharpoons}} \mathrm{III} \tag{36}$$

The overall transformation for the ECE process in Scheme X corresponds to the isomerization of the cis-monocation **I** to the trans isomer **III** reminiscent of that in Eq. 32. However it differs from the relatively straightforward thermal process in Eq. 32 in that it proceeds via a prior electron transfer. As such, this cis/trans conversion of the carbonylmanganese cation **I** is designated as an *oxidatively induced rearrangement*. Two questions arise immediately. Is it a chain process similar to the electron transfer chain or ETC catalysis that is observed in other carbonylmetal rearrangements [51]? How fast does the dication isomerize in Eq. 35 in comparison with the monocation in Eq. 32? When the bulk anodic oxidation of **I** is interrupted periodically and small aliquots of anolyte removed, IR or CV analysis indicates that the product $\mathbf{III}^+$ is always present in amounts corresponding to the fraction of charge (i.e., coulombs) passed through the solution. This result rules out any chain process such as the aforementioned ETC catalysis [51]. Accordingly the stoichiometric pathway described in Scheme X can be used to determine the rate constant for the rearrangement step in Eq. 35. Two analytic procedures are employed. In the first method, the cathodic/anodic current ratios i_c/i_a at various scan rates v are fitted to the working curves derived by Nicholson and Shain [47a]. Although conveniently independent of the electrochemical kinetics, this method requires accurate (and difficult to measure) values of the current ratio, and it is thus difficult to apply quantitatively. Nonetheless the best value of k_2 for Eq. 35 by this procedure are listed in Table 3 at two temperatures. In the second method, digital simulations of the cyclic voltammograms are performed at various sweep rates using the finite difference approach of Felberg [45]. Unlike the Nicholson and Shain analysis, the digital simulation of the cyclic voltammogram in Fig. 10 requires a knowledge of the electrochemical parameters: the reversible potential E^0, the transfer coefficient α and the heterogeneous rate constant k_s together with the diffusion coefficient. Owing to the reversible nature of both

Table 3. Oxidatively Induced Isomerization of cis-$Mn(CO)_2(DPPE)_2^{+}$[a]

Temp. (°C)	First-Order Rate Constant k_2 (s^{-1})[b] Working Curve[c]	CV Simulation[d]
0	21	23
23	103	120

[a] From 5×10^{-3} M cis-$Mn(CO)_2(DPPE)_2^+AsF_6^-$ in acetonitrile containing 0.1 M TBAP at 23 °C. [b] Reproducibility, ±15%. [c] Derivation of Nicholson and Shain in Ref. 47. [d] Digital simulation of CV according to Feldberg in Ref. 45.

couples, i.e., I $\rightleftharpoons$ I$^+$ and III $\rightleftharpoons$ III$^+$, values of E^0, α, and k_s are determined by standard methods [52, 53]. The acceptable agreement in the values of k_2 in Table 3 determined by digital simulation with those based on the Nicholson and Shain working curves reinforces our confidence in the essential correctness of the mechanism in Scheme X. These studies show that cis/trans isomerization is enhanced markedly by the oxidative conversion of cis-$Mn(CO)_2(DPPE)_2^+$ to its dication. Thus the magnitude of the first-order rate constant in Table 3 increases by seven orders of magnitude for k_1 in Eq. 32. Such a rate enchancement in the cis-trans rearrangement of $Mn(CO)_2(DPPE)_2^+$ underscores the substantial difference in lability of the 18-e monocation **I** on its oxidative conversion to the paramagnetic 17-e dication **III**.

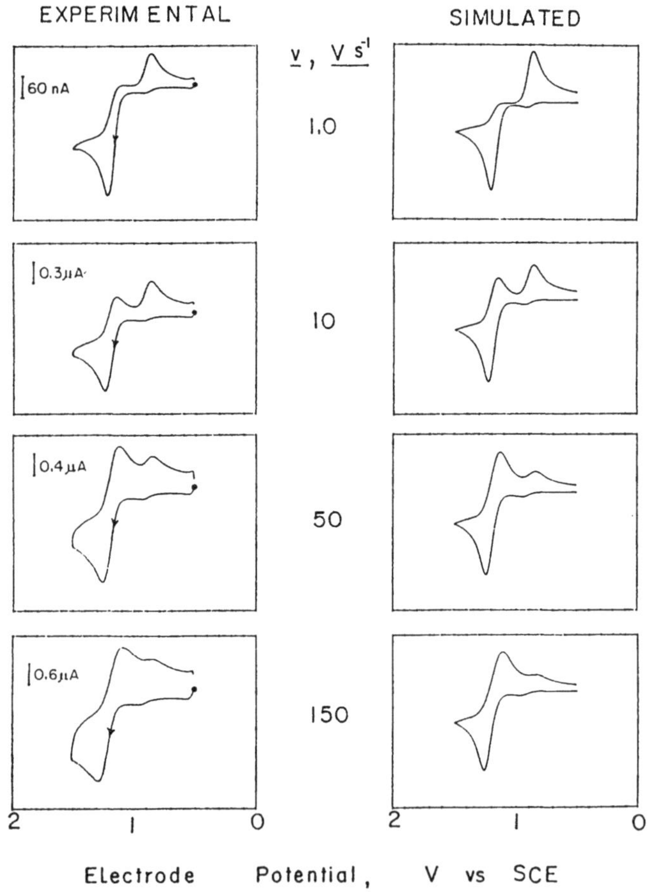

Fig. 10. Comparison of the experimental (left) and computer simulated (right) cyclic voltammograms at various scan rates (v) of cis-$Mn(CO)_2(DPPE)_2^+$ based on Feldberg's method. [Note the pronounced charging currents induced at high scan rates were not included in the CV simulations.]

3. Thermal annihilation of carbonylmetal cations by carbonylmetallate anions

The formation of metal-metal bonds in dimeric metal carbonyls derives from various reductive procedures including the use of carbon monoxide, metals, alkylmetals, etc., as reagents [54]. The oxidation of carbonylmetallate anions is also known to lead to carbonylmetal dimers [55]. Heterobimetallic carbonyls result from the interaction of carbonylmetallates with different types of metal halides and homo dimers [56]. The latter can be generally classified as nucleophilic substitution processes, although mechanistic studies of such metal-metal bond formations are generally lacking. The heterolytic coupling of carbonylmetal cations and anions is indicated in the treatment of tetracarbonylcobaltate (—I) with hexacarbonylrhenium(I) to afford the mixed carbonyl [57], i.e.

$$Co(CO)_4^- + Re(CO)_6^+ \rightarrow ReCo(CO)_9 + CO \tag{37}$$

Similarly the treatment of $Co(CO)_4^-$ with the phenanthroline-substituted tetracarbonylmanganese(I) cation leads to the substituted heterobimetallic carbonyl [58], viz.

$$Co(CO)_4^- + (\eta^2\text{-phen})Mn(CO)_4^+ \rightarrow (\eta^2\text{-phen})MnCo(CO)_7 + CO$$

and the highly efficient coupling of the carbonylmanganese anion and cation, yields the homo dimer [40],

$$Mn(CO)_5^- + Mn(CO)_6^+ \rightarrow Mn_2(CO)_{10} + CO \tag{38}$$

The availability of various substituted-carbonylmanganese cations and anions presents a unique opportunity to examine the mechanism of the coupling processes in detail, especially with regard to the products and stoichiometry for the general representation:

$$Mn(CO)_5L^+ + Mn(CO)_4P^- \rightarrow Mn_2(CO)_8(L)(P) + CO \tag{39}$$

where L and P respresent various phosphorus and nitrogen-centered ligands associated with the carbonylmanganese cation and anion, respectively.

3.1 Coupling of anionic carbonylmanganates(—I) and carbonylmanganese(I) cations

The composition of the reaction mixture from the ion-pair interaction of $Mn(CO)_5L^+$ and $Mn(CO)_4P^-$ in Eq. 39 differs according to the nature of the ligands L and P. In order to facilitate the descriptions of the product mixtures, the interactions of the carbonylmanganates with each of the cations will be described separately.

3.1.1 Pentacarbonylmanganage(—I) and hexacarbonylmanganese(I) react upon mixing (<5 min) in tetrahydrofuran solutions to afford dimanganese decacarbonyl as the sole carbonyl product. Essentially the same results are obtained in THF solution with the acetonitrile and pyridine derivatives as shown in Table 4, i.e.

$$Mn(CO)_5^- + Mn(CO)_5L^+ \rightarrow Mn_2(CO)_{10} + L \tag{40}$$

Table 4. Products and Stoichiometry from the Coupling of $Mn(CO)_5^-$ and $Mn(CO)_5L^+$[a]

$Mn(CO)_5L^+$ L	(mmol)	$Mn_2(CO)_{10}$ mmol (%)	$Mn_2(CO)_9L$ mmol (%)	$Mn_2(CO)_8L_2$ mmol (%)	$HMn(CO)_4L$ mmol (%)	Time[b] (min)	Material Balance
CO	(.05)	.040 (80)		–	–	<5	80
CH_3CN	(.05)	.044 (88)		–	–	<5	88
py	(.05)	.042 (84)		–	–	<5	84
PPh_3	(.05)	.033 (66)		.007 (14)	.008 (8)	<5	88
$P(p\text{-Tol})_3$	(.05)	.028 (56)		.007 (14)	.015 (15)	<5	85
PPh_2Et	(.05)	.023 (46)	tr	.006 (16)	.018 (18)	30	76
$PPhEt_2$	(.05)	.019 (38)	tr	.006 (12)	.022 (22)	200	72
PEt_3	(.05)	.014 (28)	[c] (10)	.006 (12)	.017 (17)	300	67
PPh_2Me	(.05)	.018 (36)	tr	.004 (8)	.017 (17)	5	61
$PPhMe_2$	(.05)	.020 (40)	[c] (5)	.010 (20)	.016 (16)	200	81

[a] In 10 mL of THF with 5×10^{-3} M $NaMn(CO)_5$ and 5×10^{-3} M $Mn(CO)_5L^+PF_6^-$ (or BF_4^-) at 22 °C. [b] Approximate time required for complete disappearance of ions. [c] Estimated, based on severe overlap of CO bands. tr = trace (<1%).

where L = CO, MeCN, and pyridine. When pentacarbonylmanganate(I) is treated with the triarylphosphine-substituted cations high yields of $Mn_2(CO)_{10}$ are formed together with minor amounts of the bis-substituted dimer $Mn_2(CO)_8L_2$ [59] and the substituted hydride $HMn(CO)_4L$ [60], i.e.

$$Mn(CO)_5^- + Mn(CO)_5L^+ \rightarrow \underset{(60\%)}{Mn_2(CO)_{10}} + \underset{(14\%)}{Mn_2(CO)_8L_2} + \underset{(10\%)}{HMn(CO)_4L} \quad (41)$$

where L = phosphines and the numbers in parentheses indicate averaged yields. This somewhat complex product mixture can be readily analyzed by quantitative IR spectroscopy (±15%) using calibration curves constructed from the characteristic ν_{CO} stretching bands of the authentic components. The analysis is confirmed by chromatographic separation of the components. On this basis, the overall material balance in excess of 80% represents the quantitative conversions of the carbonylmanganese anion and the various cations. No other carbonyl-containing manganese product is discerned from the IR spectra of the products by spectral subtraction of the components listed in Table 4. Treatment of pentacarbonylmanganate with the alkyl and aralkylphosphines derivatives also affords variable amounts of $Mn_2(CO)_{10}$ and the analogous carbonylmanganese products $Mn_2(CO)_8L_2$ and $HMn(CO)_4L$. In addition, small but distinctive amounts of the mono-substituted dimanganese carbonyl $Mn_2(CO)_9L$ are detected. Furthermore, the coupling processes of $Mn(CO)_5^-$ with $Mn(CO)_5PPh_3^+$, etc. often proceed at significantly attentuated rates, as qualitatively indicated by the reaction times listed in column 6, Table 4. The material balance in Table 4 indicates that all of the phosphine (L) included in $Mn(CO)_5L^+$ is not retained in the carbonylmanganese products. The examination of the ^{31}P NMR spectrum of the reaction mixture indicates that the deficit is present as free phosphine.

3.1.2 Tetracarbonyl(triphenyl phosphite)manganate(−I) and the cationic hexacarbonylmanganese(I) rapidly yields two manganese dimers $Mn_2(CO)_{10}$ and $Mn_2(CO)_8[P(OPh)_3]_2$ in a roughly 4 : 3 ratio, i.e.

$$Mn(CO)_4P^- + Mn(CO)_6^+ \rightarrow \underset{(40\%)}{[Mn_2(CO)_{10}} + \underset{(30\%)}{Mn_2(CO)_8P_2]} + CO \qquad (42)$$

where $P = P(OPh)_3$. When the acetonitrile and pyridine-substituted cations are employed, the same pair of dimanganese carbonyls is obtained on mixing, together with significant amounts of the mono-substituted analogue, i.e.

$$Mn(CO)_4P^- + Mn(CO)_5L^+$$

$$\rightarrow \underset{(30\%)}{[Mn_2(CO)_{10}} + \underset{(30\%)}{Mn_2(CO)_9P} + \underset{(20\%)}{Mn_2(CO)_8P_2]} + L \qquad (43)$$

where $P = P(OPh)_3$ and $L = MeCN$ and py. The absence of any other carbonyl-containing product in the reaction mixture indicates that the nitrogen ligands (MeCN and py) are rapidly replaced, as given by the approximate stoichiometry presented in Eq. 43. In marked contrast, the reaction of the phosphine-substituted cations $Mn(CO)_5L^+$ with $Mn(CO)_4[P(OPh)_3]^-$ affords no $Mn_2(CO)_{10}$ or $Mn_2(CO)_9P$ – the most important products being the other dimanganese carbonyls, i.e.

$$Mn(CO)_4P^- + Mn(CO)_5L^+$$

$$\rightarrow \underset{(20\%)}{Mn(CO)_8P_2} + \underset{(30\%)}{Mn(CO)_8L_2} + \underset{(20\%)}{Mn_2(CO)_8(L)(P)} \qquad (44)$$

where $P = P(OPh)_3$. In addition, minor amounts of the manganese hydrides $HMn(CO)_4L$ and $HMn(CO)_4P$ are observed. With $Mn(CO)_5PEt_3^+$ and $Mn(CO)_5PEt_2Ph^+$, the IR spectrum of the solution taken immediately after mixing with $Mn(CO)_4P(OPh)_3^-$ shows the presence (~20%) of the reactant ion pair, the complete disappearance of which requires about 1 h.

3.1.3 Pentacarbonyl(triphenylphosphine)manganate(−I) reacts upon mixing with hexacarbonylmanganese(I) and its acetonitrile and pyridine derivatives to produce three dimanganese carbonyls, i.e.

$$Mn(CO)_4P^- + Mn(CO)_5L^+$$

$$\rightarrow \underset{(30\%)}{[Mn_2(CO)_{10}} + \underset{(20\%)}{Mn_2(CO)_9P} + \underset{(40\%)}{Mn_2(CO)_8P_2]} + L \qquad (45)$$

where $P = PPh_3$ and $L = CO$, MeCN and py. The complete replacement of acetonitrile and pyridine respectively, according to Eq. 45 is supported by the singular absence of other carbonyl-containing products. However with the phosphine-substituted cations, the ligand **L** is completely retained to afford a mixture of the homo- and cross-coupled dimers, similar to that described in Eq. 44, where $P = PPh_3$ and $L =$ triaryl- and aralkylphosphines. Of the phosphine-substituted cations, $Mn(CO)_5PEt_3^+$ is somewhat unusual in that it also affords significant amounts of the hydrides $HMn(CO)_4PEt_3$ and $HMn(CO)_4PPh_3$.

3.1.4 The effect of added phosphine on the ion-pair couplings can be examined with the anionic carbonylmanganates(—I) and carbonylmanganese(I) cations, since they are all substitution stable. Thus, the separate solutions of these ions in acetonitrile or tetrahydrofuran remain unchanged in the presence of various amounts of added phosphine – certainly for periods far exceeding the time required for the ion-pair coupling (vide supra) to proceed to completion. Accordingly, the effect of added phosphine on the course of dimanganese formation shows three important changes that accompany the presence of triphenylphosphine on the ion-pair coupling of $Mn(CO)_5^-$ and $Mn(CO)_5(NCMe)^+$ in Eq. 40. First, with increasing amounts of PPh_3, the yield of $Mn_2(CO)_{10}$ drops precipitously, as indicated by the trend in column 2, Table 5. The deficit is made up with a pair of new products $Mn_2(CO)_8(PPh_3)_2$ and $HMn(CO)_4PPh_3$.

$$Mn(CO)_5^- + Mn(CO)_5(NCMe)^+ \xrightarrow{PPh_3} Mn_2(CO)_8(PPh_3)_2 + HMn(CO)_4PPh_3, \text{ etc.} \qquad (46)$$

Table 5. Effect of Added Phosphine on Ion Pair Annihilation[a]

Added PPh_3 mmol	$Mn_2(CO)_{10}$ mmol	(%)	$Mn_2(CO)_9P$ mmol	(%)	$Mn_2(CO)_8P_2$ mmol	(%)	$HMn(CO)_4P$ mmol	(%)	Material Balance
0[b]	.044	(88)	–		–		–		88
0.1[b]	.035	(70)	–		.004	(8)	.008	(8)	86
0.3[b]	.032	(64)	–		.007	(14)	.012	(12)	90
0.5[b]	.021	(42)	d		.009	(18)	.020	(20)	80
0[c]	.011	(22)	.009	(18)	.026	(52)			92
0.05[c]	.008	(16)	.010	(20)	.027	(54)			90
0.15[c]	.006	(12)	.009	(18)	.032	(64)			94
0.25[c]	.003	(6)	.008	(16)	.033	(66)	f		88
0.10[e]	.0		.0		.020	(40)[h]	.040	(40)[j]	g
0.10[k]	.0		0		.035	(70)[m]	.030	(30)[n]	

[a] Conditions as stated in Tables 1 and 3. P = PPh_3, unless indicated otherwise. [b] $NaMn(CO)_5$ and $Mn(CO)_5(NCMe)^+PG_6^-$. [c] $NaMn(CO)_4PPh_3$ and $Mn(CO)_5(NCMe)^+PF_6^-$. [d] Traces observed. [e] $NaMn(CO)_5$ and $Mn(CO)_5PEt_3^+PF_6^-$ with added PEt_3. [f] $HMn(CO)_4PPh_3$ detected, but not quantified. [g] Relative yield, not quantified. [h] $Mn_2(CO)_8(PEt_3)_2$. [j] $HMn(CO)_4PEt_3$ (20%) and $HMn(CO)_3(PEt_3)_2$ (20%). [k] $NaMn(CO)_5$ and $Mn(CO)_5PPh_3^+PF_6^-$ with added PEt_3. [m] Mixture of $Mn_2(CO)_8(PPh_3)_2$ (38%), $Mn_2(CO)_8(PPh_3)(PEt_3)$ (14%), $Mn_2(CO)_8(PEt_3)_2$ (18%). [n] $HMn(CO)_4PPh_3$ and $HMn(CO)_4PEt_3$.

Small amounts of the mono-substituted dimer $Mn_2(CO)_9PPh_3$ are observed only at the highest concentrations of added PPh_3.

The effect of added triphenylphosphine on the ion-pair coupling of the PPh_3-substituted anion and the MeCN-substituted cation represented in Eq. 45 is more subtle. Thus the increasing amounts of added PPh_3 lead to a gradual diminution in $Mn_2(CO)_{10}$ yield, which is solely made up by an increase in the yield of $Mn_2(CO)_8(PPh_3)_2$. The amount of the mono-substituted $Mn_2(CO)_9PPh_3$ remains relatively invariant. Hydridomanganese carbonyl $HMn(CO)_4PPh_3$ is found only at the highest concentrations of added triphenylphosphine. Added phosphine also affects the ion-pair reaction of the phosphine-substituted cations. For example, when $Mn(CO)_5PEt_3^+$ is treated with pentacarbonylmanganate (compare entry 8, Table 4) in the presence of two equiv. of added triethylphosphine, the formation of $Mn_2(CO)_{10}$ is quenched with a concomitant increase in the yield of $Mn_2(CO)_8(PEt_3)_2$ to 40%, i.e.

$$Mn(CO)_5^- + Mn(CO)_5PEt_3^+ \xrightarrow{PEt_3} Mn_2(CO)_8(PEt_3)_2 + 2CO \quad (47)$$

together with $HMn(CO)_4PEt_3$ (20%) and $HMn(CO)_3(PEt_3)_2$ (20%).

3.1.5 The effects of solvent and added salt on the ion-pair coupling are also pertinent. For example, acetonitrile and tetrahydrofuran represent aprotic media of different properties, especially as evaluated by most measures of solvent polarity [61]. The large solvent effect on the ion-pair coupling of the pentacarbonylmanganate anion **A** is indicated by the more than 100-fold difference which separates its reactivity with hexacarbonylmanganese (I) cation in tetrahydrofuran and in acetonitrile (compare entries 1 and 2, Table 6). The prolonged reaction times in the more polar acetonitrile can be related to the stabilization of the separate ions. The latter is consistent with the decreased reactivity of the ions in tetrahydrofuran containing 0.3 M tetra-n-butylammonium perchlorates (see entries 3 and 9, Table 6).

Table 6. Solvent and Salt Effects on the Ion-Pair Coupling of Pentacarbonylmanganate(−I) with Various Carbonylmanganese(I) Cations[a]

$Mn(CO)_5L^+$ L	Solvent	Salt[b] (M)	Time[c] (min)	Products
CO	THF	0	< 5	$Mn_2(CO)_{10}$[d]
CO	MeCN	0	>1000	$Mn_2(CO)_{10}$[e]
CO	THF	0.3	~ 90	$Mn_2(CO)_{10}$[e]
CO	MeCN	0.2	>2000	$Mn_2(CO)_{10}$[e]
CH_3CN	THF	0	< 5	$Mn_2(CO)_{10}$[d]
CH_3CN	MeCN	0	~ 500	$Mn_2(CO)_{10}$[e]
CH_3CN	THF	0.3	~ 30	$Mn_2(CO)_{10}$[e]
Pyridine	THF	0	< 5	$Mn_2(CO)_{10}$[d]
Pyridine	THF	0.3	~ 30	$Mn_2(CO)_{10}$[e]
PEt_3	THF	0	~ 10^2	{ $Mn_2(CO)_{10}$, $Mn_2(CO)_8(PEt_3)_2$[d]
PEt_3	THF	0.3	> 10^3	$Mn_2(CO)_8(PEt_3)_2$[f]

[a] $NaMn(CO)_5$ (1.2 M, 5 mL) added to 5 mL of 1×10^{-2} M $Mn(CO)_5L^+BF_4^-$. [b] Tetra-n-butylammonium perchlorate. [c] Required for disappearance of ions (<5 = upon mixing). [d] See Table 4. [e] Only product observed. [f] Plus $HMn(CO)_4PEt_3$.

3.2 Mechanism of dimer formation via ion-pair annihilation

The ion-pair interaction of an anionic carbonylmanganate with a carbonylmanganese cation to afford various dimanganese carbonyls formally represents an annihilation of -1 and $+1$ oxidation states to a pair of metal(0). As such the material balance for the coupling of manganese centers can be represented from among the three possible combination of ligand attachments, viz.

Scheme XI:

$$Mn^{-I}(CO)_4P^- + Mn^{I}(CO)_5L^+ \longrightarrow \begin{cases} Mn_2{}^{0}(CO)_8(P)_2 & (48) \\ Mn_2{}^{0}(CO)_8(P)(L) & (49) \\ Mn_2{}^{0}(CO)_8(L)_2 & (50) \end{cases}$$

The most direct transformation of an ion pair is included in the cross-coupled combination in Eq. 49. The corresponding stoichiometry may be expressed in Eq. 40 when P = CO and L = CO, MeCN or pyridine. On the other hand, the interactions of the other ion pairs can lead to the homo-coupled products $Mn_2(CO)_8(P)_2$ and $Mn_2(CO)_8(L)_2$. Moreover, when P and L are both phosphines, all three combinations, viz, the homo-coupled $Mn_2(CO)_8P_2$ and $Mn_2(CO)_8(L)_2$ as well as the cross-coupled $Mn_2(CO)_8(P)(L)$, are obtained. In order to account for this apparent ligand-dependent selectivity in the ion pair couplings, let us classify them in terms of the relative amounts of each combination (i.e., Eq. 48, 49 or 50) which is formed. Two clear-cut categories emerge in which the carbonylmanganese ions labelled with ligands P and L either (a) maintain their original identity or (b) they are randomly scrambled: (A) *The cross-coupled combination of carbonylmanganese moieties*, i.e.

$$Mn(CO)_5^- + Mn(CO)_5L^+ \rightarrow Mn_2(CO)_{10} + L \qquad (51)$$

corresponds to the replacement of the ligand L in the cation by a carbonylmanganese group. The extent to which the latter is an electron-rich anion then represents a nucleophilic substitution at a cationic manganese center. Such a transformation obtains in Eqs. 39 and 40 when L is CO and a nitrogen-centered ligand such as pyridine and acetonitrile, respectively. However, the situation is somewhat ambiguous since the dimanganese decacarbonyl can also derive from a pair of $Mn(CO)_5^-$ anions (via oxidation) and/or a pair of $Mn(CO)_5L^+$ cations (via reductive loss of L). (B) *The homo-coupled combination of carbonylmanganese moieties*, i.e.

$$\begin{aligned} &Mn(CO)_4P^- + Mn(CO)_5L^+ \\ &\rightarrow [\underset{(25\%)}{Mn_2(CO)_8P_2} + \underset{(25\%)}{Mn_2(CO)_8L_2} + \underset{(50\%)}{Mn_2(CO)_8(P)(L)}] + CO \end{aligned} \qquad (52)$$

represents the complete loss of anionic and cationic identities in the ion-pair precursor and the subsequent statistical reassemblage of the carbonylmanganese moieties as dimers. Such a randomization is unambiguously represented in Eq. 44 owing to the retention of either P or L (or both) in the dimanganese products. The same basic stoichiometry obtains in Eqs. 42 and 43, but in a somewhat less obvious manner owing to the ligand loss of either CO or L. The statistical mixture of manganese dimers as exemplified in Eq. 52 does provide unusual insight as to how the manganese-manganese bond is formed by ion-pair annihilation.

3.2.1 Formation of the manganese-manganese bond via radical coupling originates from electron transfer. Thus the electron balance for the ion-pair annihilation leading to the homo-coupled products $Mn_2(CO)_8(P)_2$ and $Mn_2(CO)_8(L)_2$ formally requires the oxidation of the anionic $Mn(CO)_4P^-$ and the reduction of the cationic $Mn(CO)_5L^+$, respectively. As such, the most direct pathway for the formation of these dimeric products is via the well-known coupling of carbonylmanganese radicals [62, 63], i.e.

Scheme XII:

$$Mn(CO)_4P^- \xrightarrow{-e} Mn(CO)_4P\cdot \longrightarrow 1/2\,Mn_2(CO)_8P_2 \tag{53}$$

$$Mn(CO)_5L^+ \xrightarrow{+e} Mn(CO)_5L\cdot \xrightarrow{-CO} 1/2\,Mn_2(CO)_8L_2 \tag{54a, b}$$

These radicals are the direct products of electron transfer within the ion pair, i.e.

$$Mn(CO)_4P^- + Mn(CO)_5L^+ \rightarrow Mn(CO)_4P\cdot + Mn(CO)_5L\cdot \tag{55}$$

Furthermore if the CO loss from $Mn(CO)_5L\cdot$ were rapid in Eq. 54b, the electron transfer in Eq. 55 would be tantamount to the simultaneous production of a pair of similar radicals, i.e.

$$Mn(CO)_4P^- + Mn(CO)_5L^+ \xrightarrow{-CO} Mn(CO)_4P\cdot + Mn(CO)_4L\cdot \tag{56}$$

which would afford a more or less random mixture of dimanganese carbonyls. Accordingly, the homo-coupled combination found in Eq. 52 points to an electron-transfer mechanism for the ion-pair annihilation leading to the manganese-manganese couplings. By the same token, the cross-coupling combination observed in Eqs. 39 and 40 seems to disfavor the same carbonylmanganese radicals as intermediates. In order to resolve this conundrum we recognize that the two cabonylmanganese radicals arising directly from the electron transfer in Eq. 55 are inherently different.

3.2.2 The mechanistic distinction between 19- and 17-electron carbonylmanganese radicals is represented by the carbonylmanganese radical $Mn(CO)_5L\cdot$, which is a supersaturated 19-electron species, whereas $Mn(CO)_4P\cdot$ is an unsaturated 17-electron species [26b]. The difference in the coordinative saturation in these two types of carbonylmanganese radicals immediately serves as a point of differentiation, which is modulated by the rate of ligand loss from the 19-electron species. Accordingly it is worthwhile to review the properties and behavior of 19- and 17-electron carbonylmanganese radicals when they are separately generated by independent methods.

3.2.2.1 The 19-electron radical $Mn(CO)_5L\cdot$ is a transient species, which is conveniently produced from the cathodic reduction of the carbonylmanganese(I) cation in Section 2.2, and it undergoes two basic transformations, namely, ligand dissociation and hydrogen abstraction, the competition from which is highly dependent on the nature of ligand L in the following way. (A) Ligand dissociation of the 19-electron $Mn(CO)_5L\cdot$ is rapid and can involve the loss of either L or CO to produce the 17-electron carbonylmanganese radicals [26b], i.e.

$$Mn(CO)_5L\cdot \xrightarrow{k_L} Mn(CO)_5\cdot + L \tag{57}$$

$$Mn(CO)_5L\cdot \xrightarrow{k_{CO}} Mn(CO)_4L\cdot + CO \tag{58}$$

In the case of the nitrogen-substituted radicals $Mn(CO)_5(NCMe)\cdot$ and $Mn(CO)_5(py)\cdot$, the loss of CO is not competitive with the expulsion of either MeCN or py since $Mn_2(CO)_{10}$ is obtained as the sole dimer. This selectivity is in accord with the expected difference in leaving group abilities of CO relative to MeCN and py. On the other hand, for the phosphine derivatives of $Mn(CO)_5L\cdot$ there is a competition between the loss of carbon monoxide and phosphine. Moreover the logarithm of the relative rate constants k_L/k_{CO} in Eqs. 57 and 58 shows a linear correlation with the pK_a value of the phosphine ligand [26b]. It is noteworthy that the ratio $k_L/k_{CO} = 6.7$ for L = PPh_3 is more than a factor of 20 larger than $k_L/k_{CO} = 0.3$ for the more basic PEt_3. (B) Hydrogen abstraction by $Mn(CO)_5L\cdot$ from solvent (SH) yields the hydridomanganese complexes $HMn(CO)_4L$ when followed by rapid extrusion of carbon monoxide [64, 65], i.e.

$$Mn(CO)_5L\cdot \xrightarrow[(SH)]{k_H} Mn(CO)_4(CHO)L \xrightarrow{-CO} HMn(CO)_4L \qquad (59)$$

Competition from this homolytic process is most severe with L = phosphines. For the more labile 19-electron radicals derived from L = CO, MeCN and py, the rate constant k_L in Eq. 57 appears to be sufficiently greater than k_H in Eq. 59 to obviate this competition, and hydridomanganganese carbonyls are not important products.

3.2.2.2 The 17-electron radical $Mn(CO)_4P\cdot$ can be independently generated by the anodic oxidation of the carbonylmanganate anion in Section 2.1. The ease of oxidation of $Mn(CO)_4P^-$ follows the σ-donor properties of the ligands in the order: P = $PPh_3 \gg P(OPh)_3 >$ CO; and it is an indication of the reactivity of the resulting radical $Mn(CO)_4P\cdot$, as reflected in the rates of self-dimerization [62].

$$2\,Mn(CO)_4P\cdot \xrightarrow{k_d} Mn_2(CO)_8P_2 \qquad (60)$$

For example, the second-order rate constant for $Mn(CO)_5\cdot$ is $k_d = 9\times10^8$ M^{-1} s^{-1}, and for $Mn(CO)_4PPh_3\cdot$ it is roughly two orders of magnitude slower ($k_d = 1\times10^7$ M^{-1} s^{-1}) [28, 62]. These 17-electron species are also labile and undergo rapid ligand substitution by an associative mechanism [66], e.g.

$$Mn(CO)_5\cdot + PPh_3 \xrightarrow{k_s} Mn(CO)_4PPh_3\cdot + CO \qquad (61)$$

where $k_s = 1.7\times10^7$ M^{-1} s^{-1} [28a]. The 19-electron intermediate in Eq. 61 is the same as that in Eq. 55 where L = phosphine. Thus, the presence of added phosphine during the ion-pair annihilation serves to equilibrate all carbonylmanganese radicals to a common species, as observed in Eqs. 46 and 47.

With this background established for the properties and behavior of relative 19- and 17-electron carbonylmanganese radicals, let us now reconsider the distribution among the diverse products obtained from the various ion-pair annihilations.

3.2.3 The radical-pair interactions of transient 19- and 17-electron carbonylmanganese species are pertinent to dimer formation. For example, several pathways are conceivable for the formation of $Mn_2(CO)_{10}$ in high yields from $Mn(CO)_5^-$ and $Mn(CO)_5(NCMe)^+$ in Table 4. On one hand, the most economical route for this ion-pair annihilation involves cross-coupling accompanied by the displacement of MeCN, i.e.

$$Mn(CO)_5^- + Mn(CO)_5(NCMe)^+ \rightarrow Mn_2(CO)_{10} + MeCN \qquad (62)$$

On the other hand, a prior electron-transfer involves a multistep process in which the 19- and 17-electron radical pair are involved in ligand loss (Eq. 57) and dimerization (Eq. 60), i.e.

Scheme XIII:

$$Mn(CO)_5^- + Mn(CO)_5(NCMe)^+ \longrightarrow Mn(CO)_5\cdot + Mn(CO)_5(NCMe)\cdot \quad (63)$$

$$Mn(CO)_5(NCMe)\cdot \xrightarrow{fast} Mn(CO)_5\cdot + MeCN \quad (64)$$

$$2Mn(CO)_5\cdot \xrightarrow{fast} Mn_2(CO)_{10} \quad (65)$$

The marked effect of added triphenylphosphine strongly supports the latter mechanism. Thus the presence of PPh_3 effectively diverts the course of reaction from $Mn_2(CO)_{10}$ to a pair of new products $Mn_2(CO)_8(PPh_3)_2$ and $HMn(CO)_4PPh_3$ without materially affecting the rate (see Table 5). Both of the products are readily attributed to the efficient interception of the carbonylmanganese radical, i.e.

Scheme XIV:

$$Mn(CO)_5\cdot + PPh_3 \rightleftharpoons Mn(CO)_5PPh_3\cdot \quad (66)$$

$$Mn(CO)_5PPh_3\cdot \xrightarrow{-CO} Mn(CO)_4PPh_3\cdot \quad (67)$$

$$Mn(CO)_5PPh_3\cdot \xrightarrow{SH} HMn(CO)_4PPh_3 + CO \quad (68)$$

(as described in Eqs. 61 and 59), followed by rapid dimerization (Eq. 60). Essentially the same results obtain when the ion-pair annihilation of $Mn(CO)_4PPh_3^-$/ $Mn(CO)_5(NCMe)^+$ is carried out with added PPh_3, and $Mn(CO)_5^-/Mn(CO)_5PEt_3^+$ with added PEt_3 (Eq. 47). At this juncture, it is important to re-emphasize that the control experiments establish both the precursor carbonylmanganate(—I) anion and carbonylmanganese(I) cation to be substitution stable, and no ligand exchange of the type described above can occur within the time span of these experiments.

The principal byproduct of the ion-pair coupling of $Mn(CO)_4P^-/Mn(CO)_5L^+$ is the hydridomanganese carbonyl $HMn(CO)_4L$, which most commonly arises when L = phosphines. On the other hand when L = CO, MeCN or py, the corresponding hydride is absent. These results accord with hydride formation occurring via the 19-electron radical $Mn(CO)_5L\cdot$, as described in Eq. 59. Furthermore, hydrogen atom transfer to phosphorus-substituted radicals with L = phosphine should occur more readily than that with L = CO, MeCN or py owing to the dissociation rate constant $k_L > k_{CO}$ in Eqs. 57 and 58. It is possible that the small amounts of $HMn(CO)_4P$ observed in some cases arise by protonation of the carbonylmanganate anion by adventitious moisture [67] or by multiple ligand substitutions of the 17-electron radical, i.e.

Scheme XV:

$$Mn(CO)_4P\cdot + L \rightarrow Mn(CO)_4L\cdot + P \quad (69)$$

$$Mn(CO)_5\cdot + P \rightarrow Mn(CO)_5P\cdot, \text{ etc.} \quad (70)$$

Indeed, the ion-pair annihilation of $Mn(CO)_5^-/Mn(CO)_5PEt^+$ carried out in the presence of one equiv. PEt_3 affords $Mn_2(CO)_8(PEt_3)_2$, $HMn(CO)_4PEt_3$ and the bis-substituted hydride $HMn(CO)_3(PEt_3)_2$ as the principal products, with no evidence of either $Mn_2(CO)_{10}$ or $Mn_2(CO)_9PEt_3$. These results accord with efficient interception of $Mn(CO)_5\cdot$ (compare Scheme XIV). Moreover, the appearance of $HMn(CO)_3(PEt_3)_2$ indicates the facility with which further homolytic substitution occurs [68], i.e.

$$Mn(CO)_4PEt_3\cdot + PEt_3 \xrightarrow{-CO} Mn(CO)_3(PEt_3)_2\cdot, \text{ etc.} \tag{71}$$

According to the electron-transfer formulation, the relative amounts of products are largely determined by the magnitudes of k_L, k_{CO} and k_H since they regulate the steady-state concentration of the 19-electron radical. The extent to which k_d for the coupling of the 17-electron radical is dependent on X = CO, P and L also effects the distribution among the various dimanganese carbonyls. The availability of free phosphine is responsible for raising the steady-state concentration of the 19-electron species by the microscopic reverse of Eq. 57. This results in increased hydride (Eq. 57) and phosphine incorporation into the dimanganese carbonyls (Eqs. 69 and 71]. Most importantly, the diversion of $Mn(CO)_5^-$ and $Mn(CO)_5(NCMe)^+$ by added PPh_3 in Table 5 demonstrates that a substantial portion, if not all, of the ion-pair annihilation occurs via scavengable radicals according to Scheme XIII and XIV.

3.2.4 Mechanism of Mn—Mn bond formation by ion-pair annihilation can be taken together with the foregoing analyses to show that the diverse products derived from the ion-pair annihilations can be completely accounted for by the known behavior of 19- and 17-electron carbonylmanganese radicals. Accordingly, the initiation by electron-transfer is included in the generalized formulation below for Mn-Mn bond formation.

Scheme XVI.

$$Mn(CO)_4P^- + Mn(CO)_5L^+ \rightarrow [Mn(CO)_4P\cdot, Mn(CO)_5L\cdot] \tag{72}$$

$$[Mn(CO)_4P\cdot, Mn(CO)_5L\cdot] \xrightarrow{\text{collapse}} Mn_2(CO)_8(P)(L) + CO \tag{73}$$

$$[Mn(CO)_4P\cdot, Mn(CO)_5L\cdot] \xrightarrow{\text{diffuse}} Mn(CO)_4P\cdot + Mn(CO)_5L\cdot \tag{74}$$

$$Mn(CO)_5L\cdot \xrightarrow{k_L, k_{CO}} Mn(CO)_4X\cdot + L(CO) \tag{75}$$

$$2Mn(CO)_4X\cdot \xrightarrow{k_d} Mn_2(CO)_8X_2 \tag{76}$$

where X = CO, P, L

The first-order rate constants k_L and k_{CO} represent the ligand dissociation from the 19-electron radical, and k_d is the second-order rate constant for the couplings of pairs of 17-electron radicals.

According to Scheme XVI, electron-transfer is the first step (Eq. 72) in the annihilation of the carbonylmanganese cations by anions. The nonstatistical distribution among the homo-coupled and cross-coupled dimers in Scheme XI derives primarily from the competition between cage collapse (Eq. 73) and diffusive separation (Eq. 74) of the intially formed radical pair. The extent to which cage collapse occurs faster than diffusive separation will determine the amounts of cross-coupled products obtained. Alternatively, the efficiency with which added phosphines divert the products to $Mn_2(CO)_8P_2$ (Table 5) reflects the extent to which 17-electron carbonylmanganese radicals have escaped and consequently are subject to ligand substitution (Eq. 61). In the same way the trapping of 19-electron $Mn(CO)_5L\cdot$ by hydrogen transfer reflects cage escape. Indeed the multiple substitution of phosphine in the hydride products (see Eq. 71) suggests that the 19-electron carbonylmanganese precursors are longer lived. A further indication of this difference is shown in the carbonylmanganate $Mn(CO)_4(\eta^1\text{-DPPE})^-$ in which the corresponding 17-electron radical will be susceptible to intramolecular trapping, i.e.

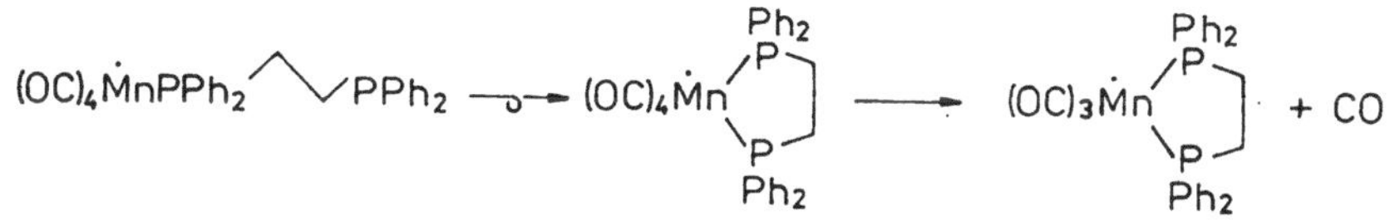

(77)

The facile annihilation of $Mn(CO)_4(\eta^1\text{-DPPE})^-$ by an equivalent amount of either $Mn(CO)_5(NCMe)^+$ or $Mn(CO)_5py^+$ affords three principal products, viz., $Mn_2(CO)_{10}$ (30%), $Mn_2(CO)_9(\eta^1\text{-DPPE})$ (15%), and $HMn(CO)_3(\eta^2\text{-DPPE})$ (35%). The formation of the chelated $HMn(CO)_3(\eta^2\text{-DPPE})$ is consistent with the trapping of the 19-electron intermediate in Eq. 77, as indicated by the mechanism for hydride formation in Eq. 59. The observation of the mono-hapto dimanganese carbonyl $Mn_2(CO)_9(\eta^1\text{-DPPE})$ suggests that the cross-coupling either involves a concerted process or a cage collapse in Eq. 73, which is faster than intramolecular trapping (Eq. 77). Be that as it may, from a purely operational point of view, there are two pathways for the formation of Mn—Mn carbonyls by ion-pair annihilation, namely, (i) a process in which pairs of carbonylmanganese radicals behave more or less independently as discrete species as a result of diffusive separation and (ii) a minor pathway in which the cation and anion more or less maintain their identity by either the cage collapse of the radical pair (Eq. 73) or by concerted action.

3.3 Coupling of Carbonylcobalt(I) Cations with Carbonylcobaltate(—I) Anions

The ready disproportionation of dicobalt octacarbonyl induced by Lewis bases has been known for more than three decades [69], but the process still presents an intriguing mechanistic story. Thus with nitrogen- and oxygen-centered bases (B), the disproportionation of $Co_2(CO)_8$ usually affords a Co(II)/(Co(—I) salt with the general stoichiometry [70]

$$3\,Co_2(CO)_8 + 12\,B \rightarrow 2[CoB_6^{2+}][Co(CO)_4^-]_2 + 8\,CO \qquad (78)$$

On the other hand, with phosphorus-centered bases P, especially in polar solvents dicobalt octacarbonyl disproportionates even at room temperature to produce the Co(I)/Co(—I) salt with a simpler stoichiometry [71, 72], i.e.

$$Co_2(CO)_8 + 2P \rightarrow [Co(CO)_3P_2^+][Co(CO)_4^-] + CO \qquad (79)$$

Interestingly, the same phosphines react with dicobalt octacarbonyl in nonpolar solvents at higher temperatures by effecting an overall ligand substitution [72b, 73a], i.e.

$$Co_2(CO)_8 + 2P \rightarrow Co_2(CO)_6P_2 + 2CO \qquad (80)$$

It is known that this disubstitution product can also be obtained by heating the Co(I)/Co(−I) salt [73], i.e.

$$[Co(CO)_3P_2^+][Co(CO)_4^-] \rightarrow Co_2(CO)_6P_2 + CO \tag{81}$$

Such an ion-pair annihilation can be formally considered as the microscopic reverse of the disproportionation in Eq. 79. As such, it is possible to envisage a reversible disproportionation that is induced by carbon monoxide, i.e.

$$Co_2(CO)_8 + CO \rightarrow [Co(CO)_5^+][Co(CO)_4^-] \tag{82}$$

Although this interchange of dicobalt octacarbonyl with its ion pair remains as yet hypothetical [74] (owing to the unknown pentacarbonylcobalt(I) cation), its viability can be established with the corresponding manganese carbonyls [75]. Indeed among the various carbonylcobalt(I) cations extant, the bis-phosphine substituted derivative $Co(CO)_3P_2^+$ is the most common – only a few examples of the mono-phosphine substituted cation $Co(CO)_4P^+$ having been reported [76]. Accordingly the previous mechanistic investigations of cobalt-centered ion pairs [77, 78] have focussed on the annihilation of the disproportionation ion pair $[Co(CO)_3P_2^+][Co(CO)_4^-]$ that is obtained directly from dicobalt octacarbonyl by treatment with phosphines [79, 80]. However this problem can be approached more generally from a consideration of carbonylcobalt cations as electrophilic electron acceptors and carbonylcobalt anions as nucleophilic electron donors in a manner similar to that established in the annihilation of carbonylmangangese ion pairs in Sections 3.1 and 3.2. Since phosphine ligands are known to strongly modulate the redox properties of carbonylmetals, [81] the substituted carbonylcobalt(I) cations $Co(CO)_3P_2^+$ and carbonylcobaltate(−I) anions $Co(CO)_3P^-$ that contain PPh_3 and PBu_3 are selected as the prototypical aryl- and alkylphosphines, respectively [82].

3.3.1 Homo Dimers from Carbonylcobalt Ion Pairs. When equimolar amounts of $Co(CO)_3(PPh_3)_2^+$ and $Co(CO)_3PPh_3^-$ in tetrahydrofuran (THF) solutions are mixed at 25 °C, a quantitative reaction results immediately (<5 min) according to the stoichiometry in Eq. 83.

$$Co(CO)_3(PPh_3)_2^+ + Co(CO)_3PPh_3^- \xrightarrow[THF]{fast} Co_2(CO)_6(PPh_3)_2 + PPh_3 \tag{83}$$

The homo dimer $Co_2(CO)_6(PPh_3)_2$ largely precipitates as a red solid in 98% yield, and it is characterized by its distinctive carbonyl band at $\nu_{CO} = 1943\ cm^{-1}$ in the IR spectrum [73b]. An equivalent amount of free triphenylphosphine is also extracted from the reaction mixture in accord with the stoichiometry in Eq. 83. When the corresponding ion pair that contains only tri-n-butylphosphine as the ligand is treated in a similar manner, high yields (93%) of the homo dimer $Co_2(CO)_6(PBu_3)_2$ are obtained. However, this coupling occurs at a significantly diminished rate (~30 min), as judged by the slow disappearance of the pair of carbonyl bands (simultaneously) at 1986 and 1826 cm^{-1} for the cation and anion, respectively. The more soluble homo dimer $Co_2(CO)_6(PBu_3)_2$ is extracted in 93% yield, and it is identified by its characteristic IR spectrum [73b].

$$Co(CO)_3(PBu_3)_2^+ + Co(CO)_3PBu_3^- \xrightarrow[THF]{slow} Co_2(CO)_6(PBu_2) + PBu_3 \tag{84}$$

No other carbonyl-containing product (<5%) is observed.

3.3.2 Cross dimers from carbonylcobalt ion pairs. When $Co(CO)_3(PPh_3)_2^+$ is treated with an equimolar amount of $Co(CO)_3PBu_3^-$ in THF solution at 25 °C, the initial IR spectrum of the reaction mixture that is recorded within 5 min shows no carbonyl band of either starting material. Immediate separation of the red solid yields the homo dimer $Co_2(CO)_6(PPh_3)_2$ in 35% yield, and the thin layer chromatography (TLC) of the mother liquor affords a pure crystalline sample of the other homo dimer $Co_2(CO)_6(PBu_3)_2$ in 28% yield. The remaining carbonyl-containing product is identified as the cross dimer $Co_2(CO)_6(PPh_3)(PBu_3)$ in 29% yield by the comparison of its IR spectrum with that of an authentic sample i.e.

$$Co(CO)_3(PPh_3)_2^+ + Co(CO)_3PBu_3^- \longrightarrow \begin{cases} Co_2(CO)_6(PPh_3)_2 + PBu_3 \\ Co_2(CO)_6(PPh_3)(PBu_3) + PPh_3 \\ Co_2(CO)_6(PBu_3)_2 + PPh_3 \end{cases} \tag{85}$$

The alternative ion-pair combination of $Co(CO)_3(PBu_3)_2^+$ and $Co(CO)_3PPh_3^-$ requires a significantly longer time (~30 min) for completion. TLC of the red solution yields only the homo dimer $Co_2(CO)_6(PBu_3)_2$ in 54% yield together with the cross dimer $Co_2(CO)_6(PPh_3)(PBu_3)$ in 26% yield. There is no evidence of the other homo dimer $Co_2(CO)_6(PPh_3)_2$, either by isolation as an insoluble red precipitate (vide supra) or from its IR spectrum.

$$Co(CO)_3(PBu_3)_2^+ + Co(CO)_3PPh_3^- \longrightarrow \begin{cases} Co_2(CO)_6(PBu_3)_2 + PPh_3 \\ Co_2(CO)_6(PPh_3)(PBu_3) + PBu_3 \end{cases} \tag{86}$$

By way of comparison, the ion pair from the disproportionation of dicobalt octacarbonyl with tri-n-butylphosphine in Eq. 79 with P = PBu_3 is also treated in THF under the same reaction conditions. However at these temperatures less than 5% of the homo dimer is produced over a 60 min-period, i.e.

$$Co(CO)_3(PBu_3)_2^+ + Co(CO)_4^- \xrightarrow[\text{slow}]{\text{very}} Co_2(CO)_6(PBu_3)_2 + CO \tag{87}$$

3.3.3 Effect of Added Phosphine on the Ion-Pair Coupling. The rate of the homo coupling of $Co(CO)_3(PPh_3)_2^+$ and $Co(CO)_3PPh_3^-$ in Eq. 83 is materially unaffected when the THF solution contains up to 15 equiv. of added tri-n-butylphosphine. However no evidence of $Co_2(CO)_6(PPh_3)_2$ can be found (<5%), the only carbonyl-containing product being the replaced homo dimer $Co_2(CO)_6(PBu_3)_2$ isolated in 84% yield, i.e.

$$Co(CO)_3(PPh_3)_2^+ + Co(CO)_3PPh_3^- \xrightarrow{PBu_3(\text{excess})} Co_2(CO)_6(PBu_3)_2 + 3\,PPh_3 \tag{88}$$

Control experiments show that neither the carbonylcobalt cation nor anion undergoes prior ligand substitution by the added PBu_3 under the reaction conditions. Moreover the homo dimer $Co_2(CO)_6(PPh_3)_2$ and the cross dimer $Co_2(CO)_6(PPh_3)(PBu_3)$ are not subject to subsequent ligand substitution by tributylphosphine to yield the homo dimer in Eq. 88.

3.3.4 Effects of solvent polarity on ion-pair annihilation. When the homo coupling of $Co(CO)_3(PPh_3)_2^+$ and $Co(CO)_3PPh_3^-$ in Eq. 83 is carried out in a more polar solvent, such as acetonitrile, the formation of $Co_2(CO)_6(PPh_3)_2$ in 84% yield is noticeably retarded. Moreover the solvent effect is unmistakable in the homo coupling of the tributylphosphine-substituted ion pair (Eq. 84), which is essentially inert in acetonitrile for periods exceeding 10 hrs. The crossed ion pair in Eq. 86 is similarly unreactive in acetonitrile, i.e.

$$Co(CO)_3(PBu_3)_2^+ + Co(CO)_3PPh_3^- \xrightarrow[MeCN]{} \text{no reaction} \tag{89}$$

with only a small amount (~10%) of the anion $Co(CO)_3PPh_3^-$ diminished after 10 hr. (The concentration of the more stable cation $Co(CO)_3(PBu_3)_2^+$ is unchanged.)

3.3.5 Inhibitory effects of added salts on the ion-pair coupling. The homo coupling of $Co(CO)_3(PPh_3)_2^+$ and $Co(CO)_3PPh_3^-$ in THF solution is visibly retarded by the presence of 0.3 M tetra-n-butylammonium perchlorate (TBAP), but the yield of the homo dimer $Co_2(CO)_6(PPh_3)_2$ is unaffected. The negative salt effect in tetrahydrofuran is dramatic in the case of the less reactive $Co(CO)_3(PBu_3)_2^+$/$Co(CO)_3PBu_3^-$ pair of ions. Thus the presence of 0.3 M TBAP is sufficient to retard the conversion in Eq. 84 to less than 5% over a 10-hr period.

$$Co(CO)_3(PBu_3)_2^+ + Co(CO)_3PBu_3^- \xrightarrow{TBAP} \text{no reaction} \tag{90}$$

Similarly, the cross-coupling pair that consists of $Co(CO)_3(PBu_3)_2^+$ and $Co(CO)_3PPh_3^-$ yields less than 5% $Co_2(CO)_6(PBu_3)_2$ in THF solution over a 10-hr period when 0.3 M TBAP is added. IR analysis indicates that the concentration of the reactant ion-pair is essentially undiminished.

3.4 Spectral study of contact ion pairs from carbonylcobalt cations and carbonylcobaltate anions

The ion-pair interactions of carbonylcobalt(I) cations with carbonylcobaltate(—I) anions is apparent in both their electronic and infrared spectra taken in solution and in the solid state.

3.4.1 The electronic spectra of both precursor salts $Co(CO)_3(PBu_3)_2^+ClO_4^-$ and $Co(CO)_4^-PPN^+$ show essentially no absorptions beyond $\lambda > 350$ nm (Fig. 11). However, the crystalline carbonylcobalt salt $[Co(CO)_3(PBu_3)_2^+][Co(CO)_4^-]$ obtained by disporportionation (Eq. 79, P = n-Bu_3P) is a bright yellow color, which appears in the diffuse reflectance spectrum as a broad absorption with a shoulder (λ_{max} ~ 400 nm) tailing out to beyond 500 nm (see inset). By comparison, the color of a solution of 3.5×10^{-3} M $[Co(CO)_3(PBu_3)_2^+][Co(CO)_4^-]$ dissolved in tetrahydrofuran is also yellow, and it shows the same absorption band, as illustrated in Fig. 11. Moreover, this band appears when the separate ions $Co(CO)_3(PBu_3)_2^+$ and $Co(CO)_4^-$ as the ClO_4^- and PPN^+ salts, respectively, are mixed in tetrahydrofuran. As such, the new spectral band with λ_{max} ~ 400 nm is assigned to the charge-transfer absorption ($h\nu_{CT}$) of the contact ion pair, i.e. [83].

$$[Co(CO)_3(PBu_3)_2^+][Co(CO)_4^-] \xrightarrow{h\nu_{CT}} [Co(CO)_3(PBu_3)_2^\bullet][Co(CO)_4^\bullet] \tag{91}$$

In order to confirm this assignment, the effects of solvent polarity and added salts are examined on the behavior of the disproportionation salt $[Co(CO)_3(PBu_3)_2^+]$ $[Co(CO)_4^-]$ in solution.

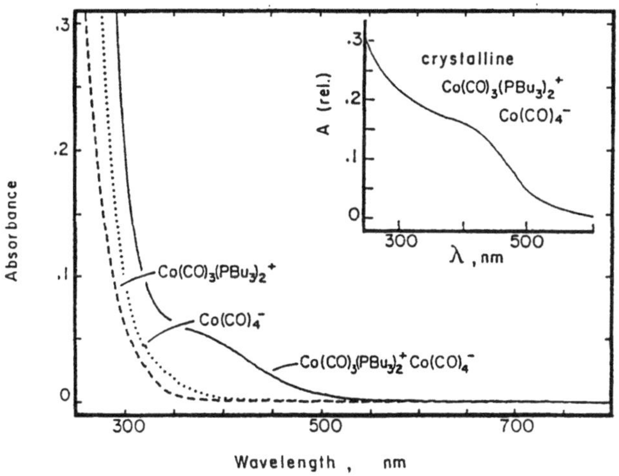

Fig. 11. Absorption spectra of 3.5×10^{-3} M solutions of $Co(CO)_3(PBu_3)_2^+$ (-------), $Co(CO)_4^-$ (·········) and $[Co(CO)_3(PBu_3)_2^+][Co(CO)_4^-]$ in THF (—). Inset: Solid-state reflectance spectrum of crystalline $[Co(CO)_3(PBu_3)_2^+][Co(CO)_4^-]$ after dilution with SiO_2.

3.4.1.1 Solvent polarity strongly affects the nature of $[Co(CO)_3(PBu_3)_2^+]$ $[Co(CO)_4^-]$ in solution as evidenced by the color change from orange in diethyl ether, yellow-orange in chlorobenzene, yellow in tetrahydrofuran, pale yellow in dichloromethane and finally to colorless in acetonitrile. The corresponding shift in the electronic absorption bands in Fig. 12A parallels the trend in the dielectric constant. As such, the complete bleaching of the color is consistent with the disappearance of the contact ion pair (and its associated charge-transfer band at ~400 nm) by stabilization of the separated ions in polar solvents such as acetonitrile, i.e.

$$[Co(CO)_3(PBu_3)_2^+][Co(CO)_4^-] \xrightarrow{MeCN} Co(CO)_3(PBu_3)_2^+ \; // \; Co(CO)_4^- \tag{92}$$

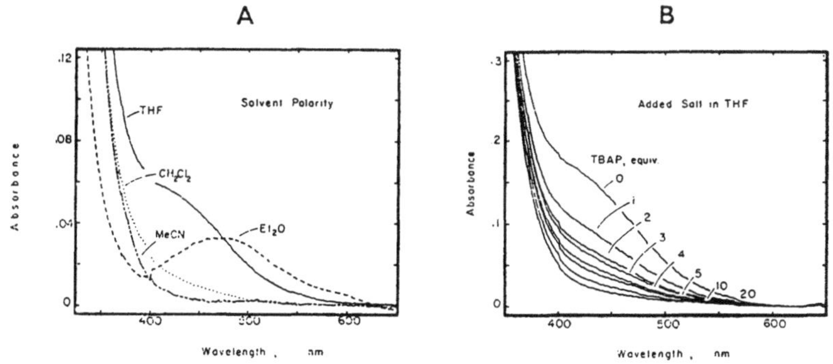

Fig. 12. Charge-transfer absorption spectrum of 3.5×10^{-3} M $[Co(CO)_3(PBu_3)_2^+]$-$[Co(CO)_4^-]$. (A) Solvent effect: THF (—), CH_2Cl_2 (·······), MeCN (—·—) and Et_2O (------). (B) Salt effect: TBAP added (top to bottom): none, 1, 2, 3, 4, 5, 10 and 20 equiv. relative to 1×10^{-2} M $[Co(CO)_3(PBu_3)_2^+][Co(CO)_4^-]$ in THF.

By the same token, the spectral red shift observed in diethyl ether can be attributed to the interionic contraction induced by this least polar solvent [84].

3.4.1.2 Added salts like tetra-n-butylammonium perchlorate (TBAP) strongly perturb the intensity of the color of $[Co(CO)_3(PBu_3)_2^+][Co(CO)_4^-]$ in THF solution. Thus Fig. 12B shows the pronounced hypochromic effect on only the charge-transfer band with $\lambda_{max} \sim 400$ nm by incremental amounts of added TBAP. Indeed, the presence of merely one equiv. of TBAP is sufficient to reduce the concentration of the contact ion pair by one-half. Ion pair exchange in Eq. 93 effectively

$$[Co(CO)_3(PBu_3)_2^+][Co(CO)_4^-] + [Bu_4N^+][ClO_4^-]$$
$$\rightleftharpoons [Co(CO)_3(PBu_3)_2^+][ClO_4^-] + [Bu_4N^+][Co(CO)_4^-] \qquad (93)$$

separates the carbonylcobalt cation from the carbonylcobaltate anion in the contact ion pair. As such, the disappearance of the CT band in Fig. 12B is reminiscent of the situation achieved with highly polar solvents.

It is important to emphasize that the equimolar combination of salts in Eq. 93 is equivalent to the mixture of the separate carbonylcobalt salts such as $[Co(CO)_3(PBu_3)_2^+][ClO_4^-]$ and $[n\text{-}Bu_4N^+][Co(CO)_4^-]$ that are employed above in ion-pair annihilation (Eq. 87) and these expectations are borne out by the presence of the well-resolved charge-transfer band with $\lambda_{CT} = 520$ nm of $[Co(CO)_3(PBu_3)_2^+]$ $[Co(CO)_3PBu_3^-]$. This contact ion pair is instantaneously generated upon mixing the separate carbonylcobalt salts that are individually transparent beyond $\lambda >$ 400 nm. Moreover the supression of the CT absorption band by added TBAP is the same as that presented for the disproportionation salt in Fig. 12B. Indeed the generality of the negative salt effect on such charge-transfer spectra emphasizes its validity as the experimental criterion for the existence of contact ion pairs [85].

3.4.2 The infrared spectra of the disproportionation salts $[Co(CO)_3(PBu_3)_2^+]$ $[Co(CO)_4^-]$ and $[Co(CO)_3(PPh_3)_2^+][Co(CO)_4^-]$ consists by and large of the spectral composite of the individual ionic components. Thus the IR absorptions of the carbonylcobalt cation in the carbonyl region occur at $\nu_{CO} = 2000 \sim 2800\ cm^{-1}$ and those for the carbonylcobaltate anionic moiety at $\nu_{CO} = 1870 \sim 1900\ cm^{-1}$. However, a closer examination of the IR absorption bands reveals a systematic variation of ν_{CO} with solvent polarity. Thus the carbonyl band of $Co(CO)_3(PPh_3)_2^+$ in acetonitrile and dichloromethane consists of a single band with $\nu_{CO} = 2010\ cm^{-1}$ for the E' mode in D_{3h} symmetry. This band in a less polar THF medium is split into two components appearing as a shoulder and a weak A_1' band which are forbidden in D_{3h} symmetry. Moreover, the salt in either a KBr disk or a Nujol mull also shows well-resolved splittings into three distinct bands symptomatic of cation distortion to C_{2v} symmetry [86]. A change in the carbonyl bands of the carbonylcobaltate anion also accompanies the transition from solution to the crystal. Thus the carbonyl band of $Co(CO)_4^-$ in either acetonitrile, dichloromethane or tetrahydrofuran consists of a single band at $\nu_{CO} \sim 1890\ cm^{-1}$, as expected for the tetrahedral anion with T_d symmetry. However in the crystal of $[Co(CO)_3(PPh_3)_2^+][Co(CO)_4^-]$ this triply degenerate band is split into three distinct components diagnostic of a configurational change to C_{3v} symmetry [87]. Owing to an adventitious overlap with the cation absorptions, the weak A_1 band, which is forbidden in T_d symmetry, is not identified. Such simultaneous distortions of both the carbonylcobalt cation and carbonylcobaltate anion (as indicated by infrared studies) thus mirror the solvent-induced changes in the amount of ion pairing (as deduced from the charge-transfer absorptions). Both spectroscopic probes point to the relevance of the intimate molecular interactions extant in contact ion pairs.

3.5 Carbonylcobalt dimers from contact ion pairs

The existence of the charge-transfer absorption bands is characteristic of important electron donor-acceptor contributions to the contact ion pair that is the direct precursor in the formation of metal-metal dimers by the mutual annihilation of carbonylcobalt(I) cations and carbonylcobaltate(−I) anions. The diverse results cannot be explained by any single process in which the metal-metal bond for the dimer is formed by the *direct* replacement of a phosphine ligand (P) on the cationic $Co(CO)_3P_2^+$ by the nucleophilic anion $Co(CO)_3P^-$. Indeed the experimental observations are (A) the homo coupling with $P = PPh_3$ in the presence of added PBu_3 produces only the substituted dimer $Co_2(CO)_6(PBu_3)_2$ under conditions in which neither the anion nor cation alone is subject to ligand substitution (Eq. 88) and (B) the rate of the substitutive coupling in Eq. 88 is essentially unaffected by the presence of 2–10 equiv. of foreign phosphine (PBu_3). Factors (A) and (B) together demand that one or more reactive intermediates intervene between the rate-limiting activation process and the product-determining step [88]. In order to facilitate the mechanistic analysis, the strong correlation of the solvent and salt effect on the *rate* of ion-pair coupling in Eqs. 89 and 90, respectively, is directly related to their effect on the *formation* of contact ion pairs in Fig. 12. Accordingly, the rapid equilibria leading up to the rate-determining step can be readily summarized as

Scheme XVII:

$$Co(CO)_3P_2^+ + Co(CO)_3P^- \overset{K}{\rightleftharpoons} [Co(CO)_3P_2^+Co(CO)_3P^-] \tag{94}$$

$$[Co(CO)_3P_2^+Co(CO)_3P^-] \xrightarrow{k_1} Co_2(CO)_6P_2 + P \tag{95}$$

where the brackets enclose the contact ion pair. Values of the formation constant $K > 10^4\ M^{-1}$ together with the configurational distortion of contact ion pairs in THF and CH_2Cl_2 point to a first-order activation process (k_1) in Eq. 95. The extensive scrambling of the ligands during ion-pair annihilation further indicates that Eq. 95 in Scheme XVII must proceed via labile intermediates, such as those formed in the rate-limiting activation by an electron-transfer process. On this basis the sequence of reactions leading to the metal-metal dimers evolve spontaneously from a pair of 19-e and 17-e carbonylcobalt radicals, as presented below for the homo-coupling process.

Scheme XVIII:

$$[Co(CO)_3P_2^+Co(CO)_3P^-] \xrightarrow{k_1} Co(CO)_3P_2^\cdot + Co(CO)_3P^\cdot \tag{96}$$

$$Co(CO)_3P_2^\cdot \rightleftharpoons Co(CO)_3P^\cdot + P \tag{97}$$

$$2\,Co(CO)_3P^\cdot \longrightarrow Co_2(CO)_6P_2 \tag{98}$$

Previous studies have indeed established both the substitutional lability and the rapid dimerization of carbonylcobalt and related radicals [89, 90]. According to

Scheme XVIII the extensive scrambling of the phosphine ligands P and P′ during cross coupling (Eqs. 85–87) derives from the facile and reversible interconversion of radical pairs, i.e.

$$\underset{(17\text{-e})}{Co(CO)_3P'\cdot} + P \rightleftharpoons \underset{(19\text{-e})}{Co(CO)_3P'P\cdot} \rightleftharpoons \underset{(17\text{-e})}{Co(CO)_3P\cdot} + P' \qquad (99)$$

In order to ascertain the viability of this mechanistic formulation, the preparative and transient electrochemical experiments relating to the cathodic reduction of $Co(CO)_3P_2^+$ and the anodic oxidation of $Co(CO)_3P^-$ [91], analogous to those presented for $Mn(CO)_5P^+$ in Section 2.2 and for $Mn(CO)_4P^-$ in Section 2.1, respectively, bear directly on the behavior of 19-e and 17-e carbonylcobalt radicals formed under the conditions of ion-pair annihilation. Thus, the high yields of homo-coupled dimers that are isolated from both the electroreduction of the cationic $Co(CO)_3P_2^+$ as well as the electrooxidation of the anionic $Co(CO)_3P^-$ confirm the rapid coupling of the carbonylcobalt radicals, as in Eq. 98. Relevant to the facile ligand substitution of carbonylcobalt radicals (Eq. 99) is the formation of the PPh_3-substituted cobalt dimer when the oxidation of carbonylcobaltate is carried out in the presence of added triphenylphosphine. Furthermore, the series of transient electrochemical studies with cyclic voltammetry establish the kinetic behavior of (A) the 19-e radical from the reduction of $Co(CO)_3P_2^+$ and (B) the 17-e radical from the oxidation of $Co(CO)_3P^-$ that are critical to the redox process (Eq. 96) as follows [92].

3.5.1 The 19-e radicals $Co(CO)_3P_2\cdot$ are generated in the initial step for the reduction of the cationic $Co(CO)_3P_2^+$. For example, the cyclic voltammogram of 5×10^{-3} M $Co(CO)_3(PPh_3)_2^+$ at a sweep rate $v = 0.5$ V s^{-1} in THF (Fig. 13A) shows a cathodic peak P_1 with $E_p = -0.69$ V vs SCE that corresponds to an uptake of 0.95 electrons by calibration relative to a ferrocene standard. Such a cathodic current leads to two reduction products. The presence of the carbonylcobaltate $Co(CO)_3PPh_3^-$ in the cyclic voltammogram is indicated by the anodic peak current P_3 with $E_p = -0.27$ V when the initial CV scan is reversed at -0.9 V. By

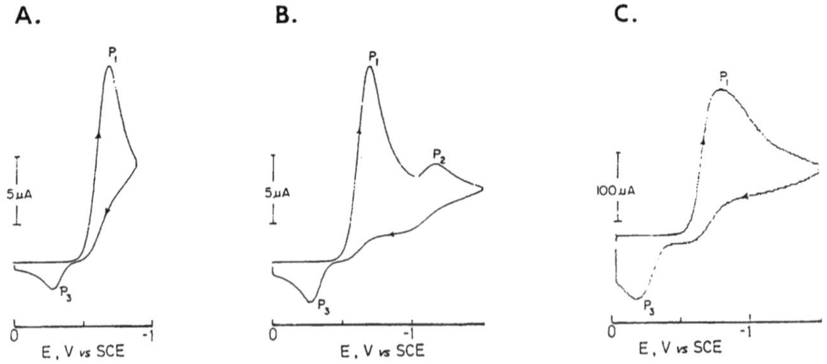

Fig. 13. Initial negative-scan cyclic voltammogram of 5×10^{-3} M $Co(CO)_3(PPh_3)_2^+$ containing 0.3 M TBAP. Switching potential at (A) -0.9 V or (B) -1.5 V at $v = 0.5$ V s^{-1}. (C) Same conditions as in (B), but at $v = 100$ V s^{-1}.

comparison with the CV behavior of an authentic sample of $Co(CO)_3PPh_3^-$, the electrode process for P_3 is

$$Co(CO)_3PPh_3^- - e \xrightarrow{[P_3]} 1/2\,Co_2(CO)_6(PPh_3)_2 \qquad (100)$$

The dimeric $Co_2(CO)_6(PPh_3)_2$ is also formed as the other product of the cathodic reduction of $Co(CO)_3(PPh_3)_2^+$; and its presence is indicated by the appearance of the cathodic peak P_2 with $E_p = -1.15$ V when the initial CV scan is carried out to a more negative potential of -1.5 V in Fig. 13B. By comparison with the CV behavior of an authentic sample of $Co_2(CO)_6(PPh_3)_2$, the electrode process for P_2 is described as

$$Co_2(CO)_6(PPh_3)_2 + 2e \xrightarrow{[P_2]} 2\,Co(CO)_3PPh_3^- \qquad (101)$$

The relative amounts of the reduction products $Co(CO)_3PPh_3^-$ and $Co_2(CO)_6(PPh_3)_2$ are indicated by the peak current ratio i_2/i_3. This ratio decreases monotonically as the CV scans of $Co(CO)_3(PPh_3)_2^+$ are increased, until P_2 disappears at $v = 100$ V s^{-1} as shown in Fig. 13C. Accordingly the electrode process for P_1 represents a composite reduction, i.e.

$$Co(CO)_3(PPh_3)_2^+ + xe \xrightarrow{[P_1]} [Co(CO)_3PPh_3^- + Co_2(CO)_6(PPh_3)_2] + PPh_3 \qquad (102)$$

The results in Fig. 13 together with the double potential-step chronoamperometry for $Co(CO)_3P_2^+$ thus delineate the transient behavior during electroreduction as an $\vec{E}C\vec{E}$ process, viz.

Scheme XIX:

$$Co(CO)_3P_2^+ \xrightarrow{e} Co(CO)_3P_2\cdot \qquad (103)$$

$$Co(CO)_3P_2\cdot \rightleftharpoons Co(CO)_3P\cdot + P \qquad (104)$$

$$Co(CO)_3P\cdot \xrightarrow{e} Co(CO)_3P^- \qquad (105)$$

The extent to which the cobalt dimer is observed by its cathodic wave P_2 during cyclic voltammetry derives in a subsequent (nonfaradaic) sacrificial destruction of $Co(CO)_3P^-$ by its cationic precursor, i.e.

$$Co(CO)_3P^- + Co(CO)_3P_2^+ \rightarrow Co_2(CO)_6P_2 + P \qquad (106)$$

The scan rate dependence of the cyclic voltammograms shows that dimer formation by Eq. 106 is rapid for P = PPh_3 but slow for P = PBu_3. Indeed this CV distinction between $Co(CO)_3(PPh_3)_2^+$ and $Co(CO)_3(PBu_3)_2^+$ mirrors the results obtained in ion-pair annihilation in Section 3.3.1 (compare Eqs. 83 and 84).

3.5.2 The 17-e radicals $Co(CO)_3P\cdot$ are generated in the initial step for the oxidation of the anionic $Co(CO)_3P^-$. For example, the cyclic voltammetry of the carbonylcobaltate anion $Co(CO)_3PPh_3^-$ in THF on the initial positive scan shows an anodic wave P_4 with $E_p = -0.27$ V vs SCE at $v = 0.5$ V s^{-1}. The anodic wave in Fig. 14A is the same as P_3 observed from $Co(CO)_3(PPh_3)_2^+$ in Fig. 13 and corresponds to the oxidative dimerization of the carbonylcobaltate, i.e.

$$Co(CO)_3PPh_3^- - e \xrightarrow{[P_4]} 1/2\,Co_2(CO)_6(PPh_3)_2 \qquad (107)$$

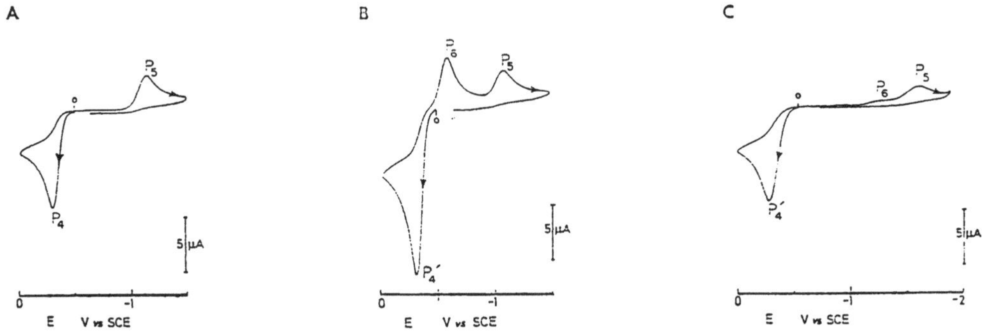

Fig. 14. Initial positive scan cyclic voltammograms of 5×10^{-3} M $Co(CO)_3PPh_3^-$ in THF containing 0.3 M TBAP at $v = 0.5$ V s^{-1} in the presence of (A) no, (B) 6 equiv. of added PPh_3 or (C) 1 equiv. of added PBu_3, relative to carbonylcobaltate.

as shown by the presence of the cathodic wave P_5 with $E_p = -1.15$ V on the return negative scan. [By comparison with the CV behavior of $Co_2(CO)_6(PPh_3)_2$, the cathodic wave P_5 is the same as P_2 in Fig. 13B.] An additional cathodic wave P_6 with $E_p = -0.69$ V is observed when the free ligand triphenylphosphine is present (Fig. 14B). The new cathodic wave corresponds to P_1 in Fig. 13A. As such, the increase in $Co(CO)_3(PPh_3)_2^+$ in the presence of added PPh_3, is associated with the electrode process

$$Co(CO)_3PPh_3^- - 2e \xrightarrow[Ph_3P]{[P_4']} Co(CO)_3(PPh_3)_2^+ \qquad (108)$$

Moreover this electrode process is also accompanied by a marked increase in the anodic peak current of P_4' owing to the change from the 1-electron process in Eq. 108 to the 2-electron process in Eq. 109. [Note that P_5 is only slightly affected by the presence of additional PPh_3.] The results in Fig. 14A thus point to the transient behavior during electrooxidation as an $\overleftarrow{E}C$ process, viz.

Scheme XX:

$$Co(CO)_3P^- \xrightarrow{-e} Co(CO)_3P\cdot \qquad (109)$$

$$2Co(CO)_3P\cdot \longrightarrow Co_2(CO)_6P_2 \qquad (110)$$

More relevant to the lability of 17-e radicals is the effect of added phosphine, as shown in Figs. 14B, i.e.

$$Co(CO)_3P\cdot + P \rightleftharpoons Co(CO)_3P_2\cdot \qquad (111)$$

$$Co(CO)_3P_2\cdot \xrightarrow{-e} Co(CO)_3P_2^+ \qquad (112)$$

In the same vein, the addition of a foreign phosphine P′, in Eq. 112 leads to the substituted cobalt dimer $Co_2(CO)_6P_2'$ and cation $Co(CO)_3(P_2')^+$ (Fig. 14C) in accord with the formulation in Eq. 99. Most importantly, the observation of these processes on the CV timescale of milliseconds confirms the rapidity with which ligand substitution occurs on the 17-e radicals $Co(CO)_3P\cdot$ [89, 90]. Therefore the extensive scrambling of phosphine ligands provides the most compelling experimental support for carbonylcobalt radicals as reactive intermediates. As such, the electron-transfer mechanism in Schemes XVII and XVIII taken together reconciles all the diverse experimental facets of [$Co(CO)_3P_2^+Co(CO)_3P^-$] annihilation leading to carbonylcobalt dimerization via contact ion pairs. Moreover the related homolytic mechanism in Scheme XIX extends the CT photochemistry to disproportionation processes.

3.6 Driving force for the thermal annihilation of contact ion pairs by electron transfer

The coupling of the carbonylmanganese cation/anion pairs in Sections 3.1–3.2 and the carbonylcobalt cation/anion pairs in Sections 3.3–3.5 both follow a common pattern involving (a) the prior association to form contact ion pairs (CIP), followed by (b) interionic electron transfer to afford (c) the carbonylmetal radical pairs that ultimately lead to (d) their coupling for the metal-metal bond. In this general formulation, the rate-limiting activation process is largely represented by the adiabatic electron transfer within the contact ion pair.

3.6.1 Electron transfer in the formation of Mn—Mn bonds by ion-pair annihilation is exemplified by the carbonylmanganese(I) cation $Mn(CO)_6^+$ and the carbonylmanganate(—I) $Mn(CO)_5$ to react upon mixing in THF solution (Eq. 40) to afford high yields of dimanganese decacarbonyl. Similarly, the substituted cations $Mn(CO)_5L^+$ with L = py, MeCN, aryl and alkylphosphines and the substituted anions $Mn(CO)_4P^-$ with P = PPh_3 and $P(OPh_3)$ lead to mixtures of dimanganese carbonyls labelled with the P and L tracers (Eq. 43). The extensive (if not complete) scrambling of the carbonylmanganese moieties during ion-pair annihilation in Scheme XVI is ascribed to the 17-e and 19-e radicals $Mn(CO)_4L\cdot$ and $Mn(CO)_5P\cdot$, respectively, as the reactive intermediates. This conclusion is strongly supported by the known behavior of both types of radicals when they are independently generated by the anodic oxidation of $Mn(CO)_4P^-$ and the cathodic reduction of $Mn(CO)_5L^+$. The reversible addition of ligands to 17-electron radicals provides a ready means for interconversion with their 19-electron counterparts. Thus, the effect of added phosphine in altering the course of Mn—Mn coupling via ligand substitution (Eq. 61) and the formation of the hydrido byproducts $HMn(CO)_4L$ via hydrogen transfer (Eq. 59) provide compelling evidence for carbonylmanganese radicals, since neither the cation nor the anion is susceptible to additives on the timescale of the coupling experiments. These experiments also rule out any type of concerted displacement mechanism as a dominant route in the formation of $Mn_2(CO)_{10}$ from ion pairs, such as [$Mn(CO)_5(NCMe)^+Mn(CO)_5^-$] or [$Mn(CO)_5py^+Mn(CO)_5^-$] in which the carbonylmanganese moieties are not labelled. The participation of specific cross-coupling is included in the unified mechanism as a minor pathway stemming from cage collapse (Eq. 73) of radicals immediately following an initial electron-transfer. The latter is consistent with the reactivity trends for carbonylmanganese ion pairs, which qualitatively parallel the differences ΔE in the oxidation and reduction potentials of $Mn(CO)_4P^-$ and $Mn(CO)_5L^+$ as follows. Since the radical-pair mechanism in Scheme XVI commences with an electron-transfer initiation, a critical part of the driving force for Mn—Mn bond formation is the oxidation-reduction of the anion/cation pair. As such, the energetics for the production of the 19- and 17-electron carbonylmanganese radical pair derives from the reversible electrode potentials for $Mn(CO)_5L^+$ and $Mn(CO)_4P^-$. (A) For the *reduction* of the parent cation $Mn(CO)_6^+$ in Eq. 45, a value of $E^0 = -1.0$ V can be estimated from the peak potentials E_c

listed in Table 2 by an additive correction of ~0.3 V [40]. The wide span in potentials indicates a large ligand dependence on the ease of reduction of $Mn(CO)_5L^+$ in the order: L = py > MeCN > CO > PPh_3 > PPh_2Et > $PPhEt_2$ > PEt_3 ≫ η^2-DPPE. (b) For the *oxidation* of the anionic $Mn(CO)_4PPh_3^-$ in Eq. 45, $E^0 = -0.40$ V vs SCE. The reversible potentials for the other carbonylmanganates in Table 1 and those estimated from irreversible peak potentials E_a in Table 7 lead to the conclusion that the ligand effect on the ease of oxidation is opposite to that in the cation in the order: L = PPh_3 ≫ $P(OPh)_3$ > CO [93]. On this basis, the driving force for electron transfer will be the most favorable for the ion pair $[Mn(CO)_5py^+Mn(CO)_4PPh_3^-]$ with $\Delta E \sim 0.5$ V and the least favorable for $[Mn(CO)_2(DPPE)_2^+Mn(CO)_5^-]$ with $\Delta E \sim 1.6$ V. Indeed, the latter is an unreactive ion pair, and it can be readily isolated as a simple salt from THF solution. i.e.

$$Na^+Mn(CO)_5^- + Mn(CO)_2(DPPE)_2^+BF_4^-$$
$$\rightarrow [Mn(CO)_2(DPPE)_2^+Mn(CO)_5^-] + NaBF_4 \quad (113)$$

owing to the insolubility of $NaBF_4$. Although the reaction rates are too fast to be measured quantitatively, the ion-pair reactivities qualitatively follow the trend in the driving forces. Thus the ion pairs $[Mn(CO)_5py^+Mn(CO)_4PPh_3^-]$ and $[Mn(CO)_5(NCMe)^+Mn(CO)_4PPh_3^-]$ are the most reactive. At the other extreme, the couplings of the parent anion $Mn(CO)_5^-$ with the phosphine-substituted cations in Table 4 require the longest times. Since $\Delta E \sim 1.2$ V for $[Mn(CO)_5PEt_3^+ Mn(CO)_5^-]$, it appears that the threshold in the driving force for ion-pair annihilation lies somewhere between 1.2 and 1.6 V in THF solution. If the rate of electron transfer in Eq. 72 is taken in the outer-sphere context of Marcus theory, [94] two other factors must also be considered. Thus the interaction of oppositely charged ions in CIPs will be aided considerably by the electrostatics. Indeed such a positive work term accords with the strong solvent dependence and the negative salt effect described in Table 6. The contribution from the reorganization energies of $Mn(CO)_4P^-$ and $Mn(CO)_5L^+$ cannot be evaluated at this juncture. However these large highly polarizable 5- and 6-coordinate ions are likely to show only minor differences in basic structural change. If so, the driving force ΔE will be the dominant factor in determining the ease with which ion pairs are annihilated to form metal-metal bonds.

3.6.2 The annihilation rates of the isomeric contact ion pairs from $[Mn(CO)_2(DPPE)_2^+Mn(CO)_2(DPPE)_2^-]$ are critical to the successful mechanistic elucidation of the reduction in Scheme IX that involves as the facile annihilation of the ion pair according to Eq. 31. For a reversible process, the second-order rate constants are related to the driving force by the equilibrium relationship [52],

$$RT \ln (k_{et}/k_d) = n\mathcal{F}\,\Delta E \quad (114)$$

in which the magnitude of ΔE is given by $(E_1{}^0 - E_2{}^0) = 0.11$ V and k_{et} is the optimized second-order rate constant for electron transfer of 2×10^2 M^{-1} s^{-1}. The calculated value from Eq. 114 of the second-order rate constant for radical disproportionation of $k_d = 0.4$ M^{-1} s^{-1} is indeed within the range predicted by the CV simulation (see Fig. 6B). The rate of ion-pair annihilation of cis-$Mn(CO)_2(DPPE)_2^+$ by $Mn(CO)_2(DPPE)_2^-$ with a second-order rate constant $k_{et} = 2 \times 10^2$ M^{-1} s^{-1} is a factor of 6 faster than the annihilation of the isomeric trans-$Mn(CO)_2(DPPE)_2^+$ with the same anion ($k_{et} = 35$ M^{-1} s^{-1}) examined earlier [39].

Such a trend in the electron-transfer rates parallels the difference in the driving force of 3.4 kcal mol^{-1}, evaluated from the reduction potentials of $E_1{}^0 = -1.61$ and -1.76 V for cis and trans-$Mn(CO)_2(DPPE)_2{}^+$, respectively. The latter together with the value of $E_2{}^0 = -1.45$ V for the oxidation of the anionic $Mn(CO)_2(DPPE)_2{}^-$, indicates that the reductions of cis- and trans-$Mn(CO)_2(DPPE)_2{}^+$ both by $Mn(CO)_2(DPPE)_2{}^-$ have driving forces of 3.7 and 7.1 kcal mol^{-1}, respectively. These driving forces are exergonic and considerably more favorable than those involved in the electron-transfer reactions of the simple, monosubstituted carbonylmanganese cations $Mn(CO)_5L^+$ and anions $Mn(CO)_4P^-$ (where L and P are both monodentate phosphines and phosphites) in Section 3.6.1. Nonetheless the rate constants k_{et} for cis- and trans-$Mn(CO)_2(DPPE)_2{}^+$ with $Mn(CO)_2(DPPE)_2{}^-$ are considerably slower than those qualitatively observed between $Mn(CO)_5L^+$ and $Mn(CO)_4P^-$ [40]. Such large rate differences that belie thermodynamics can be attributed to steric hindrance in the tetrasubstituted carbonylmanganese cations and anion, which are absent in the simpler ions. Indeed the ORTEP diagram shown in Fig. 15 visually underscores the limited access to the manganese centers arising from the ligand encumbrance. Such structural effects, even in these apparently outer-sphere electron transfers, merit a further quantitative evaluation as in the application of Marcus theory [94].

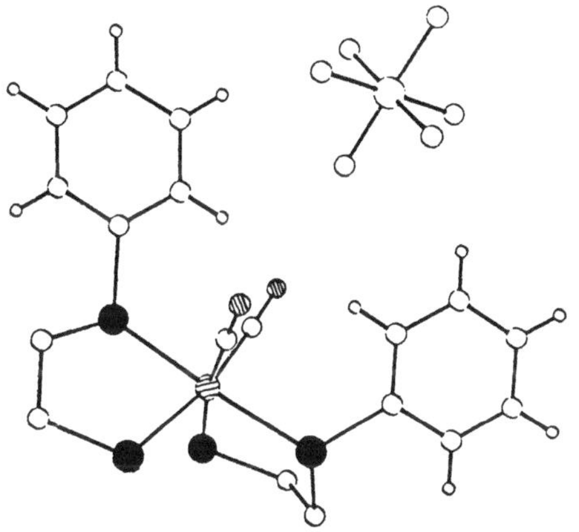

Fig. 15. ORTEP diagram of cis-$Mn(CO)_2(DPPE)_2{}^+AsF_6{}^-$ showing the ion-pair interaction. Note only 2 of 8 phenyl groups on the DPPE ligands are shown for clarity.

3.6.3 The rates of ion-pair annihilation of the carbonylcobalt ions $Co(CO)_3P_2{}^+$ and $Co(CO)_3P^-$ are also highly dependent on the phosphine ligand – the rate with $P = PPh_3$ in Eq. 83 taking place substantially faster than that with $P = PBu_3$ in Eq. 84. As such, the rate of ion-pair annihilation will parallel the driving force for electron transfer in Eq. 96 (Scheme XVIII) as given by the relationship

$$\Delta G_{et} \simeq -\mathcal{F}(E_c - E_a) \qquad (115)$$

where E_c and E_a are peak potentials in Table 7 for the irreversible CV waves P_1 and P_3 (compare Fig. 13) [95]. Evaluated in this way, the driving force for electron transfer of the PPh_3-substituted ion pair is substantially ($\sim$13 kcal mol^{-1}) greater than that for the PBu_3-substituted ion pair. This conclusion thus accounts

Table 7. Cyclic Voltammetric Potentials for Carbonylmetal Cations and Anions[a]

Electroreduction E_c[b] $Mn(CO)_5L^{+b}$	THF[c]	MeCN[c]	$Co(CO)_3P_2^{+b}$	THF[c]	MeCN[c]
$Mn(CO)_6^+$	f	−1.27	PPh_3	−0.69	−0.81
$Mn(CO)_5py^+$	−0.81	−1.12	PBu_3	−1.36	−1.42
$Mn(CO)_5[CH_3CN]^+$	−0.84	−1.19	PEt_3	−1.20	−1.36
$Mn(CO)_5PPh_3^+$	−1.05	−1.29	PPh_2Me	−0.87	−0.95
$Mn(CO)_5[P(p\text{-}Tol)_3]^+$	−1.19	−1.34	$PPhMe_2$	−1.00	−1.11
$Mn(CO)_5(PPh_2Et)^+$	−1.24	−1.39	$Co_2(CO)_6P_2$		
$Mn(CO)_5(PPhEt_2)^+$	−1.32	−1.55	PPh_3	−1.15	−1.03
$Mn(CO)_5(PEt_3)^+$	−1.36	−1.67	PBu_3	−1.66	
$Mn(CO)_5(PPh_2Me)^+$	−1.16	−1.41	PPh_2Me	−1.21	−1.15
$Mn(CO)_5(PPhMe_2)^+$	−1.34	−1.54	$PPhMe_2$	−1.25	
$Mn(CO)_2(\eta^2\text{-}DPPE)_2^+$	−1.84				
Electrooxidation E_a[c] $Mn(CO)_5^{-b}$	**THF[d]**	**MeCN[d]**	**$Co(CO)_3P^{-b}$**	**THF[d]**	**MeCN[d]**
$Mn(CO)_5^-$	−0.03	−0.11	PPh_3	−0.27	−0.28
$Mn(CO)_4[P(OPh)_3]^-$	−0.11		PBu_3	−0.35	−0.39
$Mn(CO)_4PPh_3^-$	−0.52	−0.50	PEt_3	−0.40	−0.44
$Mn(CO)_3(\eta^1\text{-}DPPE)^{-e}$	−0.48		PPh_2Me	−0.29	−0.33
			$PPhMe_2$	−0.35	−0.39
			$P(p\text{-}MeC_6H_4)_3$	−0.28	−0.31
			CO	+0.28	

[a] In either THF containing 0.3 M TBAP or MeCN containing 0.1 M TEAP at scan rates of 500 mV s^{-1} at 22°C (see Ref. 12). [b] 5×10^{-3} M. [c] Irreversible cathodic peak potential (E_p^{red}) in V vs SCE. [d] Irreversible anodic peak potential (E_p^{ox}) in V cs SCE. [e] Prepared *in situ* by cathodic (2e) reduction of $Mn(CO)_3(DPPE)Br$. [f] Insoluble.

for the reactivity pattern established from homo coupling [i.e., the homo coupling with P = PPh_3 is quantitative and rapid (Eq. 83), whereas the same process with P = PBu_3 is slow (Eq. 84)], and for cross coupling [i.e., the cross coupling with P = PPh_3 and P′ = PBu_3 is rapid, and it yields a more or less equimolar mixture of the three dicobalt carbonyls (Eq. 85). However, the same process for the reverse combination with P = PBu_3 and P′ = PPh_3 is slow, and preferentially leads to $Co_2(CO)_6P_2$ with lesser amounts of $Co_2(CO)_6PP'$ but no $Co_2(CO)_6P'_2$ (Eq. 86).]

4. Photochemical activation of carbonylmetal contact ion pairs by charge transfer

The thermal activation of carbonylmetal cation/anion pairs in Section 3 has its photochemical counterpart in the alternative charge-transfer excitation when the driving force for electron transfer $-\Delta G_{et}$ in Eq. 115 is not sufficiently exergonic.

Historically, the electronic transitions of electron donor-acceptor complexes from both neutral and charged components have been delineated as charge-transfer (CT) excitations by Mulliken [96]. The characteristic charge-transfer absorption bands have been identified in various pyridinium iodides that are constituted as acceptor/donor contact ion pairs in solution [97]. Charge-transfer bands have also been reported for carbonylmetallate salts such as $Tl^+Co(CO)_4^-$ [87], $PY^+ V(CO)_6^-$ [98] and $Co(CO)_3L_2^+ Co(CO)_4^-$ [99] where PY^+ = pyridinium cations and L = phosphines. Indeed, the strong and variable donor properties of carbonylmetallates [17] can be exploited in the quantitative study of contact ion pairs (CIP) by the close examination of their charge-transfer characteristics, expecially in a series of salts, the structures of which can be established by X-ray crystallography.

4.1 Isolation, spectral characterization and X-ray crystallography of charge-transfer salts of carbonylmetallates

Ion pairs can be comprised from three series of colorless carbonylmetallate anions, viz., tetracarbonylcobaltate [$Co(CO)_4^-$], hexacarbonylvanadate [$V(CO)_6^-$], pentacarbonylmanganate [$Mn(CO)_5^-$], together with iodide, that are prepared as the sodium or PPN salts [100–102]. Moreover three distinct classes of colorless acceptor cations, viz. cobalticenium (Cp_2Co^+) [103] dibenzenechromium [$(C_6H_6)_2Cr^+$] [104], and a series of N-methylpyridinium derivatives are available as either the halide or hexafluorophosphate salt. Thus, when a pair of colorless aqueous solutions containing 0.1 M $Na^+ Co(CO)_4^-$ and $Cp_2Co^+ Cl^-$ is poured together, dark red crystals separate immediately [105]. This remarkable visual display is also observed when an aqueous solution of $Na^+ Co(CO)_4^-$ is mixed with colorless saline solutions of cyanopyridinium (NCP^+), quinolinium (Q^+), phenylpyridinium (PP^+), and dibenzenechromium (ϕ_2Cr^+) chlorides to yield dark blue, burgundy, orange and green crystals, respectively.

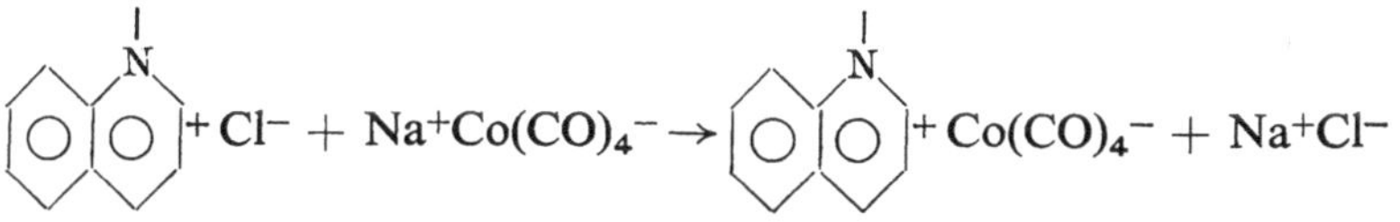

(117)

(burgundy)

Analogously, when a colorless aqueous solution of quinolinium chloride ($Q^+ Cl^-$) is added to an almost colorless aqueous solution of $Na^+V(CO)_6^-$, the well-formed dark green crystals of the vanadate salt precipitate immediately. In each case, the spontaneous separation of the highly colored salts is made even more dramatic by the absence of color in the aqueous mother liquors throughout the course of precipitation.

4.1.1 Charge-transfer spectra of contact ion pairs in the solid state and in solution. The quantitative effects in the colored salts are illustrated in Fig. 16 by the appearance of broad absorption bands in the spectral region between 350 to beyond 700 nm when they are dissolved in dichloromethane. For comparison, the inset in each of the figures shows the diffuse reflectance spectrum of the crystalline salt. The correspondence of the band maximum (λ_{CT}) and breadth (fwhm) of the absorption and reflectance spectra indicate the close similarity of the various carbonylcobaltate salts dissolved in solution with those in the crystalline solid state,

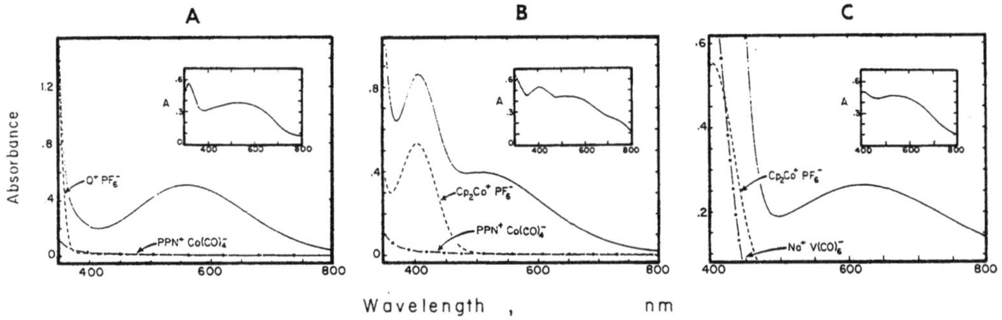

Fig. 16. Charge-transfer absorption spectrum of (A) Q^+ $Co(CO)_4^-$, (B) Cp_2Co^+ $Co(CO)_4^-$ and (C) Cp_2Co^+ $V(CO)_6^-$ as 10^{-3} M solutions in CH_2Cl_2. The diffuse reflectance spectra of the corresponding crystalline salts (10%) dispersed in silica are shown in the insets. The dashed curves are the cationic acceptors (-----) and the anionic donors (—·—) individually at the same concentrations.

respectively (Table 8). Analogously, the absorption bands of the various iodide salts in dichloromethane solution are also similar to those in the solid state (compare with Section 3.4). The absorption bands listed in Table 8 show two important trends. For the series of carbonylcobaltate salts, the maxima λ_{CT} are consistently red-shifted with cation variation in the order: $PP^+ > Cp_2Co^+ > Q^+ > \phi_2Cr^+ >$ NCP^+. Furthermore the same order applies to the corresponding carbonylvanadate, carbonylmanganate, and iodide salts. Such a progressive bathochromic shift of the absorption bands ($h\nu_{CT}$) with the decreasing cathodic potentials E_c for cation reduction (as viewed down the rows in Table 8) is illustrated in Fig. 17 [106]. The pair of lines in the figure describes the slope of unity for the linear correlation given by

$$h\nu_{CT} = -aE_c + \text{constant} \tag{118}$$

Table 8. Visible Absorption Bands of Charge-Transfer Salts in the Crystalline Solid State and in Solution[a]

Cationic Acceptor[c]	E_c (V vs SCE)[d]	Anionic Donor[b] $Co(CO)_4^-$	$V(CO)_6^-$	$Mn(CO)_5^-$	I^-
PP^+	(−1.27)	442 (494)	512 (578)	(562)[f]	(<380) (390)
Cp_2Co^+	(−0.99)	508 (520)	540 (630)	600 (570)[e]	
iQ^+	(−1.08)	510 (516)		(580)[f]	
Q^+	(−0.90)	520 (550)	520 (720)		465 (444)
ϕ_2Cr^+		516			
CMP^+	(−0.79)	(590)[f]	600 (760)	(642)[f]	490 (454)
NCP^+	(−0.67)	560 (620)[f]			

[a] Maximum (nm) of the electronic band obtained from the diffuse reflectance spectrum after 10% dispersion in silica. [b] Absorption spectrum in 10^{-3} M dichloromethane solution in parenthesis. [c] For identification of the acronym, see Ref. 106. [d] Irreversible cathodic CV peak potential at 500 mV s^{-1}. Prepared *in situ* from [e] Na^+ or [f] PPN^+ salt.

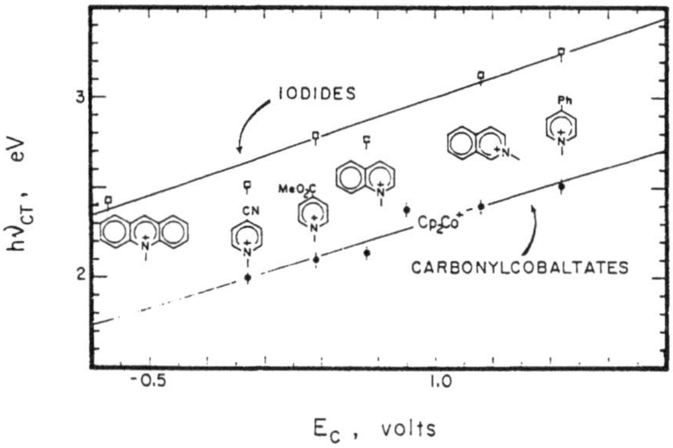

Fig. 17. Correlation of the charge-transfer absorption bands ($h\nu_{CT}$) of carbonylcobalt (⧫) and iodide (□) salts with the reduction potentials E_c of various pyridinium cations.

with $a = 1.00$ for both series of carbonylcobaltate and iodide salts. According to Mulliken charge-transfer theory, [96] the separation of $\Delta h\nu_{CT} = 0.61$ eV for the two series represents the constant difference in the donor properties (ionization potential) of $Co(CO)_4^-$ and I^- in salt pairs with the same acceptor cation. The latter is a corollary of the Mulliken relationship for a series of carbonylmetallate salts in which the charge-transfer maxima viewed across the columns in Table 8 were consistently blue-shifted in the order: $Mn(CO)_5^- \gtrsim V(CO)_6^- > Co(CO)_4^- > I^-$. Indeed such a progressive increase in the energy of the absorption bands with the increasing anodic potential E_a for anion oxidation [107] represents the linear correlation

$$h\nu_{CT} = bE_a + \text{constant} \tag{119}$$

that applies to salts derived from a common cation and a series of donor anions. The linear correlations in Eqs. 118 and 119 derive from Mulliken theory, more commonly expressed as [108]

$$h\nu_{CT} = I_D - E_A + \omega + \text{constant} \tag{120}$$

where I_D and E_A are the ionization potential and electron affinity of the donor anions and acceptor cations, respectively, in the gas phase and ω represents the ion-pair interaction. Accordingly, we will refer to these colored crystals interchangeably as charge-transfer salts in all further discussions [109]. Most importantly, the variegated crystalline salts share in common with their brightly colored solutions (Table 8), the spectral characteristics of the charge-transfer transitions that are predicted by Mulliken theory (Eq. 120), as illustrated in Fig. 17. Indeed the close relationship in Fig. 18 between the diffuse reflectance spectra of crystalline salts with the absorption spectra of the salts dissolved in dichloromethane, underscores the essential unity of the charge-transfer transitions ($h\nu_{CT}$) in the solid state and in solution. Therefore the critical interionic separations ($r_{A^+D^-}$) in the solid state are also pertinent to $h\nu_{CT}$ (Eq. 120) of these salts in solution. As such, the brightly colored solutions of charge-transfer salts derive directly from the contact ion pairs (CIP) – closely related in kind to those defined by X-ray crystallography and IR spectroscopy. Furthermore the charge-transfer spectrophotometry

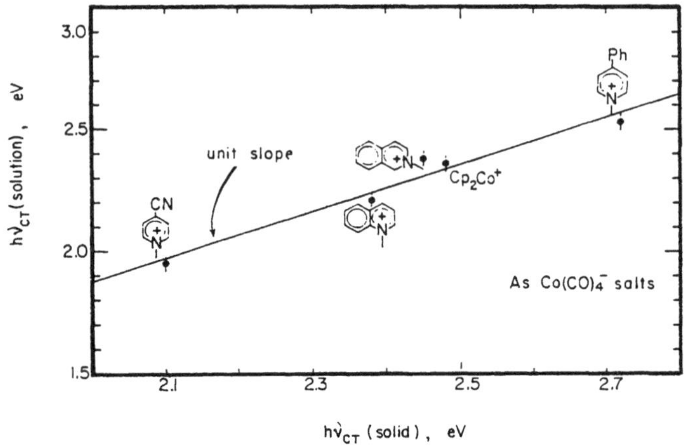

Fig. 18. The direct correspondence of the charge-transfer transitions ($h\nu_{CT}$) of contact ion pairs (as indicated) in the solid state with those in dichloromethane solution. The line is arbitrarily drawn with a slope of unity.

establishes such contact ion pairs to be the dominant species in nonpolar solvents (CH_2Cl_2, THF, etc.); and the solvatochromism in polar solvents (e.g., MeCN) mainly relates to CIP dissociation as given by the distinctive change in the carbonyl IR bands (vide infra).

4.1.2 X-ray crystallographic structures of charge-transfer salts. To establish the origin of the charge-transfer absorptions of the colored salts in Table 8, the X-ray crystallography of the tetracarbonylcobaltate salts of the representative cations Cp_2Co^+, Q^+ and NCP^+ is examined, together with that of Cp_2Co^+ I^- and Cp_2Co^+ $V(CO)_6^-$ for comparison. The X-ray diffraction patterns from $Cp_2Co^+Co(CO)_4^-$, Q^+ $Co(CO)_4^-$, and NCP^+ $Co(CO)_4^-$ as deep red, burgundy and blue crystals are all solved as the expected 1 : 1 ion pairs. In each salt, the tetracarbonylcobaltate moiety is present as a discrete tetrahedral anion with a slight distortion from ideal T_d to C_{2v} symmetry. Indeed the bond distances and angles, with only slight variations, are the same as those previously found in the colorless ionic salts of tetracarbonylcobaltate paired with alkali [110] and simple ammonium cations [111]. Likewise, the molecular structures of the acceptor moieties Cp_2Co^+ and Q^+ exist as undistorted cations in the charge-transfer salts with respect to those found in other ionic salts [112]. Most importantly, the X-ray crystallographic analyses of cobalticenium tetracarbonylcobaltate and iodide establish the interionic separations of the Cp_2Co^+/anion pairs relevant to the charge-transfer absorptions in Table 8. Thus the molecular (space-filling) models in Fig. 19 illustrate the location of the anionic donors I^- and $Co(CO)_4^-$ relative to the cobalticenium acceptor for optimal orbital overlap with the LUMO in the equatorial plane [113]. For the pyridinium salts of $Co(CO)_4^-$, the analogous charge-transfer interaction of the tetracarbonylcobaltate donor places it above the aromatic acceptor planes for optimal orbital overlap with the π-LUMOs of Q^+ and NCP^+, as illustrated by the ORTEP diagrams in Fig. 20. Such X-ray crystallographic structures indicate that these charge-transfer salts consist of contact ion pairs (CIP) that are directionally constrained for optimum CT interaction in the crystal lattice.

A B

Fig. 19. Space-filling models of (A) $Cp_2Co^+I^-$ and (B) $Cp_2Co^+Co(CO)_4^-$ based on X-ray crystallography showing the location of the anionic donors in the equatorial plane of the cobalticenium cation.

A B

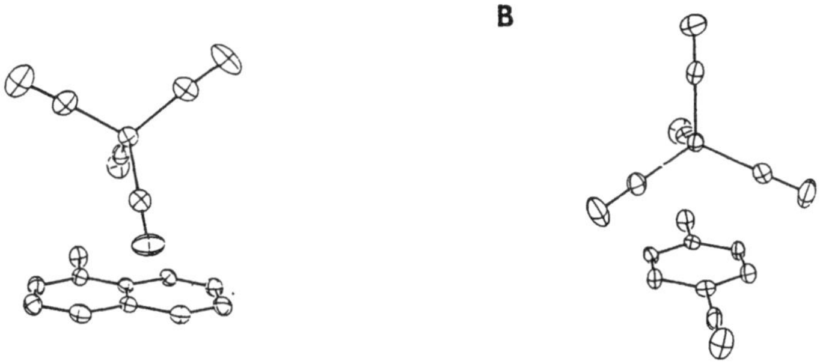

Fig. 20. ORTEP diagram of (A) Q^+ $Co(CO)_4^-$ and (B) NCP^+ $Co(CO)_4^-$ locating the relevant charge-transfer interaction of the $Co(CO)_4^-$ donor above the aromatic planes.

4.1.3 Infrared spectra of charge-transfer salts as contact ion pairs in the solid state and in solution. The X-ray crystallographic structures in Figs. 19 and 20 emphasize the intimate contact that exists between the donor anion and the acceptor cation in charge-transfer salts. Such a close proximity of the anion/cation pairs is sufficient to distort the normally tetrahedral $Co(CO)_4^-$. The resulting decrease in symmetry could be readily detected by changes in the carbonyl bands in the infrared spectrum. For example, Table 9 includes the principal carbonyl bands (ν_{CO}) in the solid-state IR spectra of three charge-transfer salts, together with those of the Na^+ and PPN^+ salts for comparison. The single band at $\nu_{CO} =$ 1883 cm^{-1} for the crystalline PPN^+ salt represents the T_2 mode of the undistorted tetrahedral $Co(CO)_4^-$, in accord with the X-ray crystallographic structure of PPN^+ $Co(CO)_4^-$ [114]. In strong contrast, the IR spectra of the charge-transfer salts all show the symmetry-forbidden A_1 band at ~2005 cm^{-1} (compare with Section 3.4.2). Most importantly, the splitting of the major T_2 band in $Q^+Co(CO)_4^-$ is akin to that previously observed in $Na^+Co(CO)_4^-$ by Edgell and coworkers, [100] and the IR spectra in Table 9 (entries 2 and 5) are both in accord with the expected three ($2A_1 + E$) bands for $Co(CO)_4^-$ in C_{3v} symmetry established by X-ray crystallography [115]. Moreover, the four carbonyl bands observed in the crystalline cobalticenium and cyanopyridinium analogues (entries 1 and 3) are consistent with the C_{2v} and C_s symmetry of $Co(CO)_4^-$ in these salts, as established by X-ray crystallography (vide supra) [116]. In other words, the splittings of the carbonyl IR bands provide reliable and sensitive measures of the tetracarbonylcobaltate distortions that are extant in crystalline charge-transfer salts. The dem-

Table 9. Solid-State and Solution IR Spectra of $Co(CO)_4^-$ Salts[a]

Cp_2Co^+	Q^+	NCP^+	Na^+	PPN^+
2007 (2006)	2007 (2004)	2006 (2003)	2025 (2007)	
1907 (1906)	1928 (1910)	1911 (1916)	1935 (1910)[b]	
1872 (1886)	1895 (1887)	1878 (1886)	1868 (1853)[c]	1883 (1887)
1858 (1870)		1865 (1870)		

[a] ν_{CO} in cm^{-1}. Solid-state spectrum (10%) in KBr disc. Tetrahydrofuran solution (10^{-2} M) in parenthesis. [b] Resolved as 2 bands at 1899 and 1906 cm^{-1} in Ref. 46. [c] Resolved as 1846 and 1856 cm^{-1} in Ref. 118, in addition to band at 1887 cm^{-1}.

onstration of the carbonyl IR bands as sensitive probes for the structure of $Co(CO)_4^-$ in crystals is also applicable to the direct interaction of oppositely charged ions, which may persist as contact ion pairs (CIP) on the dissolution of the charge-transfer salts [117]. Indeed, the comparison of the carbonyl IR bands in Table 9 shows that the charge-transfer salts dissolved in tetrahydrofuran are strikingly akin to those examined in the solid state. Furthermore, the difference between the ionic salts, $Na^+ Co(CO)_4^-$, and $PPN^+ Co(CO)_4^-$ is maintained in tetrahydrofuran solution. Since the latter derives from the undistorted $Co(CO)_4^-$, it is clear that the structures of the crystalline charge-transfer salts as established by X-ray crystallography (vide supra), are closely related to the contact ion pairs extant when these salts are dissolved in either tetrahydrofuran or dichloromethane (see Fig. 18). The X-ray crystallographic structures shown in Figs. 19 and 20 delineate (a) the interionic separations, (b) the spatial cation/anion orientations, and (c) the deviations from a tetrahedral $Co(CO)_4^-$ configuration that are pertinent to crystalline charge-transfer salts. Furthermore, the tandem applications of the X-ray crystallographic and infrared spectroscopic results reveal consistent patterns of $Co(CO)_4^-$ distortions that are inherent to such inner-sphere (CIP) salts. Therefore, the direct comparison of the carbonyl stretching bands in the IR spectra of crystalline $Co(CO)_4^-$ salts with those found in tetrahydrofuran (or dichloromethane) solution establishes the same or closely related ion-pair structures to be extant in nonpolar solvents.

4.2 Dissociation of carbonylmetallate contact ion pairs in solution

It is important to emphasize that all of these salts, regardless of the cation, when dissolved in a polar solvent such as acetonitrile show only a single band at ν_{CO} = 1892 cm^{-1} in the carbonyl IR spectrum. Such a drastic simplification of the multiple splitting of the T_2 band of the charge-transfer salts with a simple change in solvent polarity is diagnostic of the return to a tetrahedral $Co(CO)_4^-$, most likely as a solvent-separated ion pair (SSIP) previously observed with $Na^+ Co(CO)_4^-$ [118] and $Tl^+ Co(CO)_4^-$ [87], and subsequently confirmed by conductivity measurements [116, 119]. Accordingly the solvent-dependent changes in the carbonyl IR bands can be related to the displacement of the contact ion pair (CIP), e.g.

$$\underset{\text{(CIP)}}{Cp_2Co^+Co(CO)_4^-} \underset{\text{THF}}{\overset{\text{MeCN}}{\rightleftharpoons}} \underset{\text{(SSIP)}}{Cp_2Co^+ // Co(CO)_4^-} \qquad (121)$$

where // denotes the solvent separation of the ion pair in which $Co(CO)_4^-$ is sufficiently unencumbered to adopt its most symmetric structure.

4.2.1 Solvent effects on the charge-transfer salts. The marked change in the carbonyl IR bands accompanying the solvent variation from THF to MeCN coincides with the pronounced difference in color of the solutions. For example, the charge-transfer salt $Q^+ Co(CO)_4^-$ is intensely violet in THF but imperceptibly orange in MeCN at the same concentration. The quantitative effects of such a solvatochromism are illustrated in Fig. 21 by (a) the shifts of the absorption maxima and (b) the diminution in the absorbances at λ_{CT}. The concomitant bathochromic shift and hyperchromic increase in the charge-transfer bands in Fig. 21 A follows the sizeable decrease in solvent polarity from acetonitrile to tetrahydrofuran, as evaluated by the dielectric constants $\varepsilon_D = 37.5$ and 7.6, respectively [120]. The same but even more pronounced trend is apparent (Fig. 20 B) in passing from butyronitrile to dichloromethane to diethyl ether with $\varepsilon_D = 26$, 9.1 and 4.3, respectively. The marked variation in λ_{CT} with solvent polarity parallels the behavior of the carbonyl IR bands (compare with Section 3.4.1.1). This solvatochromism is thus readily ascribed to the same displacement of the CIP (Eq. 121) and its associated charge-transfer band. As such, the reversible equilibrium between CIP and SSIP is described by

$$Q^+ Co(CO)_4^- \overset{K_{CIP}}{\rightleftharpoons} Q^+ // Co(CO)_4^- \tag{122}$$

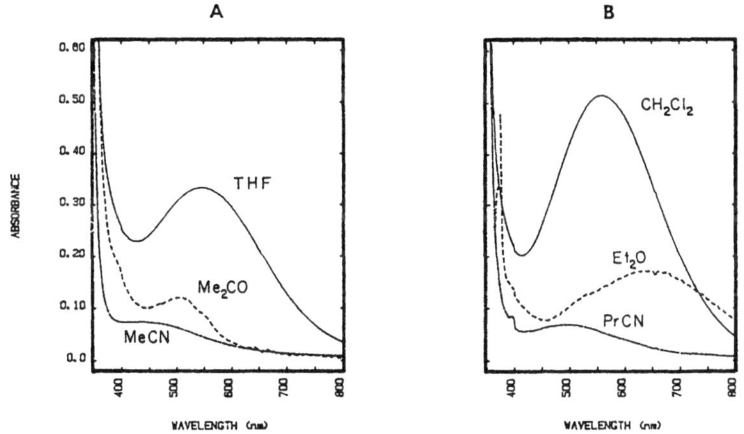

Fig. 21. Solvent dependence of the charge-transfer spectrum of $Q^+ Co(CO)_4^-$ in solutions prepared from (A) THF (10^{-3} M), MeCN (10^{-2} M) and Me_2CO (10^{-3} M) and (B) CH_2Cl_2 (10^{-3} M), n-PrCN (10^{-3} M), and Et_2O (saturated).

where the dissociation constant K_{CIP} applies to a particular solvent and temperature. The quantitative effects of solvent polarity on the dissociation constant are evaluated spectrophotometrically by measuring the change in the CT absorbance A_{CT} at various concentrations C of the charge-transfer salt from the relationship [121]

$$K_{CIP}^{-1} = \frac{A_{CT}}{\varepsilon_{CT}} + \frac{C^2 \varepsilon_{CT}}{A_{CT}} - 2C \tag{123}$$

The CIP dissociation constants evaluated according to Eq. 123 are listed in Table 10 for some typical charge-transfer salts in both polar and nonpolar solvents. The values of K_{CIP} measured conductometrically [119] for the related ionic salts, $PPN^+ Co(CO)_4^-$, $PPN^+ V(CO)_6^+$ and $Na^+ BPh_4^-$, are included in the Table for comparison. Two features in Table 10 are particularly noteworthy. First, the magnitudes of K_{CIP} in the nonpolar solvents (THF and CH_2Cl_2) are at least a 100-fold larger than those in the polar MeCN. Thus, at the concentrations employed in the IR studies (Table 9), the charge-transfer salts exist in THF primarily (>90%) as the contact ion pair. Second, extinction coefficient ε_{CIP} evaluated at the maximum of the charge-transfer band λ_{CT} is relatively invariant with solvent polarity [84]. This suggests that the same (or closely related) CIPs are always formed; and the effect of solvent polarity largely reside with changes in K_{CIP}. Such a solvent effect on the thermodynamics of CIP dissociation is indeed consistent with the influence of added salts (vide infra, Fig. 22).

Table 10. CIP Dissociation Constants of Charge-Transfer Salts in Polar and Nonpolar Solvents[a]

CT Salt	Solvent	λ_{mon}[b] (nm)	K_{CIP} (M)	ε_{CT} (M^{-1} cm^{-1})
$Cp_2Co^+Co(CO)_4^-$	MeCN	520	1.1×10^{-2}	180
	CH_2Cl_2	520	1.5×10^{-4}	230
$Q^+Co(CO)_4^-$	MeCN	458	8.1×10^{-2}	227
	THF	550	1.2×10^{-4}	420
	CH_2Cl_2	560	1.5×10^{-5}	590
$PP^+Co(CO)_4^-$	CH_2Cl_2	494	2.3×10^{-5}	380
$PPN^+Co(CO)_4^-$	THF[c]	–	9.4×10^{-5}	–
$PPN^+V(CO)_6^-$	THF[c]	–	1.2×10^{-4}	–
$PPN^+BPh_4^-$	THF[c]	–	9.1×10^{-5}	–

[a] At 25 °C in the concentration range of 10^{-4}–10^{-1} M. [b] Monitoring wavelength for CT band.
[c] From Ref. 119.

4.2.2 Salt effects on the charge-transfer salts in solution. The addition of small amounts of inert salts such as tetrabutylammonium perchlorate (TBAP) or hexafluorophosphate (TBAH) to solutions of charge-transfer salts induce large changes in the intensity of the charge-transfer absorption bands. The magnitude of the salt effect is most pronounced in nonpolar solvents (THF, CH_2Cl_2). The monotonic decrease in the CT absorbance with increasing amounts of added TBAP is characteristic of the facile competition for the contact ion pair, i.e. [122]

$$Q^+ Co(CO)_4^- + Bu_4N^+ClO_4^- \underset{}{\overset{K_{ex}}{\rightleftharpoons}} Q^+ ClO_4^- + Bu_4N^+ Co(CO)_4^- \qquad (124)$$

Indeed the single linear relationship between the charge-transfer absorbance and the mole fraction of the inert salt (TBAP) added to $Cp_2Co^+ Co(CO)_4^-$, $PP^+ Co(CO)_4^-$, $Q^+ Co(CO)_4^-$, and $\phi_2Cr^+ Co(CO)_4^-$ with the line in Fig. 22 drawn with unit slope indicates that $K_{ex} = 1.0$. Such a magnitude of K_{ex} point to the nonspecific interchange between contact ion pairs that effectively serve to disassemble the intimate charge-transfer ion pair $Q^+ Co(CO)_4^-$, without recourse to solvent variation as in SSIP formation (vide supra). As such, the primary salt effect in

Eq. 124 is largely governed by ion-pair electrostatics. This situation can be contrasted with the effect of salt on the charge-transfer absorption of $Tl^+ Co(CO)_4^-$ [87], and it is analogous to that described in Section 3.4.1.2.

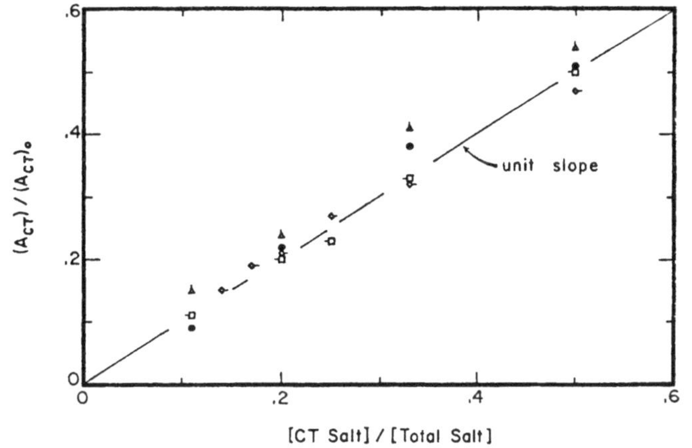

Fig. 22. Ion-pair exchange of the charge-transfer salts $Cp_2Co^+ Co(CO)_4^-$ (-□), $Q^+ Co(CO)_4^-$ (●), $\phi_2Cr^+ Co(CO)_4^-$ (▲), and $PP^+ Co(CO)_4^-$ (◇-) in 2×10^{-3} M CH_2Cl_2 containing 2–20 mM TBAP.

4.2.3 Solvatochromism of charge-transfer salts. The quantitative treatment of primary salt effects according to Fig. 22 demonstrates that the charge-transfer salts in Table 8 are purely ionic salts, indistinguishable from the more commonly used electrolytes such as tetrabutylammonium perchlorate (TBAP) or hexafluorophosphate (TBAH). Accordingly, the solvent effect on CIP dissociation to the solvent-separated ion pairs (SSIP) is of primary concern, since conductivity studies have shown that "free ions" are not particularly important in the aprotic solvents. Nonetheless "free ions" are not distinguishable from SSIPs insofar as charge-transfer absorptions are observed in purely ionic salts [117]. This spectral differentiation (or lack thereof) does not apply to other charge-transfer salts, of which $Tl^+ Co(CO)_4^-$ serves as the prime example [87]. Thus, the elegant studies of Schramm and Zink have demonstrated that a wide spectrum of charge-transfer entities exist – extending from the nonionic, highly covalent $TlCo(CO)_4$ at one extreme through a series of contact and solvent-separated ion pairs, the relative importance of which are strongly modulated by solvent polarity. In marked contrast, such a direct coordination (of the metal) to the carbonylmetallate anion is precluded in metallocenium and pyridinium salts owing to their coordinatively saturated character.

Charge-transfer salts of the carbonylmetallates $Co(CO)_4^-$, $Mn(CO)_5^-$ and $V(CO)_6^-$ bear a striking resemblance to the well-studied pyridinium iodides. Thus the solvatochromism of metallocenium and pyridinium salts of this study are fundamentally related to that of the pyridinium iodides, as seminally established by Kosower [97]. The use of carbonylmetallate donors ($Co(CO)_4^-$, $Mn(CO)_5^-$, $V(CO)_6^-$, etc.) and metallocenium acceptors (Cp_2Co^+, ϕ_2Cr^+, etc.) does successfully extend the range of donor/acceptor properties to expand the solvatochromic scale.

4.3 Photoexcitation of charge-transfer salts in solution

The red-brown solution of the contact ion pair Cp_2Co^+ $Co(CO)_4^-$ in dichloromethane shows no change, even upon prolonged irradiation (8 h) of the charge-transfer band at $\lambda > 520$ nm. However the repetition of the experiment in the presence of triphenylphosphine leads to the spontaneous evolution of carbon monoxide and the disappearance of $Co(CO)_4^-$, as judged by the diminution of its characteristic carbonyl IR band at $\nu_{CO} = 1887$ cm^{-1}. In its place a new band appears at $\nu_{CO} = 1958$ cm^{-1} for the dimeric $Co_2(CO)_6(PPh_3)_2$ [73b] that is isolated in 65% yield together with cobaltocene according to the stoichiometry

$$Cp_2Co^+Co(CO)_4^- + PPh_3 \xrightarrow{h\nu_{CT}} Cp_2Co + 1/2\,Co_2(CO)_6(PPh_3)_2 + CO \qquad (125)$$

Cobaltocene and the analogous cobalt dimer $Co_2(CO)_6(PMe_2Ph)_2$ are also obtained when triphenylphosphine is replaced with dimethylphenylphosphine, as shown in Table 11. The stoichiometry in Eq. 125 corresponds formally to an oxidation-reduction process, in which the 1-electron reduction of Cp_2Co^+ to cobaltocene is accompanied by the 1-electron oxidation of $Co(CO)_4^-$ to the carbonylcobalt(0) dimer. Moreover, the analogous photoredox dimerization of the corresponding carbonylmanganate salt occurs only when tri-n-butylphosphine is present, i.e.

$$Cp_2Co^+Mn(CO)_5^- + PBu_3 \xrightarrow{h\nu_{CT}} Cp_2Co + Mn_2(CO)_8(PBu_3)_2 + CO \qquad (126)$$

Table 11. Charge-Transfer Photochemistry of Contact Ion Pairs[a]

CT Salt	Added L (10² M)		(10² M)	Products[b]	(%)
$Cp_2Cp^+Co(CO)_4^-$	(3.8)[c]	PPh_3	(4.0)	$Co_2(CO)_6L_2$	(65)[d]
	(0.34)	PMe_2Ph	(0.8)	$Co_2(CO)_6L_2$	(50)
	(1.0)	$P(OPh)_3$	(1.0)	$Cp_2Co^+Co(CO)_3L^-$	(62)
	(1.0)	PBu_3	(1.5)	$Co_2(CO)_6L_2$	(15)
				$Co(CO)_3L_2^+Co(CO)_4^-$	(60)
$Q^+Co(CO)_4^-$	(1.2)[c]	PPh_3	(2.4)	$Co_2(CO)_6L_2$	(88)[d]
	(1.4)	PMe_2Ph	(2.0)	$Co(CO)_3L_2^+Co(CO)_4^-$	(90)
	(1.0)	PBu_3	(1.2)	$Co(CO)_3L_2^+Co(CO)_4^-$	(95)[e]
	(0.30)	$P(OPh)_3$	(0.60)	$Co_2(CO)_6L_2$	(80)
$PP^+Co(CO)_4^-$	(0.8)	PBu_3	(2.4)	$Co_2(CO)_6L_2$	(22)[f]
$\phi_2Cr^+Co(CO)_4^-$	(0.8)	PMe_2Ph	(1.8)	$Co(CO)_3L_2^+Co(CO)_4^-$	(16)[g]
	(1.0)	PBu_3	(2.0)	$Co(CO)_3L_2^+Co(CO)_4^-$	(18)[g]
$Cp_2Co^+Mn(CO)_5^-$	(1.2)[j]	PBu_3	(7.0)	$Mn_2(CO)_8L_2$	(16)[h]
	(1.2)	PPh_3	(7.0)	[i]	
$Cp_2Co^+V(CO)_6^-$	(0.4)	PMe_3	(0.7)	[i]	
	(1.0)	PPh_3	(2.0)	[i]	

[a] Irradiation with $\lambda > 550$ nm for ~2 hrs in THF at 25 °C, unless indicated otherwise. [b] Products by quantitative IR analysis of carbonyl bands, except where noted. [c] CH_2Cl_2 with $\lambda > 520$ nm. [d] Isolated. [e] Isolated 75%. [f] Maximum yield at 50% conv. [g] Maximum yield at ~30% conv. [h] Stationary state yield. [i] No products observed after 7 h irradiation. [j] Acetone.

Since these photoredox processes are specifically promoted by the CT excitation of the contact ion pair (CIP), they will be referred to hereafter as *charge-transfer dimerizations.*

When the most basic phosphine $n\text{-}Bu_3P$ is used as the additive in the charge-transfer photochemistry of the contact ion pair $Cp_2Co^+ Co(CO)_4^-$, it again leads to CO evolution and cobaltocene is isolated in 41% yield by sublimation from the reaction mixture. However a different carbonyl product is formed in 60% yield according to the stoichiometry

$$2Cp_2Co^+ Co(CO)_4^- + 2PBu_3 \xrightarrow{h\nu_{CT}} 2Cp_2Co + Co(CO)_3(PBu_3)_2^+ Co(CO)_4^- + CO \quad (127)$$

together with smaller amounts of the dimeric $Co_2(CO)_6(PBu_3)_2$ reminiscent of the stoichiometry in Eq. 125, as shown in Table 11 (entry 3). The photoinduced process in Eq. 127 (like that in Eq. 125) formally represents an oxidation-reduction of the charge-transfer salt. Accordingly this photochemical process will be referred to hereafter simply as *charge-transfer disproportionation.*

The photoinduced reaction of the contact ion pair $Cp_2Co^+ Co(CO)_4^-$ takes a third course in Table 11 (entry 4) when either the triphenylphosphine or tributylphosphine additive is replaced with triphenylphosphite, viz.

$$Cp_2Co^+ Co(CO)_4^- + P(OPh)_3 \xrightarrow{h\nu_{CT}} Cp_2Co^+Co(CO)_3[P(OPh)_3]^- + CO \quad (128)$$

which corresponds to an overall substitution of $Co(CO)_4^-$ by a single $P(OPh)_3$ ligand. Control experiments establish that (a) neither PPh_3, PBu_3, nor $P(OPh)_3$ has any effect on the charge-transfer spectrum of $Cp_2Co^+ Co(CO)_4^-$ in Fig. 16B, and (b) no thermal reaction occurs in the absence of light. In order to differentiate the photoinduced process in Eq. 128 from that in either Eq. 125 or 127, it will be simply referred to hereafter as *charge-transfer substitution.*

The photochemical behavior of the other charge-transfer salts is comparable to that of $Cp_2Co^+ Co(CO)_4^-$. Thus the orange, purple, and green solutions of $PP^+ Co(CO)_4^-$, $Q^+ Co(CO)_4^-$, and $\phi_2Cr^+ Co(CO)_4^-$, respectively, in dichloromethane are stable to visible radiation (with $\lambda > 550$ nm) for prolonged periods. Similarly the carbonylvanadate and manganate salts $Cp_2Co^+ V(CO)_6^-$ and $Cp_2Co^+ Mn(CO)_5^-$ are unaffected by visible radiation with $\lambda > 520$ nm. However, owing to the generally inert character of these charge-transfer salts, all further CT photochemical studies focus on the carbonylcobaltate salts. In each case, the actinic radiation of the charge-transfer salts in dichloromethane or tetrahydrofuran containing added ligand **L** = phosphine or phosphite results in the liberation of a mole of carbon monoxide. The stoichiometry leading to either CT dimerization, disproportionation, or substitution as given in Eqs. 125, 127 and 128, respectively, varies with the additive L. Isolation of the carbonylmetal products in Table 11 and IR spectral analyses based on the characteristic carbonyl bands indicate that one of the three processes is usually dominant for certain cationic acceptors with different phosphines (or phosphites).

At this juncture, it is important to emphasize that the photoinduced dimerization, disproportionation, and substitution always arise directly from the charge-transfer excitation of only the contact ion pair. Thus, the use of visible light with $\lambda > 520$ nm as the radiation source ensures the excitation of only the charge-transfer absorption bands in Fig. 16. Since these CT absorptions relate specifically to the contact ion pair (see Table 10), there is no ambiguity about the adventitious local excitation of either the separate anion or cation, or the photochemical generation of intermediates which do not arise from the CT excitation of the contact ion pair.

4.3.1 Quantum yields for the charge-transfer photochemistry of contact ion pairs. Owing to the critical role of the additive **L** on the charge-transfer photochemistry of contact ion pairs in Table 11, it is important to establish quantitatively the dependence of the quantum yield Φ on phosphine concentration. The charge-transfer disproportionation of $Cp_2Co^+ Co(CO)_4^-$, $Q^+ Co(CO)_4^-$ and $\phi_2Cr^+ Co(CO)_4^-$ with **L** = tri-n-butylphosphine is chosen for quantum yield measurements, since only a single, pale yellow salt $Co(CO)_3L_2^+ Co(CO)_4^-$ is formed in Table 11. Most importantly, the weak visible absorbance of $Co(CO)_3L_2^+ Co(CO)_4^-$ at $\lambda > 500$ nm causes no important interference of the CT excitation for the contact ion pairs. The quantum yields are based on Reineckate actinometry using monochromatic light ($\lambda = 550 \pm 10$ nm) [123] and carried out by quantitative IR analysis of the characteristic carbonyl bands of the cationic moiety $Co(CO)_3L_2^+$ For each of the charge-transfer salts, the value of Φ_p reaches a limiting value, which decreases in the order $\phi_2Cr^+ > Q^+ > Cp_2Co^+$ for charge-transfer disproportionation (Fig. 23). The quantum yields for charge-transfer dimerization (Eq. 125) and substitution (Eq. 128) cannot be measured reliably owing to the spectral interference by the products $Co_2(CO)_6L_2$ and $Co(CO)_3L^-$ that arise from their intense absorption bands which tail into the visible region [91, 124, 125]. However, from the product conversions in Table 11, the photoefficiences of charge-transfer dimerization and substitution are judged not to be widely divergent from the quantum yields measured in Fig. 23.

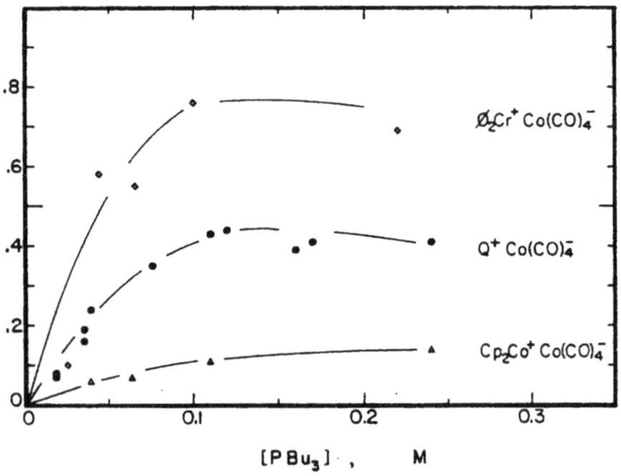

Fig. 23. Limiting quantum yields for charge-transfer photochemistry of $Co(CO)_4^-$ salts of various acceptor cations at increasing concentrations of the additive L = PBu_3.

4.3.2 Time-resolved spectra of reactive intermediates in the charge-transfer photochemistry of contact ion pairs. In order to identify the reactive intermediates in the charge-transfer photochemistry of the contact ion pairs, the time-resolved spectra are examined immediately following the application of a 10-ns pulse cosisting of the second harmonic at 352 nm of a mode-locked Nd^{3+}: YAG laser. The wavelength of this radiation source is ideally suited for the specific excitation of the contact ion pairs (see Table 10). Accordingly, the time-resolved spectra obtained from $Cp_2Co^+ Mn(CO)_5^-$ and $Cp_2Co^+ Co(CO)_4^-$ relate directly to the charge-transfer photochemistry in Table 11. Most notably, the intense band centered at $\lambda_{max} = 800$ nm in Fig. 24A is identical to the 17-electron radical $Mn(CO)_5\cdot$ that was previously generated from the direct homolytic cleavage of the dimeric $Mn_2(CO)_{10}$ [126]. Other flash photolytic studies, [62c] as well as pulse radiolysis [127] and matrix isolation [128] have verified the spectral assignment of $Mn(CO)_5\cdot$ with $\lambda_{max} \sim 800$ nm.

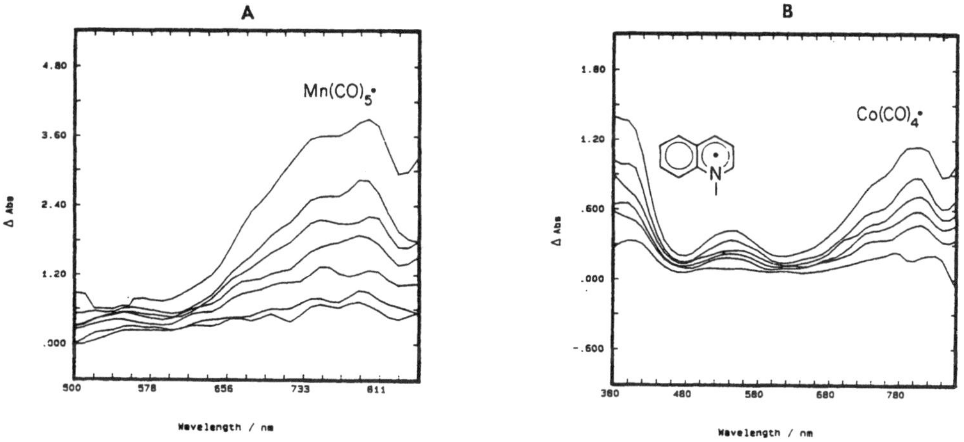

Fig. 24. Time-resolved difference spectra resulting from the 532-nm excitation of (A) 1.4×10^{-2} M $Cp_2Co^+ Mn(CO)_5^-$ at 0.9, 1.4, 1.8, 2.3, 2.8, 3.7 and 4.7 μs and (B) 1.0×10^{-2} M $Q^+ Co(CO)_4^-$ at 60. 83, 106, 128, 150, and 217 μs in Me_2CO.

The analogous observation of the spectral transient absorbing at $\lambda_{max} = 780$ nm from the CT excitation of $Cp_2Co^+ Co(CO)_4^-$ can be similarly ascribed to the 17-electron radical $Co(CO)_4\cdot$. Unfortunately the absorption spectrum of this carbonylcobalt(0) radical is not available for verification from other studies [129]. However, if one considers the location of absorption bands of their anionic precursors $Co(CO)_4^-$ and $Mn(CO)_5^-$ at $\lambda_{max} = 320$ and 330 nm, respectively, it may not be unreasonable to expect a similar relationship for the isoelectronic radicals $Co(CO)_4\cdot$ and $Mn(CO)_5\cdot$. Be that as it may, the laser-pulse excitation of the quinolonium and cyanopyridinium salts of $Co(CO)_4^-$ produces the same spectral transient with $\lambda_{max} = 780$ nm. Most revealingly, the simultaneous appearance of two bands at $\lambda_{max} = 550$ and $\sim$380 nm (end absorption) from the CT excitation of $Q^+ Co(CO)_4^-$ in Fig. 24B can be readily assigned to the quinolinyl radical (Q·) that was independently generated earlier [130]. Moreover, the high-energy tail absorption obtained from the CT excitation of $NCP^+ Co(CO)_4^-$ is in accord with the known cyanopyridinyl radical (NCP) with its intense absorption at $\lambda_{max} =$

390 nm [131]. Taken together, these time-resolved spectra provide strong support for the primary step in the photoactivation of tetracarbonylcobaltate salts, e.g.

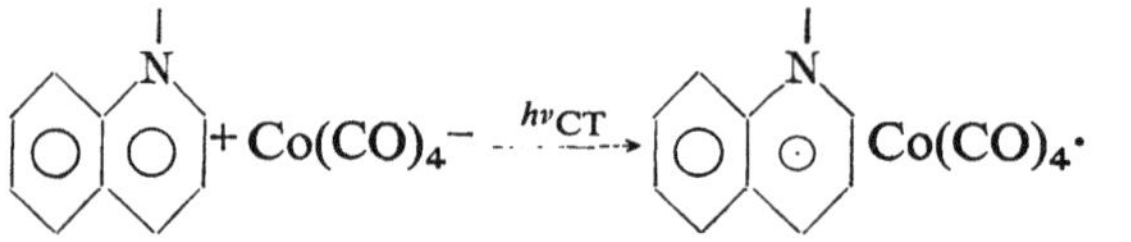

(129)

that is wholly consistent with the photoredox process for the pentacarbonylmanganate salt (vide supra). These time-resolved spectroscopic studies thus show that the charge-transfer excitation of CIPs results in the simultaneous production of the reactive 17-e carbonylmetal radicals [$Mn(CO)_5\cdot$, $Co(CO)_4\cdot$, etc.] together with the reduced acceptor radical [Cp_2Co, Q·, NCP·, etc.] Furthermore the timescale of the CT photoexcitation indicates that these radicals are initially paired within the solvent cage [132].

4.3.3 Temporal evolution of transient radical pairs from the charge-transfer excitation of CIPs. The absence of productive photochemistry upon the steady-state irradiation of contact ion pairs alone in solution (i.e. without added phosphine) is consistent with the spontaneous return of the transient radicals to the ground state intact, e.g.

$$Cp_2Co,\ Mn(CO)_5\cdot \rightarrow Cp_2Co^+\ Mn(CO)_5^-$$

Such a regeneration of the CIP from the radical pair accords with the invariance of the charge-transfer spectra even upon prolonged irradiation. Back electron transfer from the radical pair is also supported in the time-resolved spectroscopic studies by the restoration of the transient absorbances to the original baselines.

Kinetics for the return of the radical pair from $Cp_2Co^+\ Mn(CO)_5^-$ can be measured by following the absorbance change of $Mn(CO)_5\cdot$ at the monitoring wavelength of $\lambda_{mon} = 800$ nm. Fig. 25A illustrates a typical decay curve for $Mn(CO)_5\cdot$ following the CT excitation at 532 nm of $Cp_2Co^+\ Mn(CO)_5^-$ in MeCN with the 10-ns laser pulse. The reaction order of the kinetics is established by linear least-squared computer fits of the observed decrease of the absorbance with time. Thus, the second-order kinetics for the disappearance of $Mn(CO)_5\cdot$ is demonstrated by the excellent fit of the smooth computed curve to the experimental decay in Fig. 25A. The second-order rate constant k_2 evaluated in this manner is insensitive to solvent variation (Table 12, entries 1 and 2).

Disappearance of the radicals derived from such carbonylcobaltate salts as $Cp_2Co^+\ Co(CO)_4^-$ also follows the same second-order kinetics (Fig. 25B). It is noteworthy that the CT excitation of the quinolinium salt $Q^+\ Co(CO)_4^-$ allows a pair of second-order rate constants to be extracted from the absorbance decays at $\lambda_{mon} = 550$ and 780 nm (Table 12, entry 4). Assignment of the latter to the 17-e $Co(CO)_4\cdot$ (vide supra) then defines the second-order kinetics to derive from the mutual annihilation of the radicals, i.e.

$$Q\cdot + Co(CO)_4\cdot \xrightarrow{k_2} Q^+\ Co(CO)_4^- \qquad (130)$$

Furthermore the same kinetic relationship holds for the pair of radicals from the cyanopyridinium salt $NCP^+\ Co(CO)_4^-$ in Table 12 (entry 7). Since the absorbance at $\lambda_{mon} = 380$ nm does not return completely to the baseline, it is possible that the highly persistent NCP [132] does not always undergo mutual annihilation as presented in Eq. 130. The kinetic results in Table 12 also show that radical decay suffers only a minor retardation in the presence of added phosphine (entry 6).

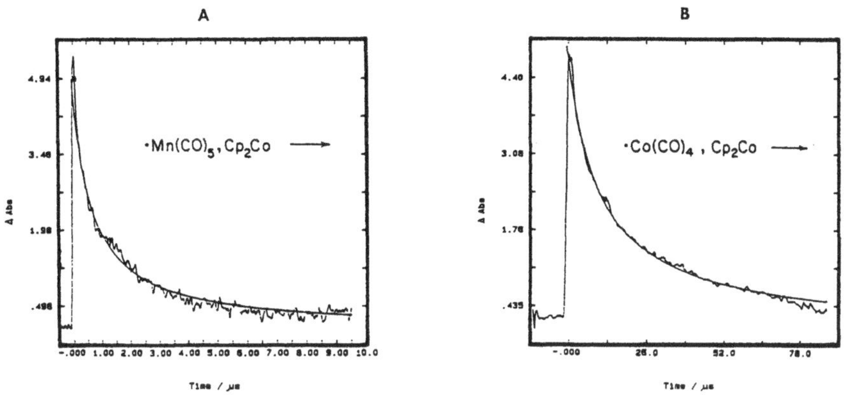

Fig. 25. Decay of the CT transients from (A) 2.0×10^{-2} M $Cp_2Co^+ Mn(CO)_5^-$ and (B) 1.0×10^{-1} M $Cp_2Co^+ Co(CO)_4^-$ in Me_2CO at 25 °C by following the absorbance change at λ_{mon} = 800 and 780 nm, respectively.

Table 12. Kinetics for the Radical-Pair Return to the Contact Ion Pair[a]

Contact Ion Pair	(10^3 M)	Solvent	Radical	[nm][b]	τ^c (μs)	k^d (A^{-1} s^{-1})
$Cp_2Co^+Mn(CO)_5^-$	(10)	MeCN	$Mn(CO)_5\cdot$	[790]	0.4	2.7×10^7
$Cp_2Co^+Mn(CO)_5^-$	(20)[e]	Me_2CO	$Mn(CO)_5\cdot$	[790]	0.3	4.4×10^7
$Cp_2Co^+Co(CO)_4^-$	(13)	Me_2CO	$Co(CO)_4\cdot$	[780]	10	2.0×10^6
$Q^+Co(CO)_4^-$	(8)	Me_2CO	$Co(CO)_4\cdot$	[780]	8	1.6×10^7
			$Q\cdot$	[550]	6	0.9×10^7
$Q^+Co(CO)_4^-$	(0.6)	CH_2Cl_2	$Q\cdot$	[550]	14	1.4×10^7
$Q^+Co(CO)_4^-$ /PBu_3	[f]	CH_2Cl_2	$Q\cdot$	[550]	17	3.7×10^6
$NCP^+Co(CO)_4^-$	(11)	Me_2CO	$Co(CO)_4\cdot$	[780]	21	1×10^6
			$NCP\cdot$	[380]	42	5×10^5

[a] As charge-transfer salt dissolved in 3 mL solvent at 25 °C and irradiated at 532 nm, unless indicated otherwise. [b] λ_{max}. [c] Apparent halflife. [d] Second-order rate constant in absorbance units, reliability ±20%. [e] From an equimolar mixture $PPN^+Mn(CO)_5^-$ and $Cp_2Co^+PF_6^-$. [f] $Q^+Co(CO)_4^-$ (1.55 mM) plus added PBu_3 (10 mM).

4.4 Mechanism of charge-transfer photochemistry of contact ion pairs

The charge-transfer salts of carbonylmetallates are distinguished from the iodide salts in the breadth of photochemistry that obtains upon exposure to visible light. For example, the continuous irradiation of the charge-transfer band ($\lambda \sim$ 350 to 500 nm) of the pyridinium iodides inflicts no permanent change of either the crystalline salt or that dissolved in CH_2Cl_2 or THF [133]. Moreover the highly colored $Co(CO)_4^-$, $V(CO)_6^-$ and $Mn(CO)_5^-$ salts are similarly unaffected by visible light for prolonged periods in the solid state or in solution. However, merely the presence of small amounts of additives **L** are sufficient to activate the carbonylmetallate salts to undergo a variety of distinctive photochemical reactions in solution.

At first glance, all of the photochemical processes described in Table 11 are unique, with none showing any apparent stoichiometry relationship to the others. For example, the carbonylcobalt dimerization (Eq. 125) and disproportionation (Eq. 127) represent one-electron oxidation processes of $Co(CO)_4^-$, whereas the formation of $Co(CO)_3L^-$ in Eq. 128 relates to a nonredox ligand substitution. The incorporation of the additive L into the carbonylmetal product, whether it be $Co_2(CO)_6L_2$, $Co(CO)_3L_2^+$ or $Co(CO)_3L^-$ in Eqs. 125, 127 and 128, respectively, is the sole feature that these photochemical processes have in common. Since the carbonylmetallates $Co(CO)_4^-$, $V(CO)_6^-$, and $Mn(CO)_5^-$ are all thermally substitution-stable anions, [134] the introduction of **L** into the carbonylmetallate moiety must occur in some reactive intermediate.

The time-resolved spectroscopic studies of the transient intermediates provide the keys to understanding the charge-transfer photochemistry of contact ion pairs. Thus the unambiguous identification of the carbonylmetal radical $Mn(CO)_5\cdot$ immediately following the 10-ns laser-flash excitation of the contact ion pair $Cp_2Co^+ Mn(CO)_5^-$ at $\lambda = 532$ nm in Fig. 24A relates directly to the photoredox process,

$$Cp_2Co^+ Mn(CO)_5^- \xrightarrow{h\nu_{CT}} Cp_2Co, Mn(CO)_5\cdot \tag{131}$$

in accord with Mulliken theory [96]. The concomitant production of cobaltocene in Eq. 131 could not be observed owing to its spectral overlap with the local absorptions of the separate ions Cp_2Co^+ and $Mn(CO)_5^-$ [102, 103]. Nonetheless the persistence of the transient 800-nm band at times as short as 100 ns following the laser excitation coupled with the isolation of cobaltocene from the steady-state experiments (Table 11), represents the reasonable experimental verification of the Mulliken charge-transfer formulation. Most importantly, the enhanced reactivity of the 17-e radical $Mn(CO)_5\cdot$ leads to rapid ligand substitution by tributylphosphine and homolytic dimerization, i.e.

$$Mn(CO)_5\cdot + PBu_3 \xrightarrow{k_s} Mn(CO)_4PBu_3\cdot + CO \tag{132a}$$

$$2\,Mn(CO)_4PBu_3\cdot \xrightarrow{k_d} Mn_2(CO)_8(PBu_3)_2 \tag{132b}$$

The large magnitudes of the rate constants k_s and $k_d = 1.0 \times 10^9$ and $1.0 \times 10^8\ M^{-1}\ s^{-1}$, respectively, established by Brown ahd coworkers [28, 62c] are critical for the description of CT dimerization in Eq. 126, as given in the mechanistic scheme below.

Scheme XXI:

$$Cp_2Co^+Mn(CO)_5^- \underset{k_{et}}{\overset{h\nu_{CT}}{\rightleftharpoons}} Cp_2Co + Mn(CO)_5\cdot \tag{133}$$

$$Mn(CO)_5\cdot + L \xrightarrow{k_s} Mn(CO)_4\cdot + CO \tag{134}$$

$$2\,Mn(CO)_4L\cdot \xrightarrow{k_d} Mn_2(CO)_8L_2 \tag{135}$$

It is important to emphasize that the photoefficiency of CT dimerization according to Scheme XXI is generally dependent on the competition between back electron transfer (k_{et}) and ligand substitution (k_sL). As such, the absence of photochemistry without added **L** derives from the facile back electron transfer of the initial radical pair, $Mn(CO)_5\cdot$ and Cp_2Co, to regenerate the CIP in Eq. 133. Thus in order to account for the observed CT dimerization, a value of $k_{et} < 10^8$ s^{-1} is required to allow competition for phosphine scavenging of $Mn(CO)_5\cdot$ with $k_s = 1.0 \times 10^9$ M^{-1} s^{-1} at roughly 0.1 M concentrations of added PBu_3. Furthermore, the absence of CT photochemistry when the added PBu_3 is replaced with the less nucleophilic PPh_3 (see entry 13, Table 11) is a direct consequence of the mechanism in Scheme XXI, since the second-order rate constant for substitution $k_s = 1.7 \times 10^7$ M^{-1} s^{-1} with L = PPh_3 is roughly two orders of magnitude slower than that with PBu_3 [28].

The kinetic results from the time-resolved experiments show that back electron transfer leading to annihilation of $Mn(CO)_5\cdot$ actually proceeds by *second*-order kinetics, with the rate constant of $k_{et} = 3 \times 10^{10}$ M^{-1} s^{-1}, as determined from the data in Table 12 (entries 1 and 2) together with the measured extinction coefficient of $\varepsilon = 980$ M^{-1} cm^{-1} [62c]. As such, any description of the photostationary state attained in the absence of additives must include the diffusive separation of the initial radical pair, [135] e.g.

$$Cp_2Co^+Mn(CO)_5^- \underset{\varkappa_3}{\overset{h\nu_{CT}}{\rightleftharpoons}} Cp_2Co, Mn(CO)_5\cdot \underset{\varkappa_2}{\overset{\varkappa_1}{\rightleftharpoons}} Cp_2Co + Mn(CO)_5\cdot \qquad (136)$$

where the measured rate constant $k_{et} \cong \varkappa_2\varkappa_3/\varkappa_1$. Furthermore the phosphine substitution of $Mn(CO)_5\cdot$ to yield $Mn(CO)_4L\cdot$ in Eq. 134 serves to enhance the photochemical efficiency by minimizing the energy-wasting back electron transfer, i.e.

$$Cp_2Co + Mn(CO)_4L\cdot \xrightarrow{\text{slow}} Cp_2Co^+Mn(CO)_4L^- \qquad (137)$$

owing to the significantly attenuated reduction potentials of the phosphine-substituted radicals [27b].

4.5 Selectivity in the charge-transfer photochemistry of contact ion pairs

The time-resolved spectra also accord with the Mulliken formulation for the photochemical activation of carbonylcobaltate salts. As such, the charge-transfer mechanism outlined for $Mn(CO)_5\cdot$ should be generally applicable to the kindred 17-e radical, i.e.

Scheme XXII:

$$Cp_2Co^+Co(CO)_4^- \underset{\varkappa_3}{\overset{h\nu_{CT}}{\rightleftharpoons}} Cp_2Co, Co(CO)_4\cdot \underset{\varkappa_2}{\overset{\varkappa_1}{\rightleftharpoons}} Cp_2Co + Co(CO)_4\cdot \qquad (138)$$

$$Co(CO)_4\cdot + L \xrightarrow{k_s} Co(CO)_3L\cdot + CO \qquad (139)$$

Indeed the cage mechanism in Eq. 138 is confirmed by the limiting value of the quantum yield, Φ_p (limit) that is observed at high concentrations of added phosphine, as illustrated in Fig. 23. According to Scheme XXII, the efficiency of CT photochemistry arises from the carbonylcobalt radicals $Co(CO)_4\cdot$ that successfully undergo cage escape ($\varkappa_1$) in competition with back electron transfer ($\varkappa_3$). Importantly, the trend in the values of Φ_p (limit) in the order $\Phi_2Cr^+ > Q^+ > Cp_2Co^+$

derives from the relative ease of back electron transfer with the acceptor radicals $\phi_2Cr < Q\cdot < Cp_2Co$, as based on the values of $E_c = -0.80$, 0.90 and -0.95 V, respectively. However, there are specific elaborations to Scheme XXII that must be included to account for the unique product selectivities in charge-transfer dimerization, substitution and disproportionation as they obtain with different phosphines described separately below.

4.5.1 A. Charge-transfer dimerization. The second-order rate constant for back electron transfer (k_{et}) with $Co(CO)_4\cdot$ appears to be an order of magnitude slower in Table 12 than that for $Mn(CO)_5\cdot$ in Scheme XXI [105]. Secondly, the formation of the dimeric $Co_2(CO)_6L_2$ with L = PPh_3 and PMe_2Ph in Table 11 (entries 1 and 2) undoubtedly arises from the efficient scavenging of the carbonylcobalt radical (Eq. 139), followed by dimerization, i.e.

$$2Co(CO)_3L\cdot \xrightarrow{k_d} Co_2(CO)_6L_2 \tag{140}$$

with the second-order rate constant $k_s > 10^5$ M^{-1} s^{-1} and $k_{et} \sim 10^7$ M^{-1} s^{-1}. Moreover, the energy-wasting back electron transfer of $Co(CO)_4\cdot$ (with Cp_2Co) is obviated by the conversion to $Co(CO)_3PPh_3^-$ whose annihilation, i.e.

$$Cp_2Co + Co(CO)_3PPh_3\cdot \xrightarrow{\text{slow}} Cp_2Co^+Co(CO)_3PPh_3 \tag{141}$$

is minimized as the result of its significantly reduced potential for reduction [136]. Indeed the quantum yield of $\Phi_p = 0.1$ for the CT dimerization of $Cp_2Co^+Co(CO)_4^-$ with PPh_3 in Fig. 23 indicates that $Co(CO)_4\cdot$ undergoes ligand substitution roughly ten times slower than back electron transfer in Eq. 138. Thus the mechanism of charge-transfer dimerization is essentially the same as that presented in Scheme XVIII for ion-pair annihilation – the difference merely lying in the electron transfer initiated photochemically in Eq. 138 and thermally in Eq. 96, respectively.

4.5.2 B. Charge-Transfer Substitution. Carbonylcobalt radicals are also critical intermediates when the phosphines PPh_3 and PMe_2Ph are replaced by the phosphite $P(OPh)_3$ in Table 11 (entry 4). The formation of $Co(CO)_3[P(OPh)_3]^-$ does not derive in Eq. 128 from direct ligand substitution, since $P(OPh)_3$ is too weak a nucleophile to effect CO replacement in the substitution-inert $Co(CO)_4^-$. However, the additives L = PPh_3, PMe_2Ph and $P(OPh)_3$ are strongly differentiated in the redox properties of the 17-e intermediates $Co(CO)_3L\cdot$ as formed in Eq. 139 (Scheme XXII). For example, the phosphite-substituted radical $Co(CO)_3[P(OPh)_3]$ is significantly more readily reduced than its phosphine analogue $Co(CO)_3(PMe_2Ph)\cdot$, the cathodic potentials at $E_c = -0.05$ and -0.45 V, respectively, [91] representing a difference of 9 kcal mol^{-1}. Coupled with the oxidation potential of -0.95 V for cobaltocene, [137] the driving force for back electron transfer

$$Cp_2Co + Co(CO)_3[P(OPh)_3]\cdot \xrightarrow{k_{et}} Cp_2Co^+Co(CO)_3[P(OPh)_3]^- \tag{142}$$

is estimated to be exergonic by $\Delta G \sim -20$ kcal mol^{-1}. Although the rate of electron transfer could not be measured directly by time-resolved spectroscopy, it is estimated that $k_{et} > 10$ M^{-1} s^{-1} from the large driving force for Eq. 142. Such large values of k_{et} leading to Cp_2Co^+ $Co(CO)_3[P(OPh)_3]^-$ could thus be sufficient to obviate the formation of $Co_2(CO)_6[P(OPh)_3]_2$ via the bimolecular coupling of

$Co(CO)_3[P(OPh_3]\cdot$ in Eq. 140. The extent to which any of the dimeric $Co_2(CO)_6$-$[P(OPh)_3]_2$ is formed will be subsequently reduced to the salt, i.e.

$$Co_2(CO)_6[P(OPh)_3]_2 + 2Cp_2Co \rightarrow 2Cp_2Co^+Co(CO)_3[P(OPh)_3]^- \quad (143)$$

as independently established by control experiments.

4.5.3 C. Charge-transfer disproportionation. The role of carbonylcobalt radicals in the conversion of $Co(CO)_4^-$ to $Co(CO)_3L_2^+$ was delineated in earlier electrochemical studies [138]. Therefore the charge-transfer disproportionation of $Cp_2Co^+Co(CO)_4^-$ can be considered in two distinct parts, namely (a) the 2-oxidation of the anionic donor, i.e.

$$Co(CO)_4^- + 2PBu_3 \xrightarrow{-2e} Co(CO)_3(PBu_3)_2^+ + CO \quad (144)$$

coupled with (b) the reduction of 2 equiv. of Cp_2Co^+, so that the combination of (a) and (b) represents the experimental stoichiometry in Eq. 127. The facile ligand substitution of carbonylcobalt radicals constitutes the pathway for the 2e transformation in Eq. 144, i.e.

Scheme XXIII:

$$Co(CO)_4^- \xrightarrow{-e} Co(CO)_4\cdot \quad (145)$$

$$Co(CO)_4\cdot + L \longrightarrow Co(CO)_3L\cdot + CO \quad (146)$$

$$Co(CO)_3L\cdot + L \xrightarrow{-e} Co(CO)_3L_2^+ \quad (147)$$

which represents an electrochemical ECE mechanism [138]. As such, transient electrochemical techniques differentiate phosphine ligands $L = PPh_3$ and PBu_3 by their effect on $Co(CO)_3L_2^+$ – the redox potential of $Co(CO)_3(PBu_3)_2^+$ being 0.61 V or 14 kcal mol^{-1} more favorable than that for $Co(CO)_3(PPh_3)_2^+$. The importance of such a redox property is also illustrated in the photodissociation of $Co_2(CO)_6(PBu_3)_2$ in the presence of both $Q^+PF_6^-$ and PBu_3. Separately, neither Q^+ nor PBu_3 has any perceptible effect on the photostationary state of $Co_2(CO)_6(PBu_3)_2$ upon irradiation of the σ, σ^* band [124] with $\lambda \sim 380$ nm. However when both are present, essentially quantitative yields of $Co(CO)_3(PBu_3)_2^+$-PF_6^- are obtained by a photoredox process than can be described as

Scheme XXIV:

$$Co_2(CO)_6(PBu_3)_2 \xrightarrow{h\nu} 2Co(CO)_3(PBu_3)\cdot \quad (148)$$

$$Co(CO)_3(PBu_3)_2\cdot + PBu_3 \xrightarrow{[e]} Co(CO)_3(PBu_3)_2^+ \quad (149)$$

where [e] represents the redox couple $[A^+ \rightarrow A\cdot]$ for the pyridinium acceptor A^+ [139, 140]. The importance of Eq. 149 is underscored in Table 11 by the acceptor cations which favor CT disproportionation in the order: $A^+ = \phi_2Cr^+ > Q^+ > Cp_2Co^+ > PP^+$. It is noteworthy that this also represents the decreasing trend in the reduction potentials of the cations in the same order, namely, $E_c = -0.80$, -0.90, -0.95 and -1.22 V, respectively, A particularly dramatic influence of the cation is shown by the entirely different products obtained from the CT activation of Cp_2Co^+ $Co(CO)_4^-$ and Q^+ $Co(CO)_4^-$ in the presence of the same phosphine additive PMe_2Ph (compare entries 2 and 6, Table 11).

Finally, care must be exercised in the photochemical activation of contact ion pairs to irradiate only the charge-transfer absorption bands, and not those of the (colored) products. For example, the irradiation of either $Cp_2Co^+ Co(CO)_4^-$ or $Q^+ Co(CO)_4^-$ in the presence of PBu_3 at $\lambda > 510$ nm leads only to the dimeric $Co_2(CO)_6(PBu_3)_2$, despite the fact that the CT photochemistry of the same solution at $\lambda > 550$ nm leads smoothly to only the disporportionation products (Table 11, entries 4 and 7). In fact, control experiments demonstrate that the carbonylcobalt dimer arises from a secondary process by the adventitious excitation of the disproportionation salt, i.e.

$$Co(CO)_3(PBu_3)_2{}^+Co(CO)_4{}^- \xrightarrow{h\nu_{CT}} Co_2(CO)_8(PBu_3)_2 + CO \qquad (150)$$

It is also important to emphasize that the mechanism of charge-transfer disproportionation is basically the same as that presented in Scheme XIX for ion-pair annihilation – the difference merely lying in the electron transfer initiated photochemically in Eq. 138 and thermally in Eq. 96, respectively.

4.5.4 Unifying theme in the charge-transfer photochemistry of carbonylmetallate salts. The charge-transfer photochemistry of carbonylmetallate salts $A^+Mn(CO)_5^-$ and $A^+Co(CO)_4^-$ derives from contact ion pairs (CIP), whose structures are established by X-ray crystallography. These crystallographic structures can be successfully interrelated with contact ion pairs in solution by their characteristic carbonyl IR bands. Photoexcitation of the charge-transfer absorption band of the CIPs leads to the radical pairs [A·, $Mn(CO)_5\cdot$] and [A·, $Co(CO)_4\cdot$] that can be observed directly by time-resolved spectroscopy. Although prone to undergo facile back electron transfer, the radical pair can be scaveneged by phosphine additives **L** in solution. The rapid ligand substitution of the 17-e radicals $Mn(CO)_5\cdot$ and $Co(CO)_4\cdot$ gives rise to quantum efficient photochemistry that is generally described as charge-transfer dimerization, disproportionation and substitution according to the stoichiometries in Eqs. 125, 127 and 128, respectively. The critical role of the phosphine ligand on the substituted carbonylmetal radicals, $Mn(CO)_4L\cdot$ and $Co(CO)_3L\cdot$, especially with regard to oxidation-reduction and coupling, is wholly consistent with their properties, principally delineated by Brown and coworkers from independent sources [11, 28]. Moreover, the absence of significant CT photochemistry from carbonylvanadate salts $A^+ V(CO)_6^-$ accords with ligand substitution rates of the corresponding 17-e radical $V(CO)_6$ that are substantially slower [141] than those of $Mn(CO)_5\cdot$ and $Co(CO)_4\cdot$ in Table 12.

5. Epilogue

Metal carbonyls are among the oldest organometals, and their reactions are among the most fundamental processes in organometallic chemistry. Traditionally, the thinking in contemporary organometallic chemistry has relied heavily on the 16- and 18-electron rule [142], which is widely held and consistently applied [3–6]. However, many organometals including metal carbonyls are excellent electron donors and acceptors, and their properties are clearly delineated in their charged analogues. Thus, the focus has been directed here on electron-transfer processes with carbonylmetal cations and carbonylmetallate anions that are known to be excellent electrophiles and nucleophiles, respectively.

Basic to electron transfer of diamagnetic, 18-electron carbonylmetal cations and anions is the existence of 19-e and 17-e carbonylmetal radicals, and their facile interconversion, as presented in Section 2. Taken together, the carbonylmetal cations and carbonylmetallate anions undergo mutual annihilation to afford the corresponding radical pair comprised of the paramagnetic 19-e and 17-e species. Thermal (Schemes XVI and XVIII) and photochemical (Schemes XXI and XXII) activation of such an organometallic ion pair is modulated by the driving force (Section 3.6) for interionic electron transfer evaluated from the electrode potentials for reduction (E_c) and oxidation (E_a) of the carbonylmetal cation and carbonylmetallate anion, respectively. Otherwise, the thermal and photochemical pathways are largely indistinguishable. In both cases, the ionic association to form contact ion pairs (CIP) is preliminary to electron transfer. The tandem application of X-ray crystallography and carbonyl IR spectroscopy in Sections 4.1 and 4.2 serves to delineate the structural features relevant to the charge-transfer photoactivation of CIPs.

The behavior of 19-e and 17-e carbonylmetal radicals is seminal to defining the products of ion-pair annihilation – independent of whether they derive via thermal or photochemical activation. Owing to their enhanced lability, the magnitudes of the rate processes leading to back electron transfer, homolytic dimerization, and substitution of these carbonylmetal radicals are shown to be determinant in the competition.

In a more general context, the facile annihilation of carbonylmetal cations and carbonylmetallate anions relates to nucleophile-electrophile interactions that form the basis of many organometallic reactions [143]. As such, the critical role of electron transfer merits further attention [144].

6. Acknowledgments

I gratefully acknowledge the numerous contributions of my collaborators mentioned in the references, and especially D. J. Kuchynka, K. Y. Lee and T. M. Bockman for their outstanding efforts. I also thank the National Science Foundation and Robert A. Welch Foundation for financial support, and the National Science Foundation and Texas Advanced Research Program for funds to construct the laser flash spectrometer.

7. References

1. I. Wender and P. Pino (Eds.), Organic Synthesis via Metal Carbonyls, Vol. **1**, Wiley, New York, 1968. Vol. **2**, 1976.
2. R. A. Sheldon, Chemicals from Synthesis Gas, Reidel, Dordrecht, 1983.
3. C. M. Lukehart, Fundamental Transition Metal Organometallic Chemistry, Brooks/Cole, Monterey, California, 1985.
4. F. Basolo and R. G. Pearson, Mechanisms of Inorganic Reactions, 2nd edn., Wiley, New York, 1967, p. 533.
5. (a) G. R. Dobson, Acc. Chem. Res. **9** (1976) 300. (b) A. J. Deeming, Inorg. React. Mech. **7** (1981) 275 and related review in this series.
6. J. D. Atwood, Inorganic and Organometallic Reaction Mechanisms, Brooks/Cole, Monterey, California, 1985.

7. G. J. Bezems, P. H. Rieger and S. Visco, J. Chem. Soc. Chem. Commun. (1981) 265.
8. J. W. Hershberger, R. J. Klingler and J. K. Kochi, J. Am. Chem. Soc. **104** (1982) 3034.
9. D. P. Summers, J. L. Luong and M. S. Wrighton, J. Am. Chem. Soc. **103** (1981) 5238.
10 (a).B. H. Byers and T. L. Brown, J. Am. Chem. Soc. **97** (1975) 947. (b) M. Absi-Halabi and T. L. Brown, J. Am. Chem. Soc. **99** (1977) 2982. (c) D. R. Kidd and T. L. Brown, J. Am. Chem. Soc. **100** (1978) 4095.
11. T. L. Brown, Ann. N. Y. Acad. Sci. **80** (1980) 333.
12. J. K. Kochi, J. Organomet. Chem. **300** (1986) 139.
13. (a) B. A. Narayanan and J. K. Kochi, J. Organomet. Chem. **272** (1984) C49. (b) B. A. Narayanan, C. Amatore and J. K. Kochi, Organometallics **5** (1986) 926.
14. (a) H. H. Ohst and J. K. Kochi, J. Chem. Soc. Chem. Commun. (1986) 121; J. Am. Chem. Soc. **108** (1986) 2897. (b) M. G. Richmond and J. K. Kochi, Inorg. Chem. **26** (1987) 541. (c) H. H. Ohst and J. K. Kochi, Inorg. Chem. **25** (1986) 2066; Organometallics **5** (1986) 1359. (d) T. M. Bockman and J. K. Kochi, J. Am. Chem. Soc. **109** (1987) 7725.
15. J. W. Hershberger, C. Amatore and J. K. Kochi, J. Organomet. Chem. **250** (1983) 345.
16. (a) C. J. Pickett and D. Pletcher, J. Chem. Soc. Dalton Trans. (1975) 879; Ibid., (1976) 636. (b) D. Pletcher, J. Chem. Soc. Rev. **4** (1975) 471. (c) R. J. Klingler and J. K. Kochi, J. Am. Chem. Soc. **102** (1980) 4790.
17. For a recent review, see N. G. Connelly and W. E. Geiger, Adv. Organometal. Chem. **23** (1984) 18.
18. (a) R. E. Dessy and L. A. Bares, Acc. Chem. Res. **5** (1972) 415. (b) R. E. Dessy and R. L. Pohl, J. Am. Chem. Soc. **90** (1968) 2005 and related papers.
19. J. P. Collman and L. S. Hegedus, Principles and Applications of Organotransition Metal Chemistry, University Science Books, Mill Valley, California, 1980, p. 199ff.
20. J. K. Kochi, Organometallic Mechanisms and Catalysis, Academic, New York, 1978, Chap. 8.
21. (a) K. Theopold, Nachr. Chem. **34** (1986) 876. (b) M. C. Baird, Chem. Revs. **88** (1988) 1217.
22. (a) A. E. Stiegman and D. R. Tyler, Comments Inorg. Chem. **5** (1986) 215. (b) A. E. Stiegman, M. Stieglitz and D. R. Tyler, J. Am. Chem. Soc. **105** (1983) 6032.
23. M. Chanon (Ed.), Paramagnetic Organometallic Species in Activation, Selectivity, and Catalysis, Reidel Publ., Dordrecht, 1989.
24. J. O. Howell and R. M. Wightman, J. Phys. Chem. **88** (1984) 3915.
25. (a) D. O. Wipf, E. W. Kristensen, M. R. Deakin and R. M. Wightman, Anal. Chem. **60** (1988) 306. (b) C. A. Amatore, A. Jutand and F. Pflüger, J. Electroanal. Chem. **21′** (1987) 361.
26. (a) B. A. Narayanan, C. Amatore and J. K. Kochi, Organometallics **6** (1987) 129. (b) D. J. Kuchynka, C. Amatore and J. K. Kochi, Inorg. Chem. **25** (1986) 4087.
27. (a) C. Amatore, D. J. Kuchynka and J. K. Kochi, J. Electroanal. Chem. **241** (1988) 181. (b) D. J. Kuchynka, C. Amatore and J. K. Kochi, J. Organometal. Chem. **328** (1987) 133.
28. (a) T. R. Herrinton and T. L. Brown, J. Am. Chem. Soc. **107** (1987) 5700. (b) A. Fox, J. Malito and A. Poe, J. Chem. Soc. Chem. Commun. (1981) 1052. (c) Q. Z. Shi, T. G. Richmond, W. C. Trogler and F. Basolo, J. Am. Chem. Soc. **104** (1982) 4032. See however R. M. Kowalewski, F. Basolo and W. C. Trogler, J. Am. Chem. Soc. **108** (1986) 6046.
29. (a) For example in the cyclopentadienyl and related derivatives of metal carbonyls [22] there is some ambiguity of $\eta^{3,4}$-bonding (i.e. Cp ring slippage) [30] in the absence of X-ray crystallographic structures of the 19-e radicals. (b) C. E. Philbin, C. A. Granatir and D. R. Tyler, Inorg. Chem. **25** (1986) 4886.
30. Cf. M. E. Rerek and F. Basolo, J. Am. Chem. Soc. **106** (1984) 5908.
31. J. Heinze, Angew. Chem. Int. Ed. Eng. **23** (1984) 831. See also Ref. 16b.
32. J. A. Armstead, D. J. Cox and R. J. Davis, J. Organomet. Chem. **236** (1982) 213.
33. H. W. Walker, R. S. Herrick, R. J. Olsen and T. L. Brown, Inorg. Chem. **23** (1984) 3748.
34. (a) S. B. McCullen and T. L. Brown, J. Am. Chem. Soc. **104** (1982) 7496. See also (b) D. R. Kidd, C. P. Cheng and T. L. Brown, J. Am. Chem. Soc. **100** (1978) 1978 and Ref. 11.
35. (a) In cyclic voltammetry, only the observation of the peak current on the reverse scan is needed for chemical reversibility. In this situation, the reversible potential E^0 is adequately

expressed as the average of the anodic and cathodic peak potentials $(E_a + E_c)/2$. (b) J. O. Howell, J. M. Goncalves, C. Amatore, L. Klasinc, R. M. Wightman and J. K. Kochi, J. Am. Chem. Soc. **106** (1984) 3968.
36. See, e.g. Wightman et al. in Ref. 25a and C. P. Andrieux, D. Garreau, P. Hapiot, J. Pinson and J. M. Saveant, J. Electroanal. Chem. **243** (1988) 321.
37. M. L. Olmstead, R. G. Hamilton and R. S. Nicholson, Anal. Chem. **41** (1969) 260.
38. See (a) P. M. Zizelman, C. Amatore and J. K. Kochi, J. Am. Chem. Soc. **106** (1984) 3771. (b) M. N. Golovin, M. M. Rahman, J. E. Belmonte and W. P. Giering, Organometallics **4** (1985) 1981.
39. D. J. Kuchynka and J. K. Kochi, Inorg. Chem. **27** (1988) 2574.
40. K. Y. Lee, D. J. Kuchynka and J. K. Kochi, Organometallics **6** (1987) 1886.
41. R. R. Gagne, C. A. Koval and G. C. Lisensky, Inorg. Chem. **19** (1980) 2854.
42. M. G. Richmond and J. K. Kochi, Inorg. Chem. **25** (1986) 656; M. G. Richmond and J. K. Kochi, Organometallics in Ref. 14.
43. D. J. Kuchynka and J. K. Kochi, Inorg. Chem., **27** (1988) 2574.
44. (a) J. G. Gaudiello, T. C. Wright, R. A. Jones and A. J. Bard, J. Am. Chem. Soc. **107** (1985) 888. See also (b) D. F. Untereker and S. Bruckenstein, Anal. Chem. **44** (1972) 1009. D. F. Untereker and S. Bruckenstein, J. Electroanal. Chem. **57** (1974) 77.
45. (a) S. W. Feldberg in A. J. Bard (Ed.), Electroanalytical Chemistry, Vol. **3**, Dekker, New York, 1969, p. 199. (b) S. W. Feldberg in H. B. Mark (Ed), Computer Applications in Analytical Chemistry, Dekker, New York, 1972, p. 185.
46. L. Nadjo, J. M. Saveant, J. Electroanal. Chem. **48** (1973) 113.
47. (a) R. A. Nicholson and I. Shain, Anal. Chem. **36** (1964) 706. (b) R. A. Nicholson and I. Shain, Anal. Chem. **37** (1965) 178. (c) J. M. Saveant, Electrochim. Acta **12** (1967) 753.
48. For the successful application of CV simulation, see (a) J. W. Hershberger, R. J. Klingler and J. K. Kochi, J. Am. Chem. Soc. **105** (1983) 61. (b) T. M. Bockman and J. K. Kochi in Ref. 14e.
49. (a) For the square-pyrimidal structure of the 17-e radical, see Ref. 39. With other systems, both theory [R. Hoffman, A. R. Rossi, Inorg. Chem. **14** (1975) 365; M. Elian and R. Hoffman, Inorg. Chem. **14** (1975) 1058] as well as experiment [M. C. R. Symons and R. L. Sweany, Organometallics **1** (1982) 834; S. P. Church, R. L. Poliakoff, J. A. Timney ad J. J. Turner, J. Am. Chem. Soc. **103** (1981) 7515 and D. R. Kidd, C. P. Cheng and T. L Brown, J. Am. Chem. Soc. **100** (1978) 4103] also favor the square pyramidal structure. (b) For the fluxional behavior of the d^9 complex $Rh(DPPE)_2$, see K. T. Mueller, A. J. Kunin, S. Greiner, T. Henderson, R. W. Kreilick, and R. Eisenberg, J. Am. Chem. Soc. **109** (1987) 6313.
50. D. J. Kuchynka and J. K. Kochi, Organometallics, **8** (1989) 677.
51. (a) A. M. Bond, R. Colton and K. McGregor, Inorg. Chem. **25** (1986) 2378.
(b) A. M. Bond, R. Colton, and J. E. Kevekordes, Inorg. Chem. **25** (1986) 749.
(c) A. M. Bond, S. W. Carr and R. Colton, Organometallics **3** (1984) 541.
52. A. J. Bard and L. R. Faulkner, Electrochemical Methods, Wiley, New York, 1980.
53. J. O. Howell and R. M. Wightman, Anal. Chem. **56** (1984) 524. See also Nadjo in Ref. 46.
54. (a) H. E. Podall, J. H. Dunn and H. Shapiro, J. Am. Chem. Soc. **82** (1960) 1325. (b) W. Hieber and H. Fuchs, Z. Anorg. Allgem. Chem. **248** (1941) 256. (c) A. Davison, J. A. McCleverty and G. Wilkinson, J. Chem. Soc. (1963) 1133. (d) W. Hieber, H. Schulten and R. Marin, Z. Anorg. Allgem. Chem. **240** (1939) 261. (e) J. A. McCleverty and G. Wilkinson, Inorg. Synth. **8** (1966) 211. (f) C. Ungurenasu and J. Palie, J. Chem. Soc., Chem. Commun. (1975) 388.
55. (a) R. B. King and F. G. A. Stone, Inorg. Synth. **7** (1963) 193. (b) R. Birdwhistell, P. Hackett and A. R. Manning, J. Organomet. Chem. **157** (1978) 239. (c) J. A. Armstead, D. J. Cox and R. Davis, J. Organomet. Chem. **236** (1982) 213. (d) R. E. Dessy, P. M. Weissman and R. L. Pohl, J. Am. Chem. Soc. **88** (1966) 5117.
56. (a) R. E. Dessy and P. M. Weissman, J. Am. Chem. Soc. **88** (1966) 5129. (b) V. N. Pandey, Inorg. Chim. Acta **23** (1977) L26. (c) S. Fontana, O. Orama, E. O. Fischer, U. Schubert and F. R. Kreissl, J. Organomet. Chem. **149** (1978) C57. (d) A. Davison and J. E. Ellis, J. Or-

ganomet. Chem. **36** (1972) 113. (e) D. A. Roberts, W. C. Mercer, S. M. Zakurak, G. L. Geoffroy, C. W. DeBrosse, M. E. Cass and C. G. Pierpont, J. Am. Chem. Soc. **104** (1982) 910. (f) M. A. Bennett and D. J. Patmore, Inorg. Chem. **10** (1971) 2387. (g) U. Anders and W. A. G. Graham, J. Am. Chem. Soc. **89** (1967) 539. (h) J. K. Ruff, Inorg. Chem. **7** (1968) 1818. (i) E. O. Fischer, T. L. Lindner, F. R. Kreissl and P. Braunstein, Chem. Ber. **110** (1977) 3139. (j) G. D. Michels and H. J. Svec, Inorg. Chem. **20** (1981) 3445. (k) C. Laurence, W. E. Lindsell, K. I. McCullough and P. N. Preston, J. Chem. Soc., Chem. Commun. (1982) 1001. (l) Compare also J. A. Hriljac, P. N. Swepston and D. Shriver, Organometallics **4** (1985) 158. (m) R. E. Dessy and P. M. Weissman, J. Am. Chem. Soc. **88** (1966) 5124. (n) See also D. L. Morse and M. S. Wrighton, J. Organomet. Chem. **125** (1977) 71.

57. (a) T. Kruck and M. Höfler, Angew. Chem., Int. Eng. Ed. **3** (1964) 701. See also (b) K. K. Joshi and P. L. Pauson, Z. Naturforsch. **17b** (1962) 565. (c) A. N. Nesmeyanov, K. N. Anisimov, N. E. Kolobova and I. S. Kolomnokov, Izvest. Akad. Nauk SSSR, Otdel, Khim. Nauk **194** (1963). (d) P. Szabo, L. Fekete, G. Bor, Z. Nagy-Magos and L. Marko, J. Organomet. Chem. **12** (1968) 245.
58. T. Kruck and M. Höfler, Chem. Ber. **97** (1964) 2289.
59. A. G. Osborne and M. H. B. Stiddard, J. Chem. Soc. (1964) 634.
60. (a) J. A. Gladysz, W. Tam, G. M. Williams, D. L. Johnson and D. W. Parker, Inorg. Chem. **18** (1979) 1163. (b) J. D. Atwood, R. J. Ruszczyk and B. L. Huang, J. Organomet. Chem. **299** (1986) 205.
61. (a) E. M. Kosower, Introduction to Physical Organic Chemistry, Wiley, New York, 1968. (b) C. Reichardt, Liebig Ann. **752** (1971) 64.
62. (a) J. L. Hughey, C. P. Anderson and T. J. Meyer, J. Organomet. Chem. **125** (1977) C49. (b) R. W. Wegman, R. J. Olsen, D. R. Gard, L. R. Faulkner and T. L. Brown, J. Am. Chem Soc. **103** (1981) 6089. (c) H. W. Walker, R. S. Herrick, R. J. Olsen and T. L. Brown, Inorg. Chem. **23** (1984) 3748.
63. (a) H. Yesaka, T. Kobayashi, K. Yasufuku and S. Nagakura, J. Am. Chem. Soc. **105** (1983) 6249. (b) W. L. Waltz, O. Hackelberg, L. M. Dorfman and A. Wojcicki, J. Am. Chem. Soc. **100** (1978) 7259.
64. B. A. Narayanan, C. A. Amatore and J. K. Kochi, Organometallics **3** (1984) 802. See also Ref. 13.
65. J. M. Hanckel, K.-W. Lee, P. Rushman and T. L. Brown, Inorg. Chem. **25** (1986) 1852.
66. (a) A. Fox, J. Malito, A. Poë, J. Chem. Soc. Chem. Commun. (1981) 1052. (b) M. S. Wrighton, D. S. Ginley, J. Am. Chem. Soc. **97** (1975) 2065. (c) D. R. Kidd and T. L. Brown in Ref. 10c. (d) S. B. McCullen, H. W. Walker and T. L. Brown, J. Am. Chem. Soc. **104** (1982) 4007. (e) B. H. Byers and T. L. Brown, J. Am. Chem. Soc. **99** (1977) 2527. (f) N. W. Hoffman and T. L. Brown, Inorg. Chem. **17** (1978) 613. (g) M. Absi-Halabi and T. L. Brown in Ref. 10b. (h) Q.-Z. Shi, T. G. Richmond, W. C. Trogler and F. Basolo in Ref. 28c.
67. G. Zotti, S. Zeechin and G. Pilloni, J. Organomet. Chem. **246** (1983) 61.
68. Compare (a) D. R. Kidd, C. P. Cheng and T. L. Brown, J. Am. Chem. Soc. **100** (1978) 4103. (b) S. B. McCullen and T. L. Brown in Ref. 34a. (c) D. J. Kuchynka, C. Amatore and J. K. Kochi, J. Organomet. Chem. in Ref. 27b.
69. (a) J. E. Ellis, J. Organomet. Chem. **86** (1975) 1. (b) T. A. Manuel, Adv. Organomet. Chem. **3** (1965) 181. (c) R. D. W. Kemmitt and D. R. Russell, in G. Wilkinson (Ed.), Comprehensive Organometallic Chemistry, Vol. **5**, 1, Pergamon, Oxford, 1982. (d) A. L. Pidcock, in C. A. McAuliffee (Ed.), Transition Metal Complexes of Phosphorus, Arsenic and Antimony Ligands, John Wiley & Sons, New York, 1973, p. 88.
70. (a) I. Wender, H. W. Sternberg and M. Orchin, J. Am. Chem. Soc. **74** (1952) 1216. (b) E. R. Tucchi and G. H. Gwynn, J. Am. Chem. Soc. **86** (1964) 4838. (c) W. Hieber, J. Sedlmeier and W. Abeck, Chem. Ber. **86** (1953) 700. (d) W. Hieber and J. Sedlmeier, Chem. Ber. **87** (1954) 25. (e) W. Hieber and R. Wiesboeck, Chem. Ber. **91** (1958) 1146.
71. (a) A. Sacco and M. Freni, J. Inorg. Nucl. Chem. **8** (1958) 566. (b) A. Sacco and M. Freni, Annali. Chim. **48** (1958) 218. (c) O. Vohler, Chem. Ber. **91** (1958) 1235. (d) S. Attali and R. Poilblanc, Inorg. Chim. Acta. **6** (1972) 475.

72. (a) W. Hieber and F. Freyer, Chem. Ber. **93** (1960) 462. (b) W. Hieber and F. Freyer, Chem. Ber. **91** (1958) 1230.
73. (a) J. A. McCleverty, A. Davison and G. Wilkinson, J. Chem. Soc. **A** (1969) 2610. (b) A. R. Manning, J. Chem. Soc. **A** (1968) 1135.
74. However, see S. Metlin, I. Wender and H. W. Sternberg, Nature **183** (1959) 457.
75. T. Kruck and M. Höfler, Chem. Ber. in Ref. 58.
76. (a) W. Hieber and H. Duchatsch, Chem. Ber. **98** (1965) 1745. (b) See Wender et al. in Ref. 70a.
77. (a) M. F. Mirbach, M. J. Mirbach and R. W. Wegman, Organometallics **3** (1984) 900. (b) Compare Ref. 73b and R. B. King, Inorg. Chem. **2** (1963) 936.
78. J. D. Atwood, Inorg. Chem. **26** (1987) 2918.
79. R. F. Heck, J. Am. Chem. Soc. **85** (1963) 657.
80. M. Absi-Halabi, J. D. Atwood, N. P. Forbus and T. L. Brown, J. Am. Chem. Soc. **102** (1980) 6248. (b) See also R. L. Sweany and T. L. Brown, Inorg. Chem. **16** (1977) 415.
81. (a) J. W. Hershberger and J. K. Kochi, Polyhedron **2** (1983) 919. (b) See also Ref. 26 and 27.
82. (a) R. B. King, Inorg. Chem. **2** (1963) 936. (b) Manning (Ref. 73b). (c) McCleverty et al. (Ref. 73a). (d) See M. S. Arabi, A. Maisonnat, S. Attali and R. Poilblanc, J. Organomet. Chem. **67** (1974) 109.
83. T. M. Bockman and J. K. Kochi, J. Am. Chem. Soc. **110** (1988) 1294.
84. Compare K. M. C. Davis, Mol. Assoc. **1** (1975) 151.
85. M. Szwarc (Ed.), Ions and Ion Pairs in Organic Reactions, Vol. **1** and **2**, Wiley, New York, 1972, 1974.
86. W. Hieber and H. Duchatsch, Chem. Ber. **98** (1965) 2530.
87. C. Schramm and J. I. Zink, J. Am. Chem. Soc. **19** (1980) 2854.
88. E. S. Lewis and C. E. Boozer, in A. Weissberger (Ed.), Investigation of Rates and Mechanisms of Reactions, Vol. **VIII** Tech. Org. Chem., Wiley, New York, 1963.
89. (a) Y. Mugnier, P. Reeb, C. Moise and E. Laviron, J. Organomet. Chem. **254** (1983) 111. (b) P. Reeb, Y. Mugnier, C. Moise and E. Laviron, J. Organomet. Chem. **273** (1984) 247.
90. R. W. Wegman and T. L. Brown, Inorg. Chem. **22** (1983) 183. Compare also R. W. Wegman, R. W. Olsen, D. R. Gard, L. R. Faulkner and T. L. Brown in Ref. 62b. H. W. Walker, R. S. Herrick, R. J. Olsen and T. L. Brown in Ref. 33. M. Absi-Halabi and T. L. Brown in Ref. 10b. For steric effect of phosphines, see C. A. Tolman, Chem. Rev. **77** (1977) 313.
91. K. Y. Lee and J. K. Kochi, Inorg. Chem., **28** (1989) 567.
92. (a) C. Amatore, D. J. Kuchynka and J. K. Kochi, J. Electroanal. Chem. **241** (1988) 181, Ref. 26b. (b) D. J. Kuchynka and J. K. Kochi in Ref. 43.
93. (a) A. G. Osborne and M. H. B. Stiddard, J. Chem. Soc. (1965) 700. (b) M. R. Snow and M. H. B. Stiddard, J. Chem. Soc. **A** (1966) 777.
94. R. A. Marcus, J. Chem. Phys. **24** (1956) 966; **26** (1957) 867; **43** (1965) 679. For a review, see R. D. Cannon, Electron Transfer Mechanisms, Butterworths, London, 1980.
95. (a) Eq. 115 is based on the use of irreversible CV peak potentials in parallel behavior to the reversible electrode potentials. See, e.g. (b) R. J. Klingler and J. K. Kochi, J. Am. Chem. Soc. **102** (1980) 4790; J. Phys. Chem. **85** (1981) 1731, (c) L. L. Miller, G. D. Nordblom and E. A. Mayeda, J. Org. Chem. **37** (1972) 917. (d) A. J. Bard and L. R. Faulkner in Ref. 52.
96. (a) R. S. Mulliken, J. Am. Chem. Soc. **74** (1952) 811. (b) R. S. Mulliken and W. B. Person, Molecular Complexes: A Reprint and Lecture Volume, Wiley, New York, 1969.
97. (a) E. M. Kosower, J. Am. Chem. Soc. **80** (1958) 3253. (b) E. M. Kosower and J. A. Skorcz, J. Am. Chem. Soc. **82** (1960) 2195. (c) E. M. Kosower and M. Mohammad, J. Phys. Chem.**74** (1970) 1153.
98. (a) F. Calderazzo, G. Pampaloni, M. Lanfranchi and G. Pelizzi, J. Organometal. Chem. **1** (1985) 296. (b) F. Calderazzo, G. Pampaloni, G. Pelizzi and F. Vitali, Organometallics **7** (1988) 1083.
99. A. Volger and H. Kunkeley, Organometallics **7** (1988) 1449.
100. D. P. Schussler, W. R. Robinson and W. F. Edgell, Inorg. Chem. **13** (1974) 153. See also Ref. 87.
101. D. Rehder, J. Organometal. Chem. **37** (1972) 303.

102. (a) R. A. Faltynek and M. S. Wrighton, J. Am. Chem. Soc. **100** (1978) 2701. (b) PPN = bis-triphenylphosphoranyideneammonium, $(Ph_3P)_2N^+$.
103. Y. S. Sohn, D. N. Hendrickson and H. B. Gray, J. Am. Chem. Soc. **93** (1971) 3603.
104. E. O. Fischer, F. Scherer and H. O. Stahl, Chem. Ber. **93** (1960) 2065.
105. T. M. Bockman and J. K. Kochi, J. Am. Chem. Soc. **111** (1989) 4669.
106. Specifically, NCP^+ = 4-cyano-N-methylpyridinium, Q^+ = N-methyl-1-quinolinium, PP^+ = 4-phenyl-N-methylpyridinium, iQ^+ = N-methyl-2-quinolinium, CMP^+ = 4-carbomethoxy-N-pyridinium, NCP^+ = 4-cyano-N-Methylpyridinium, $\phi_2Cr^+ = (C_6H_6)_2Cr^+$ generally isolated and used as the iodide or chloride salts.
107. E_a for $Mn(CO)_5^-$, $V(CO)_6^-$, $Co(CO)_4^-$ and I^- are −0.11, 0.06, 0.33 and 0.36 V vs SCE at a scan rate of $v = 500$ mV s^{-1}.
108. M. W. Hanna and J. L. Lippert, in R. Foster (Ed.), Molecular Complexes, Vol. **I**, Elek Science, London, 1973, p. 1 ff.
109. See R. Foster, Organic Charge-Transfer Complexes, Academic, New York, 1969.
110. (a) P. Klüfers, Z. Kristallogr. **167** (1984) 275. (b) P. Klüfers, Z. Kristallogr. **167** (1984) 253. (c) P. Klüfers, Z. Kristallogr. **165** (1983) 217.
111. F. Calderazzo, G. Fachinetti, F. Marchetti and P. F. Zanazzi, J. Chem. Soc., Chem. Commun. (1981) 181.
112. (a) Compare, P. E. Riley and R. E. Davis, J. Organomet. Chem. **152** (1978) 209. (b) H. Kobayashi, F. Marumo and Y. Saito, Acta Cryst. **B27** (1971) 373.
113. T. A. Albright, J. K. Burdett and M. H. Whangbo, Orbital Interactions in Chemistry, Wiley, New York, 1985.
114. H. B. Chin and R. Bau, J. Am. Chem. Soc. **98** (1976) 2434 (see footnote 11).
115. W. F. Edgell, J. Lyford, A. Barbetta and C. I. Jose, J. Am. Chem. Soc. **93** (1971) 6403.
116. (a) P. S. Braterman, Metal Carbonyl Spectra, Academic, New York, 1975. For other examples of distorted carbonylmetallates, see (b) K. H. Pannell, D. Jackson, J. Am. Chem. Soc. **98** (1976) 4443. (c) G. B. McVicker, Inorg. Chem. **14** (1975) 2087. (d) M. Y. Darensbourg, D. J. Darensbourg and H. L. C. Barros, Inorg. Chem. **17** (1978) 297. (e) M. Y. Darensbourg, Prog. Inorg. Chem. **13** (1985) 221.
117. J. E. Gordon, Organic Chemistry of Electrolyte Solutions, Wiley, New York, 1975.
118. W. F. Edgell, S. Hegde and A. Barbetta, J. Am. Chem. Soc. **100** (1978) 1406.
119. M. Darensbourg, H. Barros and C. Borman, J. Am. Chem. Soc. **99** (1977) 1647.
120. (a) J. A. Dean (Ed.), Lange's Handbook of Chemistry, McGraw Hill, 1979, 12th ed. (b) For the related E_T values, see C. Reichardt, Solvent Effects in Organic Chemistry, Verlag Chemie, New York, 1979.
121. (a) R. S. Drago and N. J. Rose, J. Am. Chem. Soc. **81** (1959) 6138. (b) Note that this measurement makes no distinction between SSIP and free ions, or of ion triplets, etc. See e.g., H. C. Wang, G. Levin, M. Szwarc, J. Am. Chem. Soc. **100** (1978) 6137.
122. See J. M. Masnovi and J. K. Kochi, J. Am. Chem. Soc. **107** (1985) 7880.
123. E. E. Wegner and A. W. Adamson, J. Am. Chem. Soc. **88** (1966) 394.
124. G. L. Geoffroy and M. S. Wrighton, Organometallic Photochemistry, Academic, New York, 1979.
125. T. J. Meyer and J. V. Caspar, Chem. Rev. **85** (1985) 187.
126. L. J. Rothberg, N. J. Cooper, K. S. Peters and V. Vaida, J. Am. Chem. Soc. **104** (1982) 3536.
127. W. L. Waltz, O. Hackelberg, L. M. Dorfman and A. Wojcicki, J. Am. Chem. Soc. **100** (1978) 7259.
128. (a) S. P. Church, M. Poliakoff, J. A. Timney and J. J. Turner, J. Am. Chem. Soc. **103** (1981) 7515. (b) A variety of other 17-e carbonylmetal radicals commonly show low-energy absorptions in the spectral region $\lambda \sim 600$–800 nm. See M. C. Baird in Ref. 21 b and references therein.
129. The spectral observation of $Co(CO)_4$ was unfortunately not extended beyond $\lambda = 700$ nm by (L. A. Hanlan, H. Huber, E. P. Kundig, B. R. McGarvey, G. A. Ozin, J. Am. Chem. Soc. **97** (1975) 7054. (b) For the complex photochemistry of $Co_2(CO)_8$, see R. W. Wegman and T. L. Brown, in Ref. 90.

130. (a) R. F. Cozzen, and T. A. Gover, J. Phys. Chem. **74** (1970) 3003. In this study, the high-energy band of Q. at λ 400 nm is obscured by the very strong absorption by I_2^-. (b) In CT salts, the observation of the maximum of the high-energy band is complicated by tail absorptions of $Mn(CO)_5^-$ and $Co(CO)_4^-$.
131. M. Itoh and S. Nagakura, Bull. Chem. Soc. Jpn. **39** (1966) 369.
132. E. M. Kosower, Topic Curr. Chem. **112** (1983) 117.
Compare (a) E. M. Kosower and L. Lindqvist, Tetrahedron Lett. (1965) 4481; (b) T. W. Ebbesen and G. Ferraudi, J. Phys. Chem. **87** (1983) 3717. (c) Cozzens and Giver in Ref. 130.
134. (a) J. A. S. Howell and P. M. Burkinshaw, Chem. Revs. **83** (1983) 557. (b) A. Davison and J. E. Ellis, J. Organomet. Chem. **31** (1971) 239. (c) F. Ungvary and A. Wojcicki, J. Am. Chem. Soc. **109** (1987) 6848.
135. Compare, e.g. T. Koenig and H. Fischer, in J. K. Kochi (Ed.), Free Radicals, Vol. **1**, Wiley, New York, 1973, p. 157ff.
136. (a) From cyclic voltammetry $E_a = 0.33$ and -0.28 V at 500 mV s^{-1} for $Co(CO)_4^-$ and $Co(CO)_3PPh_3^-$, respectively. (b) See K. Y. Lee and J. K. Kochi in Ref. 91.
137. U. Koelle, J. Organomet. Chem. **152** (1978) 225.
138. (a) P. Reeb, Y. Mugnier, C. Moise and E. Laviron in Ref. 89p. (b) See also K. Y. Lee in Ref. 91.
139. See also A. E. Stiegman and D. R. Tyler, Coord. Chem. Rev. **217** (1985) 63.
140. (a) I. Carelli, M. E. Cardinali, A. Casini and A. Arnone, J. Org. Chem. **41** (1976) 3967. (b) S. Kato, J. Nakaya and E. Imoto, J. Electrochem. Soc. Jpn. **46** (1972) 708. (c) Pyridinyl radicals formed in the CT photochemistry presumably dimerized [132], although their fate was not established.
141. Q. Z. Shi, T. G. Richmond, W. C. Trogler and F. Basolo, J. Am. Chem. Soc. **106** (1984) 71.
142. C. A. Tolman, Chem. Soc. Rev. **1** (1972) 337.
143. (a) J. P. Collman and L. S. Hegedus in Ref. 19. (b) A. Yamamoto, Organotransition Metal Chemistry, Wiley, New York, 1986. (c) L. Brandsma and H. Verkruijsse, Preparative Polar Organometallic Chemistry, Springer-Verlag, Berlin, 1987.
144. See J. K. Kochi, Angew. Chem. Int. Ed. Eng. **27** (1988) 1227.

CHAPTER 8

Electron Spin Resonance Studies of Organometallic Radicals

Philip H. Rieger

Department of Chemistry, Brown University, Providence, RI 02912, U.S.A.

Table of Contents

1. Introduction

Electron spin resonance (ESR) spectroscopy has been used for over 40 years to study organic radicals, transition metal complexes, and a variety of other paramagnetic species, but most of the applications to organotransition metal radicals have been in the last decade. Although ESR spectroscopy is supposed to be a mature field with a fully developed theory [1], there have been some surprises as organometallic problems have explored new domains in ESR parameter space. This review will present a synopsis of the fundamentals of ESR spectroscopy, emphasizing those aspects which are particularly relevant to organometallic radicals and those which have presented surprises to the spectroscopist. Most of the examples are taken from work in which the author has been involved. Inevitably, there will be points left untouched and theoretical developments which start at the wrong point for the reader. In such cases, please refer to one of the excellent texts on ESR spectroscopy [2–7].

The electron spin resonance spectrum of a free radical with one unpaired electron is the simplest of all forms of spectroscopy. The observed transitions are always between the electron spin energy levels characterized by the quantum numbers $m_s = \pm 1/2$. If unpaired electrons in radicals were indistinguishable from free electrons, the only information content of an ESR spectrum would be the integrated intensity, proportional to the radical concentration. Fortunately, an unpaired electron interacts with its environment, and the details of ESR spectra depend on the nature of those interactions.

There are two kinds of environmental interactions which are important in the ESR spectrum of a free radical: (i) To the extent that the unpaired electron has unquenched orbital angular momentum, the total magnetic moment is different from the spin-only moment (either larger or smaller, depending on how the angular momentum vectors couple). It is customary to lump the orbital and spin angular momenta together in an effective spin and to treat the effect as a shift in the energy of the spin transition. (ii) The electron spin energy levels may be split by nuclear spin magnetic moments – the nuclear hyperfine interaction. Each interacting nucleus of spin I splits the electron spin energy levels into $(2I + 1)$ sublevels. Since the transitions are observed between sublevels with the same values of m_I, nuclear spin splitting of energy levels is mirrored by splitting of the resonance line.

2. Isotropic ESR Spectra

An isotropic ESR spectrum – the spectrum of a freely tumbling radical in liquid solution – can contain several kinds of chemically useful information: (i) The hyperfine coupling pattern provides information on the numbers and kinds of magnetic nuclei with which the unpaired electron interacts. (ii) The spacing of the lines and the center of gravity of the spectrum yield the hyperfine coupling constants, $\langle A_i \rangle$, and g-value, $\langle g \rangle$, which are related to the way in which the unpaired electron spin density is distributed in the molecule. (iii) The integrated intensity of the spectrum is proportional to the concentration of radicals in the sample. (iv) The spectral linewidths are related to the rate of rotational motions, which average anisotropies in the g- and hyperfine tensors (see § 3), and to the rates of fluxional processes, which average nuclear positions in a radical. (v) The saturation behavior of a spectrum – the variation of integrated intensity with microwave power – is related to the spin-lattice relaxation time, a measure of the rate of energy transfer between the electron spin and its surroundings.

The degree to which these kinds of information have been exploited varies widely. Although experimentalists often make qualitative observations relating "strong" or "weak" spectra to chemical circumstances, to the best of my knowledge, there have been no quantitative applications of integrated intensities in the study of organometallic radicals. This could be a useful analytical approach in some cases. Lacking precedents, the interested reader is referred to Wertz and Bolton [5], who discuss the technique, point out the variables which must be controlled, and suggest intensity standards.

Largely because spin-orbit coupling results in spin state admixture, electronic spin-lattice relaxation times are normally very short in organometallic radicals. This has two consequences. Short relaxation times mean that saturation problems, which plague ESR spectroscopists studying organic radicals and NMR spectroscopists in general, are largely absent in organometallic ESR studies. Thus spectra ordinarily can be recorded at full microwave power with salutary consequences for sensitivity. On the other hand, relaxation times are most easily determined by measuring spectral intensity as a function of microwave power in the saturation region. If relaxation times are short, very high power is required, out of the range of operation of most spectrometers. To the best of my knowledge, there has been no serious attempt either to use ESR to measure relaxation times in organometallic radical systems or to exploit such measurements to gain insights into electronic structure and dynamics.

2.1 Line Positions in Isotropic Spectra

ESR spectra of radicals in liquid solution are usually interpreted in terms of a spin Hamiltonian

$$\mathscr{H} = \mu_B \langle g \rangle \vec{B} \cdot \vec{S} + \sum_i \langle A_i \rangle \vec{I}_i \cdot \vec{S} \tag{1}$$

where μ_B is the Bohr magneton. The first term of Eq. (1) represents the Zeeman interaction of the effective electron spin, $\vec{S}$, and the applied magnetic field, $\vec{B}$, and the second represents the nuclear hyperfine interaction of $\vec{S}$ and the nuclear spins, $\vec{I}_i$. The spectral information is contained in the parameters, $\langle g \rangle$, the effective gyromagnetic ratio or g-value, and $\langle A_i \rangle$, the electron-nuclear hyperfine coupling constant for nucleus i. With the use of spin functions based on the quantum numbers m_s and m_i, Eq. (1) can be used to compute energy levels. Equating energy differences for the allowed transitions ($\Delta m_s = \pm 1, \Delta m_i = 0$) with the microwave photon energy,

$$E(m_s = 1/2) - E(m_s = -1/2) = h\nu \tag{2}$$

the resonant magnetic field can be predicted. To first-order in perturbation theory, the resonant field is

$$B = B_0 - \sum_i \langle a_i \rangle m_i \tag{3}$$

where $B_0 = h\nu/\langle g\rangle \mu_B$ represents the center of the spectrum and $\langle a_i\rangle = \langle A_i\rangle/\langle g\rangle \mu_B$ is the hyperfine coupling constant in field units.

The coupling constant in Eq. (1) has energy units, but the energies are very small so that frequency (MHz) or wave number (10^{-4} cm^{-1}) units are more commonly used. Even more often, however, isotropic coupling constants are given in units of magnetic field strength, usually gauss, though *SI* purists sometimes use millitesla (1 mT = 10 G). Conversions from units of gauss to frequency or wave number units involve the *g*-value

$$\langle A\rangle/\text{MHz} = 2.8025(\langle g\rangle/g_e)\,\langle a\rangle/\text{G} \tag{4a}$$

$$\langle A\rangle/10^{-4}\,\text{cm}^{-1} = 0.93480(\langle g\rangle/g_e)\,\langle a\rangle/\text{G} \tag{4b}$$

Note that coupling constants in 10^{-4} cm^{-1} are comparable in magnitude to those expressed in gauss. Although the units used for isotropic hyperfine coupling constants is largely a matter of taste, the components of an anisotropic hyperfine coupling tensor (see § 3) should be given in frequency or wave number units unless the *g*-tensor is virtually isotropic.

2.1.1 Hyperfine Coupling Patterns. Nuclear hyperfine coupling results in a multi-line ESR spectrum, analogous to the spin-spin coupling multiplets of NMR spectra. ESR spectra are simpler to understand than NMR spectra in that second-order effects normally do not alter the intensities of components; on the other hand, ESR multiplets can be much more complex when the electron interacts with several high-spin nuclei, and, as we will see in Section 2.2, there can be considerable variation in linewidth within a spectrum. However, just as in NMR, a multiplet pattern gives an important clue to the identity of a radical.

For example, the series of radicals derived from the nucleophilic displacement of $Co(CO)_4^-$ from $[(RCCR')CO_2(CO)_6]^-$ by CO, phosphines, or phosphites was identified in large part from the resulting hyperfine patterns [8]. Depending on the nucleophile, spectra were observed which showed coupling to 0, 1, 2, or 3 ^{31}P nuclei, but always to a single ^{59}Co nucleus. The parameters $\langle g\rangle$ and $\langle a^{\text{Co}}\rangle$ were found to depend slightly on the substituents on the acetylene moiety, which suggests that the acetylene was retained in the radical, but otherwise the parameters were nearly constant over the series, which suggests a single family of radicals. The appearance of 0–3 phosphorus couplings suggested three additional ligands, either CO or a phosphine or phosphite. Taken together, this information identified the radicals as $(RCCR')CoL_3$, where L = CO, PR_3, or $P(OR)_3$. Fig. 1a shows the experimental first-derivative spectrum of $(Ph_2C_2)Co(CO)_2P(OMe)_3$, *1*, and Fig. 1c shows a first-order "stick spectrum" ($\langle g\rangle = 2.061$, $\langle a^{\text{Co}}\rangle = 45.2$ G, $\langle a^{\text{P}}\rangle = 176.2$ G). Hyperfine coupling to ^{59}Co ($I = 7/2$) and ^{31}P ($I = 1/2$) nuclei results in an octet of doublets. The spectrum is somewhat complicated by the obvious variation in linewidth, but the assignments are quite straightforward.

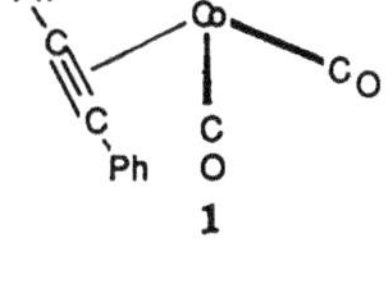

1

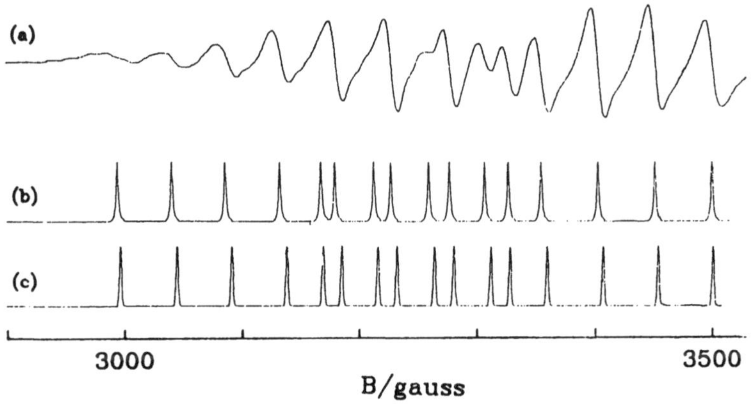

Fig. 1. (a) Isotropic ESR spectrum of $(Ph_2C_2)Co(CO)_2P(OMe)_3$ in THF solution at 260 K (from Reference 7; reproduced with permission); (b) Second-order "stick spectrum"; (c) First-order "stick spectrum".

When several magnetically equivalent nuclei are present in a radical, some of the multiplet lines are degenerate, which results in variations in component intensity. Equivalent spin 1/2 nuclei such as ^1H, ^{19}F, or ^{31}P result in multiplets with intensities given by binomial coefficients (1 : 1 for two nuclei, 1 : 2 : 1 for two, 1 : 3 : 3 : 1 for three, etc.). Nuclei with $I > 1/2$ give less familiar multiplet intensity ratios. Thus, for example, three equivalent ^{59}Co nuclei ($I = 7/2$) give (to first order) 22 lines with intensity ratios 1:3:6:10:15:21:28:36:42:46:48:48:46 ..., as shown in Fig. 2b. The experimental spectrum of $[PhCCo_3(CO)_9]^-$, *2* [9], which shows this coupling pattern, is given in Fig. 3.

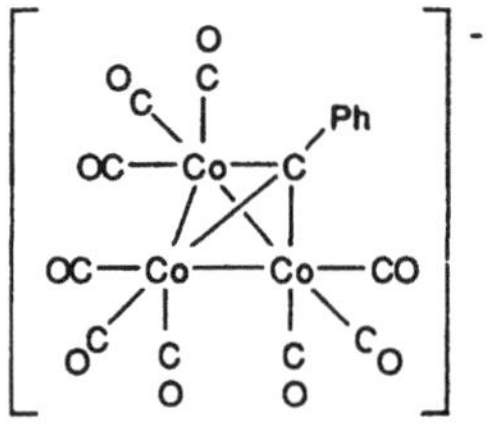

2

2.1.2 Spin Hamiltonian Parameters from Spectra. Once a hyperfine pattern has been recognized, the line position information can be summarized by the spin Hamiltonian parameters, $\langle g \rangle$ and $\langle a_i \rangle$. These parameters can be extracted from spectra by a linear least-squares fit of experimental line positions to Eq. (3). However, for high-spin nuclei and/or large couplings, one soon finds that the lines are not evenly spaced as predicted by Eq. (3) and second-order corrections must be made. Solving the spin Hamiltonian, Eq. (1), to second order, Eq. (3) becomes

$$B = B_0 - \sum_i \{\langle a_i \rangle m_i + [\langle a_i \rangle^2/2B] [I_i(I_i + 1) - m_i^2] + \ldots\} \quad (5)$$

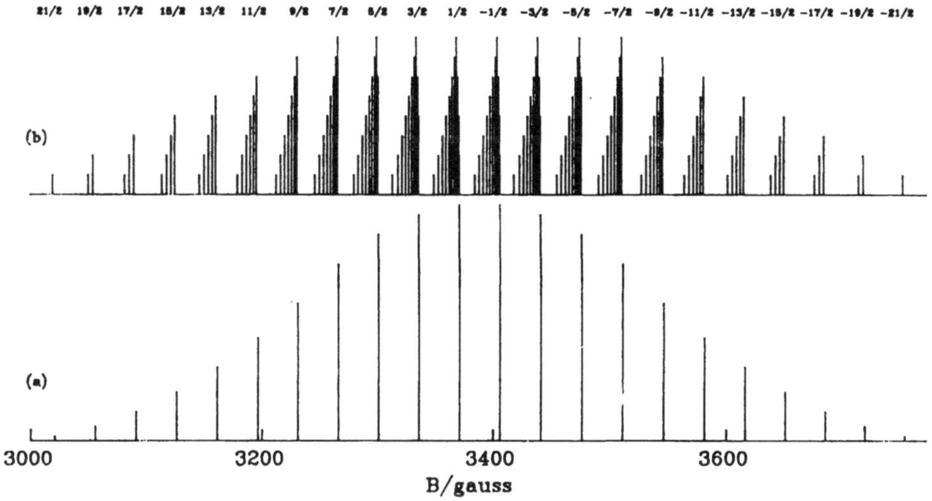

Fig. 2. "Stick spectrum" showing hyperfine pattern for coupling to three equivalent ^{59}Co nuclei ($I = 7/2$) computed to (a) second-order and (b) first-order in perturbation theory.

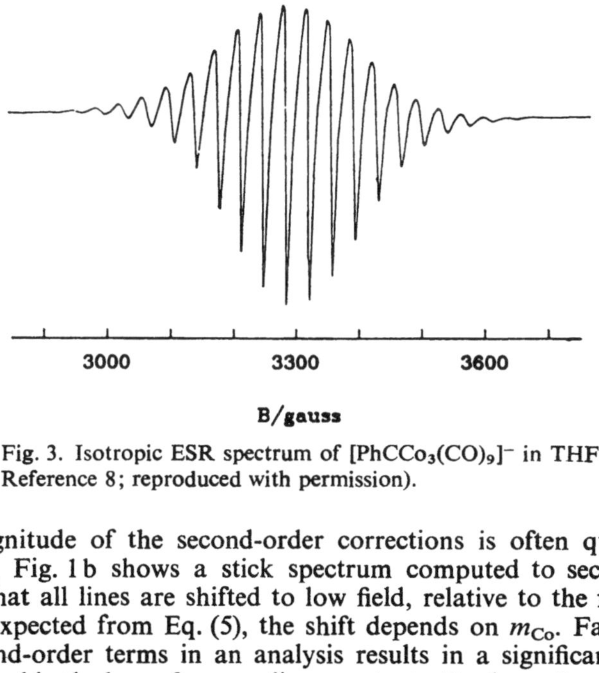

Fig. 3. Isotropic ESR spectrum of $[PhCCo_3(CO)_9]^-$ in THF solution at 40 °C (from Reference 8; reproduced with permission).

The magnitude of the second-order corrections is often quite significant. For example, Fig. 1b shows a stick spectrum computed to second-order for *1*. [8] Notice that all lines are shifted to low field, relative to the first-order spectrum, but, as expected from Eq. (5), the shift depends on m_{Co}. Failure to account for the second-order terms in an analysis results in a significant error in both the g-value and in the hyperfine coupling constants. For large I_i and $\langle a_i \rangle$, well-resolved spectra may warrant inclusion of third- or higher-order corrections. Since the third-order corrections involve cross terms among coupling constants, in principle they permit determination of the relative signs of the coupling constants. However, in the example of Fig. 1, the third-order corrections amount to ca. 0.1–0.2 G, too small to be significant given the resolution of the experimental spectrum. When coupling constants are really large, it is better to fit line positions to the exact energy levels given by the Breit-Rabi equation [10, 11].

2.1.3 Second-order Splittings. Equation (5) describes line positions correctly for spectra with hyperfine coupling to two or more nuclei provided that the nuclei are not magnetically equivalent. When two or more nuclei are completely equivalent, i.e., both instantaneously equivalent and equivalent over a time average, then the nuclear spins should be described in terms of the total nuclear spin quantum numbers J and m_J rather than the individual I_i and m_i. In this "coupled representation", the degeneracies of some multiplet lines are lifted when second-order shifts are included. This can lead to extra lines and/or asymmetric line shapes. For example, the three equivalent spin 7/2 ^{59}Co nuclei in tricobalt-carbon radical anions should be described in terms of 11 J-states with J ranging from 21/2 to 1/2. The $m_J = 17/2$ feature, for example, has three components with $J = 21/2$, 19/2, and 17/2, degeneracies of 1, 2, and 3, and second-order shifts of 97 $\langle a \rangle^2/4B$, 55 $\langle a \rangle^2/4B$, and 17 $\langle a \rangle^2/4B$, respectively. The shifts are too small to be resolved, but they lead to an asymmetric absorption line envelope with apparent broadening on the low-field side, as shown in Fig. 2a and as is observed in the experimental spectrum of *2*, shown in Fig. 3 [10].

2.1.4 Interpretation of Isotropic Parameters. Proton coupling constants in the spectrum of an organic π-radical are directly proportional to the spin density in the p-orbital of the adjacent carbon atom,

$$\langle a^H \rangle = Q_{CH}{}^H \rho_C{}^\pi \quad (6)$$

It is sometimes assumed that metal or ligand hyperfine couplings are similar measures of spin densities in organometallic radicals. Unfortunately, such an assumption is usually unwarranted. An isotropic hyperfine coupling has three contributions: (i) Fermi contact interaction between the nuclear spin and electron spin density in the valence-shell s-orbital; (ii) Fermi contact interaction between the nuclear spin and spin density in inner-shell s-orbitals arising from spin polarization by unpaired electron density in valence-shell p- or d-orbitals; and (iii) A contribution from spin-orbit coupling. The first contribution is positive (for a positive nuclear magnetic moment), the second is normally negative, and the third may be of either sign. Because direct involvement of hydrogen 1s character in the SOMO of an organic π-radical is symmetry-forbidden and spin-orbit coupling is negligible in carbon-based radicals, proton couplings in such radicals result solely from spin polarization and thus are proportional to the polarizing spin density. All three contributions are usually significant for organometallic radicals. Although there are a few cases where polarization constants, analogous to $Q_{CH}{}^H$, have been estimated [12], they are of use only in a more complete analysis based on the results of a solid state ESR study.

As we will see below, g-tensors are often difficult to interpret reliably. The interpretation of isotropic g-values is even more dangerous. Thus, isotropic ESR spectra should be used to characterize a radical by means of the hyperfine coupling pattern, to study its dynamical properties through linewidth effects, or to measure its concentration by integration of the spectrum and comparison with an appropriate standard, but considerable caution should be exercised in interpreting the g-value or nuclear hyperfine coupling constants.

2.2 Linewidths in Isotropic Spectra

2.2.1 Incomplete Averaging of Anisotropies. Isotropic ESR spectra of organotransition metal radicals often show substantial variations in linewidth from one hyperfine component to another (see, for example, Figs. 1a and 3). The widths can often be expressed as a power series in the nuclear spin quantum numbers,

$$W = \alpha + \sum_i [\beta_i m_i + \gamma_i m_i^2 + \ldots] \tag{6}$$

The major linewidth contribution in solution spectra usually arises from incomplete averaging of anisotropies in the *g*- and hyperfine tensors (see § 3); the parameters of Eq. (6) then depend on $\Delta g = g_{\parallel} - g_{\perp}$, $\Delta A_i = A_{i,\parallel} - A_{i,\perp}$ and τ_R, the rotational correlation time,

$$\alpha - \alpha_0 \propto (\Delta g)^2 \tau_R \tag{7a}$$

$$\beta_i \propto \Delta g \, \Delta A_i \, \tau_R \tag{7b}$$

$$\gamma_i \propto (\Delta A_i)^2 \tau_R \tag{7c}$$

Since these terms are proportional to τ_R, they increase with decreasing temperature. There are several linewidth contributions, included in α_0, which do not depend on m_i. These include spin-rotation interaction which increases with $1/\tau_R$ and thus with increasing temperature. These and other linewidth effects have been studied in some detail and are discussed elsewhere [10, 13, 14].

If the *g*- and hyperfine tensor anisotropies are known from analysis of a solid-state spectrum, the linewidth parameters β_i and γ_i can be used to compute the rotational correlation time, τ_R, which can be a useful measure of freedom of motion. Linewidths in ESR spectra of nitroxide spin labels, for example, have been used to probe the motional freedom of biological macromolecules [15]. Since τ_R is related to the molecular hydrodynamic volume, V_h, and the solution viscosity, η, by

$$\tau_R = \eta V_h / kT \tag{8}$$

ESR linewidths can provide a measure of the effective size of a paramagnetic molecule or ion, useful information when there is a suspicion that a radical has polymerized. For example, vanadium(IV) shows an eight-line ESR spectrum in basic aqueous solution, suggesting a monomeric unit, $VO(OH)_3(H_2O)_2^-$ [16], although previous work on the system had indicated polymers such as $V_3O_7^{2-}$ or $V_4O_9^{2-}$. Comparison of values of τ_R from the basic solution linewidths with those obtained from spectra of acidic solutions containing $VO(H_2O)_5^{2+}$ showed that the hydrodynamic volume of the aquo cation is actually about twice that of the basic solution species. This effectively rules out the presence of ESR-active polymers in solution [17].

2.2.2 Rates of Fluxionality from Linewidths. ESR linewidths are also sensitive to processes that modulate the *g*-value or hyperfine coupling constants or limit the lifetime of the electron spin state. The effects are closely analogous to those observed in NMR spectra of dynamical systems. However, since ESR linewidths are typically on the order of 0.1–10 G (0.3–30 MHz), rate processes which give observable increases in linewidths must be fast. Bimolecular processes which contribute to ESR linewidths have mostly been nearly diffusion-controlled, e.g., intermolecular electron exchange between naphthalene and its anion radical [18] and reversible axial ligation of square planar copper(II) complexes [19].

The effect of rate processes on linewidths can be understood quantitatively in terms of the modified Bloch equations [20], or, more accurately, in terms of density matrix [21] or relaxation matrix [13, 22] formalisms. If a rate process modulates a line position through an amplitude ΔB ($\Delta\omega$ in angular frequency units) by fluctuating between two states, each with lifetime τ, a single line is observed with an excess width proportional to $(\Delta\omega)^2 \tau$ when $\tau^{-1} \gg \Delta\omega$ – the fast exchange limit. As the lifetime increases, the line broadens to indetectability and then re-emerges as two broad lines. These shift apart and sharpen until, in the slow exchange limit ($\tau^{-1} \ll \Delta\omega$), two lines are observed with widths proportional to τ^{-1}.

The ESR spectrum of (η^3-cyclooctenyl)-Fe[P(OMe)$_3$]$_3$, *3*, reported by Krusic and coworkers [23], provides a good illustration of dynamical linewidth effects. This radical, produced by sodium amalgam reduction of the corresponding Fe(0) cation, showed a 1:3:3:1 quartet spectrum ($\langle a^P \rangle = 20$ G) when dissolved in tridecane at 140 °C, Fig. 4. Under these conditions, the central lines were substantially broader than the outer lines, as expected if a fluxional process rapidly exchanges instantaneously nonequivalent phosphorus nuclei. As the temperature was lowered, the central lines broadened further until, at 50 °C, only a doublet remained. In pentane solution at 23 °C, the doublet lines had developed satellites which grew until, at −56 °C, the spectrum showed a doublet of 1:2:1 triplets ($\langle a_1^P \rangle = 100.5$ G, $\langle a_2^P \rangle = -19.6$ G). Since these couplings average to 20 G, they must be of opposite sign. On further cooling the central lines of the triplets broadened, virtually disappearing at −100 °C. Finally, in 1:1 cyclopropane: ethane at −140 °C, eight lines were observed, a doublet of doublets of doublets, indicating

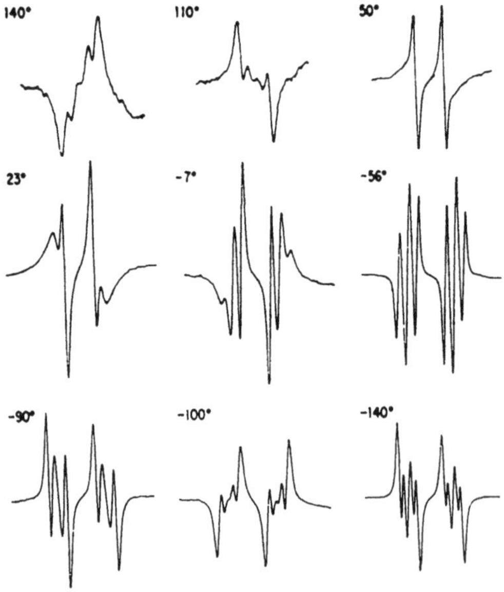

Fig. 4. ESR spectra of (η^4-cyclooctenyl)-Fe[P(OMe)$_3$]$_3$ in tridecane (140°, 110, and 50 °C), pentane (23°, −7°, −56°, −90°, and −100 °C), and cyclopropane/ethane (−140 °C) (from Reference 22; reproduced with permission).

three nonequivalent phosphorus nuclei ($\langle a^P \rangle = 99.6, -27.9, -10.5$ G). These results suggest two fluxional processes, a fast process which averages the two phosphorus nuclei with negative couplings, and a slower one which averages all three phosphorus nuclei. Detailed analysis gave (at 298 K) $k_1 = 2 \times 10^8$ s^{-1} ($E_a =$ 21 kJ mol^{-1}), $k_2 = 1 \times 10^{10}$ s^{-1} ($E_a = 17$ kJ mol^{-1}). This is a somewhat unusual case in that the radical was sufficiently stable that it could be studied over a wide temperature range in several solvents; in the absence of contributions to the widths from incomplete averaging of metal hyperfine anisotropies, the nonkinetic linewidths were relatively sharp and constant from one component to another. Kinetic linewidth contributions are not always so obvious.

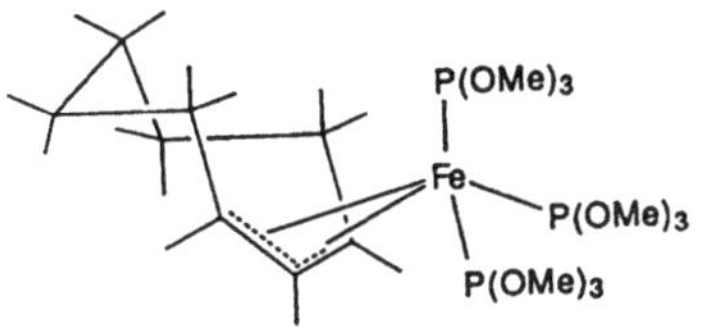

3

Casagrande, et al. [8], used linewidth effects to study the rate of fluxionality in $(Ph_2C_2)Co(CO)[P(OMe)_3]_2$, *4*. The experimental spectrum, shown in Fig. 5a, can be described as a 1:2:1 triplet of octets; the spectrum is complicated by a large linewidth dependence on m_{Co}, but, as demonstrated in Figs. 5b and 5c, the central lines of the triplets are much broader than the outer lines. This radical

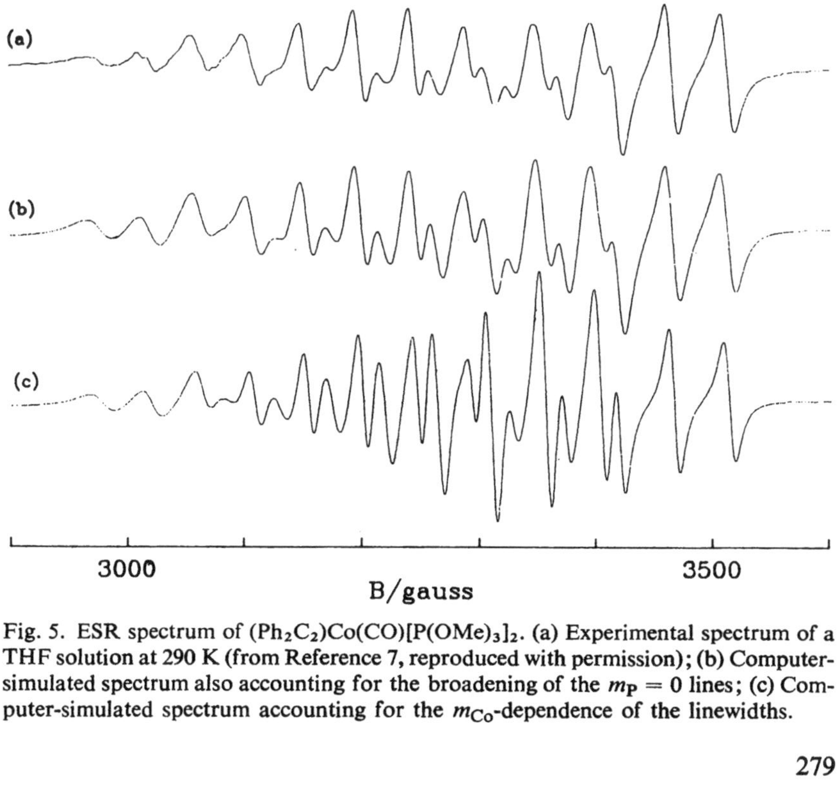

Fig. 5. ESR spectrum of $(Ph_2C_2)Co(CO)[P(OMe)_3]_2$. (a) Experimental spectrum of a THF solution at 290 K (from Reference 7, reproduced with permission); (b) Computer-simulated spectrum also accounting for the broadening of the $m_P = 0$ lines; (c) Computer-simulated spectrum accounting for the m_{Co}-dependence of the linewidths.

has a distorted tetrahedral structure with the semi-occupied molecular orbital (SOMO) largely cobalt $3d_{z^2}$ in character [24]. Thus the ligand sites can be described as axial or equatorial relative to the unique z-axis. Several isomers are possible, but the ^{31}P couplings distinguish between the isomer with an axial phosphite (ax, eq) and those with either CO or the acetylene axial and both phosphites equatorial (eq, eq). The rate of interconversion between eq, eq and ax, eq isomers was estimated from the relative widths of the $M_P = \pm 1$ and 0 lines, given the isotropic coupling constants for the various ^{31}P nuclei (which were determined from the frozen solution spectrum [24]). The average rate of conformational change was found to be about 2×10^{10} s^{-1} (E_a = 17 kJ mol^{-1}) at 298 K.

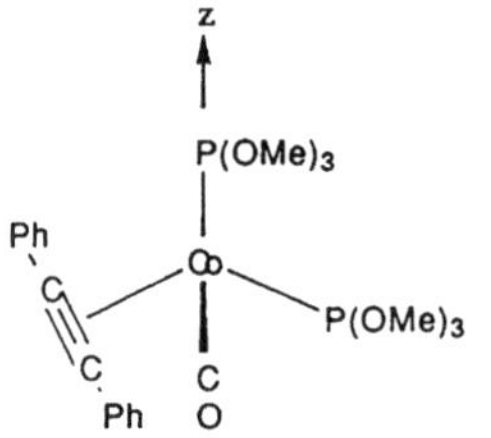

4 (ax,eq)

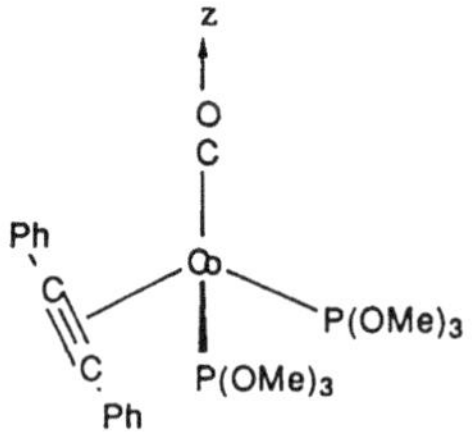

4 (eq,eq)

3. Anisotropic ESR Spectra

The anisotropies which lead to line broadening in isotropic ESR spectra influence solid-state spectra more directly. Accordingly, a more complex spin Hamiltonian is required to interpret such spectra

$$\mathscr{H} = \mu_B \vec{B} \cdot \boldsymbol{g} \cdot \vec{S} + \sum_i \vec{I}_i \cdot \boldsymbol{A}_i \cdot \vec{S} \tag{6}$$

In Eq. (6), $\boldsymbol{g}$ and $\boldsymbol{A}_i$ are 3×3 matrices representing the anisotropic Zeeman and nuclear hyperfine interactions. We will follow common practice and refer to these as the g- and hyperfine tensors, although strictly speaking, they are not proper second-rank tensors. In general, a coordinate system can be found – the tensor principal axes – in which $\boldsymbol{g}$ (or $\boldsymbol{A}_i$) is diagonal. If $\boldsymbol{g}$ and $\boldsymbol{A}_i$ are diagonal in the same coordinate system, we say that their principal axes are coincident.

In species with two or more unpaired electrons, a fine structure term must be added to the spin Hamiltonian to represent electron spin-spin interactions. We will avoid this complication by confining our attention to radicals with one unpaired electron ($S = 1/2$).

Nuclear quadrupole interactions introduce line shifts and forbidden transitions in spectra of radicals with nuclei having $I > 1/2$. In practice, quadrupolar effects are observable only in very well-resolved spectra or in spectra of radicals with nuclei having small magnetic moments and large quadrupole moments. The most extreme case of a small magnetic moment to quadrupole moment ratio is that of ^{191}Ir, ^{193}Ir, and spectra of $Ir(CN)_5^{3-}$ [25], $Ir(CN)_5Cl^{4-}$ and $Ir(CN)_4Cl_2^{4-}$ [26], and $[Ir_2(CO)_2(PPh_3)_2(\mu\text{-}RNNR)_2]^+$, R = p-tolyl [27], show easily recognizable quadrupolar effects. Other nuclei for which quadrupolar effects might be expected include $^{151}Eu/^{153}Eu$, $^{155}Gd/^{157}Gd$, ^{175}Lu, ^{181}Ta, ^{189}Os, and ^{197}Au. When quadrupolar effects are important, it is usually necessary to take account of the nuclear Zeeman interaction as well. The nuclear quadrupole and nuclear Zeeman interactions add two more terms to the spin Hamiltonian. Since these terms considerably complicate an already complex situation, we will confine our attention here to nuclei for which quadrupolar effects can be neglected.

When a radical is oriented such that the magnetic field direction is located by the polar and azimuthal angles, θ and φ, relative to the g-tensor principal axes, the resonant field is given, to first order in perturbation theory, by [28]

$$B = B_0 - \sum_i \frac{A_i m_i}{g\mu_B} \tag{7}$$

where

$$B_0 = h\nu/g\mu_B \tag{8}$$

$$g^2 = g_x^2 \sin^2\theta \cos^2\varphi + g_y^2 \sin^2\theta \sin^2\varphi + g_z^2 \cos^2\theta \tag{9}$$

$$A_i^2 = A_{iz}^2 S_{ix}^2 + A_{iy}^2 S_{iy}^2 + A_{iz}^2 S_{iz}^2 \tag{10}$$

$$S_{ik} = [g_x \sin\theta \cos\varphi l_{ixk} + g_y \sin\theta \cos\varphi l_{iyk} + g_z \cos\theta l_{izk}]/g \tag{11}$$

and the l_{ijk} are direction cosines indicating the orientation of the kth principal axis of the ith hyperfine tensor relative to the jth g-tensor principal axis. When the tensor principal axes are coincident, only one of the l_{ijk} of Eq. (11) will be nonzero. When the hyperfine tensor components are large, second-order terms [28] must be added to Eq. (7); these result in down-field shifts, proportional to m_i^2.

3.1 Solid-State ESR Spectra

So long as they are dilute (to avoid line broadening from intermolecular spin exchange), radicals can be studied in the solid state as solutes in single crystals, powders, glasses or frozen solutions. Radicals can be produced *in situ* by UV- or γ-irradiation of a suitable precursor in a crystalline or glassy matrix. While many organometallic radicals have been studied in this way [29], it is often easier to obtain solid state ESR spectra by freezing the liquid solution in which the radical is formed. A variety of techniques then can be used to generate radicals, e.g., chemical reactions, electrochemical reduction or oxidation, or photochemical methods. Furthermore, the radical is studied under conditions more closely approximating those in which its reaction chemistry is known.

3.1.1 Spectra of Dilute Single Crystals. Spectra of radicals in a dilute single crystal are obtained for a variety of orientations, usually with the field perpendicular to one of the crystal axes. Spectra usually can be analyzed as if they were isotropic to obtain an effective g-value and hyperfine coupling constants. Since the g- and hyperfine tensor principal axes are not necessarily the same as the crystal axes, the tensors, written in the crystal axis system usually will have off-diagonal elements. Thus, for example, if spectra are obtained for a variety of orientations in the crystal xy-plane, the effective g-value is

$$g_\varphi^2 = (g_{xx}\cos\varphi + g_{yx}\sin\varphi)^2 + (g_{xy}\cos\varphi + g_{yy}\sin\varphi)^2 + (g_{xz}\cos\varphi + g_{yz}\sin\varphi)^2 \tag{12}$$

or

$$g_\varphi{}^2 = K_1 + K_2 \cos 2\varphi + K_3 \sin 2\varphi \tag{13}$$

where

$$K_1 = (g_{xx}{}^2 + g_{yy}{}^2 + g_{xz}{}^2 + g_{yz}{}^2 + 2g_{xy}{}^2)/2 \tag{14a}$$

$$K_2 = (g_{xx}{}^2 - g_{yy}{}^2 + g_{xz}{}^2 - g_{yz}{}^2)/2 \tag{14b}$$

$$K_3 = g_{xy}(g_{xx} + g_{yy}) + g_{xz}g_{yz} \tag{14c}$$

A sinusoidal plot of $g_\varphi{}^2$ vs φ can be analyzed to determine K_1, K_2, and K_3. Exploration of another crystal plane gives another set of K's which depend on other combinations of the g_{ij}; eventually enough data are obtained to determine the six independent g_{ij} ($\boldsymbol{g}$ is a symmetric matrix so that $g_{ij} = g_{ji}$). The g^2-matrix then is diagonalized to obtain the principal values and the transformation matrix, elements of which are the direction cosines of the g-tensor principal axes relative to the crystal axes. An analogous treatment of the effective hyperfine coupling constants leads to the principal values of the A^2-matrix and the orientation of its principal axes in the crystal coordinate system.

3.1.2 Analysis of Powder Spectra. Since ESR spectra are normally recorded as the first derivative of absorption vs. field, dA/dB, observable features in the spectrum of a powder correspond to molecular orientations for which dA/dB is large in magnitude or changes in sign. For any spin Hamiltonian, there will be minimum and maximum resonant fields at which the absorption changes rapidly from zero, leading to a large value of the derivative and features which resemble positive-going and negative-going absorption lines. Peaks in the absorption envelope correspond to derivative sign changes and lead to features resembling isotropic derivative lines. The interpretation of a powder spectrum thus depends on the connection of the positions of these features to the g- and hyperfine tensor components.

Early treatments of powder patterns attempted to deal with the spatial distribution of resonant fields by analytical mathematics [30]. This approach led to some valuable insights but the algebra is much too complex when nonaxial hyperfine tensors are involved. Consider the simplest case: a single resonance line without hyperfine structure. The resonant field is given by Eq. (8). Features in the first derivative spectrum correspond to discontinuities or turning points in the absorption spectrum which arise when $\partial B/\partial\theta$ or $\partial B/\partial\varphi$ are zero,

$$\frac{\partial B}{\partial \theta} = \frac{h\nu}{\mu_B} \frac{g_z{}^2 - g_\perp{}^2}{g^3} \sin\theta \cos\theta = 0 \tag{15a}$$

or

$$\frac{\partial B}{\partial \varphi} = \frac{h\nu}{\mu_B} \frac{g_x{}^2 - g_y{}^2}{g^3} \sin^2\theta \sin\varphi \cos\varphi = 0 \tag{15b}$$

These equations have three solutions: (i) $\theta = 0$; (ii) $\theta = 90°$, $\varphi = 0$; and (iii) $\theta = \varphi = 90°$. Since θ and φ in Eqs. (9) and (15) are in the g-tensor axis system, observable features are expected for those fields corresponding to orientations along the principal axes of the g-tensor. This being the case, the principal values of the g-tensor are obtained from a straightforward application of Eq. (8).

Powder spectra with hyperfine structure often can be interpreted similarly with spectral features identified with orientation of the magnetic field along one of the *g*- and hyperfine tensor principal axes. However, this simple situation often breaks down. Using a first-order theory and one hyperfine coupling, Ovchinnikov and Konstantinov [31] have shown that Eqs. (15) may have up to six solutions corresponding to observable spectral features. Three of these correspond to orientation of $\boldsymbol{B}$ along principal axes, but the "extra lines" correspond to less obvious orientations. Even more extra lines may creep in when the spin Hamiltonian is treated to second order or when more than one hyperfine coupling is involved. The problem is illustrated by the resonant field vs. cos θ and φ surface shown in Fig. 6. This surface corresponds to the $m_{\mathrm{Cu}} = -3/2$ "line" in the spectrum of $\mathrm{Cu(acac)_2}$ ($g = 2.0527, 2.0570, 2.2514$; $A^{\mathrm{Cu}} = 27.0, 19.5, 193.4 \times 10^{-4}\ \mathrm{cm^{-1}}$) [31]. The minimum resonant field, $B = 3290.7$ G, corresponds to $\boldsymbol{B}$ along the *z*-axis ($\cos\theta = \pm 1$). With $\boldsymbol{B}$ along the *x*-axis ($\cos\theta = 0, \varphi = 0°$), the surface shows a saddle point at 3344.3 G, and with $\boldsymbol{B}$ along the *y*-axis ($\cos\theta = 0, \varphi = 90°$), there is a local minimum at 3325.5 G. In addition, another saddle point occurs in the *yz*-plane at $B = 3371.2$ G ($\cos\theta = \pm 0.482, \varphi = 90°$); the only maximum is in the *xz*-plane at $B = 3379.0$ G ($\cos\theta = \pm 0.459, \varphi = 0°$). Thus five features are expected and indeed are shown in the computer-simulated spectrum of $\mathrm{Cu(acac)_2}$ shown in Fig. 7. The two high-field features correspond to off-axis field orientations and thus are "extra lines". The situation is more complex when the *g*- and hyperfine tensor principal axes are noncoincident (see § 3.4.2); in this case, none

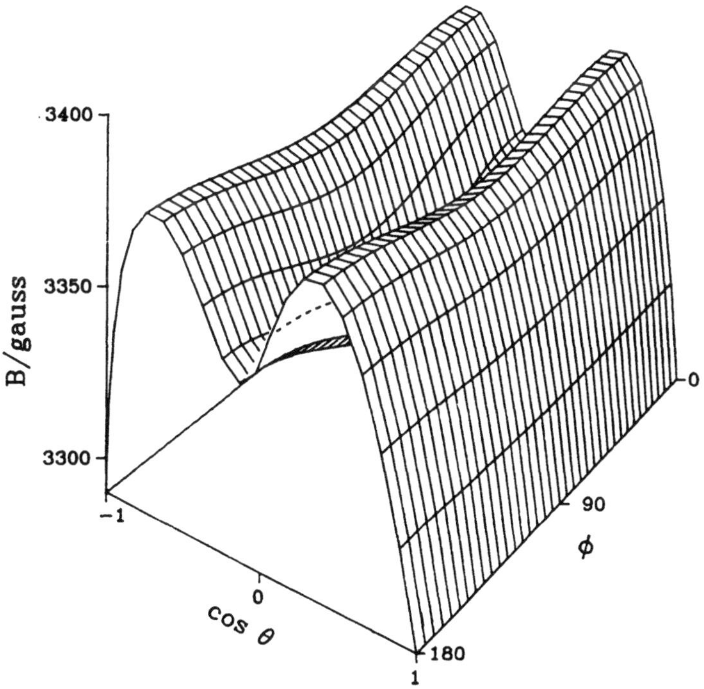

Fig. 6. Resonant field as a function of cos θ and φ for the $m_{\mathrm{Cu}} = -3/2$ "line" of the frozen solution spectrum of $\mathrm{Cu(acac)_2}$; ESR parameters from Reference 31.

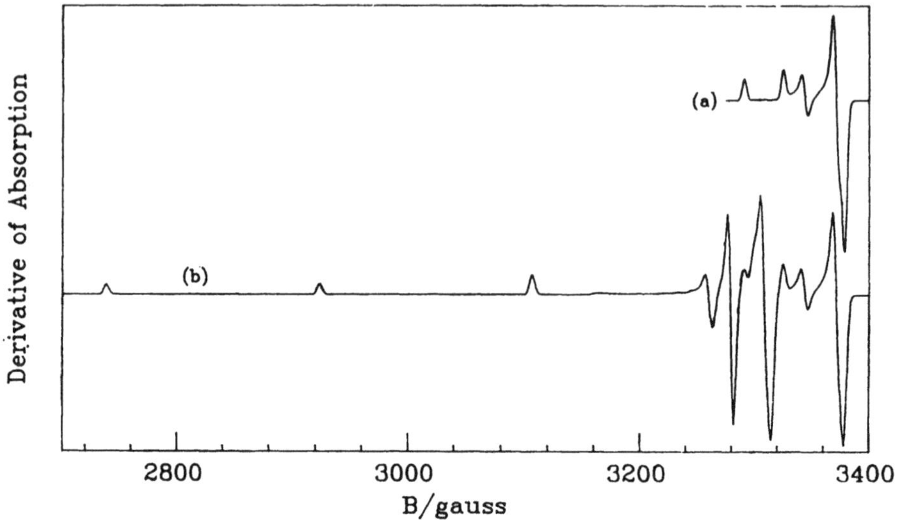

Fig. 7. Computer-simulated first-derivative ESR powder spectrum of $Cu(acac)_2$. (a) Features corresponding to $m_{Cu} = -3/2$; (b) Complete spectrum.

of the features need correspond to orientation of *B* along a principal axis direction.

Since the analytical approach is so complicated, powder patterns have usually been analyzed by comparing the experimental spectrum with a spectrum, computer-simulated using estimates of the *g*- and hyperfine tensor components. Parameters are then adjusted and another simulation computed until the fit is satisfactory (or the experimentalist loses patience with the problem). The most straightforward computer simulation method [32] involves computation of the resonant magnetic field using Eq. (7) for ca. 10^5 values of cos θ and φ for each set of nuclear spin quantum numbers. The field is then divided into equal increments and the number of times the resonant field falls between B_i and B_{i+1} is plotted vs. *B* to give an approximation to the unbroadened absorption spectrum. The absorption spectrum is then broadened by numerical convolution with a line shape function and differentiated to give the desired simulation. Although the "cut and try" approach to spectrum analysis works reasonably well when there are small numbers of parameters, analysis of a complex spectrum is exceedingly tedious.

DeGray and Rieger [33] have developed a computer algorithm to locate powder pattern features in spectra, given estimates of the spin Hamiltonian parameters. The method employs a search of the resonant field surface in cos θ, φ space for maxima, minima, and saddle points. Since the search procedure requires computation of *B* for ~100 orientation, the method is much faster than a complete simulation. The predicted locations of spectral features then are compared with the experimental values and the parameters refined using a nonlinear least-squares method. Using this method, relatively complex powder patterns can be analyzed, provided that the spectrum is sufficiently well resolved that enough features can be located and identified to determine the parameters. Even with least-squares fitting, however, comparison of the experimental spectrum with a computer simulation is required to check the assignments of spectral features.

3.2 Interpretation of the *g*-Tensor

The *g*-value of a free electron is a scalar, $g_e = 2.00232$. In a radical species, *g* becomes a tensor because of the admixture of orbital angular momentum into *S* through spin-orbit coupling. The components of the *g*-tensor thus differ from g_e to the extent that *p*- or *d*-orbital character has been incorporated and differ from one another, depending on which *p*- or *d*-orbitals are involved.

In general, the components of the *g*-tensor are given by

$$g_{ij} = g_e\,\delta_{ij} + \sum_{k,k'} \sum_{m \neq 0} \frac{\zeta_k \langle m \,|\, l_{ki} \,|\, 0\rangle \langle 0 \,|\, l_{k'j} \,|\, m\rangle}{E_0 - E_m} \tag{16}$$

where the indices *i* and *j* refer to molecular coordinate axes (*x*, *y*, *z*), *k* and *k'* sum over atoms with unpaired electron density, and *m* sums over filled and empty molecular orbitals with energy E_m (E_0 is the energy of the SOMO); ζ_k is the spin-orbit coupling constant for atom *k*, and l_{ki} is the *i*-component orbital angular momentum operator for atom *k*. The integrals $\langle m \,|\, l_{ki} \,|\, n\rangle$ are easily computed if the MO's are written as linear combination of real *p* or *d* atomic orbitals. The results of operation by l_i on these functions is shown in Table 1. Thus, for example, $\langle z \,|\, l_x \,|\, y\rangle = i$ and $\langle z^2 \,|\, l_y \,|\, xz\rangle = -\sqrt{3}\,i$.

Table 1. Angular momentum operations on the real *p* and *d* orbitals

	l_x	l_y	l_z
$\|x\rangle$	0	$-i\|z\rangle$	$iy\rangle$
$\|y\rangle$	$i\|z\rangle$	0	$-i\|x\rangle$
$\|z\rangle$	$-i\|y\rangle$	$i\|x\rangle$	0
$\|x^2 - y^2\rangle$	$-i\|yz\rangle$	$-i\|xz\rangle$	$2i\langle xy\rangle$
$\|xy\rangle$	$i\|xz\rangle$	$-i\|yz\rangle$	$-2i\|x^2 - y^2\rangle$
$\|yz\rangle$	$i\|x^2 - y^2\rangle + \sqrt{3}\,i/z^2\rangle$	$i\|xy\rangle$	$-i\|xz\rangle$
$\|xz\rangle$	$-i\|xy\rangle$	$i\|x^2 - y^2\rangle - \sqrt{3}\,i\|z^2\rangle$	$i\|yz\rangle$
$\|z^2\rangle$	$-\sqrt{3}\,i\|yz\rangle$	$\sqrt{3}\,i\|xz\rangle$	0

Notice that d_{z^2} is unique among the *d*-orbitals in that l_z does not couple it to any other orbital. Thus if the major metal contribution to the SOMO is d_{z^2}, g_z will be close to the free electron value. Accordingly, when one *g*-tensor component is found close to the free electron value, it is often taken as evidence for a d_{z^2}-based SOMO; such reasoning should be applied with caution, however, since cancellation of negative and positive terms in Eq. (16) could have the same effect.

Spin-orbit coupling to empty MO's ($E_0 - E_m < 0$) gives a negative contribution to g_{ij}, whereas coupling to filled MO's has the opposite effect. Thus, ESR spectra of d^1 vanadium(IV) complexes generally have *g*-values less than g_e (admixture of empty MO's) whereas d^9 copper(II) complexes have *g*-values greater than g_e (admixture of filled MO's).

Since the *g*-tensor has only three principal values and there are almost always many potentially interacting molecular orbitals, there is rarely sufficient information to interpret a *g*-tensor with complete confidence. When a well resolved and reliably assigned optical spectrum is available, the energy differences, $E_0 - E_m$ are known and can be used in Eq. (16). Extended Hückel MO calculations can be useful, but one is most commonly reduced to arguments designed to show that the observed g-tensor is consistent with the interpretation placed on the hyperfine tensor.

3.3 Interpretation of the Hyperfine Tensor

Electron-nuclear hyperfine coupling arises mainly through two mechanisms: (i) The Fermi contact interaction between the nuclear spin and s-electron spin density; this contribution is isotropic and has been discussed above. (ii) The electron spin-nuclear spin dipolar interaction; this contribution is almost entirely anisotropic, i.e., neglecting spin-orbit coupling, the average dipolar contribution to the hyperfine coupling is zero.

3.3.1 The Dipolar Part of the Hyperfine Tensor. The general form of the dipolar contribution to the hyperfine term of the Hamiltonian is

$$\mathscr{H}_{\text{dipolar}} = g_e g_N \mu_B \mu_N \left\langle \psi_{\text{SOMO}} \left| \frac{\vec{S}\cdot\vec{I}}{r^3} - \frac{3(\vec{S}\cdot\vec{r})(\vec{I}\cdot\vec{r})}{r^5} \right| \psi_{\text{SOMO}} \right\rangle \tag{17}$$

where g_e and g_N are the electron and nuclear g-values, μ_B and μ_N are the Bohr and nuclear magnetons, and the matrix element is evaluated by integration over the spatial coordinates, leaving the spins as operators. Equation (17) can then be written

$$\mathscr{H}_{\text{dipolar}} = \vec{S}\cdot A_d \cdot \vec{I} \tag{18}$$

where A_d is the dipolar contribution to the hyperfine tensor,

$$A = \langle A \rangle\, 1 + A_d \tag{19}$$

(1 is the unit matrix). In evaluating the matrix element of Eq. (17), the integration over the angular variables is quite straightforward. The integral over r, however, requires a good atomic orbital wavefunction. Ordinarily, the integral is combined with the constants as a parameter

$$P = g_e g_N \mu_B \mu_N \langle r^{-3} \rangle \tag{20}$$

P has been computed using Hartree-Fock atomic orbital wavefunctions and can be found in several published tabulations [34–36]. Because of the r^{-3} dependence of P, dipolar coupling of a nuclear spin with electron spin density on another atom is usually negligible.

If an atom contributes p_x, p_y, and p_z atomic orbitals to the SOMO

$$c_x|x\rangle + c_y|y\rangle + c_z|z\rangle \tag{21}$$

the total p-orbital spin density is

$$\varrho^p = c_x^2 + c_y^2 + c_z^2 \tag{22}$$

and the dipolar contribution to the hyperfine tensor can be written

$$(A_d)_{ij} = (2/5)\, P l_{ij} \tag{23}$$

where the l_{ij} are

$$l_{xx} = 2c_x^2 - c_y^2 - c_z^2 \tag{24a}$$

$$l_{yy} = -c_x^2 + 2c_y^2 - c_z^2 \tag{24b}$$

$$l_{zz} = -c_x^2 - c_y^2 + 2c_z^2 \tag{24c}$$

$$l_{ij} = -3c_i c_j \quad (i \neq j) \tag{24d}$$

The dipolar hyperfine tensor for p-orbitals can always be written

$$\boldsymbol{A}_d = (2/5)\, P\varrho^p \begin{pmatrix} 2 & 0 & 0 \\ 0 & -1 & 0 \\ 0 & 0 & -1 \end{pmatrix} \tag{25}$$

The p-orbital axis corresponds to the positive principal value of the tensor. When the p-orbitals are written as hybrids, the orbital shape is unchanged, but the principal axes of the hyperfine tensor, which reflect the spatial orientation of the hybrid p-orbital, differ from those in which the SOMO was formulated. Thus, for example, a p-hybrid with $c_x = c_z$ ($c_y = 0$) corresponds to a p-orbital with the major axis in the xz-plane and halfway between the x-and z-axes (Euler angle $\beta = 45°$).

Similarly, if an atom contributes d atomic orbitals to the SOMO,

$$c_{z^2}|z^2\rangle + c_{yz}|yz\rangle + c_{xz}|xz\rangle + c_{x^2-y^2}|x^2 - y^2\rangle + c_{xy}|xy\rangle \tag{26}$$

the total d-orbital spin density is

$$\varrho^d = (c_{z^2})^2 + (c_{yz})^2 + (c_{xz})^2 + (c_{x^2-y^2})^2 + (c_{xy})^2 \tag{27}$$

and the dipolar contribution to the hyperfine tensor is [37]

$$(A_d)_{ij} = (2/7)\, P l_{ij} \tag{28}$$

where P is given by Eq. (20) and the l_{ij} are:

$$l_{xx} = -c_{z^2}^2 - 2c_{yz}^2 + c_{xz}^2 + c_{x^2-y^2}^2 + c_{xy}^2 - 2\sqrt{3}\, c_{z^2}c_{x^2-y^2} \tag{29a}$$

$$l_{yy} = -c_{z^2}^2 + c_{yz}^2 - 2c_{xz}^2 + c_{x^2-y^2}^2 + c_{xy}^2 + 2\sqrt{3}\, c_{z^2}c_{x^2-y^2} \tag{29b}$$

$$l_{zz} = 2c_{z^2}^2 + c_{yz}^2 + c_{xz}^2 - 2c_{x^2\ y^2}^2 - 2c_{xy}^2 \tag{29c}$$

$$l_{xy} = -2\sqrt{3}\, c_{z^2}c_{xy} + 3c_{yz}c_{xz} \tag{29d}$$

$$l_{yz} = \sqrt{3}\, c_{z^2}c_{yz} + 3c_{xz}c_{xy} - 3c_{yz}c_{x^2-y^2} \tag{29e}$$

$$l_{xz} = \sqrt{3}\, c_{z^2}c_{xz} + 3c_{xz}c_{x^2-y^2} + 3c_{yz}c_{xy} \tag{29f}$$

The dipolar contribution to the hyperfine tensor for a pure d-orbital is

$$\boldsymbol{A}_d = (\pm 2/7)\, P\varrho^d \begin{pmatrix} 2 & 0 & 0 \\ 0 & -1 & 0 \\ 0 & 0 & -1 \end{pmatrix} \tag{30}$$

where the positive sign applies for d_{z^2} and the lower sign to the other four orbitals. Hybrid combinations of d_{yz}, d_{xz}, and d_{xy} or $d_{x^2-y^2}$ and d_{xy} give a d-orbital of the same shape and the same dipolar tensor, though the principal axes in general are different from the axes in which the SOMO was formulated. Other hybrid orbitals are generally of different shape, reflected by different principal values of the dipolar tensor, usually with different principal axes.

Spin-orbit coupling perturbs these results, adding terms to the diagonal tensor components on the order of $P(g_i - g_e)$. These can be neglected only when the g-tensor anisotropy is small. Calculation of the spin-orbit coupling correction is fairly straightforward for mononuclear complexes where the SOMO is composed mainly of d-orbitals from a single metal [6, 12, 38]. In radicals with two or more transition metals, the spin-orbit coupling calculation is seriously nontrivial. A major part of the problem is that the solution must be gauge-invariant, that is, it must not depend on the choice of coordinate system. This problem was addressed in the context of spin-orbit coupling corrections to the g-tensor [39], with Eq. (16) as the result, but it has received only cursory examination with regard to spin-orbit contributions to hyperfine tensors [40]. Fortunately, polynuclear radicals containing first-row transition metals generally have g-tensor components sufficiently close to g_e that the problem can be ignored. As organometallic radicals with second- and third-row transition metals appear, the problem will become more urgent; it is to be hoped that some theoretician will deem the problem worthy of attention.

3.3.2 Reconstruction of SOMO's from Dipolar Coupling Tensors. The AO composition of the SOMO can often be deduced from the dipolar hyperfine tensor, particularly when the radical has enough symmetry to restrict possible hybridization. Thus an axial hyperfine tensor can usually be interpreted in terms of coupling to a single p- or d-orbital. A departure from axial symmetry may be due to spin-orbit coupling effects, if (for example) $A_{\parallel} = A_z$ and $A_x - A_y \approx P(g_x - g_y)$. If the departure from axial symmetry is larger, it is usually caused by d-orbital hybridization. The procedure is best illustrated by example.

3.3.3 Example: $[Mn(dppe)_2(CO)(CNBu)]^{2+}$. The spectrum of the low-spin manganese(II) complex, $[Mn(dppe)_2(CO)(CNBu)]^{2+}$, *5* [41], in a CH_2Cl_2/THF glass is shown in Fig. 8a. The spin Hamiltonian parameters, obtained from a least-squares fit of the field positions of the spectral features [42], are given in Table 2, and a computer simulation based on those parameters is shown in Fig. 8b.

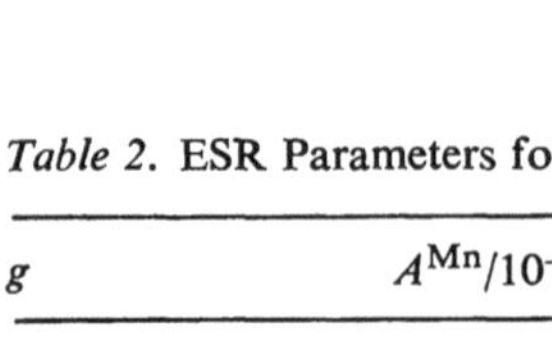

5

Table 2. ESR Parameters for $[Mn(dppe)_2(CO)(CNBu)]^{2+}$

g	$A^{Mn}/10^{-4}$ cm^{-1}	$A^{P}/10^{-4}$ cm^{-1}
2.107	30.2	27.2
2.051	20.6	25.3
1.998	146.9	26.4

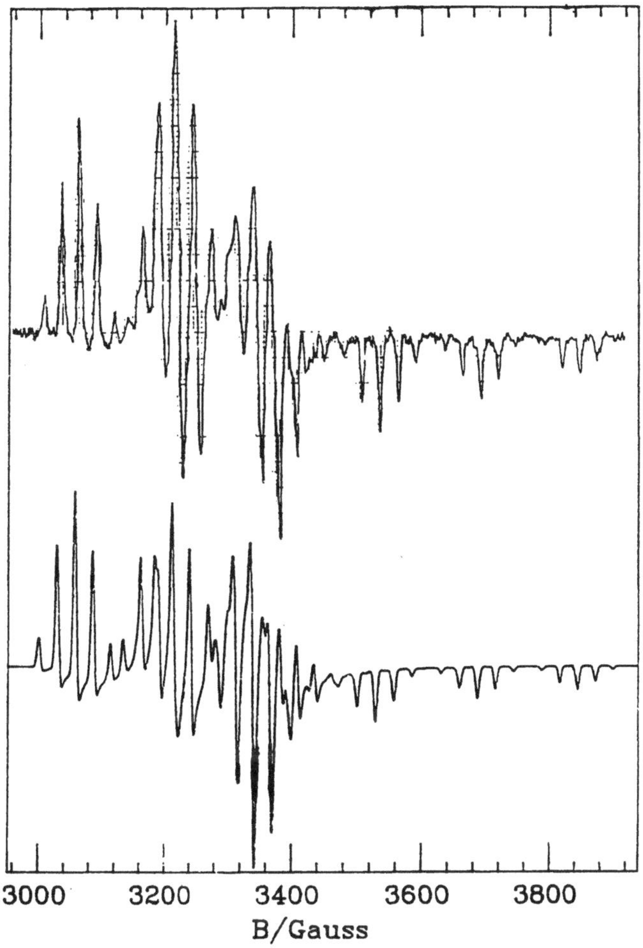

Fig. 8. (a) ESR spectrum of $[Mn(dppe)_2(CO)(CNBu)]^{2+}$ in a CH_2Cl_2/THF glass (from Reference 41, reproduced by permission); (b) Computer simulation using the parameters of Table 2.

Complex *5* has approximate C_{2v} symmetry, although the actual symmetry is reduced to C_2 or C_s, depending on the conformation of the ethylene bridges of the dppe ligands. Since *5* has a nominal d^5 configuration, the SOMO is expected to be one of the "t_{2g}" orbitals of an idealized octahedral complex – d_{xz} (b_1), d_{yz} (b_2), or $d_{x^2-y^2}$ (a_1), where the representations refer to C_{2v}. The energies of the d_{xz} and d_{yz} orbitals are expected to be lowered by back-donation into the π^* orbitals of the CO ligand so that the SOMO is most likely based on $d_{x^2-y^2}$, possibly with some symmetry-allowed d_{z^2} admixture

$$|\mathrm{SOMO}\rangle = a|z^2\rangle + b|x^2 - y^2\rangle + \dots \tag{31}$$

Although the isotropic spectrum of *5* was not sufficiently resolved to unambiguously determine $\langle A^{\mathrm{Mn}}\rangle$, other closely related species give isotropic couplings on the order of 60–70 G [42]; if we assume an isotropic coupling in this range, all three tensor components must above the same sign. If the isotropic hyperfine coupling is negative, as expected if it arises mostly through polarization of inner-shell s orbitals, we have $\langle A \rangle = -65.9 \times 10^{-4}$ cm^{-1}. Assuming that the SOMO is mostly $d_{x^2-y^2}$, ($b \gg a$) and including the appropriate spin-orbit coupling corrections, we have [24, 38]

$$A_z - \langle A \rangle = P\left[\frac{4}{7}(a^2 - b^2) - \frac{2}{3}\Delta g_z - \frac{5}{42}(\Delta g_x + \Delta g_y)\right] \tag{32}$$

With $\Delta g_z = -0.004$, $(\Delta g_x + \Delta g_y) = 0.154$, and $P = 207.6 \times 10^{-4}$ cm^{-1} [35], we get $(a^2 - b^2) = -0.655$. The departure from axial symmetry is due to spin-orbit coupling and/or $d_{x^2-y^2}/d_{z^2}$ hybridization,

$$A_x - A_y = P\left[-\frac{8\sqrt{3}}{7}ab + \frac{17}{14}(\Delta g_x - \Delta g_y)\right] \tag{33}$$

Substituting the parameters, we have $ab = \pm 0.058$. (The upper sign applies if the components are listed in the order x, y, z in Table 2, the lower sign if the order is y, x, z.) Finally, we get $b^2 = 0.660$, $a^2 = 0.005$. The d_{z^2} component is not really significant, given the accuracy of the data and the theory, i.e., most of the departure from axial symmetry can be explained by the spin-orbit coupling correction.

Using Eq. (16), the g-tensor components are found to be

$$\Delta g_{xx} = 2\zeta_{\mathrm{Mn}} \sum_k \frac{b^2 c_{yz,k}{}^2}{E_0 - E_k} \tag{34a}$$

$$\Delta g_{yy} = 2\zeta_{\mathrm{Mn}} \sum_k \frac{b^2 c_{xz,k}{}^2}{E_0 - E_k} \tag{34b}$$

$$\Delta g_{zz} = 2\zeta_{\mathrm{Mn}} \sum_k \frac{4b^2 c_{xy,k}{}^2}{E_0 - E_k} \tag{34c}$$

If we assume coupling with single pure d_{yz}, d_{xz}, and d_{xy} orbitals, we have $\Delta E_{yz} = 16\zeta$, $\Delta E_{xz} = 19\zeta$, $\Delta E_{xy} = -1\,100\zeta$, qualitatively consistent with the expected MO energy level scheme.

3.3.4 Example: Some Cobalt(0) *Radical Anions.* ESR spectra of [CpCo(1,3-COD)]$^-$, *6*, in frozen THF solution and [CpCo(1,5-COD)]$^-$, *7*, in frozen DMF were reported by Geiger and coworkers [43] and are reproduced in Figs. 9a and 10a. These spectra have been reinterpreted to give the parameters shown in Table 3; computer-simulated spectra based on these parameters are shown in Figs. 9b and 10b. Also shown in the table are the ESR parameters for [(C_5Ph_5) $Co(CO)_2$]$^-$, *8* [12].

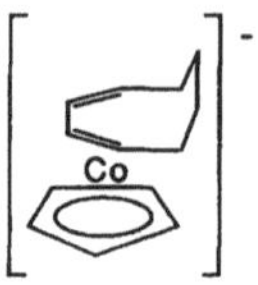

6

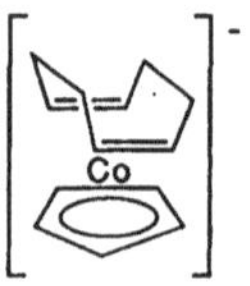

7

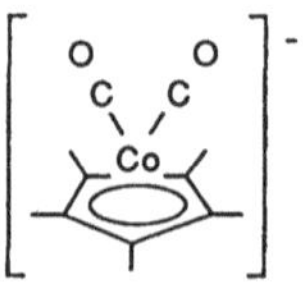

8

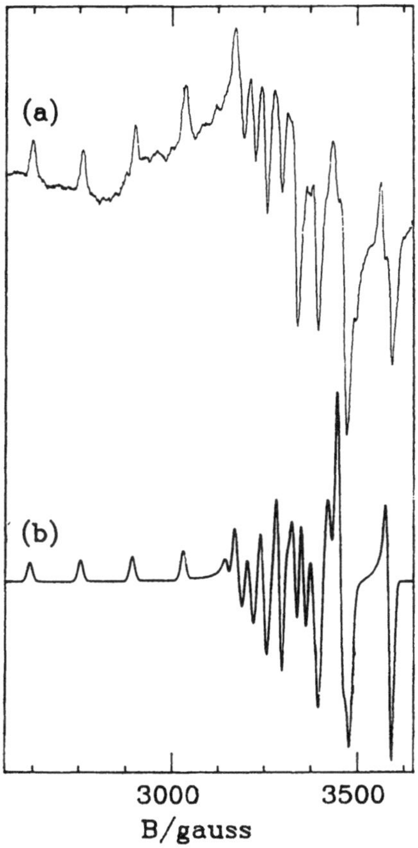

Fig. 9. ESR spectrum of $[CpCo(1,3\text{-}COD)]^-$; (a) experimental spectrum in frozen THF solution (from Reference 43; reproduced with permission); (b) computer-simulation, based on the parameters of Table 3.

Table 3. ESR parameters for Cobalt(0) Radical Anions

Radical Anion	g_x	g_y	g_z	A_x[a]	A_y[a]	A_z[a]
$[CpCo(1,5\text{-}COD)]^-$	2.171	2.027	1.985	(−)158.6	(−)36.7	(−)45.8
$[CpCo(1,3\text{-}COD)]^-$	2.151	2.027	1.997	(−)139.2	(−)36.4	(−)38.2
$[(C_5Ph_5)Co(CO)_2]^-$ [b]	2.018	2.041	1.995	(−)157.9	(−)16.8	(−)44.1

[a] Units of 10^{-4} cm^{-1}. [b] From Reference 12.

The hyperfine tensor components must have identical signs in order that the average values match the observed isotropic couplings; we assume the signs are negative since the isotropic couplings almost certainly arise from polarization of inner shell *s* orbitals (see below).

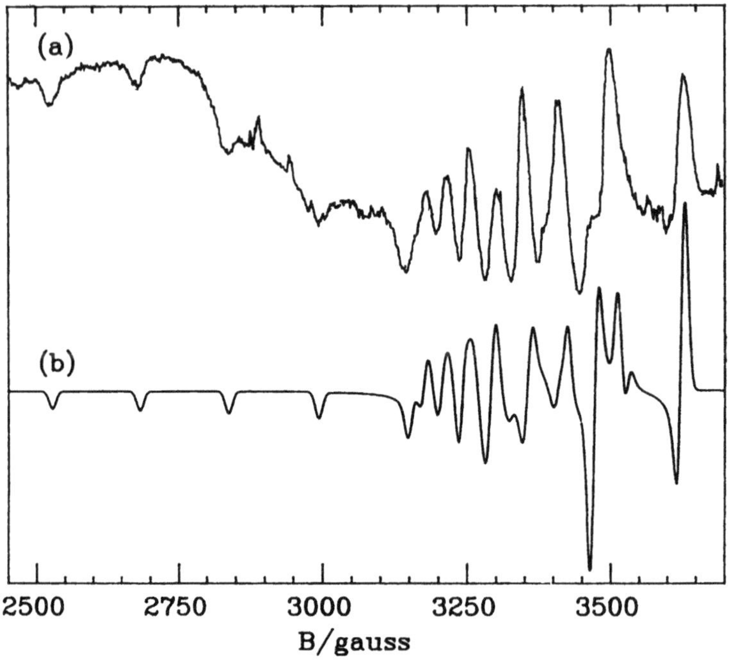

Fig. 10. ESR spectrum of $[CpCo(1,5\text{-}COD)]^-$; (a) experimental spectrum in frozen DMF solution (from Reference 43, reproduced with permission); (b) computer-simulation, based on the parameters of Table 3.

The SOMO in these radicals is expected from extended Hückel MO calculations [43, 44] to be primarily cobalt $3d_{yz}$ in character. In the $\boldsymbol{C}_s$ symmetry of the radicals, d_{yz} belongs to the a'' representation and d-hybridization is possible only with d_{xy}. Assuming that such hybridization is negligible, the g-tensor components are given by [12]

$$\Delta g_{xx} = 2\zeta_{Co}a^2 \sum_k \frac{c_{x^2-y^2,k}{}^2 + 3c_{z^2,k}{}^2}{E_0 - E_k} \tag{35a}$$

$$\Delta g_{yy} = 2\zeta_{Co}a^2 \sum_k \frac{c_{xy,k}{}^2}{E_0 - E_k} \tag{35b}$$

$$\Delta g_{zz} = 2\zeta_{Co}a^2 \sum_k \frac{c_{xz,k}{}^2}{E_0 - E_k} \tag{35c}$$

The dipolar contribution to the hyperfine tensor is given by Eq. (25), but spin-orbit coupling contributions are significant. These often can be expressed in terms of the g-tensor components (as in the Mn(II) example discussed above), but here spin-orbit coupling with the four other d-orbitals contributes somewhat differently to the g-tensor and to the hyperfine tensor. The simplest way of expressing the hyperfine tensor is in terms of the isotropic coupling, the x-component, and the departure from axial symmetry,

$$\langle A\rangle = A_s + \frac{1}{3} P(\Delta g_x + \Delta g_y + \Delta g_z) \tag{36a}$$

$$A_x - \langle A\rangle = P\left[-\frac{4}{7}a^2 + \frac{2}{3}\Delta g_x - \frac{5}{42}(\Delta g_y + \Delta g_z)\right] \tag{36b}$$

$$A_y - A_z = P\left[\frac{17}{14}(\Delta g_y + \Delta g_z) + \frac{6a^2\zeta}{7}\left(\frac{1}{\Delta E_{x^2-y^2}} - \frac{1}{\Delta E_{z^2}}\right)\right] \tag{36c}$$

With the assumed signs of the hyperfine components of Table 3, Eq. (36b) can be used unambiguously to compute $a^2 = \varrho^d$ with the results shown in Table 4.

Table 4. d-Electron spin densities in cobalt(0) radical anions

Radical Anion	ϱ^d	A_s[a]	A_s/Q_d
$[CpCo(1,5\text{-}COD)]^-$	0.681	−97.0	0.740
$[CpCo(1,3\text{-}COD)]^-$	0.591	−87.2	0.666
$[(C_5Ph_5)Co(CO)_2]^-$	0.540	−77.4	0.591

[a] In units of 10^{-4} cm^{-1}.

Since $3d_{yz}/4s$ admixture is symmetry-forbidden for these radicals, the Fermi contact contribution to the isotropic coupling must be entirely from spin polarization,

$$A_s = Q_d\varrho^d \tag{37}$$

Thus we can obtain an independent estimate of the d-electron spin density from the values of A_s, taking $Q_d = -131 \times 10^{-4}$ cm^{-1}, estimated from the isotropic cobalt coupling in $[PhCCo_3(CO)_9]^-$. The results are shown in the last column of Table 4. The spin densities estimated from the isotropic couplings are consistently about 10% higher than those from the dipolar coupling tensor, suggesting a systematic error in one of the parameters, but a reliable ordering of the spin densities.

The g-tensor presents an interesting problem in these cases. EHMO calculations [43, 44] suggest that the SOMO is the highest-energy MO primarily cobalt $3d$ in character. At lower energy is an orbital with d_{xz} character and still lower, but grouped at about the same energy, are MO's with $d_{x^2-y^2}$, d_{xy}, and d_{z^2} contributions. Equations (35) then would suggest that $\Delta g_{xx}/4 \approx \Delta g_{yy} < \Delta g_{zz}$. With the assignments of Table 4, the first relationship is approximately correct for *6* and *7*, but very poor for *8*. The second relationship is not found for any of the anions. Reversing the y and z assignments makes the agreement worse. In discussing this problem for *8* [12], we postulated admixture of some cobalt $4p_y$ character in the SOMO.

$$|\text{SOMO}\rangle = a|yz\rangle + b|y\rangle + \ldots \tag{38}$$

This would result in additional contributions to g_{xx} and g_{zz},

$$g_{xx}(p) = 2\zeta_p b^2/\Delta E_z \quad \text{and} \quad g_{zz}(p) = 2\zeta_p b^2/\Delta E_x \tag{39}$$

where ζ_p is the cobalt $4p$ spin-orbit coupling parameter ($\zeta_p \approx \zeta_d/3$). If MO's with significant p_z or p_x character lie just above the SOMO, then g_{xx} and g_{zz} would be less positive than expected from Eqs. (35), possibly even negative. The component g_{xx} is indeed smaller than expected for *8* and EHMO calculations suggest a MO with significant p_z character just above the SOMO in energy; this orbital is apparently substantially higher in energy in *6* and *7*. A MO with significant p_x character, at about the same energy for all three anions, is implied by these results but is unsubstantiated by the reported EHMO calculations.

3.4 Noncoincident Tensor Axes

In general, the g- and nuclear hyperfine coupling tensors, $\boldsymbol{g}$ and $\boldsymbol{A}_i$, can be written in diagonal form with three principal values – g_x, g_y, g_z and A_{ix}, A_{iy}, A_{iz}. In textbooks on ESR [2–7], it is usually assumed that the same set of principal axes diagonalizes all the relevant tensors. While this is sometimes true, there are many instances where the principal axes are noncoincident [45].

3.4.1 Symmetry Considerations. Kneubühl [46, 47] has given a detailed group theoretical analysis of symmetry restrictions on the orientations of g- and hyperfine tensor principal axes. His results are summarized in Table 5.

Table 5. Symmetry restrictions on g-tensor components

Symmetry	Point Groups	Restrictions on Diagonal Elements	Restrictions on Off-Diag. Elements	Required Tensor Axes
Triclinic	$\boldsymbol{C}_1, \boldsymbol{C}_i$	none	none	
Monoclinic	$\boldsymbol{C}_2, \boldsymbol{C}_s, \boldsymbol{C}_{2h}$	none	$g_{xz} = g_{yz} = 0$	z
Orthorhombic	$\boldsymbol{C}_{2v}, \boldsymbol{D}_2, \boldsymbol{D}_{2h}$	none	$g_{xz} = g_{yz} = g_{xy} = 0$	x, y, z
Axial	$\boldsymbol{C}_n, \boldsymbol{C}_{nv}, \boldsymbol{C}_{nh}, \boldsymbol{D}_n, \boldsymbol{D}_{nd}, \boldsymbol{D}_{nh}, n > 2$	$g_{xx} = g_{yy}$	$g_{xz} = g_{yz} = g_{xy} = 0$	x, y, z

For a nucleus sharing all the molecular symmetry elements (e.g., the metal nucleus in a mononuclear complex), the hyperfine tensor is subject to the same restrictions as the g-tensor. In orthorhombic or axial symmetry, such nuclear hyperfine tensors necessarily share principal axes with the g-tensor. In monoclinic symmetry, one hyperfine axis is also a g-tensor axis, but the other two may be different. In triclinic symmetry ($\boldsymbol{C}_1$ or $\boldsymbol{C}_i$), none of the three principal axes need be shared by the g-tensor and hyperfine tensor. The hyperfine tensor for a ligand atom (or for a metal in polynuclear complexes) is constrained only by the symmetry elements which the nucleus shares with the molecule.

Although symmetry considerations often permit g- and hyperfine tensor principal axes to be noncoincident, there are relatively few cases of such noncoincidence reported in the literature. Most of the examples discussed by Pilbrow and Lowrey in their 1980 review [45] are cases of transition metal ions doped into a host lattice at sites of low symmetry and none are relevant to the present review. This is not to say that tensor axis noncoincidence is rare but that the effects have only rarely been recognized.

3.4.2 Experimental Determination of Tensor Axis Orientations. We have seen in Section 3.2 that spectra of dilute single crystals are analyzed in a way that gives the orientations of the g- and hyperfine tensor principal axes relative to the crystal axes. Historically, most of the information on noncoincident tensor axes is derived from such studies.

At first glance, it would appear that all orientation dependence should be lost in the spectrum of a randomly oriented sample and that location of the g- and hyperfine tensor principal axes would be impossible. While it is true that there is no way of obtaining tensor axes relative to molecular axes from a powder pattern, it is frequently possible to find the orientation of a set of tensor axes relative to those of another tensor.

The observable effects of tensor axis noncoincidence on powder patterns range from blatantly obvious to negligible. In general, the effects of axis noncoincidence will be more noticeable if two (or more) tensors have large anisotropies, which are comparable in magnitude, e.g., $\Delta g \mu_B B \approx \Delta A$. This follows from the fact that minimum and maximum resonant fields are determined by a competition between extrema in the angle-dependent values of g and A. Consider the case of noncoincident g- and hyperfine tensor axes. For large values of $|m_I|$, the field extrema will be determined largely by the extrema in the effective hyperfine coupling and will occur at angles close to the hyperfine tensor axes, but for small $|m_I|$, the extrema will be determined by extrema in the effective g-value and will correspond to angles close to the g-tensor axes. The result of such a competition is that a series of features which would be equally spaced (to first order), acquires markedly uneven spacings.

There are two corollaries stemming from this generalization. Since spin 1/2 nuclei give only two hyperfine lines, there can be no variation in spacings. Thus powder spectra cannot be analyzed to extract the orientations of hyperfine tensor axes for such important nuclei as ^{1}H, ^{13}C, ^{19}F, ^{31}P, ^{57}Fe, and ^{103}Rh. Secondly, since the observable effects in powder spectra depend on the magnitude of the tensor anisotropies, the principal axes of the hyperfine tensor for a nucleus with small hyperfine coupling generally cannot be located from a powder spectrum, even though the relative anisotropy may be large.

A good example of the effect of g- and hyperfine tensor axis noncoincidence is the ESR spectrum of $[CpCr(CO)_2NO]^-$, *9*, studied by Geiger and coworkers [48]. The g- and ^{14}N hyperfine tensors are approximately axial for this radical, but the $g_\parallel$ axis lies close to the perpendicular plane of the hyperfine tensor. If the $g_\parallel$ axis was exactly in the $A_\perp$ plane, the three positive-going $g_\parallel$, $A_\perp$ features, corresponding to resonant field maxima, would be evenly spaced. In fact, the spacings are noticeably uneven by an amount far greater than can be explained by second-order shifts. The effect can be understood, and the spectrum simulated virtually exactly, by assuming that the $g_\parallel$ axis is about 15° out of the $A_\perp$ plane.

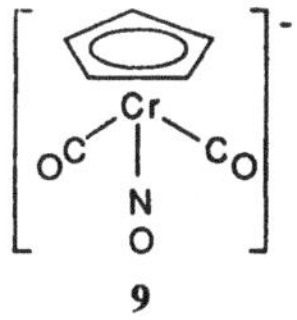

9

3.4.3 Example of Tensor Axis Noncoincidence: $Fe(CO)_5^+$ and $Fe(CO)_5^-$. Two particularly interesting organometallic examples were reported by Morton, Preston and coworkers [49, 50]. Spectra of single crystals of $Cr(CO)_6$, doped with ^{13}C- or ^{57}Fe-enriched $Fe(CO)_5$ and γ-irradiated at 77 K, showed two different radicals. One species, identified as $Fe(CO)_5^+$, showed coupling to ^{57}Fe and to a unique ^{13}C nuclei with axial hyperfine tensors sharing principal axes with the g-tensor [49]. Coupling was also observed to four other ^{13}C nuclei with identical coupling tensors but with the major axis approximately along the g-tensor x-axis for two nuclei and along the g-tensor y-axis for the other two. The parameters are listed in Table 6. If the radical is square pyramidal (C_{4v}) $Fe(CO)_5^+$, *10*, the principal axes of the g-tensor must be the molecular axes (the C_4 axis and normals to the reflection planes). The iron atom and the carbon of the axial CO group have the full symmetry of the group and so these hyperfine tensors must share principal axes with the g-tensor. The four equatorial carbonyl carbons, on the other hand, lie in reflection planes, but not on the C_4-axis and so are symmetry-required to share only one principal axis with the g-tensor. In fact, the major tensor axes for the equatorial carbons are tilted slightly in the $-z$ direction from the ideal locations along the $\pm x$ and $\pm y$ axes. The g-tensor suggests that the metal contribution is d_{z^2} and the iron hyperfine tensor then can be used to estimate about 55% iron $3d$ and 34% axial carbon $2p_z$ spin density. The spin density on the equatorial carbons then is mostly negative and due to spin polarization.

Table 6. ESR Parameters for $Fe(CO)_5^+$ and $Fe(CO)_5^-$ [a]

$Fe(CO)_5^+$:				
	$g_\parallel = 2.001$	$A_\parallel^{Fe} = (+)9.4$	$A_\parallel^{C1} = (+)19.6$	$A_\parallel^{C2-C5} = (+)6.4$
	$g_\perp = 2.081$	$A_\perp^{Fe} = (-)6.2$	$A_\perp^{C1} = (-)17.6$	$A_\perp^{C2-C5} = (+)8.6$
$Fe(CO)_5^-$:				
	$g_1 = 1.989$	$A_1^{Fe} = (+)6.7$	$A_1^{C1} = (+)87.4$	
	$g_2 = 2.003$	$A_2^{Fe} = (-)4.5$	$A_2^{C1} = (+)70.7$	
	$g_3 = 2.010$	$A_3^{Fe} = (-)3.2$	$A_3^{C1} = (+)65.7$	

[a] Coupling constants in units of 10^{-4} cm^{-1}. Data from references 49 and 50.

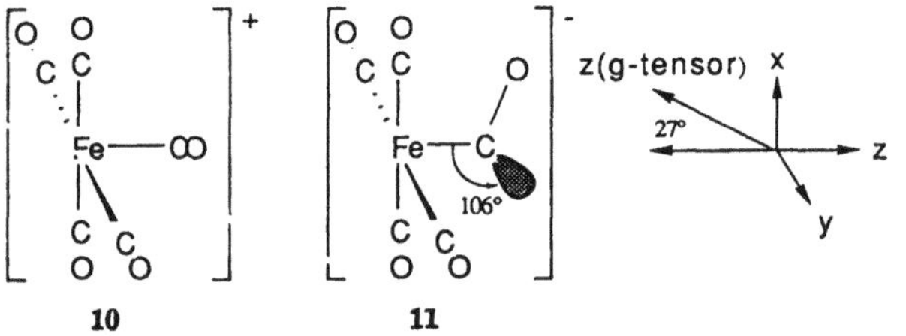

10 11

The other species observed in irradiated $Fe(CO)_5$-doped crystals of $Cr(CO)_6$ also shows coupling to ^{57}Fe, to a unique ^{13}C, and to four other carbons. However, in this case $\boldsymbol{g}$, $\boldsymbol{A}^{Fe}$, and $\boldsymbol{A}^{C1}$ have only one tensor axis in common (that corresponding to the third component of each tensor listed in Table 6). The other ^{57}Fe hyperfine axes are rotated by about 27° and those of the ^{13}C hyperfine tensor by about 48° relative to the g-tensor axes. Insufficient data were accumulated to determine the complete hyperfine tensors for the other four carbons, but the components are considerably smaller ($4\text{–}15 \times 10^{-4}$ cm^{-1}). The hyperfine tensors suggest about 38% iron $3d_{z^2}$, 18% carbon $2p$, and 6% carbon $2s$ spin densities. Using detailed arguments regarding the orientation of the g-tensor axes relative to the crystal axes, the authors conclude that the carbon $2p$ axis is oriented at about 106° relative to the Fe—C bond axis and that the Fe—C—O bond angle is about 119°.

The most striking feature of these results is the orientation of the unique ^{13}C hyperfine tensor axes, relative to those of the ^{57}Fe hyperfine axes. This orientation led Fairhurst, et al. [50], to assign the spectrum to $Fe(CO)_5^-$, *11*, and to describe the species as a substituted acyl radical. However, these authors did not discuss the orientation of the g-tensor axes. The y-axis, normal to the reflection plane, is common to all three tensors. The x- and z-axes of the g-tensor, on the other hand are oriented about 27° away from the corresponding ^{57}Fe hyperfine tensor axes. Since the iron d-orbital contribution to the SOMO appears to be nearly pure d_{z^2}, the ^{57}Fe hyperfine tensor major axis must correspond to the local z-axis, assumed to be essentially the Fe—C bond. Thus we must ask: Why are the g-tensor axes different? The SOMO can be written

$$|\mathrm{SOMO}\rangle = a|z^2, \mathrm{Fe}\rangle + b_x|x, C\rangle + b_z|z, C\rangle \tag{40}$$

where $a = 0.62$, $b_x = -0.41$, and $b_z = 0.12$. Spin-orbit coupling will mix the SOMO with MO's having iron d_{yz} or d_{xz} character, but d_{yz} is involved in the π orbitals of the C=O group,

$$|\pi\rangle = c_{yz}|yz, \mathrm{Fe}\rangle + c_y|y, C\rangle \tag{41}$$

Assuming that there is only one π orbital close enough in energy to couple significantly, Eq. (16) gives the g-tensor components:

$$\Delta g_{xx} = 2\zeta_{\mathrm{Fe}} \frac{3a^2 c_{yz}^2 + \sqrt{3}\, a b_z^2 c_{yz} c_y}{\Delta E_\pi} \tag{42a}$$

$$\Delta g_{yy} = 2\zeta_{\mathrm{Fe}} \frac{3a^2 c_{xz}^2}{\Delta E_{xz}} \tag{42b}$$

$$\Delta g_{zz} = 0 \tag{42c}$$

$$g_{xz} = -2\zeta_{\mathrm{Fe}} \frac{\sqrt{3}\, a b_x c_{yz} c_y}{\Delta E_\pi} \tag{42d}$$

The g-tensor can be diagonalized by rotation about the y-axis by the angle β,

$$\tan 2\beta = \frac{2g_{xz}}{g_{xx} - g_{zz}} = \frac{-2\sqrt{3}\, a b_x}{3a^2(c_{yz}/c_y) + \sqrt{3}\, a b_z} \tag{43}$$

With $\beta = 27°$, this expression gives $c_{yz}/c_y \approx 0.5$, a reasonable result.

This may be a rather general effect; if the unpaired electron in a radical is delocalized asymmetrically, and other MO's are similarly delocalized, the g-tensor will have off-diagonal elements, which may be large enough to shift the principal axes away from the molecular coordinate system.

3.4.4 Example of Tensor Axis Noncoincidence: $[(RCCR')Co_2(CO)_6]^-$ *and* $[SFeCo_2(CO)_9]^-$. Noncoincident tensor axis effects are seen in the frozen solution spectra of $[(RCCR')Co_2(CO)_6]^-$, *12* [37], and $[SFeCo_2(CO)_9]^-$, *13* [51], but the effects are rather more subtle than those discussed above for $[CpCr(NO)(CO)_2]^-$.

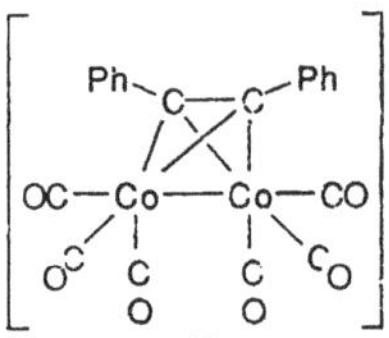

12

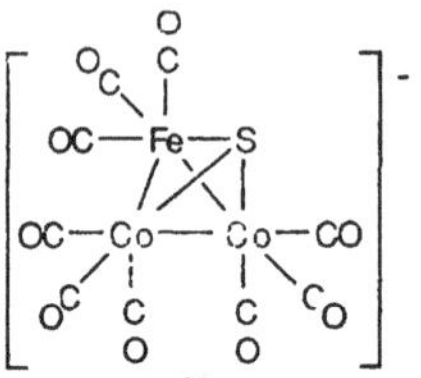

13

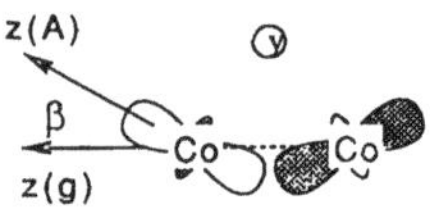

14

In these cases, the g-tensor is nearly isotropic, but the principal axes of the ^{59}Co hyperfine tensors are noncoincident. The largest hyperfine tensor component ($a_y = 66.0$ G in the case of *13*) results in 15 features, evenly spaced (apart from small second-order shifts). Another series of features, less widely spaced, shows some variation in spacing and, in a few cases, resolution into components. This behavior can be understood as follows: Suppose that the hyperfine tensor y-axes are coincident and consider molecular orientations with the magnetic field in the xz-plane. To first order, the resonant field then is

$$B = B_0 - m_1 a_+ - m_2 a_- \tag{44}$$

where $B_0 = h\nu/g\mu_B$ and

$$a_{\pm}{}^2 = a_z^2 \cos^2(\theta \pm \beta) + a_x^2 \sin^2(\theta \pm \beta) \tag{45}$$

where β describes the orientation of the hyperfine tensor z-axes relative to the g-tensor z-axis. Since $\boldsymbol{g}$ is nearly isotropic, the extrema in B are determined mostly by the hyperfine term. When $m_1 = m_2$, a_+ and a_- are equally weighted and the extrema occur at $\theta = 0$ or 90°, but when $m_1 \neq m_2$, the extrema correspond to other angles. Consider, for examples, the five components of the $m_1 + m_2 = +3$ feature. With $|a_z| = 53.6$ G, $|a_x| = 15.5$ G, $\beta = 18°$, the hyperfine contributions to the field extrema and the corresponding values of θ are given in Table 7. In the experimental spectrum of *13*, two resolved field maximum features were seen, which correspond to the first two and the last three of the above components. Since the resolution is sensitive to the noncoincidence, it was possible to fit the experimental spectrum to obtain β quite accurately.

Table 7. Splitting of M = 3 features in $[SFeCo_2(CO)_9]^-$ Spectrum

(m_1, m_2)	θ_{min}	$(B - B_0)_{min}/G$	θ_{max}	$(B - B_0)_{max}/G$
+7/2, −1/2	−20°	37.0	+67°	166.1
−1/2, +7/2	+20°	37.0	−67°	166.1
+5/2, +1/2	−15°	55.3	+78°	156.9
+1/2, +5/2	+15°	55.3	−78°	156.9
+3/2, +3/2	0°	66.5	90°	153.6

It is relatively easy to understand the significance of the noncoincident tensor axes in these cases. For *12*, the C_{2v} molecular symmetry permits a specific prediction of the possible tensor axis orientations. The g-tensor principal axes must be coincident with the molecular symmetry axes. The two cobalt nuclei are located in a reflection plane (which we label xz) so that symmetry requires the y-axis to be a principal axis for all three tensors. The other two axes may be rotated, relative to the molecular x- and z-axes, by $\pm\beta$. (Since the two nuclei are symmetrically equivalent, the rotations must be equal and opposite.)

Because the magnitudes and probable signs of the cobalt hyperfine tensors suggest a SOMO predominantly d_{z^2} in character, the major axes of the hyperfine tensors approximate the local z-axes at the cobalt atoms and the angular displacement indicates a bent Co—Co anti-bonding interaction, as shown in structure *14*. Here the C—C or Fe—S bond axis (the molecular y-axis) is perpendicular to the plane of the page. Comparison with the crystal structure of neutral $(Ph_2C_2)Co_2(CO)_6$ [52] shows that these local axes are roughly in the direction of the axial carbonyl ligands (the Co—CO bond is tilted 28° away from the Co—Co vector). Thus, it seems reasonable to say that the local axes on a metal are determined primarily by the more strongly interacting ligands and that bonds to other atoms can be described as bent.

3.4.5 Example of Tensor Axis Noncoincidence: (o-xylylene)-Mn(dmpe)$_2$. As part of a study of Mn(II) dialkyls, Wilkinson, Hursthouse, and coworkers [53] reported the ESR spectrum of the approximately octahedral (o-xylylene)Mn(dmpe)$_2$, *15*, Fig. 11a. The spectrum was interpreted assuming coincident g- and hyperfine tensor axes, but a simulation based on the reported parameters gave a very poor fit to the published spectrum [54]. On closer examination, it was realized that this is a rather extreme example of a spectrum influenced by noncoincident g- and hyperfine tensor principal axes. The clue evident in the spectrum is the large gap between the $m_{Mn} = -1/2$ and $+1/2$ "parallel" features, suggesting one or more extra features. Fig. 12 shows a set of simulated spectra for a hypothetical low-spin Mn(II) species; all seven spectra correspond to the same g- and hyperfine tensors, but the angle β, between the g- and hyperfine tensor z-axes, varies from 0 to 90°. As shown in Fig. 12, it is possible to obtain spectra with more than six resolved "parallel" features. Indeed, the spectrum is sufficiently sensitive to the angle that β can be evaluated quite precisely. The final parameters for *15*, based on least-squares fitting of the positions of the resolved features and the isotropic parameters, [53] are given in Table 8; a computer simulation using these parameters is shown in Fig. 11b.

Complex *15* has approximate C_{2v} symmetry, but the ethylene bridges of the dmpe ligands destroy the reflection planes; in addition, the x-ray structure [53] shows a small fold at the methylene groups of the o-xylylene ligand, which destroys the C_2 axis. Thus, the molecule can be regarded, with increasing accuracy, as C_{2v}, C_s, C_2, or C_1. In order to explain the tensor axis noncoincidence, the metal contribution to the SOMO must be a d-hybrid. Since *15* is a d^5 low-spin Mn(II) species, the SOMO is expected to be dominated by one of the orbitals of the octahedral t_{2g} set – d_{xz}, d_{yz}, or $d_{x^2-y^2}$. This is consistent with a clue contained in the ESR parameters. The four ^{31}P couplings are apparently equivalent, and all are relatively small. Thus no lobe of the SOMO can be directed toward a phosphorus atom, and major d_{xy} or d_{z^2} contributions can be ruled out. Consider the twelve binary hybrids based on the "t_{2g}" orbitals which are listed in Table 9. Since we know that the hyperfine tensor major axis is not a g-tensor principal axis, we can immediately reject hybrids 1, 3, 4, 7, 8, and 10 for which the major axis is a common axis. Hybrids 2 and 6, 5 and 9, and 11 and 12 differ only in the x- and y-labels

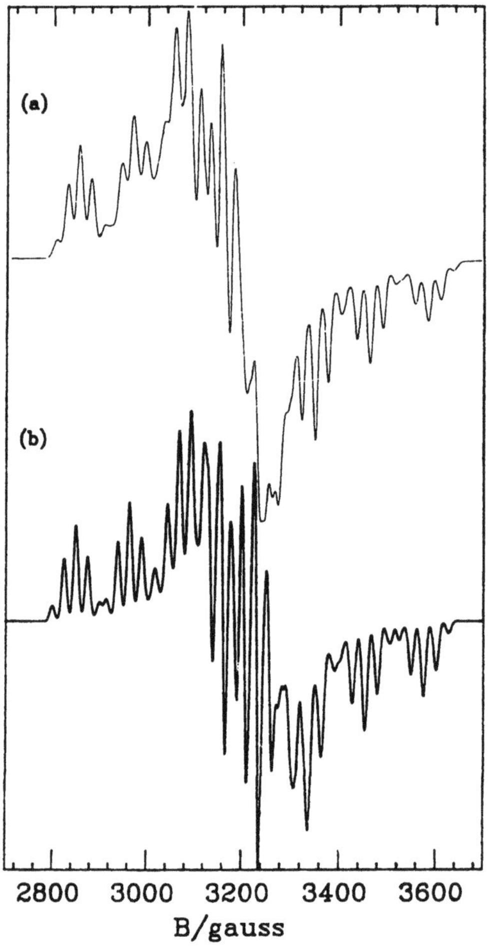

Fig. 11. (a) ESR spectrum of (o-xylylene)Mn(dmpe)$_2$ in frozen toluene solution (from Reference 53, reproduced with permission); (b) Computer simulation of spectrum using parameters of Table 8.

and are essentially indistinguishable, so there are only three cases to consider in detail.

Hybrids 11 and 12 can be written

$$|\text{SOMO}\rangle = a|yz\rangle + b|xz\rangle \tag{46}$$

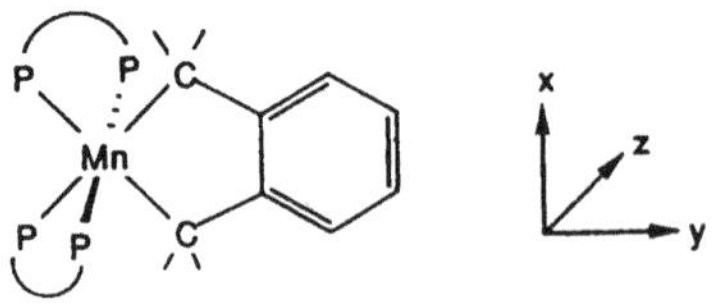

15

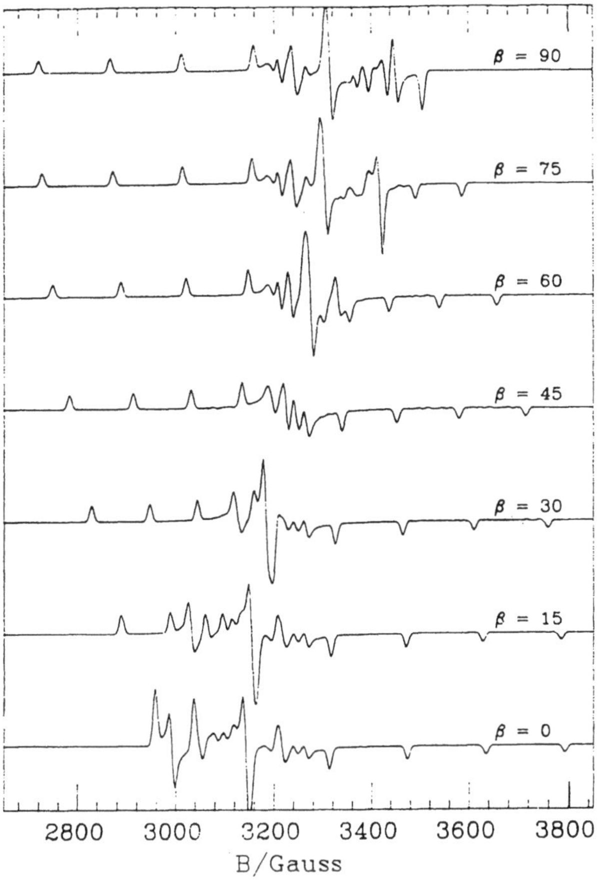

Fig. 12. Computer-simulated ESR spectra for a hypothetical low-spin Mn(II) radical with $g = (2.100, 2.050, 2.000)$, $A^{Mn} = (150, 25, 25) \times 10^{-4}$ cm^{-1}, for various values of β, the Euler angle between the g-tensor and hyperfine tensor z-axes.

Table 8. ESR Parameters for (o-xylylene)Mn(dmpe)$_2$[a]

g	A^{Mn}	A^{P}
2.110	27	24.5
2.035	27	24.5
2.000	125	24.5

[a] Hyperfine tensor components in units of 10^{-4} cm^{-1}; $\beta = 41°$.

Straightforward application of Eqs. (29) shows that the hyperfine tensor remains axial, independent of the hybridization ratio, although the principal axes are rotated in the xy-plane by an angle α given by

$$\tan 2\alpha = -\frac{2b/a}{1 - b^2/a^2} = \frac{2a/b}{1 - a^2/b^2} \tag{47}$$

Table 9. SOMO Candidates: Binary *d*-Hybrids

Number	Representation C_3	C_2	Major *d*-AO	Minor *d*-AO	Approx. Major HF Axis	Common Axis
1	a'	a	$d_{x^2-y^2}$	d_{z^2}	z	x, y, z
2	a'	–	$d_{x^2-y^2}$	d_{yz}	z	x
3	a'	–	d_{yz}	$d_{x^2-y^2}$	x	x
4	a'	–	d_{yz}	d_{z^2}	x	x
5	a''	–	d_{xz}	d_{xy}	y	x
6	–	a	$d_{x^2-y^2}$	d_{xz}	z	y
7	–	a	d_{xz}	$d_{x^2-y^2}$	y	y
8	–	a	d_{xz}	d_{z^2}	y	y
9	–	b	d_{yz}	d_{xy}	x	y
10	–	–	$d_{x^2-y^2}$	d_{xy}	z	z
11	–	–	d_{xz}	d_{yz}	y	z
12	–	–	d_{yz}	d_{xz}	x	z

Assuming that only MO's with predominantly d_{xz} or $d_{x^2-y^2}$ character are close enough in energy to contribute significantly, the g-tensor components are

$$\Delta g_{xx} = 2a^2\zeta_{\text{Mn}}/\Delta E_{x^2-y^2} \tag{48a}$$

$$\Delta g_{yy} = 2b^2\zeta_{\text{Mn}}/\Delta E_{x^2-y^2} \tag{48b}$$

$$\Delta g_{zz} = 2a^2\zeta_{\text{Mn}}/\Delta E_{xz} \tag{48c}$$

$$g_{xy} = -2ab\zeta_{\text{Mn}}/\Delta E_{x^2-y^2} \tag{48d}$$

Rotation in the xy-plane by the angle α' diagonalizes the tensor, but, at this level of approximation, α' and α are identical. Thus for d_{xz}/d_{yz} hybrids, the g- and hyperfine tensor principal axes are coincident for all hybridization ratios.

Hybrid 9 (or, interchanging x and y, hybrid 5),

$$|\text{SOMO}\rangle = a|d_{yz}\rangle + b|d_{xy}\rangle \tag{49}$$

gives an axial hyperfine tensor with principal axes rotated in the xz-plane,

$$\tan 2\beta = -\frac{2b/a}{1 - b^2/a^2} \tag{50}$$

If only MO's with mainly d_{xz} or $d_{x^2-y^2}$ character are coupled, the g-tensor components are

$$\Delta g_{xx} = 2a^2\zeta_{\text{Mn}}/\Delta E_{x^2-y^2} + 2b^2\zeta_{\text{Mn}}/\Delta E_{xz} \tag{51a}$$

$$\Delta g_{yy} = 0 \tag{51b}$$

$$\Delta g_{zz} = 2a^2\zeta_{\text{Mn}}/\Delta E_{xz} + 8b^2\zeta_{\text{Mn}}/\Delta E_{x^2-y^2} \tag{51c}$$

$$g_{xz} = -2ab\zeta_{\text{Mn}}/\Delta E_{xz} - 4ab\zeta_{\text{Mn}}/\Delta E_{x^2-y^2} \tag{51d}$$

The tensor is diagonalized by rotation about the y-axis. With another parameter, $\Delta E_{x^2-y^2}/\Delta E_{xz}$, to play with, virtually any angle can be rationalized, in particular, when the energy differences are equal, we get $\tan 2\beta' = 2a/b$ so that for small b/a, $\beta \approx 0°$, $\beta' \approx 45°$. However, we then have $\Delta g_x \approx \Delta g_z$, $\Delta g_y \approx 0$; these are important discrepancies, which are difficult to resolve.

Hybrid 6 or (interchanging x and y, hybrid 2) can be written

$$|\text{SOMO}\rangle = a|d_{x^2-y^2}\rangle + b|d_{xz}\rangle \tag{52}$$

In this case, the hyperfine tensor is diagonalized by rotation about the y-axis by the angle β,

$$\tan 2\beta = 2b/a \tag{53}$$

Assuming that the only MO's close enough in energy to undergo significant spin-orbit mixing are those with d_{xz} and d_{yz} character, we obtain the g-tensor components.

$$\Delta g_{xx} = 2a^2\zeta_{\text{Mn}}/\Delta E_{yz} \tag{54a}$$

$$\Delta g_{yy} = 2a^2\zeta_{\text{Mn}}/\Delta E_{xz} \tag{54b}$$

$$\Delta g_{zz} = 2b^2\zeta_{\text{Mn}}/\Delta E_{yz} \tag{54c}$$

$$g_{xz} = -2ab\zeta_{\text{Mn}}/\Delta E_{yz} \tag{54d}$$

The principal values of the g-tensor are

$$\Delta g_x = 2(a^2 + b^2)\,\zeta_{\text{Mn}}/\Delta E_{yz} \tag{55a}$$

$$\Delta g_z = 0 \tag{55b}$$

Note that g_z is expected to be nearly equal to the free electron value in agreement with experiment. Rotation about the y-axis by the angle β', given by Eq. (50), diagonalizes the g-tensor. For small b/a, β and β' are comparable in magnitude but differ in sign. The experimentally observed angle is the difference, $|\beta - \beta'| = 41°$, which is obtained for $b/a = \pm 0.40$ ($\beta = \pm 22°$, $\beta' = \pm 19°$). This ratio results in a significant departure of the hyperfine tensor from axial symmetry; however, the inclusion of a small amount of d_{z^2} character in the SOMO restores axial symmetry without significantly altering the above arguments. The resulting hybrid can best be thought of as $d_{x^2-y^2}$ in shape, but rotated about the y-axis so that the x^2-lobes are 22° away from the molecular x-axis. Inclusion of spin-orbit coupling corrections to the hyperfine tensor [55] only slightly modifies this result; with spin-orbit corrections, the hybrid which fits the g-tensor, hyperfine tensor and noncoincidence angle has 80% $d_{x^2-y^2}$, 15% d_{xz}, and 5% d_{z^2} and a total d-spin density of 0.59.

An extended Hückel MO calculation [54] on *15* supports this analysis. The three "t_{2g}" orbitals are indeed close together in energy and remain nearly non-bonding metal-based d-orbitals. Furthermore, the SOMO is predicted to be primarily $d_{x^2-y^2}$ with a small d_{xz} admixture, in good agreement with our analysis of the ESR results.

4. References

1. A. Hudson, Electron Spin Resonance **11 A** (1988) 55–76.
2. P. B. Ayscough, Electron Spin Resonance in Chemistry, Methuen, London, 1967.

3. P. W. Atkins and M. C. R. Symons, The Structure of Inorganic Radicals, Elsevier, Amsterdam, 1967.
4. A. Abragam and B. Bleaney, Electron Paramagnetic Resonance of Transition Ions, Clarendon Press, Oxford, 1970.
5. J. E. Wertz and J. E. Bolton, Electron Spin Resonance, McGraw-Hill, New York, 1972.
6. N. M. Atherton, Electron Spin Resonance, Ellis Horwood, Chicester, 1973.
7. W. Gordy, Theory and Applications of Electron Spin Resonance, Wiley, New York, 1980.
8. L. V. Casagrande, T. Chen, P. H. Rieger, B. H. Robinson, J. Simpson, and S. J. Visco, Inorg. Chem. **23** (1984) 2019–2025.
9. B. M. Peake, P. H. Rieger, B. H. Robinson and J. Simpson, Inorg. Chem. **18** (1979) 1000–1005.
10. G. Breit and I. I. Rabi, Phys. Rev. **38** (1931) 2082–2083.
11. A. R. Boate, J. R. Morton, and K. F. Preston, J. Mag. Reson. **24** (1976) 259–268.
12. N. G. Connelly, W. E. Geiger, G. A. Lane, S. J. Raven, and P. H. Rieger, J. Am. Chem. Soc. **108** (1986) 6219–6224.
13. G. K. Fraenkel, J. Phys. Chem. **71** (1967) 139–171.
14. W. B. Lewis and L. O. Morgan, in R. L. Carlin (Ed.), Transition Met. Chem., Vol. **4**, Marcel Dekker, New York, 1968, pp. 33–112.
15. B. H. Robinson and A. H. Beth, Electron Spin Resonance **8** (1983) 346–377.
16. M. M. Iannuzzi and P. H. Rieger, Inorg. Chem. **14** (1975) 2895–2899.
17. M. M. Iannuzzi, C. P. Kubiak, and P. H. Rieger, J. Phys. Chem. **80** (1976) 541–545.
18. R. L. Ward and S. I. Weissman, J. Am. Chem. Soc. **79** (1957) 2086–2090.
19. B. J. Corden and P. H. Rieger, Inorg. Chem. **10** (1971) 263–272.
20. H. S. Gutowsky, D. W. McCall, and C. P. Slichter, J. Chem. Phys. **21** (1953) 279–292. See also Reference 6, pp. 309–313.
21. J. I. Kaplan, J. Chem. Phys. **28** (1958) 278–282; **29** (1958) 462; S. Alexander, ibid. **37** (1962) 967–974, 974–980. See also Reference 6, pp. 313–323.
22. See also Reference 6, pp. 323–327.
23. S. D. Ittel, P. J. Krusic, and P. Meakin, J. Am. Chem. Soc. **100** (1978) 3264–3266.
24. J. A. DeGray, Q. Meng, and P. H. Rieger, J. Chem. Soc. Faraday I, **83** (1987) 3565–3573.
25. N. V. Vugman, A. O. Caride, and J. Danon, J. Chem. Phys. **59** (1973) 4418–4422.
26. N. V. Vugman and N. M. Pinhal, Mol. Phys. **49** (1983) 1315–1319.
27. J. A. DeGray, P. H. Rieger, N. G. Connelly, and G. Garcia, J. Mag. Reson., in press.
28. P. H. Rieger, J. Mag. Reson. **50** (1982) 485–489.
29. M. C. R. Symons, J. R. Morton, and K. F. Preston, in K. S. Suslick (Ed.), High-Energy Processes in Organometallic Chemistry, American Chemical Society, Washington, DC, 1987, pp. 169–190.
30. P. C. Taylor, J. F. Baugher, and H. M. Kriz, Chem. Rev. **75** (1975) 203–240.
31. I. V. Ovchinnikov and V. N. Konstantinov, J. Mag. Reson. **32** (1978) 179–190.
32. P. C. Taylor and P. J. Bray, J. Mag. Reson. **2** (1970) 305–331.
33. J. A. DeGray and P. H. Rieger, Bull. Mag. Reson. **8** (1987) 95–101.
34. B. A. Goodman and J. B. Raynor, Adv. Inorg. Chem. Radiochem. **13** (1971) 136–362.
35. J. R. Morton and K. F. Preston, J. Mag. Reson. **30** (1978) 577–582.
36. A. K. Koh and D. J. Miller, Atom. Data Nucl. Data Tab. **33** (1985) 235–253.
37. B. M. Peake, P. H. Rieger, B. H. Robinson, and J. Simpson, J. Am. Chem. Soc. **102** (1980) 156–163.
38. B. R. McGarvey, in T. F. Yen (Ed.), Electron Spin Resonance of Metal Complexes, Plenum Press, New York, 1969, pp. 1–11.
39. A. F. Stone, Proc. Roy. Soc. (London) **A 271** (1963) 424–434.
40. T. Kawamura, K. Fukamachi, T. Sowa, S. Hayashida, and T. Yonezawa, J. Am. Chem. Soc. **103** (1981) 364–369.
41. G. A. Carriedo, V. Riera, N. G. Connelly, and S. J. Raven, J. Chem. Soc., Dalton Trans. (1987) 1769–1773.
42. N. G. Connelly, A. L. Rieger, and P. H. Rieger, to be published.

43. T. A. Albright, W. E. Geiger, J. Moraczweski, and B. Tulyathan, J. Am. Chem. Soc. **103** (1981) 4787–4794.
44. T. A. Albright and R. Hoffmann, Chem. Ber. **111** (1978) 1578–1590.
45. J. R. Pilbrow and M. R. Lowrey, Rep. Prog. Phys. **43** (1980) 433–495.
46. F. K. Kneubühl, Phys. Kondens. Mat. **1** (1963) 410–447; **4** (1965) 50–62.
47. For a summary of Kneubühl's methods, see Reference 45.
48. W. E. Geiger, P. H. Rieger, B. Tulyathon, and M. C. Rausch, J. Am. Chem. Soc. **106** (1984) 7000–7006.
49. T. Lionel, J. R. Morton, and K. F. Preston, J. Chem. Phys. **76** (1982) 234–239.
50. S. A. Fairhurst. J. R. Morton, and K. F. Preston, J. Chem. Phys. **77** (1982) 5872–5875.
51. B. M. Peake, P. H. Rieger, B. H. Robinson, and J. Simpson, Inorg. Chem. **20** (1981) 2540–2543.
52. F. A. Cotton, J. D. Jamerson, and B. R. Stults, J. Am. Chem. Soc. **98** (1976) 1774–1779.
53. C. G. Howard, G. S. Girolami, G. Wilkinson, M. Thornton-Pett, and M. B. Hursthouse, J. Chem. Soc., Dalton Trans. (1983) 2631–2637.
54. P. H. Rieger, unpublished work.
55. R. D. Pike, A. L. Rieger, and P. H. Rieger, J. Chem. Soc., Faraday Trans. I, in press.

CHAPTER 9

Mechanisms for Substitution, Disproportionation, and Insertion in Seventeen Electron Organometallic Radicals

William C. Trogler

Department of Chemistry, D-006, University of California at San Diego, La Jolla, California 92093, U.S.A.

Table of Contents

1. Introduction

Organometallic radicals that contain 17 valence electrons about the metal exhibit a broad range of reactivities [1–11]. Some of these, such as radical recombination and atom abstraction, generally are unique to such compounds. This article will focus on the mechanisms of ligand substitution, insertion, and disproportionation reactions, which are common to both 18- and 17-electron complexes. Here the unique aspect of radical reactivity derives from the 10^6–10^{10} enhancement in reaction rates that occur in a 17-electron complex as compared to the 18-electron analogue. Although the information is less complete, similar enhancements appear possible in polynuclear radicals when a metal atom in the cluster acquires a 17-electron configuration.

Organometallic radicals that dimerize efficiently often exhibit competing atom abstraction or electron transfer chemistry. The observation of simultaneous ligand substitution or insertion reactivity occurs less frequently since the lifetime of the radical intermediate is short. Dimerization rates may approach the diffusion limit.

Thus, in the electrochemical oxidation of $Co(CO)_4^-$ or $Mn(CO)_5^-$ to produce the corresponding radicals dimerization usually occurs before other chemistry can occur. Using the faster technique of photolysis to break the metal-metal bond in $Mn_2(CO)_{10}$ the substitution chemistry of 17-electron $\cdot Mn(CO)_5$ can be observed directly. Sterically hindered 17-electron complexes, such as $\cdot V(CO)_6$ or $\cdot V(\eta\text{-}C_5H_5)_2CO$, can not dimerize and provide examples of stable radicals that can be studied more easily. Cationic radicals (e.g. $Fe(CO)_3(PPh_3)_2^+$) often resist dimer formation. Finally, because electron transfer can be rapid, disproportionation provides an efficient path for decomposition of 17-electron complexes, which may limit the time available for other reactions to occur. One must consider all the available reaction paths, since the inefficiency of one form of reactivity is essential for observing another.

2.1 Substitution Reactions

The exceptional lability of a CO ligand in a 17-electron complex was suggested by observations [12] that CO substitution by PPh_3 in photochemically generated $\cdot Mn(CO)_5$ competes with recombination to $Mn_2(CO)_{10}$. An associative mechanism for substitution was proposed. Shortly thereafter it was shown [13] that up to two CO ligands are rapidly replaced by PPh_3 ($Ph = C_6H_5$) in $HRe(CO)_5$. After careful purification of reagents and rigorous exclusion of O_2 for mechanistic work, $HRe(CO)_5$ was found to be surprisingly unreactive. The addition of $Re_2(CO)_{10}$ to the reaction mixture and the generation of $\cdot Re(CO)_5$ by irradiation of the $\sigma \rightarrow \sigma^*$ absorption in the dimer catalyzed the substitution process. A radical chain mechanism shown in Fig. 1 was proposed to explain these results. Since the $\cdot Re(CO)_5$ intermediate occurs in low (undetectable) concentrations, and since $HRe(CO)_5$ is unreactive for days in its absence, the rate of substitution in $\cdot Re(CO)_5$ must be tremendously accelerated and an enhanced lability toward CO dissociation was postulated [13]. We now address the origin of the rate acceleration in 17-electron complexes.

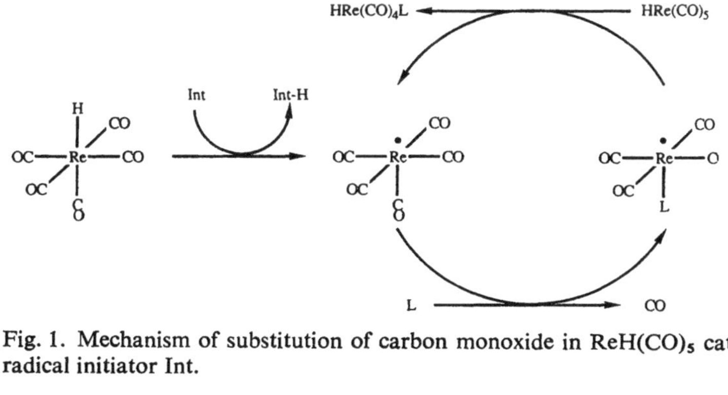

Fig. 1. Mechanism of substitution of carbon monoxide in $ReH(CO)_5$ catalyzed by a radical initiator Int.

2.1.1 *Four Coordinate Radicals*

The prototypical four coordinate radical is $\cdot Co(CO)_4$. Unfortunately, quantitative studies of its substitutional reactivity are not available. Matrix isolation IR and EPR studies [14] suggest that $\cdot Co(CO)_4$ ($t_2^6\, e^3$) adopts a Jahn-Teller distorted tetrahedral geometry (C_{3v}). Since photolysis of $Co_2(CO)_8$ results in CO dissociation [15, 16], as well as metal-metal bond cleavage, flash photolysis studies have not yet been used to isolate substitution processes in the radical. Thermal substitution reactions of $Co_2(CO)_8$ with $AsPh_3$, $P(t\text{-}Bu)_3$ or $P(CH_2CH_2CN)_3$ proceed by CO dissociation [17–19] and do not involve radical intermediates. Disproportionation products, such as $[Co(CO)_3L_2][Co(CO)_4]$, form on the addition of basic phosphine ligands, L = PBu_3, PPh_3, and $P(i\text{-}Pr)_3$ [18, 19]. The kinetics of the latter processes show nonintegral reaction orders in $Co_2(CO)_8$ (1.0–1.5) and an apparent order in L less than one. A radical chain mechanism shown in Eq. 1a–1i was proposed. This scheme is modified slightly from the original mechanism to favor associative substitution for $\cdot Co(CO)_4$ as found (vide infra) in well-characterized examples.

$$Co_2(CO)_8 + L \rightleftharpoons Co_2(CO)_8L \tag{1a}$$

$$Co_2(CO)_8L \rightarrow \cdot Co(CO)_3L + \cdot Co(CO)_4 + CO \tag{1b}$$

$$\cdot Co(CO)_3L + Co_2(CO)_8 \rightarrow Co(CO)_3L^+ + Co_2(CO)_8^{-\cdot} \tag{1c}$$

$$Co_2(CO)_8^{-\cdot} \rightarrow Co(CO)_4^- + \cdot Co(CO)_4 \tag{1d}$$

$$Co(CO)_3L^+ + L \rightarrow Co(CO)_3L_2^+ \tag{1e}$$

$$\cdot Co(CO)_4 + L \rightarrow \cdot Co(CO)_3L + CO \tag{1f}$$

$$\cdot Co(CO)_3L + \cdot Co(CO)_4 \rightarrow Co_2(CO)_7L \tag{1g}$$

$$\cdot Co(CO)_3L + \cdot Co(CO)_3L \rightarrow Co_2(CO)_6L_2 \tag{1h}$$

$$\cdot Co(CO)_4 + \cdot Co(CO)_4 \rightarrow Co_2(CO)_8 \tag{1i}$$

The high substitution lability of the $\cdot Co(CO)_4$ radical has been confirmed directly. Irradiation of the contact ion pair species $[Cp_2Co][Co(CO)_4]$ into its charge transfer band reversibly generates the neutral radicals [20]. Back electron transfer is not as favorable when a CO is substituted by L = PPh_3 or PMe_2Ph and an 85% yield of dimer forms according to Eq. 2.

$$\begin{array}{c} \text{18e} \quad \text{18e} \qquad\qquad \text{17e} \quad \text{17e} \\ [Cp_2Co][Co(CO_4)] \xrightarrow{h\nu} \{Cp_2CoCo(CO)_4\} \xrightarrow{L} \\ \cdot Co(CO)_3L + CO \xrightarrow{k_d} 1/2[Co_2(CO)_6L_2] \end{array} \tag{2}$$

Since the rate constant of dimerization, k_d, for $\cdot Co(CO)_4$ is 4×10^8 M^{-1} s^{-1} and since k_d for $\cdot Co(CO)_3(PBu_3)$ is 9×10^7 M^{-1} s^{-1} [21], the substitution rate must be high to account for the results of Eq. 2.

2.1.2 *Five Coordinate Radicals*

As noted in the beginning of this section the $\cdot M(CO)_5$ 17-electron systems, where M = Mn or Re, provided the first evidence for high substitutional lability in a metal radical, although the mechanism was uncertain. Initial work [22] on the thermal CO replacement reactions of Mn—Mn and Re—Re single bonded complexes was interpreted as proceeding by metal-metal bond homolysis, followed by associative substitution in the radical intermediates. Although the conclusion of associative substitution in the radicals is correct, the assumption of metal-metal bond homolysis has been shown to be incorrect for most dinuclear Mn and Re complexes [23]. Mononuclear radical intermediates are generally not involved in these reactions. The first sound experimental evidence for associative substitution was obtained [24] in a study of the competition between atom abstraction and CO substitution in photochemically generated $\cdot Re(CO)_5$ as depicted in Eq. 3. In the presence of CCl_4 and L = PBu_3

$$Re_2(CO)_{10} \underset{}{\overset{h\nu}{\rightleftharpoons}} 2\cdot Re(CO)_5 \tag{3a}$$

$$\cdot Re(CO)_5 + CCl_4 \xrightarrow{k_b} ClRe(CO)_5 \tag{3b}$$

$$\cdot Re(CO)_5 + L \xrightarrow{k_c} \cdot Re(CO)_4L + CO \tag{3c}$$

$$\cdot Re(CO)_4L + CCl_4 \xrightarrow{k_d} ClRe(CO)_4L \tag{3d}$$

or PPh_3 UV irradiation of $Re_2(CO)_{10}$ produces $ClRe(CO)_5$ and $ClRe(CO)_4L$ as stable products. The ratio $[ClRe(CO)_5]/[ClRe(CO)_4L] = [U]/[S]$ is directly proportional to $[CCl_4]$ at constant $[PPh_3]$, which confirms the expected second order nature of step b. If step c exhibits a two term rate law $k_c = k'_c + k''_c[L]$ with CO dissociative and associative components then $[S]/[U] = (k'_c + k''_c[L])/(k_b[CCl_4])$. A plot of [S]/[U] vs $[PPh_3]/[CCl_4]$ is linear with a zero intercept $k'_c \simeq 0$ and the substitution of CO in $\cdot Re(CO)_5$ must therefore be associative. For L = PPh_3 the ratio k''_c/k_b equals 22.1 and for L = PBu_3 it equals 34. When combined with a literature value [25] of $k_b = 4.9 \times 10^7\ M^{-1}\ s^{-1}$ at 22 °C one obtains estimates of $k''_c = 1.1 \times 10^9\ M^{-1}\ s^{-1}$ and $1.6 \times 10^9\ M^{-1}\ s^{-1}$ for second order rates of substitution of CO in $\cdot Re(CO)_5$ by PPh_3 and PBu_3, respectively, The small difference in rates for PPh_3 and PBu_3, in spite of their considerable difference in size and basicity [26], may reflect weak Re—P bonding in the transition state and a compression of reactivity near the diffusion limit. The mechanism proposed [24] favors nucleophilic attack at the open coordination site of square pyramidal $\cdot Re(CO)_5$ as had been suggested earlier for $\cdot Mn(CO)_5$ [12].

The reactivity of $Mn_2(CO)_{10}$ was probed directly by nanosecond flash photolysis of $Mn_2(CO)_{10}$ in the presence of L = PBu_3 [27]. Kinetics were complicated by competing CO dissociation and Mn—Mn bond cleavage on photolysis as shown in Fig. 2. When $[L] = 2 \times 10^{-3}$ M only $Mn_2(CO)_9L$ formed, but at higher $[L] = 2 \times 10^{-2}$ M both $Mn_2(CO)_9L$ and $Mn_2(CO)_8L_2$ formed in a 1:3 ratio. This provides qualitative evidence for associative substitution in $\cdot Mn(CO)_5$.

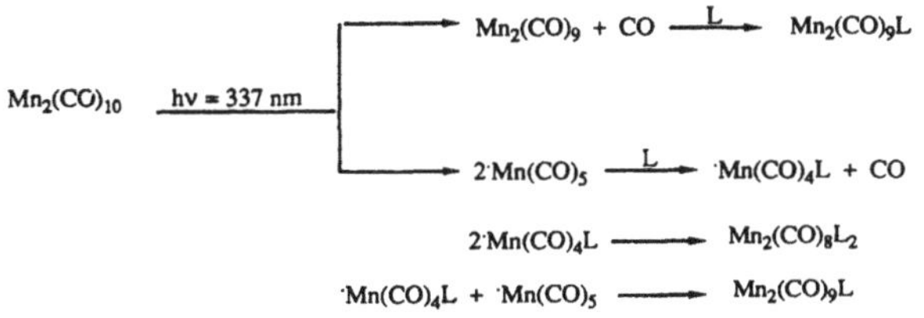

Fig. 2. Scheme for photochemically induced ligand substitution in $Mn_2(CO)_{10}$.

The reactions of photochemically produced $\cdot Mn(CO)_5$ with CCl_4 and L = PPh_3, $P(i\text{-}Pr)_3$, $P(O\text{-}i\text{-}Pr)_3$, and PBu_3 have also been examined by the competition technique [28]. For L = $AsPh_3$, which reacts slowly, the competing halogen abstraction substrate CH_2Br_2 was used. As in the studies [24] of $\cdot Re(CO)_5$ only an associative mechanism for substitution obtains with k as low as $6.5 \pm 0.8 \times 10^4\ M^{-1}\ s^{-1}$

for $AsPh_3$ and as high as $1.0 \pm 0.1 \times 10^9$ M^{-1} s^{-1} for PBu_3 as the nucleophile. The low reactivity with PPh_3 ($k = 1.7 \pm 0.2 \times 10^7$ M^{-1} s^{-1}) is somewhat surprising since only a slight difference in reactivity between PPh_3 and PBu_3 was observed for $\cdot Re(CO)_5$ [24]. This may reflect a more dominant role for steric effects in $\cdot Mn(CO)_5$ and an intrinsically lower susceptibility toward nucleophilic attack, which results in an increased selectivity. The reduction of $Mn(CO)_6^+$, and of the substituted complexes $Mn(CO)_5L^+$, has been examined as a method for generating 19-electron intermediates that could be involved in associative reactions of $\cdot Mn(CO)_5$. Such species have not been directly observed, presumably because of rapid dissociative loss of a ligand to form a 17-electron species. These studies are discussed in greater detail elsewhere in this volume [29].

Persistent substituted radicals $\cdot M(CO)_3L_2$, where M = Mn or Re and L = bulky phosphine, have been generated by irradiation of $M_2(CO)_8L_2$ in the presence of excess L [30–32]. These radicals, which are stable for days in hydrocarbon solutions, have been assigned a square pyramidal structure with trans L in the basal plane. Recently $\cdot Re(CO)_3(PCy_3)_2$ was isolated by addition of PCy_3 to $(\eta^5\text{-}CPh_3)$ $Re(CO)_3$ in benzene solution and a square pyramidal structure has been confirmed crystallographically [33]. Kinetics studies of the reaction sequence shown in Eq. 4 for L = $P(i\text{-}Bu)_3$ show [32] a first order dependence on [CO] thereby suggesting that substitution is associative even with a weak nucleophile such as CO.

$$\cdot Mn(CO)_3L_2 + CO \xrightarrow{k} \cdot Mn(CO)_4L + L \quad (4a)$$

$$2\cdot Mn(CO)_4L \xrightarrow{\text{fast}} Mn_2(CO)_8L_2 \quad (4b)$$

The $\cdot Fe(CO)_3(PPh_3)_2^+$ radical cation isoelectronic to $\cdot Mn(CO)_5$ has been isolated by one-electron oxidation of the corresponding neutral precursor [34–37] with $[N(4\text{-}Br\text{-}C_6H_4)_3][PF_6]$ in CH_2Cl_2 solvent. Other derivatives with L = $AsPh_3$, $PMePh_2$, $P(OPh)_3$, $1/2(PPh_2)_2CH_2$, or $1/2(PPh_2CH_2)_2$ can be generated in solution. The IR spectra of these complexes suggest a trigonal bipyramidal structure about iron with apical L and this has been supported by results of a ^{13}C labeling study [38]. The range of $Fe(CO)_3L_2^+$ cations has been extended to L = PMe_3, PCy_3, and $SbPh_3$ in solution electrochemical studies [39, 40]. On oxidation of $Fe(CO)_3L_2$ species with $Ag[PF_6]$ an Ag—Fe bonded intermediate $Ag[Fe(CO)_3L_2]_2^+$ can be detected in the IR spectra [34]. In studies of the oxidation of $Fe(CO)_3L_2$ complexes (L = PPh_3, $AsPh_3$, and $SbPh_3$) at Pt, Ag, or Hg electrode surfaces [39] the stabilities of the radical cations formed on oxidation at Ag or Hg electrodes increase markedly. The presence of Ag—Fe and Hg—Fe bonded species was proposed to explain these results.

Electrochemically generated $\cdot Fe(CO)_3(PPh_3)_2^+$ decomposes [41] in the cavity of an EPR spectrometer in the presence of acetone or acetonitrile by a second order process. The reaction between $Fe(CO)_3(PPh_3)_2^+$ and py quantitatively produces [40] $Fe(CO)_3(PPh_3)_2$ and Fe^{2+}. The kinetics of this disproportionation reaction were examined for $Fe(CO)_3L_2^+$ (L = PMe_3, PPh_3, PBu_3, and PCy_3) in CH_2Cl_2 using pulsed electrochemical techniques [40] and the mechanism shown in Eq. 5 is consistent with the experimental data. An associative mechanism obtains for the

$$[Fe(CO)_3L_2]^+ + py \xrightarrow{k_1} [Fe(CO)_2(py)L_2]^+ + CO \quad (5a)$$

$$[Fe(CO)_2(py)L_2]^+ + [Fe(CO)_3L_2]^+ \xrightarrow{\text{fast}} Fe(CO)_3L_2 + [Fe(CO)_2(py)L_2]^{2+} \quad (5b)$$

$$[Fe(CO)_2(py)L_2]^{2+} + 5py \xrightarrow{\text{fast}} 2CO + 2L + [Fe(py)_6]^{2+} \quad (5c)$$

rate limiting ligand substitution step. The large decrease in k from 85 $M^{-1}\ s^{-1}$ for $L = PMe_3$ to $1.5 \pm 0.1 \times 10^{-4}\ M^{-1}\ s^{-1}$ at 25 °C for $L = PCy_3$ suggests a strong dependence on steric effects at the metal center. This agrees with the proposed associative mechanism, which requires a six-coordinate intermediate or transition state. Variation of the rate of substitution with electronic properties of the incoming pyridine nucleophile leads to a negative slope ($\varrho = -3.25$) in the Hammett plot. This indicates a strong dependence on the electron donor properties of the nucleophile [40].

2.1.3 (a) *Substitution in Six Coordinate Radicals*

With phosphorus, arsenic, nitrogen, and oxygen donor nucleophiles, $V(CO)_6$ substitutes CO (Eq. 6) by an associative process [42–44]. For nitrogen and oxygen

$$V(CO)_6 + L \xrightarrow{k} V(CO)_5L + CO \qquad (6)$$

donor ligands subsequent electron transfer reactions lead to disproportionation as considered later. A linear correlation between the basicity of L and $\log k$ for phosphorus donors, as well as the low $\Delta H\ddagger$ and negative $\Delta S\ddagger$ observed, favor an associative mechanism for CO substitution. Replacement of a second CO ligand (Eq. 7) also

$$V(CO)_5L + L \rightarrow V(CO)_4L_2 + CO \qquad (7)$$

occurs by an associative mechanism; however, the second order rates are at least 10^{-3} that for the first substitution. The rate enhancement for associative CO replacement in 17-electron $\cdot V(CO)_6$ as compared to 18-electron $Cr(CO)_6$ was estimated [42] as 10^{10}. This resembles the 10^9 rate acceleration noted [40] for the $\cdot Fe(CO)_3L_2^+$ system.

In contrast to the wealth of kinetic data available for $\cdot V(CO)_6$ most observations of the reactions of $\cdot Cr(CO)_6^+$ have been qualitative. Initially the lifetime of $\cdot Cr(CO)_6^+$ was shown to be $\sim$1–20 s based on cyclic voltammetric studies of $Cr(CO)_6$ ($E_{1/2}^{ox} = 1.53$ V vs SCE) in CH_3CN solvent [45]. By contrast, the lifetimes of $\cdot Mo(CO)_6^+$ and $W(CO)_6^+$ are too short to observe directly. Given the estimated half-life of 2.4 sec for $\cdot V(CO)_6$ in 1 M CH_3CN [44] the kinetic stabilities of $\cdot Cr(CO)_6^+$ and $\cdot V(CO)_6$ appear to be comparable. When $\cdot Cr(CO)_6^+$ is generated in the presence of trifluoroacetate anion the reduction wave disappears in the cyclic voltammogram [46] and the reaction of Eq. 8 is suggested to occur by a second order process. Later it was shown [47] that the situation is complicated

$$\cdot Cr(CO)_6^+ + O_2CCF_3^- \rightarrow \cdot Cr(CO)_5(O_2CCF_3) + CO \qquad (8)$$

because the oxidation wave shifts 200 mV to more positive potentials and transforms from a one-electron to a two-electron process. It was proposed [47] that after the substitution of Eq. 8 the $Cr(CO)_5(O_2CCF_3)$ produced undergoes further oxidation at the electrode surface to Cr^{2+}.

Seventeen electron $\cdot Cr(CO)_5X$ complexes (X = F, Cl, Br, I) have been generated by electrochemical oxidation of the corresponding 18-electron anions [47]. There is only a slight difference (0.05 V range) in oxidation potentials; however, $\cdot Cr(CO)_5F$ exhibits the greatest reactivity toward disproportionation. The *trans*-$Cr(CO)_4(PPh_3)_2^+$ radical ion has also been generated electrochemically in solution and the PF_6^- and ClO_4^- salts can be isolated from the reaction between the neutral complex and $NOPF_6$ or $AgClO_4$ [48]. Addition of CH_3CN or $[NBu_4]Br$ as nucleophiles results in disproportionation.

Electrocatalytic substitution of carbon monoxide in $Cr(CO)_5(NCCH_3)$, $W(CO)_5(py)$, $W(CO)_5(NCCH_3)$, *cis*-$W(CO)_4(NCCH_3)_2$, *cis*-$W(CO)_4(py)_2$, *cis*-$Mo(CO)_4(py)_2$, *fac*-$Mo(CO)_3(NCCH_3)_3$, and *fac*-$W(CO)_3(NCCH_3)_3$ occurs on passing an anodic current through solutions containing the metal complex and ligands such as PPh_3 or RNC [49]. Catalytic factors of 2–100 turnovers per mole of charge passed are obtained, which suggests a high substitutional lability in the radical cations. Electrocatalytic substitution mechanisms receive coverage elsewhere [29] in this volume.

(*b*) *Isomerization in Six Coordinate Radicals*

Although the relation to CO substitution is not clear, an extensive body of electrochemical data [50–58] documents rapid structural isomerization (either trans $\leftrightarrow$ cis or fac $\leftrightarrow$ mer) on oxidation of the following 18-electron complexes to 17-electron radicals: *trans or cis*-$Mo(CO)_2(diphos)_2$, *cis*-$Cr(CO)_2(diphos)_2$, and *cis*-$W(CO)_2(diphos)_2$ [51]; *cis* or *trans*-$Cr(CO)_2(dppm)_2$, *cis* or *trans*-$Mo(CO)_2(dppm)_2$, and *cis*-$W(CO)_2(dppm)_2$ [52]; *fac*-$Mn(CO)_3(dppm)X$ for X = Cl or Br [54]; *fac*-$Mn(CO)_3(SbPh_3)_2Br$ [55]; *fac*-$Cr(CO)_3L_3$ for L = $P(OMe)_3$, $P(OEt)_3$, $P(OPh)_3$, $P(OMe)_2Ph$, and PMe_2Ph [56, 57]; *fac*-$Mo(CO)_3L_3$ for L = PMe_2Ph, $P(OPh)_3$, $P(OMe)_2Ph$, $P(OMe)Ph_2$, and $PPh_2(CH_2Ph)$ [53]; *fac*-$M(CO)_3(\eta^1$-diphos$)(\eta^2$-diphos) for M = Cr, Mo, and W [50]; diphos = $Ph_2PCH_2CH_2PPh_2$ and dppm = $Ph_2PCH_2PPh_2$. Several generalizations emerge from these studies. First, the rate of isomerization in the 17-electron systems is tremendously (qualitative half-lives of seconds vs hours) accelerated as compared to the 18-electron analogues. An intramolecular mechanism, whereby a trigonal coordination face of the octahedron twists, is favored [55]. Second, in a series of Cr, Mo, and W complexes the rates of isomerization increase in the order Cr < Mo < W. Quantitative measurements [57] for the *fac/mer*-$Cr(CO)_3[P(OMe)_3]_3^{+/0}$ redox couples show that the equilibrium constant K° for the fac $\leftrightarrow$ mer equilibrium in the 18-electron complex is 4.0 ± 0.5 at 22 °C, while the corresponding value in the radical cation $K^+ = 640 \pm 130$. Corresponding values for the isomerization rate are $k^\circ = 1.8 \pm 0.3 \times 10^{-4}\ s^{-1}$ and $k^+ = 0.11 \pm 0.01\ s^{-1}$ at 22 °C. One electron oxidation of *trans*-(carbene)$_2M(CO)_4$ complexes (M = Cr, Mo, and W) also results in rapid isomerization to the cis isomer by a unimolecular process [58]. Thus, structural isomerization as well as associative substitution becomes accelerated in six coordinate carbonyl radicals.

2.1.4 Radicals Containing Pi Bound Organic Ligands

Studies of CO substitution in η^6-arene or η^5-cyclopentadienyl complexes are complicated by the possibility of ring slippage. Extensive kinetics studies of associative substitution of carbon monoxide in 18-electron systems have defined ring slippage, which makes an orbital on the metal available for nucleophilic attack (e.g. η^5-$C_5H_5 \rightarrow \eta^3$-$C_5H_5$ or η^6-$C_6H_6 \rightarrow \eta^4$-$C_6H_6$), as an important mechanistic path [59]. That ring slipped intermediates are viable species in reactions of 17-electron complexes was shown [60] by the isolation of such a complex on addition of CO to bis(indenyl)vanadium, Fig. 3. Recent work [61] implies that ring slippage may be kinetically accelerated in 17-electron systems. While the 18-electron complex

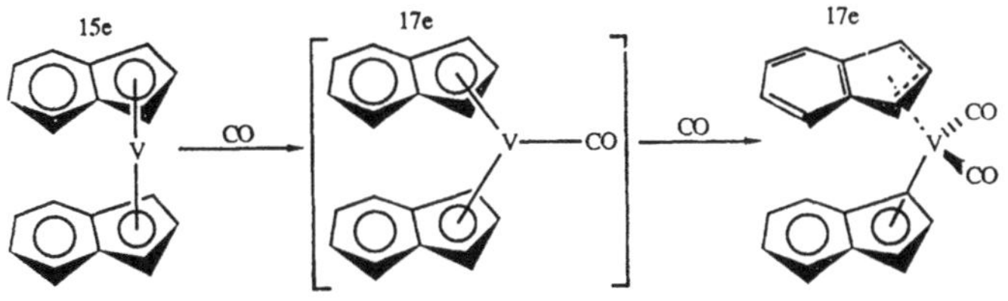

Fig. 3. Proposed scheme for slippage of the indenyl ring on addition of CO to bis-(indenyl)vanadium.

$(H_3C)_4C_2(\eta^5\text{-}C_5H_4)_2Cr(CO)$ slowly adds a second CO ligand under high CO pressure to form $(H_3C)_4C_2(\eta^5\text{-}C_5H_4)(\eta^3\text{-}C_5H_4)\,Cr(CO)_2$ the corresponding reaction of the 17-electron radical cation occurs rapidly. Since the transition state for this process may involve initial nucleophilic attack at Cr by CO the acceleration of ring slippage could have the same origin as the rapid associative substitution processes of 17-electron radicals discussed in the previous sections. One key question is whether ring slippage precedes nucleophilic attack. Either of the mechanisms depicted in Eq. 9 and 10 would exhibit second order kinetics. Intuitively the mechanism of Eq. 9

$$\underset{17e}{\cdot M(\eta^5\text{-}C_5H_5)_2(CO) + L} \xrightarrow{k} \underset{19e}{\text{“}\cdot M(\eta^5\text{-}C_5H_5)_2(CO)L\text{”}} \quad \text{or} \tag{9a}$$

$$\underset{17e}{\text{“}\cdot M(\eta^5\text{-}C_5H_5)(\eta^3\text{-}C_5H_5)(CO)L\text{”}} \longrightarrow \underset{17e}{\cdot M(\eta^5\text{-}C_5H_5)_2L} + CO \tag{9b}$$

$$\underset{17e}{\cdot M(\eta^5\text{-}C_5H_5)_2(CO)} \underset{}{\overset{\text{fast}}{\rightleftharpoons}} \underset{15e}{\cdot M(\eta^5\text{-}C_5H_5)(\eta^3\text{-}C_5H_5)(CO)} \tag{10a}$$

$$\xrightarrow{k'} \underset{17e}{\cdot M(\eta^5\text{-}C_5H_5)(\eta^3\text{-}C_5H_5)(CO)L} \rightarrow \underset{17e}{\cdot M(\eta^5\text{-}C_5H_5)_2L} + CO \tag{10b}$$

seems more likely since only 17- and 19-electron intermediates are involved, whereas in Eq. 10 the initial preequilibrium generates a 15-electron species. If the mechanism of Eq. 10 were operative one would expect 18-electron complexes to react more rapidly than analogous 17-electron complexes, since a 16-electron intermediate would form in the former system vs a 15-electron intermediate in the latter. As noted above [61] the opposite trend obtains, which favors ring slippage concomitant with nucleophilic attack.

Reactivity comparisons of 17-electron cyclopentadienyl and pentadienyl complexes of vanadium also suggest a secondary role for ring slippage [62, 63]. In 18-electron complexes the reactivity of compounds containing the η^5-pentadienyl ligand ($\eta^5\text{-}C_5H_7$) usually exceeds that of cyclopentadienyl analogues and facile $\eta^5\text{-}\eta^3\text{-}\eta^1$ interconversions have been invoked to explain these differences [64]. By contrast, rates of associative substitution of CO by ^{13}CO in $\cdot V(\eta^5\text{-}C_5H_7)_2(CO)$; $\cdot VCp(\eta^5\text{-}C_5H_7)(CO)$, or $\cdot VCp(\eta^5\text{-}2,4\text{-}Me_2C_5H_5)(CO)$ occur about 10^4 slower than in $\cdot VCp_2(CO)$ or $\cdot VCp_2{}^*(CO)$, where $Cp = \eta^5\text{-}C_5H_5$ and $Cp^* = \eta^5\text{-}C_5Me_5$. The pentadienyl complexes exhibit such low reactivity that two terms rate laws ($k_1 + k_2[^{13}CO]$) are observed and the dissociative path (k_1) dominates the substitution process [62]. This was the first case where a dissociative mechanism could be observed in a 17-electron complex. A similar rate of CO dissociation was observed in the pentadienyl and cyclopentadienyl complexes, as well as in the 18-electron analogue $CrCp(\eta^5\text{-}C_5H_7)(CO)$ [65]. Thus, CO dissociation is not accelerated in the 17-electron analogues of 18-electron complexes. One might expect this since a highly electron deficient 15-electron coordinatively unsaturated intermediate must be involved if a 17-electron complex reacts dissociatively.

The kinetics of substitution for several nitrogen donor ligands (L = NCMe, py, 3-F-py, 3-Cl-py, and 4-X-py for X = NO_2, OMe, $C(O)CH_3$, and Me) in $\cdot Mn(\eta^5\text{-}C_5H_4Me)(CO)_2L^+$ has been explored for a range of trivalent phosphorus nucleophiles [66, 67]. Through the use of transient electrochemical techniques, described in detail elsewhere in this volume [68], the reaction was shown to be associative and the low $\Delta H\ddagger$ (4.4 kcal/mole) and negative $\Delta S\ddagger$ (−25 cal/mole K) for L = PPh_3 nucleophile [66] resemble those for the reaction between $\cdot V(CO)_6$ and PPh_3 nucleophile ($\Delta H\ddagger$ = 10.0 kcal/mole and $\Delta S\ddagger$ = −28 cal/mole K) [42]. As discussed later the large negative entropic term seems characteristic for associative substitutions in organometallic radicals. The nucleophile dependence parallels that found for $\cdot V(CO)_6$, so the labilizing effect of ligands in 17- electron complexes may be observed for either CO or nitrogen donors as leaving groups.

In an experiment, similar to that used to define associative reactivity in $\cdot Re(CO)_5$ [24], the competition between chlorine abstraction from $CClPh_3$ and substitution of CO by PR_3 (R = Bu, OPh, and Ph) in 17-electron $\cdot W(\eta\text{-}C_5H_5)(CO)_3$ was used to establish an associative mechanism for ligand substitution [69]. The 17-electron species may be generated by rapid hydrogen atom abstraction according to Fig. 4.

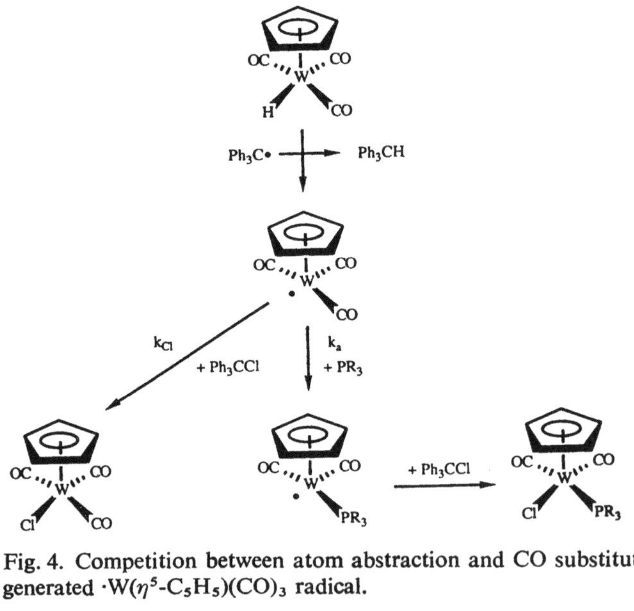

Fig. 4. Competition between atom abstraction and CO substitution in the thermally generated $\cdot W(\eta^5\text{-}C_5H_5)(CO)_3$ radical.

The product ratio for this kinetic scheme follows Eq. 11 if the substitution step is associative.

$$\frac{[W(\eta^5\text{-}C_5H_5)(CO)_2LCl]}{[W(\eta^5\text{-}C_5H_5)(CO)_3Cl]} = \frac{k_a[L]}{k_{Cl}[RCl]} \tag{11}$$

Under a CO atmosphere the amount of substitution product formed decreases, which can be attributed to reversibility of the ligand substitution step or to formation of $\cdot W(\eta^5\text{-}C_5H_5)(CO)_4$ and $\cdot W(\eta^3\text{-}C_5H_5)(CO)_4$.

A study of the electrochemical oxidation of a series of (η^6-arene)chromium and (η^6-arene)tungsten tricarbonyl complexes originally suggested [70] that coordination sphere expansion occurs when nucleophiles are added to the 17-electron tungsten radical cations. Oxidation of $(\eta\text{-}C_6Me_6)W(CO)_3$ in the presence of CH_3CN, DMF, or PR_3 (R = Bu, Ph, OPh, OBu), produces a new reduction wave that can be attributed to a 19-electron complex; however, recent work finds the initial oxidation wave to be a more complex two-electron process [71] and a 19-electron species can not be observed. Although (arene)tricarbonylchromium radical cations were known to be capable of being formed in a one-electron oxidation process [72 73], long term stability has only been achieved in solvents of low nucleophilicity (e.g. trifluoroacetic acid, propylene carbonate, or benzene [71, 74, 75]) or at low temperatures (e.g. −90 °C in CH_2Cl_2 [76]). The use of tetra-*n*-butylammonium hexfluorophosphate as supporting electrolyte results in superior stability of the radical cation when compared to corresponding perchlorate, tetrafluoroborate, or trifluomethanesulfonate salts [76]. In DMF a net irreversible 3-electron oxidation of $Cr(\eta^6\text{-}XC_6H_5)(CO)_3$ to XC_6H_5, Cr(III), and CO occurs [77]. Partial CO substitution by $P(OBu)_3$ also has been detected [71] during a one-electron oxidation-reduction cycle. The two-electron oxidation of $Cr(\eta^6\text{-}C_6H_6)(CO)_3$ in CH_3CN has been proposed to follow an ECE_h mechanism (E denotes redox at the electrode, C denotes a chemical reaction step, and E_h denotes a homogeneous redox reaction [78–81]). Activation parameters of $E_a = 10.3 \pm 0.1$ kcal/mole and $\Delta S^{\ddagger} = -15.2 \pm 0.4$ cal/mol K measured for the decomposition of $\cdot Cr(\eta^6\text{-}1,3,5\text{-}Me_3C_6H_3)(CO)_3{}^+$ would be consistent with a mechanism that initially involves associative substitution at Cr.

A comprehensive study [71] of the electrochemistry of $Cr(\eta^6\text{-}C_6H_6)(CO)_3$, $Cr(\eta^6\text{-mesitylene})(CO)_3$, and the tungsten analogues defines the stability of 17-electron species formed on oxidation. For example, rotating disk voltammetry studies of $Cr(\eta^6\text{-}C_6H_6)(CO)_3$ show a one-electron oxidation process in CH_2Cl_2 −0.1 M $[NBu_4]PF_6$; however, in CH_2Cl_2 −0.1 M $[NBu_4]ClO_4$ a net two-electron oxidation transpires. The kinetic scheme shown in Eq. 12 was

$$\cdot Cr(\eta^6\text{-}C_6H_6)(CO)_3{}^+ \xrightarrow[k]{Nu} \cdot Cr(\eta^6\text{-}C_6H_6)(CO)_3Nu^+ \xrightarrow{fast} C_6H_6 + 3CO + Cr(I) \quad (12a)$$

$$Cr(I) + \cdot Cr(\eta^6\text{-}C_6H_6)(CO)_3{}^+ \rightarrow Cr(II) + Cr(\eta^6\text{-}C_6H_6)(CO)_3 \quad (12b)$$

$$Cr(I) \xrightarrow{-e} Cr(II) \quad (12c)$$

proposed to explain the decomposition process. In CH_2Cl_2 the rate k depends linearly on $[PF_6{}^-]$, which functions as the nucleophile. Thus, an associative mechanism is suggested and the activation parameters for decomposition ($\Delta H^{\ddagger} \sim 10$ kcal/mol and $\Delta S^{\ddagger} \sim -17$ cal/mol K, based on rates at two temperatures) support this view.

Substitution of the carbon monoxide ligand in $\cdot Cr(\eta^5\text{-}C_5H_5)(CO)_3$ by PEt_3, PBu_3, PMe_2Ph, and PEt_2Ph proceeds rapidly at room temperature; however, heating is necessary with PCy_3 and PPh_3 nucleophiles [82]. The radical, generated by thermal bond homolysis of the metal-metal bonded complex, conforms qualitatively to an associative mechanism for substitution.

Flash photolysis and time resolved IR spectroscopic studies have been applied to the photochemistry of $[Fe(\eta^5\text{-}C_5H_5)(CO)_2]_2$ [83]. Slow addition of L = MeCN, PBu_3 and PPh_3 to a triplet-intermediate [k (25 °C) $\sim 10^5$–10^6 M^{-1} s^{-1}], formulated as $Fe_2(\eta^5\text{-}C_5H_5)_2(CO)_3$, can be readily monitored. This intermediate arises from photodissociation of CO. The 17-electron species $\cdot Fe(\eta^5\text{-}C_5H_5)(CO)_2$ also forms (by photochemical homolysis of the Fe—Fe bond), but recombination occurs too rapidly for the aforementioned L to attack at iron. With L = $P(OMe)_3$ a rapid substitution reaction to form $\cdot Fe(\eta^5\text{-}C_5H_5)(CO)[P(OMe)_3]$ could be detected. This substitution process follows a second order rate law, consistent with an associative process, and at 25 °C the rate is 9×10^8 M^{-1} s^{-1}. It is somewhat surprising that the better nucleophile PBu_3 reacts too slowly to see direct substitution with the radical intermediate. This may suggest an important role for steric effects.

Besides inducing substitution, electron transfer can promote isomerization or coupling of π-bound organic ligands [9]. The 17-electron radical cations $\cdot Fe(\eta^5\text{-}C_5H_5)[\eta^5\text{-}C_6Me_5(CH_2)]^+$, $\cdot Fe(\eta^4\text{-}C_8H_8)(CO)_3^+$, $\cdot Fe(\eta^5\text{-}C_5H_5)(CO)_2[CH_2C(R){=}CR_2]^+$, $\cdot Fe(\eta^5\text{-}C_5H_5)(CO)_2SR$, $\cdot Rh(\eta^5\text{-}C_5H_5)(\eta^4\text{-}C_8H_8)^+$, $\cdot Rh(\eta^5\text{-}C_5H_5)(CO)(PPh_3)^+$, $\cdot Ru(CO)_2L(\eta^4\text{-}C_8H_8)^+$ for L = CO or PPh_3, $\cdot Fe(\eta^5\text{-}C_5H_5)(C{=}CR_2)(Ph_2PCH_2CH_2PPh_2)^{2+}$, and $\cdot Rh(\eta^5\text{-}C_5H_5)(PPh_3)_2^+$ [84–91] all dimerize by carbon-carbon or sulfur-sulfur bond formation through the organic ligand. The $\cdot Rh(\eta^5\text{-}C_5H_5)(CO)L^+$ cations exhibit dual reactivity. For L = PPh_3 dimerization leads to formation of the fulvalene bridged dication [88]; however, for L = PMe_3 or $P(OPh)_3$ metal-metal bond formation predominates [92] (Fig. 5). Delocalization of unpaired spin density onto unsaturated organic ligands must therefore be considered in an analysis of the reactivity of 17-electron π-complexes. Isomerization before ligand-ligand coupling may be important, since one-electron oxidation of $Rh(\eta^5\text{-}C_5H_5)(\eta^4\text{-}C_4H_8)$ is suggested [87] to initially produce $Rh(\eta^5\text{-}C_5H_5)(\eta^5\text{-}C_8H_8)^+$, which maintains an 18-electron count at the metal and localizes the radical onto the C_8 ring. Dimerization then occurs by C—C bond formation through the allyl radical portion of the C_8 ring. A caveat on the literal correlation of product structures with the spin density distribution in the radical precursor has been noted [11].

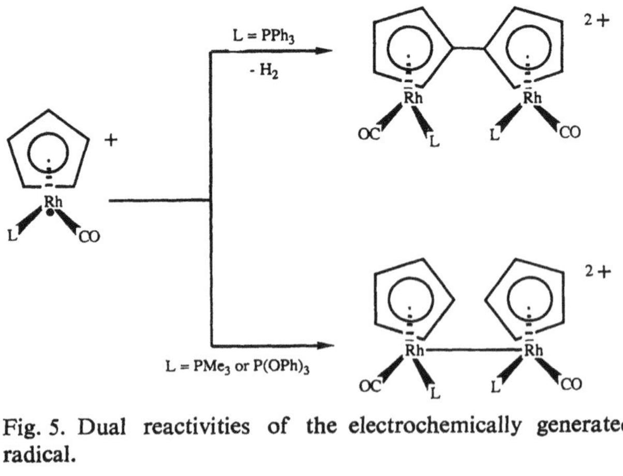

Fig. 5. Dual reactivities of the electrochemically generated $\cdot Rh(\eta^5\text{-}C_5H_5)(CO)L^+$ radical.

2.1.5 Cluster Compounds Containing an Unpaired Electron

The cluster analogue of a 17-electron mononuclear transition metal complex is the one-electron oxidized product of a closed shell cluster complex. Alternatively, a 17-electron center can be generated by slippage of a multielectron ligand or by dissociation of a metal-metal bond (e.g. $M{-}M\cdot^{-} \rightarrow M{:}^{-} + \cdot M$) on one-electron reduction of a cluster species. Therefore, one needs to consider both one-electron oxidations and reductions as mechanisms for introducing a 17-electron radical center in a cluster. For one-electron reduced clusters one must also recognize the enhanced rate of CO dissociation possible in an electron rich species. Since the redox induced lability of cluster complexes is catalogued elsewhere in this volume [93], our discussion here will be limited to the few kinetically documented examples.

Although most examples of redox initiated lability in clusters involve reduction of a coordinatively saturated complex [93], a 10^6 increase in associative reactivity has been established for the 33-electron dinuclear radical, $\cdot Fe_2(CO)_7(\mu\text{-}PPh_2)$, as compared to 34-electron $Fe_2(CO)_6(\mu\text{-}CO)(\mu\text{-}PPh_2)^-$ and $FeCo(CO)_7(\mu\text{-}PPh_2)$ [94]. Kinetics measurements were determined using double potential step chronocoulometry and agree with the EC mechanism shown in Eq. 13. The structural isomerization that happens on oxidation of $Fe_2(CO)_6(\mu\text{-}CO)(\mu\text{-}PPh_2)^-$ is

$$\text{(E)}\ Fe_2(CO)_6(\mu\text{-}CO)(\mu\text{-}PPh_2)^- \underset{+e}{\overset{-e}{\rightleftharpoons}} \cdot Fe_2(CO)_7(\mu\text{-}PPh_2) \qquad (13a)$$

$$\text{(C)}\ \cdot Fe_2(CO)_7(\mu\text{-}PPh_2) + PR_3 \xrightarrow{k_1} \cdot Fe_2(CO)_6(PR_3)(\mu\text{-}PPh_2)^- \qquad (13b)$$

$PR_3 = PCy_3, PPh_3, PPh_2Me, PPhMe_2, PMe_3, PPh(OMe)_2, P(OMe)_3$

rapid on the time scale of the electrochemical experiments. All metal complexes involved in Eq. 13 have been isolated and crystallographically characterized, to provide definitive structures for the molecular species involved. For $PR_3 = PPh_3$ and PCy_3 the rate of product formation follows second order kinetics with $k_1 = 44.2 \pm 2.4$ and $23.3 \pm 2.5\ M^{-1}\ s^{-1}$, respectively, at 20 °C. Associative substitution can be attributed to attack at the 5-coordinate iron radical center in the unsymmetrical dimer as shown in Fig. 6. This requires concomitant loss of a CO ligand on the adjacent metal to account for the structure of the product. The rate enhancement of 10^5–10^6 in the electron deficient cluster is less than the 10^9 observed in mononuclear $\cdot Fe(CO)_3(PR_3)_2^+$ radicals. The attenuated reactivity may be attributed to increased steric hindrance and decreased electrophilicity in the cluster. It should be noted that the dinuclear system contains a radical localized on one metal because of the structural isomerization on oxidation (μ-CO → terminal-CO) of the 34-electron anion. If the unpaired spin density were delocalized over several metal centers then its associative reactivity might be further attenuated.

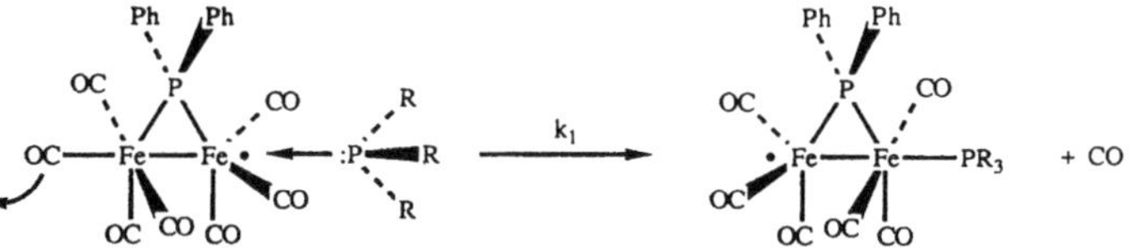

Fig. 6. Proposed mechanism of attack of a phosphine ligand on dinuclear 33-electron $\cdot Fe_2(CO)_7(PPh_2)$.

Cluster radicals formed by one-electron reduction of even electron cluster complexes show an enhanced ability to undergo ligand substitution reactions. Several origins for enhanced reactivity have been identified. For example, $Co_4(CO)_{10}(\mu_4\text{-}PPh)_2$ slowly dissociates CO on one-electron reduction to the radical anion [95]. This is suggested by qualitative observations that the amount of $\cdot Co_4(CO)_9(\mu_4\text{-}PPh]_2L^-$ formed on the cyclic voltammetry time scale does not depend on the amount of $L = PPh_3$ added and its formation decreases on the addition of an atmosphere of CO above the solution. Such behavior has a parallel in mononuclear complexes where reduction of an 18-electron complex to a 19-electron species results in ligand loss to form a 17-electron complex [29, 96].

The $Fe_3(\mu_3\text{-}E)_2(CO)_9$ clusters, where E = PPh, S, Se, and NPh, undergo rapid substitution of a CO ligand after one-electron reduction in the presence of phosphorus donor ligands [97, 98]. The E = S, Se, and NPh derivatives react by dissociative loss of CO with respective first order rates of 6.6, 1.0, and 3×10^3 s^{-1} at 25 °C [98]. However, the E = PPh cluster reacts by rate limiting opening of the cluster where the 3-electron donor μ_3-PPh group isomerizes to a 2-electron donor $\mu_2\text{-}PPh^-$ ligand on reduction [97] (Fig. 7).

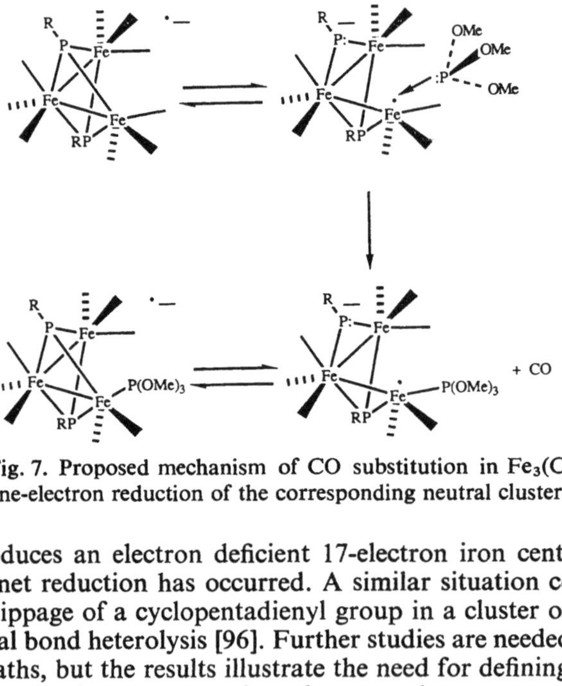

Fig. 7. Proposed mechanism of CO substitution in $Fe_3(CO)_9(\mu^3\text{-}PR)_2^{-\cdot}$ formed by one-electron reduction of the corresponding neutral cluster.

This introduces an electron deficient 17-electron iron center in the cluster even though a net reduction has occurred. A similar situation could be created by an $\eta^5 \rightarrow \eta^3$ slippage of a cyclopentadienyl group in a cluster or by a process such as metal-metal bond heterolysis [96]. Further studies are needed to assess the viability of these paths, but the results illustrate the need for defining the electron count of the reactive metal atom in a polynuclear complex.

2.1.6 *Lability Trends*

Tables 1–3 summarize kinetic data for rates of associative substitution of carbon monoxide and other ligands in the 17-electron complexes discussed in the preceding sections. Several comparisons can be drawn among the various entries. First, steric effects at the metal center play a major rule in influencing radical reactivity.

Table 1. Rates for Associative Substitution of Carbon Monoxide in 17-Electron Radicals

Radical	Nucleophile	$T°$ (C)	k (M^{-1} s^{-1})	Ref.
$\cdot Re(CO)_5$	PPh_3	22	1.1×10^9	see 2.1.2
$\cdot Re(CO)_5$	PBu_3	22	1.6×10^9	see 2.1.2
$\cdot Mn(CO)_5$	$AsPh_3$	24	6.5×10^4	28
$\cdot Mn(CO)_5$	PPh_3	24	1.7×10^7	28
$\cdot Mn(CO)_5$	$P(i\text{-}Pr)_3$	24	6.7×10^7	28
$\cdot Mn(CO)_5$	$P(O\text{-}i\text{-}Pr)_3$	24	3.1×10^7	28
$\cdot Mn(CO)_5$	PBu_3	24	1.0×10^9	28
$\cdot Fe(CO)_3(PPh_3)_2^+$	$(CH_3)_2CO$	20	3.5×10^{-3}	41
$\cdot Fe(CO)_3(PMe_3)_2^+$	py	25	85	40
$\cdot Fe(CO)_3(PPh_3)_2^+$	py	25	13.6	40
$\cdot Fe(CO)_3(PBu_3)_2^+$	py	25	8.1	40
$\cdot Fe(CO)_3(PCy_3)_2^+$	py	25	1.5×10^{-4}	40
$\cdot Fe(CO)_3(PPh_3)_2^+$	3,4-Me_2-py	25	100	40
$\cdot Fe(CO)_3(PPh_3)_2^+$	4-Me-py	25	64	40
$\cdot Fe(CO)_3(PPh_3)_2^+$	3-Me-py	25	52	40
$\cdot Fe(CO)_3(PPh_3)_2^+$	4-CH_3CO-py	25	4.9	40
$\cdot Fe(CO)_3(PPh_3)_2^+$	3-F-py	25	0.76	40
$\cdot V(CO)_6$	PMe_3	25	132	42
$\cdot V(CO)_6$	PBu_3	25	50.2	42
$\cdot V(CO)_6$	$PMePh_2$	25	3.99	42
$\cdot V(CO)_6$	$P(O\text{-}i\text{-}Pr)_3$	25	0.94	42
$\cdot V(CO)_6$	$P(OMe)_3$	25	0.70	42
$\cdot V(CO)_6$	PPh_3	25	0.25	42
$\cdot V(CO)_6$	$P(i\text{-}Pr)_3$	25	0.11	42
$\cdot V(CO)_6$	$AsPh_3$	17	0.018	42
$\cdot V(CO)_6$	py	25	1.22	44
$\cdot V(CO)_6$	Et_3N	25	0.765	44
$\cdot V(CO)_6$	MeCN	25	0.284	44
$\cdot V(CO)_6$	MeOH	25	2.29×10^{-2}	44
$\cdot V(CO)_6$	$(CH_3)_2CO$	25	1.93×10^{-2}	44
$\cdot V(CO)_6$	THF	25	4.35×10^{-3}	44
$\cdot V(CO)_6$	2,5-Me_2THF	25	1.06×10^{-3}	44
$\cdot V(CO)_6$	DME	21	2.8×10^{-4}	44
$\cdot V(CO)_6$	$MeNO_2$	21	1.05×10^{-4}	44
$\cdot V(CO)_6$	Et_2O	25	9.63×10^{-4}	44
$\cdot V(CO)_5(PBu_3)$	PBu_3	25	5.42×10^{-4}	42
$\cdot V(CO)_5(PMePh_2)$	$PMePh_2$	25	2.28×10^{-3}	42
$\cdot V(CO)_5[P(O\text{-}i\text{-}Pr)_3]$	$P(O\text{-}i\text{-}Pr)_3$	25	7.38×10^{-4}	42
$\cdot V(CO)_5[P(OMe)_3]$	$P(OMe)_3$	25	1.49×10^{-3}	42
$\cdot V(CO)_5(PPh_3)$	PPh_3	25	2.81×10^{-4}	42
$\cdot V(CO)_5(PBu_3)$	THF	25	2.7×10^{-6}	44
$\cdot V(CO)_5(PBu_3)$	Et_2O	25	1.8×10^{-6}	44
$\cdot V(CO)_5(PPh_3)$	Et_2O	25	3.0×10^{-6}	44
$\cdot VCp^*_2(CO)$	MeO_2CCCCO_2Me	25	7.67×10^{-3}	62
$\cdot VCp^*_2(CO)$	^{13}CO	10	21.5	62
$\cdot VCp^*_2(CO)$	^{13}CO	60	260	62

Table 1.

Radical	Nucleophile	T° (C)	k (M^{-1} s^{-1})	Ref.
$\cdot VCp_2(CO)$	^{13}CO	60	800	63
$\cdot VCp(\eta^5\text{-}2,4\text{-}Me_2C_5H_5)(CO)$	^{13}CO	60	0.013	63
$\cdot VCp(\eta^5\text{-}C_5H_7)(CO)$	^{13}CO	60	0.006	63
$\cdot V(\eta^5\text{-}C_5H_7)_2(CO)$	^{13}CO	60	0.0038	63
$\cdot WCp(CO)_3$	PPh_3	24	8.6×10^2	69
$\cdot WCp(CO)_3$	$P(OPh)_3$	24	1.3×10^4	69
$\cdot FeCp(CO)_2$	$P(OMe)_3$	25	9×10^8	83
$\cdot Fe_2(CO)_7(\mu\text{-}PPh_2)$	PCy_3	20	23	94
$\cdot Fe_2(CO)_7(\mu\text{-}PPh_2)$	PPh_3	20	44	94
$\cdot Fe_2(CO)_7(\mu\text{-}PPh_2)$	PPh_2Me	20	350	94
$\cdot Fe_2(CO)_7(\mu\text{-}PPh_2)$	$PPhMe_2$	20	950	94
$\cdot Fe_2(CO)_7(\mu\text{-}PPh_2)$	PMe_3	20	1.1×10^3	94
$\cdot Fe_2(CO)_7(\mu\text{-}PPh_2)$	$PPh(OMe)_2$	20	225	94
$\cdot Fe_2(CO)_7(\mu\text{-}PPh_2)$	$P(OMe)_3$	20	240	94

py = pyridine, Cp = $\eta^5\text{-}C_5H_5$, Cp* = $\eta^5\text{-}C_5Me_5$, Me = CH_3, Et = C_2H_5, Ph = C_6H_5, Cy = $c\text{-}C_6H_{11}$, Pr = $n\text{-}C_3H_7$, Bu = $n\text{-}C_4H_9$.

Table 2. Activation Parameters for Associative Substitution of Carbon Monoxide in 17-Electron Radicals

Radical	Nucleophile	ΔH‡ (kcal/mol)	ΔS‡ (cal/mol K)	Ref.
$\cdot Fe(CO)_3(PPh_3)_2^+$	py	9.8 ± 0.3	-21 ± 1	40
$\cdot Fe(CO)_3(PCy_3)_2^+$	py	14 ± 1.5	-27 ± 5	40
$\cdot V(CO)_6$	PMe_3	7.6 ± 0.7	-23 ± 3	42
$\cdot V(CO)_6$	PBu_3	7.6 ± 0.4	-25 ± 2	42
$\cdot V(CO)_6$	$PMePh_2$	8.9 ± 0.3	-26 ± 1	42
$\cdot V(CO)_6$	$P(OMe)_3$	10.9 ± 0.2	-22.6 ± 0.8	42
$\cdot V(CO)_6$	PPh_3	10.0 ± 0.4	-28 ± 2	42
$\cdot V(CO)_6$	$(CH_3)_2CO$	14 ± 1	-22 ± 4	44
$\cdot V(CO)_5[P(OMe)_3]$	$P(OMe)_3$	13.2 ± 0.4	-27.6 ± 1.8	42
$\cdot VCp^*_2(CO)$	MeO_2CCCCO_2Me	10.8 ± 1.1	-32 ± 4	62
$\cdot VCp^*_2(CO)$	^{13}CO	8.9 ± 0.7	-21 ± 3	62
$\cdot V(C_5H_7)_2(CO)$	^{13}CO	22.7 ± 0.5	-2 ± 1	62
$\cdot Fe_2(CO)_7(\mu\text{-}PPh_2)$	PPh_3	9.9 ± 1.0	-18 ± 4	94

For example, six coordinate $\cdot V(CO)_6$ is about 10^7 less reactive toward PBu_3 nucleophile than five-coordinate $\cdot Mn(CO)_5$ or $\cdot Re(CO)_5$. The reactivity of $\cdot Fe(CO)_3(PCy_3)_2^+$ with pyridine nucleophile is nearly 10^6 less than $\cdot Fe(CO)_3(PMe_3)_2^+$. In the substitution of phosphine ligands by CO the hindered complex $\cdot Mn(CO)_3[P(i\text{-}Bu)_3]_2$ exhibits a 10^2 decrease in reactivity as compared to $\cdot Mn(CO)_3(PBu_3)_2$. Carbonyl complexes that contain cyclopentadienyl rings may show reduced reactivity, although that is not always the case, as evidenced by the entry for $\cdot FeCp(CO)_2$.

Table 3. Rates for Associative Substitution of Noncarbonyl Ligands in 17-Electron Radicals

Complex	Nucleophile	T (°C)	k (M^{-1} s^{-1})	Ref.
$\cdot Mn(CO)_3[P(i\text{-}Bu)_3]_2$	CO	20	0.32	32
$\cdot Mn(CO)_3(PBu_3)_2$	CO	20	42.	32
$\cdot V(CO)_5(PBu_3)$	$P(OMe)_3$	25	7.00	42
$\cdot Mn(\eta^5\text{-}C_5H_4Me)(CO)_2(NCMe)^+$	PPh_3[a]	25	1.3×10^4	66
$\cdot Mn(\eta^5\text{-}C_5H_4Me)(CO)_2(NCMe)^+$	$P(OPh)_3$	25	12	66
$\cdot Mn(\eta^5\text{-}C_5H_4Me)(CO)_2(NCMe)^+$	$PMePh_2$	25	$>10^5$	66
$\cdot Mn(\eta^5\text{-}C_5H_4Me)(CO)_2(NCMe)^+$	$P(OMe)Ph_2$	25	2.5×10^4	66
$\cdot Mn(\eta^5\text{-}C_5H_4Me)(CO)_2(NCMe)^+$	$P(4\text{-}Cl\text{-}C_6H_4)_3$	25	9.5×10^2	66
$\cdot Mn(\eta^5\text{-}C_5H_4Me)(CO)_2(NCMe)^+$	$P(4\text{-}Me\text{-}C_6H_4)_3$	25	3.0×10^4	66
$\cdot Mn(\eta^5\text{-}C_5H_4Me)(CO)_2(py)^+$	PPh_3	25	5.0	66
$\cdot Mn(\eta^5\text{-}C_5H_4Me)(CO)_2(py)^+$	PEt_3	25	1.3×10^3	66
$\cdot Mn(\eta^5\text{-}C_5H_4Me)(CO)_2(py)^+$	$P(OMe)Ph_2$	22	130	67
$\cdot Mn(\eta^5\text{-}C_5H_4Me)(CO)_2(4\text{-}OMe\text{-}py)^+$	$P(OMe)Ph_2$	22	75	67
$\cdot Mn(\eta^5\text{-}C_5H_4Me)(CO)_2(4\text{-}Me\text{-}py)^+$	$P(OMe)Ph_2$	22	85	67
$\cdot Mn(\eta^5\text{-}C_5H_4Me)(CO)_2(4\text{-}C(O)Me\text{-}py)^+$	$P(OMe)Ph_2$	22	220	67
$\cdot Mn(\eta^5\text{-}C_5H_4Me)(CO)_2(3\text{-}Me\text{-}py)^+$	$P(OMe)Ph_2$	22	125	67
$\cdot Mn(\eta^5\text{-}C_5H_4Me)(CO)_2(3\text{-}F\text{-}py)^+$	$P(OMe)Ph_2$	22	345	67
$\cdot Mn(\eta^5\text{-}C_5H_4Me)(CO)_2(3\text{-}Cl\text{-}py)^+$	$P(OMe)Ph_2$	22	410	67
$\cdot Mn(\eta^5\text{-}C_5H_4Me)(CO)_2(4\text{-}NO_2\text{-}py)^+$	$P(OMe)Ph_2$	22	410	67
$\cdot Mn(\eta^5\text{-}C_5H_4Me)(CO)_2(4\text{-}NO_2\text{-}py)^+$	PEt_3	22	$2.\times10^4$	67
$\cdot Mn(\eta^5\text{-}C_5H_4Me)(CO)_2(4\text{-}NO_2\text{-}py)^+$	$PPhEt_2$	22	$2.\times10^3$	67
$\cdot Mn(\eta^5\text{-}C_5H_4Me)(CO)_2(4\text{-}NO_2\text{-}py)^+$	PPh_2Et	22	170.	67
$\cdot Mn(\eta^5\text{-}C_5H_4Me)(CO)_2(4\text{-}NO_2\text{-}py)^+$	PPh_3	22	17	67
$\cdot Mn(\eta^5\text{-}C_5H_4Me)(CO)_2(4\text{-}NO_2\text{-}py)^+$	$P(4\text{-}Me\text{-}C_6H_4)_3$	22	55	67
$\cdot Mn(\eta^5\text{-}C_5H_4Me)(CO)_2(4\text{-}NO_2\text{-}py)^+$	$P(4\text{-}MeO\text{-}C_6H_4)_3$	22	150	67
$\cdot Mn(\eta^5\text{-}C_5H_4Me)(CO)_2(4\text{-}NO_2\text{-}py)^+$	PPh_2Me	22	1100	67

[a] $\Delta H^\ddagger = 4.4$ kcal/mol, $\Delta S^\ddagger = -25$ cal/mol K.

A second general observation is that phosphine substitution for CO produces a radical less susceptible toward nucleophilic attack. For example, $\cdot V(CO)_5(PBu_3)$ is 10^5 less reactive than $\cdot V(CO)_6$ toward PBu_3 nucleophile. The radicals $\cdot Fe(CO)_3L_2^+$ have a finite lifetime in solution, while the parent $\cdot Fe(CO)_5^+$ can not be observed. This stabilizing effect may arise from a combination of increased steric bulk at the metal and decreased electrophilicity on replacing the π-acceptor CO ligand with a strong σ-donor phosphine. Electronic effects must be significant since the large reactivity difference between $\cdot V(CO)_6$ and $\cdot V(CO)_5(PBu_3)$ can not be explained on steric effects alone. One-half of the octahedron should remain unhindered in a monosubstituted octahedral complex, although crowding might be intensified in a seven-coordinate transition state.

Data in Table 3 illustrate that associative lability is not limited to CO as the leaving group because phosphorus and nitrogen donor ligands are also labile to substitution in 17-electron compounds. This lability also extends to complexes that contain cyclopentadienyl or pentadienyl ligands. These systems provide the first examples of CO dissociation in a 17-electron system (Table 4); however, the dissociative process does not exhibit any unusual acceleration, as we noted earlier.

Table 4. Rates and Activation Parameters for Dissociative Substitution of CO in 17-Electron Radicals

Complex	T (°C)	k (s^{-1})	$\Delta H\ddagger$ (kcal/mol)	$\Delta S\ddagger$ (cal/mol K)	Ref.
$\cdot VCp_2(CO)$	60	$\sim 10^{-4}$	–	–	62
$\cdot VCp^*_2(CO)$	60	$\sim 10^{-4}$	–	–	62
$\cdot VCp(\eta^5\text{-}2,4\text{-}Me_2C_5H_5)(CO)$	60	3.1×10^{-4}	27.9 ± 0.8	9 ± 2	63
$\cdot VCp(\eta^5\text{-}C_5H_7)(CO)$	60	2.7×10^{-4}	28.8 ± 0.3	11.2 ± 0.8	63
$\cdot V(\eta^5\text{-}2,4\text{-}Me_2C_5H_5)(CO)$	60	2.1×10^{-3}	27	11	63
$\cdot V(\eta^5\text{-}C_5H_7)_2(CO)$	60	8.1×10^{-6}	28.1 ± 0.4	2 ± 1	63

Activation parameters for the associative substitution path in 17-electron radicals (Table 2) exhibit a remarkable similarity for a range of first row metal-ligand systems. Generally $\Delta H\ddagger$ is low (9 to 14 kcal/mole) and $\Delta S\ddagger$ is large and unfavorable (−20 to −30 cal/mol K). The exception to this rule, $\cdot V(\eta^5\text{-}C_5H_7)_2(CO)$, may be suspect because the dominant reaction path is CO dissociation and the data obtained for the associative path (by difference) are prone to error. The $\Delta H\ddagger$, which are less than the 25–30 kcal/mole required for CO dissociative processes [99, 100], suggest advanced metal-nucleophile bond formation in the transition state for CO substitution. This derives further support from the highly unfavorable $\Delta S\ddagger$, which suggests a highly organized transition state. A key question is whether the associative reaction process involves a 19-electron complex as a transition state (A) or intermediate (B) as depicted in Fig. 8. This seems likely to depend on the

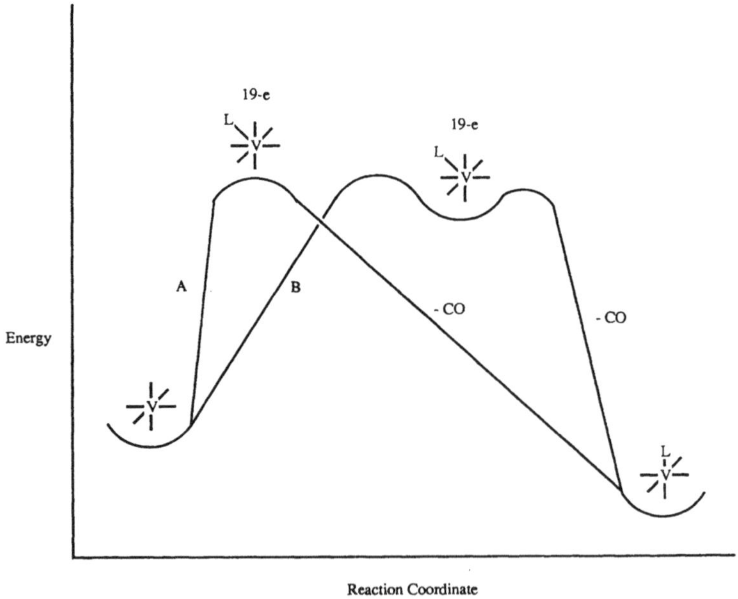

Fig. 8. Alternative roles of a 19-electron complex as a transition state (A) or an intermediate (B) depicted for ligand substitution in $\cdot V(CO)_6$.

specific system as noted in a following section on electronic structure-reactivity correlations.

2.1.7 *Gas Phase 17-Electron Species*

It is noteworthy that the first observation of ligand substitution in a gas phase organometallic species was for 17-electron $\cdot CoCp(CO)^-$ reacting with PF_3 to form $\cdot CoCp(PF_3)^-$ [101]. The 19-electron $\cdot CoCp(CO)_2^-$ was also detected as a long-lived species. In considering gas phase reactions one must recognize that a simple dissociative process of a thermalized complex can not occur, however, a collision activated dissociation can occur.

Studies of the 17-electron complexes $\cdot Fe(CO)(NO)_2^-$, $\cdot Co(CO)_2(NO)^-$, and $\cdot Co(\eta^3\text{-}C_3H_5)(CO)_2$ show [102] that associative addition of PF_3 occurs at rates approaching the diffusion limit. In the presence of He buffer gas a 19-electron intermediate persists as a stable species with relative fractional yields of formation as shown above the arrows in Eq. 14–16.

$$\cdot Fe(CO)(NO)_2^- + PF_3 \xrightarrow[He]{0.79} \cdot Fe(CO)(NO)_2(PF_3)^- \tag{14a}$$

$$\xrightarrow{0.21} \cdot Fe(NO)_2(PF_3)^- + CO \xrightarrow[PF_3]{He} \cdot Fe(NO)_2(PF_3)_2^- \tag{14b}$$

$$\cdot Co(CO)_2(NO)^- + PF_3 \xrightarrow[He]{0.38} \cdot Co(CO)_2(NO)(PF_3)^- \xrightarrow{0.62} \tag{15a}$$

$$\cdot Co(CO)(NO)(PF_3)^- + CO \xrightarrow[PF_3]{He} \cdot Co(CO)(NO)(PF_3)_2^- \rightarrow \tag{15b}$$

$$\cdot Co(NO)(PF_3)_2^- + CO \xrightarrow[PF_3]{He} \cdot Co(NO)(PF_3)_3^- \tag{15c}$$

$$\cdot Co(\eta^3\text{-}C_3H_5)(CO)_2^- + PF_3 \xrightarrow{0.48} \cdot Co(\eta^1\text{-}C_3H_5)(CO)_2(PF_3)^- \tag{16a}$$

$$\xrightarrow{0.52} \cdot Co(\eta^3\text{-}C_3H_5)(CO)(PF_3)^- + CO \xrightarrow[PF_3]{He} \tag{16b}$$

$$\cdot Co(\eta^1\text{-}C_3H_5)(CO)(PF_3)_2^- \rightarrow \cdot Co(\eta^3\text{-}C_3H_5)(PF_3)_2^- + CO \tag{16c}$$

The stability of complexes of expanded coordination number was attributed to the presence of NO or the π-bound allyl ligands, which can undergo geometrical isomerization (e.g. linear NO $\rightarrow$ bent NO or $\eta^3\text{-}C_3H_5 \rightarrow \eta^1\text{-}C_3H_5$) and thereby maintain a 17-electron count. Of complexes that lack a suitable ligand, such as $\cdot Fe(CO)_4^-$, $\cdot Cr(CO)_5^-$, and $\cdot MnH(CO)_4^-$, only $\cdot Fe(CO)_4^-$ undergoes CO substitution by an associative process (Eq. 17). Attempts to detect a species of expanded

$$\cdot Fe(CO)_4^- + PF_3 \rightarrow \cdot Fe(CO)_3(PF_3)^- + CO \xrightarrow{PF_3} \tag{17a}$$

$$\cdot Fe(CO)_2(PF_3)_2^- + CO \xrightarrow{PF_3} \cdot Fe(CO)(PF_3)_3^- + CO \tag{17b}$$

coordination number (e.g. $\cdot Fe(CO)_4(PF_3)^-$) in the presence of a buffer gas were not successful [93]. When PMe_3 was substituted for PF_3 in the above studies no reaction occurred [102, 103]. Thus, the limited data available suggest that gas phase 17-electron complexes undergo facile associative substitution and that 19-electron intermediates are only stable with flexible ligands capable of removing the excess electron density.

2.2 Substitution – Disproportion Reactions

As noted in preceding sections the thermal substitution of CO groups in 17-electron radicals (e.g. $\cdot V(CO)_6$, $\cdot Cr(\eta\text{-}C_6H_6)(CO)_3^+$, or $\cdot Fe(CO)_3(PR_3)_2^+$) can induce disproportionation. Photochemically induced disproportionation reactions are reviewed elsewhere [96, 104], and the fundamental question of electron transfer addressed [29]. Here we summarize two mechanisms favored for such processes.

For $\cdot V(CO)_6$ the disproportionation observed in the presence of oxygen or nitrogen donor ligands is attributed to a substitution-electron transfer sequence as shown in Eq. 18 [43, 44].

$$\cdot V(CO)_6 + L \xrightarrow{slow} \cdot V(CO)_5L + CO \tag{18a}$$

$$\cdot V(CO)_6 + \cdot V(CO)_5L \xrightarrow{fast} V(CO)_6^- + V(CO)_5L^+ \tag{18b}$$

$$V(CO)_5L^+ + \cdot V(CO)_6 + 5L \xrightarrow{fast} VL_6^{2+} + V(CO)_6^- + 5CO \tag{18c}$$

Destabilization of a t_{2g} orbital in $\cdot V(CO)_6$ by replacing a CO ligand with the weak-field base, L, provides a driving force for electron transfer to ultimately afford the stable 18-electron vanadium hexacarbonyl anion and V(II). Thus the exceptional lability of radicals toward substitution ultimately aids the disproportionation process. The susceptibility of these radicals to associative attack may also accelerate the electron transfer process if the CO ligand acts as an inner sphere electron transfer bridge [105]. Circumstantial evidence for this derives support from the observation of an isocarbonyl bridged species $(OC)_5V(CO)VL_4(OC)V(CO)_5$ as a stable product for L = THF or Et_2O [106]. When L = pyridine such a species was detected by FTIR spectroscopy as an intermediate in the net reaction of Eq. 18 [44]. An illustration of this reactivity is the rapid electron transfer that occurs between the one-electron reductant $V(\eta^5\text{-}C_5Me_5)_2$ and $\cdot V(CO)_6$ to form the isocarbonyl bridged product $V(\eta^5\text{-}C_5Me_5)_2(OC)V(CO)_5$ [107]. Electron transfer appears to be reversible in forming the $\cdot V(CO)_6$ disproportionation products since they partially revert to $\cdot V(CO)_6$ when coordinating agents are removed [44, 108].

An alternative disproportionation mechanism has been advanced [96] to explain certain photochemically induced electron transfer processes. In this mechanism the entering ligand L not only substitutes CO, but electron transfer proceeds from a proposed 19-electron intermediate present in a small steady state concentration. The scheme suggested for photoinduced disproportionation of $Mn_2(CO)_{10}$ in the presence of nitrogen donor ligands, L, is shown in Eq. 19. The distinction between the two mechanisms for disproportionation of radicals is difficult to establish experimentally and more quantitative studies are needed.

$$\cdot Mn(CO)_5 + L \rightarrow \cdot Mn(CO)_4L + CO \tag{19a}$$

$$\cdot Mn(CO)_4L + L \rightarrow \cdot Mn(CO)_3L_2 + CO \tag{19b}$$

$$\cdot Mn(CO)_3L_2 + L \rightleftharpoons \cdot Mn(CO)_3L_3 = 19\text{-e} \tag{19c}$$

$$\cdot Mn(CO)_3L_3 + \cdot Mn_2(CO)_{10} \rightarrow [Mn(CO)_3L_3]^+ + \cdot Mn(CO)_5 + [Mn(CO)_5]^- \tag{19d}$$

2.3 Insertion Reactions

Electron deficient Cr(III) alkyl radicals $[Cr(\eta^5\text{-}C_5Me_5)ClMe]_2$ have been shown to be labile toward insertion of nitriles and olefins [109, 110]. Oxidatively induced migratory insertion of a metal carbonylalkyl complex to form a metal acyl complex has been established in several cases [111–125] and a reductively induced example has been described [126]. One of the earliest studies [111] showed that oxidation of $Fe(\eta^5\text{-}C_5H_5)(CO)_2$(*cis*-4-methylcyclohexane) and $Fe(\eta^5\text{-}C_5H_5)(CO)_2$ (*trans*-4-methylcyclohexane) with Cu^{2+} in ethanol solvent produced *cis* and *trans*-

4-methyl-ethylcyclohexanate, respectively, which show retention of stereochemistry by the migrating alkyl group. Similar results were obtained on oxidation of $Fe(\eta^5\text{-}C_5H_5)(CO)_2$(*erythro*-3,3-dimethylbutyl-1,2-d_2) with a wide range of oxidants [113]. Thus, insertion proceeds with retention of configuration as observed for normal migratory insertion reactions [127, 128] of 18-electron compounds.

Mechanistic studies of migratory insertion have focused on $Fe(\eta^5\text{-}C_5H_5)(CO)_2Me$, $Fe(\eta^5\text{-}C_5H_5)(CO)(PPh_3)Me$, and derivatives [121–125]. Oxidation of $Fe(\eta^5\text{-}C_5H_5)(CO)_2Me$ with $Cu(O_3SCF_3)_2$ at 20 °C in acetonitrile immediately yields $\cdot Fe(\eta^5\text{-}C_5H_5)(CO)(NCCH_3)(COCH_3)^+$, which persists with a half-life of 22 sec before decomposing [121]. Two factors seem important in the oxidatively induced insertion process. First, oxidation is estimated to shift the equilibrium constant for CO insertion by 10^{11}–10^{12} in favor of the insertion product [122]. Second, this change is predominantly attributed to a 10^{11} increase in the forward rate constant for insertion in the 17-electron complex [121]. Later studies [123] of $Fe(\eta^5\text{-}C_5H_5)(CO)(PPh_3)CH_3$ show that the redox-catalyzed insertion follows the rate law of Eq. 20.

$$\text{rate} = k[Fe(\eta^5\text{-}C_5H_5)_2{}^+][CO] \tag{20}$$

Here ferrocenium serves as oxidant and at 2 °C in CH_2Cl_2 solvent $k = 500 \pm 100\ M^{-1}\ s^{-1}$.

Two potential mechanisms for the insertion process in $\cdot Fe(\eta^5\text{-}C_5H_5)(CO)(PPh_3)CH_3{}^+$ were considered [125]. One possibility shown in Fig. 9 is that oxidation induces reversible insertion followed by trapping with a nucleophile such as acetonitrile or pyridine. This resembles the mechanism favored for 18-electron systems [128]. If this were the case it is difficult to explain why a 17-electron complex should go to a 15-electron acyl intermediate faster than the corresponding

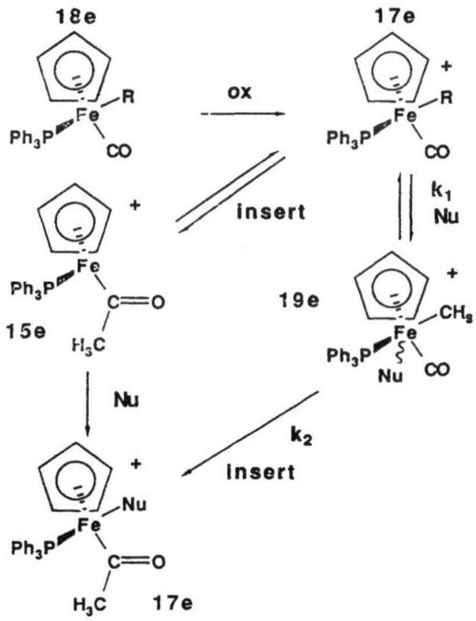

Fig. 9. Alternative mechanisms for the oxidatively induced CO insertion reaction for $Fe(\eta^5\text{-}C_5H_5)(CO)(PPh_3)CH_3$, where Nu = a nucleophile such as acetonitrile or pyridine.

18-electron complex would go to a 16-electron acyl intermediate. Given the susceptibility of 17-electron systems to undergo nucleophilic attack a more likely mechanism would be a nucleophile induced insertion that proceeds through a 19-electron intermediate or transition state as shown in the right hand column of Fig. 9. Both mechanisms, however, should exhibit a nucleophile dependence so that kinetics alone can not distinguish between the two possibilities.

The key role of a nucleophile in the oxidatively induced insertion reaction of $Fe(\eta^5\text{-}C_5H_5)(CO)(PPh_3)CH_3$ is demonstrated by the observation of its persistent 17-electron radical cation produced on electrochemical oxidation at 20 °C in CH_2Cl_2 solvent with 0.1 M $(NBu_4)PF_6$ electrolyte [125]. Addition of a 30-fold excess of pyridine to such a solution results in CO insertion on oxidation at the electrode surface as observed for experiments in CH_3CN solvent [123]. At low temperatures (−20 to 0 °C) the cyclic voltammograms of $Fe(\eta^5\text{-}C_5H_5)(CO)(PPh_3)CH_3$ reveal an intermediate species to the acyl insertion product [125]. By adding excess pyridine at −20 °C the intermediate can be formed exclusively. Since the amount of this complex depends directly on the amount of pyridine and is unperturbed by the addition of free PPh_3 the equilibrium of Eq. 21 was

$$\cdot Fe(Cp)(CO)(PPh_3)CH_3^+ + py \overset{K}{\rightleftharpoons} \cdot Fe(Cp)(CO)(PPh_3)(py)CH_3^+ \tag{21}$$

postulated. The value of K obtained (~7) suggests that a 19-electron species exists as a stable intermediate. Alternatively, the $\eta^5\text{-}C_5H_5$ ring in the pyridine adduct could slip to η^3 bonding and preserve a 17-electron configuration. Spectroscopic support for the structure of the intermediate derives from FTIR studies of $\cdot FeCp(CO)(PPh_3)CH_3^+$ generated in CH_2Cl_2 by chemical oxidation with $AgPF_6$ [125]. The terminal CO stretch at 1966 cm^{-1} shifts to 1916 cm^{-1} on adding pyridine-d_5, which is consistent with the formation of an electron-rich 19-electron adduct. Slow formation of the insertion product $\cdot FeCp(PPh_3)(py)(COCH_3)^+$ then follows (ν_{CO} = 1645 cm^{-1}) on warming. At 20 °C the k_1 step (Fig. 9) limits the rate and the second order rate constants obtained by the use of transient electrochemical techniques are shown in Table 5. The Hammet ϱ of −4.69 obtained resembles that of −3.25 observed for nucleophilic attack of pyridine at the iron center in 17-electron $\cdot Fe(CO)_3(PPh_3)_2^+$ [40, 125]. At 0° the k_2 for the insertion step, measured for several nucleophiles (Table 6), reveals that electron donation accelerates the insertion process. Here, one has the interesting paradox that oxidation to a 17-electron complex favors insertion of CO because a nucleophile can add to drive the reaction via a 19-electron intermediate.

Table 5. Rate Constants[a] for Nucleophilic Attack at $\cdot FeCp(CO)(PPh_3)CH_3^+$ by Substituted Pyridines at 20 °C

nucleophile	k (M^{-1} s^{-1})	nucleophile	k (M^{-1} s^{-1})
3,4-dimethylpyridine	200	3-fluoropyridine	0.27
4-methylpyridine	105	3-chloropyridine	0.14
3-methylpyridine	93.7	4-cyanopyridine	0.10
pyridine	33.8 ± 2.4	3-cyanopyridine	0.04

[a] Experimental conditions: $[FeCp(CO)(PPh_3)CH_3] = 1 \times 10^{-3}$ M, TBAHFP = 0.15 M, solvent = CH_2Cl_2, temperature = 20 °C. Rates measured by double potential step chronocoulometry from Ref. 125.

Table 6. Rate Constants for the Conversion of $\cdot FeCp(CO)(PPh_3)(Nu)CH_3^+$ to $\cdot FeCp(PPh_3)(Nu)(COCH_3)^+$

nucleophile, Nu	chemical step	rate constant	k
3,4-dimethylpyridine[a]	insertion	k (s^{-1})	1.10 ± 0.12
4-methylpyridine	insertion	k (s^{-1})	0.47 ± 0.04
3-methylpyridine	insertion	k (s^{-1})	0.20

[a] Experimental conditions: [$FeCp(CO)(PPh_3)CH_3$] = 1×10^{-3} M, TBAHFP = 0.15 M, solvent = CH_2Cl_2, temperature = 0.0°C. Rates measured by double potential step chronocoulometry from Ref. 125.

The above mechanism has not received universal acceptance [124]. Oxidation of $Fe(\eta^5\text{-}C_5H_4Me)(CO)_2CH_3$ in methanol/acetonitrile mixtures led to the formation of $\cdot Fe(\eta^5\text{-}C_5H_4Me)(CO)(OHMe)(COMe)^+$ and $\cdot Fe(\eta^5\text{-}C_5H_4Me)(CO)(NCCH_3)(COMe)^+$ with equal rate constants [124]. This is consistent with the presence of a 15-electron species $\cdot Fe(\eta^5\text{-}C_5H_5Me)(CO)(COMe)^+$, which is trapped by solvent. Such a species could be stabilized by η^2-coordination of the acyl group to effectively maintain a 17-electron count [124]. A lower limit of 1.4 M^{-1} s^{-1} could be set for the rate of insertion at −65 °C [124]. Two alternative explanations might also be considered. Since $\cdot Fe(\eta^5\text{-}C_5H_4Me)(CO)_2Me^+$ should be more electrophilic than $\cdot Fe(\eta^5\text{-}C_5H_4Me)(CO)(PPh_3)Me^+$ a high reactivity could result in a low selectivity for nucleophiles (i.e. any one induces insertion on collision). Another possibility would be a noninnocent role for the $LiClO_4$ electrolyte with the cationic electrophile.

An observation, which must be considered, is the lack of insertion chemistry for the stable species [$\cdot Fe(\eta^5\text{-}C_5Me_5)(CO)(\eta^1\text{-}Ph_2PCH_2CH_2PPh_2)CH_3$]$PF_6$ [129]. This complex is unreactive in the presence of nucleophiles. This could arise from unfavorable thermodynamics for the insertion process or from an inability of the sterically crowded metal center to add a nucleophile to form a 19-electron precursor for insertion. In this context we mention that free PPh_3 is not effective for inducing insertion in $\cdot FeCp(CO)(PPh_3)CH_3^+$. Better definition of steric and electronic requirements for the nucleophile is needed.

The viability of 19-electron complexes as intermediates in CO insertion reactions derives support from a report that CO insertion in $FeCp(CO)_2Me$ can be catalyzed [126] by reduction to an anion radical (19-electron) in the presence of PPh_3 to produce $FeCp(CO)(PPh_3)(COMe)$. This could be explained by reversible. insertion in the 19-electron complex to form a 17-electron acyl species, which is trapped by PPh_3. Further studies will be needed to determine the generality of the reaction mechanisms discussed.

2.4 Electronic Structure and Reactivity Correlations

Preceding examples suggest that 17 to 19-electron interconversions play the same role for organometallic radicals as the 16 to 18-electron cycle [130] for closed shell organometallic complexes. A question then arises concerning the theoretical basis for the 10^{10} acceleration in associative substitution for 17-electron systems and the unexpected stability of electron-rich 19-electron intermediates. The simplest explanation is that the interaction between an electron pair from an attacking nucleophile and the half-filled orbital on the metal leads to a doubly occupied bonding orbital and a half-filled antibonding orbital. A half-order bond provides electronic stabilization for a 19-electron species [5, 7, 12, 22, 24, 42]. Depending on the balance between steric repulsions in the 19-electron species versus the strength of the half-order bond these complexes can be expected to be either intermediates or transition states as depicted in Fig. 8. Given these considerations one expects 19-electron complexes to react by a dissociative path and several mechanistic studies support this conclusion [131–133].

The three-electron bonding interaction, which can occur between an organic free radical and a nucleophile, has been long known and its rediscovery has been a subject for recent debate [134–138]. We now consider the evidence for similar bonding in organometallic radicals. It is interesting that the first computational study bearing on this problem focused on explaining the lack of reactivity of the 17-electron complexes $\cdot Mn(\eta^4\text{-}C_4H_6)_2L$ for L = PMe_3, PEt_3, $P(OMe)_3$, and CO [139] toward substitution of L under forcing conditions. Extended Hückel calculations on $\cdot Mn(\eta^4\text{-}C_4H_6)_2CO$ suggest that the odd electron in this complex occupies an a_1 orbital (shown in Fig. 10A), which is sterically inaccessible to nucleophilic attack. This can be contrasted to $\cdot Mn(CO)_5$, which is extremely labile to CO substitution and where the singly occupied molecular orbital (SOMO) is sterically accessible to attack by a nucleophile as shown in Fig. 10B. Qualitative observations of substitutional lability and a tendency toward disproportionation for 17-electron $[\cdot Co(\eta^5\text{-}C_5H_5)(PEt_3)_2]BF_4$ has been explained [140] by nucleophilic attack on the SOMO as shown in Fig. 10C.

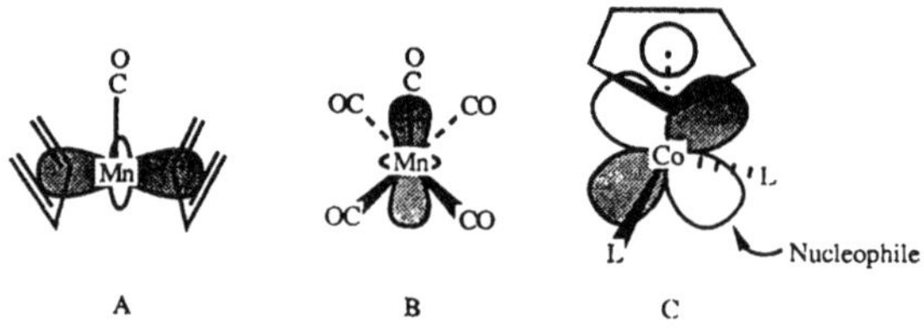

Fig. 10. Sketch of the singly occupied molecular orbitals (SOMO) calculated for 17-electron $\cdot Mn(\eta^4\text{-}C_4H_6)_2CO$-(A), $\cdot Mn(CO)_5$-(B), and $\cdot Co(\eta^5\text{-}C_5H_5)(PH_3)_2$-(C).

Addition of a model PH_3 nucleophile to a variety of four-, five and six-coordinate metal carbonyl radicals has been studied by SCF-Xα-DV calculations [141] to see whether the postulated three-electron bond forms. For an octahedral d^5 radical, such as $V(CO)_6$, nucleophilic attack at an octahedral face (Fig. 11) leads to formation of a doubly-occupied $10a_1$ orbital of V—P σ bonding character and a half-occupied $11a_1$ V—P σ^* orbital. Fig. 12 shows energies for these orbitals as a function of the V—P separation, as well as the energy for the nonbonding $12e$ orbital (t_{2g} splits into $11a_1$ and $12e$ in C_{3v} symmetry as the PH_3 nucleophile attacks along an open face of $V(CO)_6$). Thus, the theoretical calculations provide strong support for the three-electron bonding model of radical reactivity.

The calculations were also used to qualitatively evaluate the most favorable mode of nucleophilic attack at metal carbonyl radicals. For an octahedral d^5

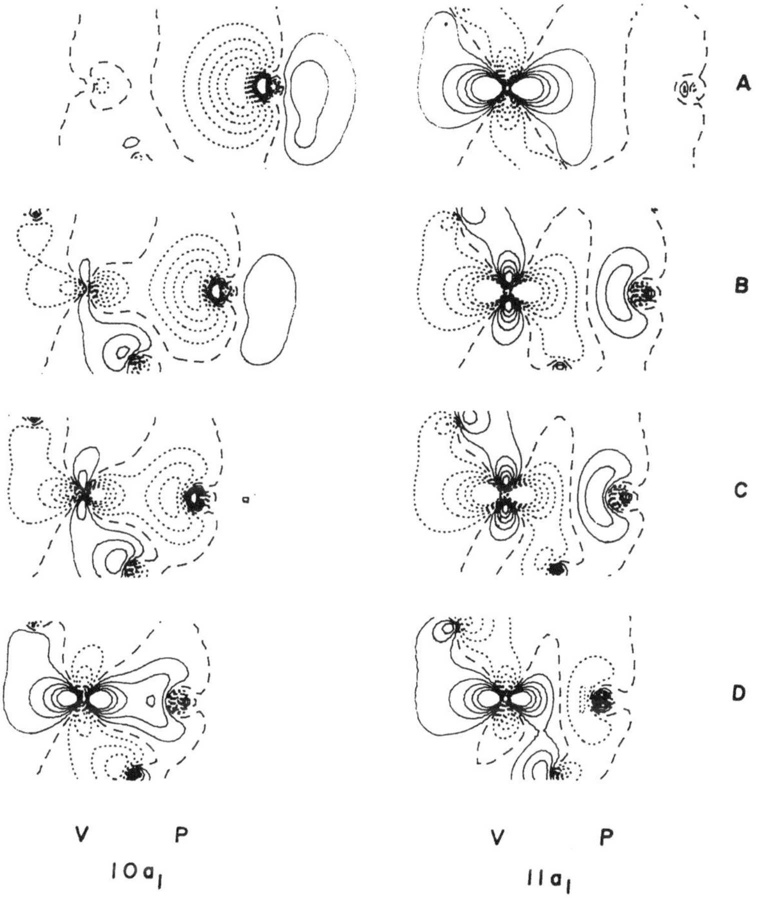

Fig. 11. Orbital plots of the interaction that develops as a PH_3 nucleophile approaches the $\cdot V(CO)_6$ 17-electron radical along an open trigonal face of the octahedron. The left column shows development of the $10a_1\sigma$-bonding orbital between the lone pair orbital on PH_3 and a d orbital on vanadium. The right column shows the corresponding singly occupied σ-antibonding $11a_1$ orbital that develops as the SOMO on $\cdot V(CO)_6$ interacts with the phosphorus-centered lone pair. Reproduced with permission from Reference 7.

complex face attack is preferred, for a square pyramidal d^7 complex attack at the open square face is preferred, for a trigonal bipyramidal d^7 complex edge attack in the equatorial plane is preferred, and for a tetrahedral d^9 radical face attack, as in an organic S_N2 reaction is favored [141]. These attack modes are summarized in Fig. 13.

Electronic structure calculations and EPR spectroscopic studies of $\cdot V(\eta^5\text{-}C_5H_5)_2CO$, $\cdot V(\eta^5\text{-}C_5Me_5)_2CO$, $\cdot V(\eta^5\text{-}C_5H_7)_2CO$, and $\cdot V(2,4\text{-}Me_2C_5H_5)_2CO$ have been used to probe the distribution of the odd electron [142]. Recall that an unusual aspect of associative CO substitution in these complexes is a million-fold

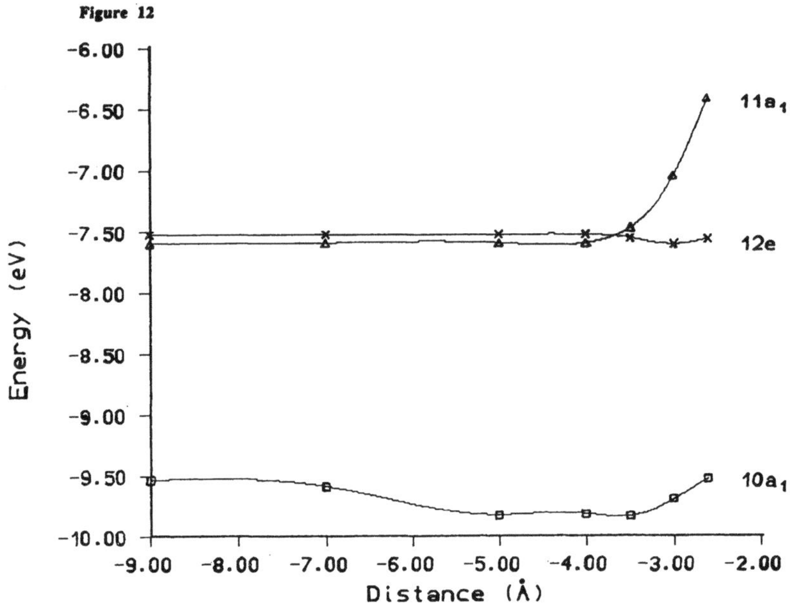

Fig. 12. Variation of the $10a_1$ and $11a_1$ orbital energies of Fig. 11 with the V-P separation. Reproduced with permission from Reference 141.

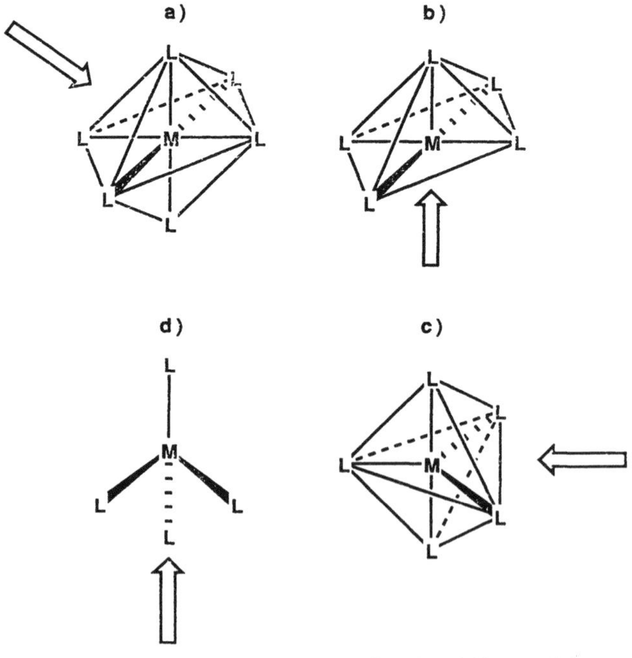

Fig. 13. Calculated preferred modes of nucleophilic attack in octahedral d^5 (a), square pyramidal d^7 (b), trigonal bipyramidal d^7 (c), and tetrahedral d^9 (d) 17-electron metal carbonyl radicals.

decrease in reactivity observed for the pentadienyl complexes. This can not be attributed to an enhanced donor ability of the pentadienyl ligand, since the C—O stretching frequency shifts 55–120 cm^{-1} to higher energy in the pentadienyl complexes. Electron density considerations would thus predict enhanced electrophilicity at vanadium and an increase in the rate of nucleophilic attack at vanadium in the pentadienyl complexes, which contradicts experiment. SCF-Xα-DV calculations show a similar MO scheme for the two systems (Fig. 14) with identical symmetry SOMO's. Little spin density delocalizes onto the CO ligands as judged by ^{13}C hyperfine splittings in isotopically enriched complexes. Calculations, however, reveal a difference in the primary direction of hybridization of the $12a_1$ orbital for the cyclopentadienyl and pentadienyl complexes (Fig. 15). Shielding of the primary lobe of the SOMO from an approaching nucleophile by the organic rings in the pentadienyl complex may account for the marked decrease in reactivity.

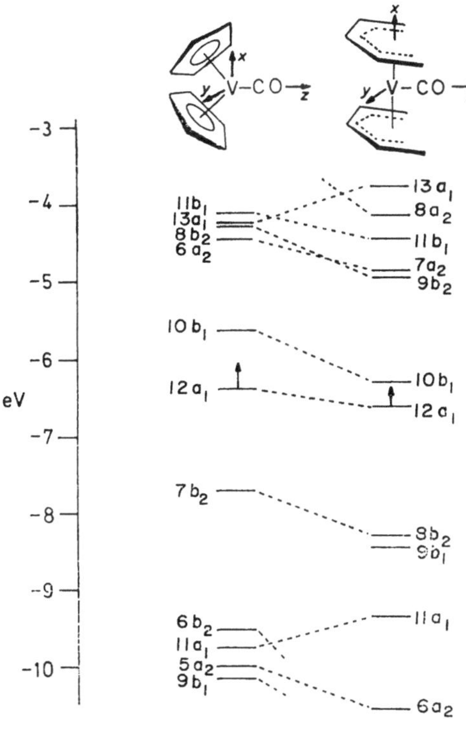

Fig. 14. Comparison of the orbital energy diagrams for $\cdot V(\eta^5\text{-}C_5H_5)_2CO$ and $\cdot V(\eta^5\text{-}C_5H_7)_2CO$ from SCF-Xα-DV calculations. Reproduced with permission from Reference 142.

We discussed earlier the similar labilities of five-coordinate iron radical centers in $Fe(CO)_3L_2^+$ and in $Fe_2(CO)_7(\mu\text{-}PPh_2)$ toward associative substitution of carbon monoxide [40, 94]. Spectroscopic and theoretical studies of the former complexes suggest that the odd electron occupies an e' orbital (probably subject to a splitting from Jahn-Teller distortion) in the equatorial plane of the trigonal bipyramid. This is consistent with the postulate of nucleophilic attack at an equatorial edge

Fig. 15. Sketch of the SOMO's in $\cdot V(\eta^5\text{-}C_5H_5)_2CO$ (a) and $\cdot V(\eta^5\text{-}C_5H_7)_2CO$ (b) from SCF-Xα-DV calculations.

of a trigonal bypyramidal d^7 complex [141]. Single crystal EPR studies of $Fe_2(CO)_7(\mu\text{-}PPh_2)$ doped into the diamagnetic host $CoFe(CO)_7(\mu\text{-}PPh_2)$ and extended Hückel MO calculations suggest the odd electron in the dimer localizes on the five-coordinate iron atom (Fig. 6) [143]. The SOMO is an iron localized orbital in the equatorial plane of the trigonal bipyramid similar to that postulated in the mononuclear complex. EPR parameters are also consistent with the analogy between $Fe(CO)_3L_2^+$ ($g = 2.053\text{–}2.057$, $A_p = 17\text{–}19$ G) and $Fe_2(CO)_7(\mu\text{-}PPh_2)$ ($g = 2.052$, $A_p = 30$ G).

There have been several studies of quantitative linear free energy correlations for steric and electronic effects when nucleophiles attack 17-electron radicals [40, 42, 67, 94, 125, 144, 145]. For example, in the reaction between $\cdot V(CO)_6$ and PR_3 a linear correlation holds between the log of the substitution rate and ΔHNP, a measure of phosphorus donor basicity relative to an organic reference base [42]. Of the seven ligands examined only $P(i\text{-}Pr)_3$ deviates from the correlation and its low reactivity may be attributed to the steric bulk of isopropyl substituents. Another study of the reaction between $\cdot Mn(\eta^5\text{-}C_5H_4Me)(CO)_2L^+$ and a series of 20 phosphorus donor nucleophiles provides a unified treatment of steric and electronic effects for the entering PR_3 ligand, as well as leaving group electronic effects for L, a para-substituted pyridine [67]. The fit to Eq. 22 shows a correlation co-

$$\log k = A + B[pK_a^{\mathrm{L}}] + C[pK_a^{\mathrm{PR_3}}] + D[\theta] \tag{22}$$

efficient of 0.99 for the 20 systems studied, where A, B, C, and D are the fitting constants, pK_a^{L} is the pK_a of the leaving group, $pK_a^{\mathrm{PR_3}}$ is the pK_a of the entering group, and θ the cone angle steric factor tabulated [26] for phosphorus donor ligands. The electronic parameter $C = 0.28$ resembles that of $C = 0.42$ when the data for $\cdot V(CO)_6$ are fit to a similar equation. An excellent linear correlation exists between the log of rates for nucleophilic attack by PR_3 on $\cdot Mn(\eta^5\text{-}C_5H_4Me)(CO)_2L^+$ and the log of rates for nucleophilic attack by PR_3 on EtI [67]. This analogy provides strong support for the S_N2 model of organometallic radical reactivity depicted in Fig. 13.

The correlation of reactivities has been further refined by separating the electronic parameters for the PR_3 nucleophile into σ and π components ($\pi^a + \pi^d$ for donor or acceptor behavior) [144]. Regression analysis of rates of nucleophilic attack at the cluster radical $\cdot Fe_2(CO)_7(\mu\text{-}PPh_2)$ for seven PR_3 donors could be fit to Eq. 23 with a correlation coefficient of 0.89 [94]. Addition of a term $E\pi_a$ to

$$\log k = A + C[pK_a^{\mathrm{PR_3}}] + D\theta \tag{23}$$

include the PR_3 π-acceptor parameter improves the correlation to an index of 0.98. Most important the least squares values of A, C, and D change little in the new fit. The pK_a coefficient of $C = 0.06$ is reduced considerably from that of 0.2–0.4 in the mononuclear metal centered radicals discussed above. A reduced electronic parameter agrees with the lowered reactivity enhancement observed for the cluster system.

Plots of ($\log k - \log k_\sigma$) vs. θ, where k_σ is the calculated σ-donor contribution to k, have been used to construct steric profiles of radical reactivity [144]. These plots display steric thresholds for reactions (i.e. a break in the plot when ligand size exceeds a threshold value). Thus, $\cdot V(CO)_6$ shows a steric threshold ($\theta \sim 145°$) similar to the 18-electron complexes $Mo(CO)_6$ and $W(CO)_6$, which have slow associative paths attributed to an I_d mechanism [144]. The key difference derives from the greater degree of metal phosphorus bonding in the electronic term for $\cdot V(CO)_6$.

Steric and electronic effects for associative reactions of metal carbonyls have been surveyed using Eq. 24 [145, 146]. The electronic discrimination of $\cdot V(CO)_6$ (reflected by the value of C) toward nucleophiles is higher than for any of the

$$\log k = A + C[pK_a^{PR_3}] + D\theta \qquad (24)$$

18-electron complexes analyzed. In this context we note that the Hammett rho values obtained for nitrogen donor ligands attacking 17-electron metal centers are large and negative (−3 to −5), which implies a strong donation from the nucleophile in the transition state [40, 125]. Thus the high sensitivity of radicals toward electronic properties of the nucleophile also extends to nitrogen donors. A particularly interesting comparison can be made between 17-electron $\cdot V(CO)_6$ and 18-electron $V(CO)_5(NO)$ [147]. Both complexes exhibit a similar high associative reactivity toward nucleophiles, which in $V(CO)_5(NO)$ has been attributed to weakening of the V—CO bond trans to NO [147]. SCF-Xα-DV calculations were performed for $V(CO)_5(NO)$ and compared with those for $V(CO)_6$. In $V(CO)_5(NO)$, the fully occupied π-bonding d-t_{2g} orbitals split into $b_2(d_{xy})$ and $e(d_{xz,yz})$ orbitals. The e level, which can π-bond to NO, is preferentially stabilized, and the NO π-orbital contribution is 4–5 times that of the trans CO ligand. The superior π-acceptor character of NO should weaken V—C π-bonding to the axial group. This explanation receives support from the linear free energy analysis, which suggests a high reactivity for $\cdot V(CO)_6$ based on bond formation to the nucleophile. For $V(CO)_5NO$ an intrinsic high reactivity (i.e. not very nucleophile dependent) stems from an intrinsic weakness of the bond to the leaving group [146]. Although, the above data is sketchy it suggests that associative reactions of 17-electron organometallic radicals and those of 18-electron organometallic complexes differ in a greater sensitivity to electronic properties of the nucleophile for the former systems. This is consistent with an important role for the metal-nucleophile three-electron bond.

3. Acknowledgment

Support of research in my laboratory on organometallic radicals by the U.S. National Science Foundation is gratefully acknowledged (CHE-8815958).

4. References

1. T. L. Brown, Ann. N. Y. Acad. Sci. **333** (1980) 80.
2. J. Halpern, Pure Appl. Chem. **58** (1986) 575; Angew. Chem. **97** (1985) 308.
3. J. K. Kochi, Organometallic Mechanisms and Catalysis, Academic Press, New York, 1978.
4. M. F. Lappert and P. W. Lednor, Adv. Organomet. Chem. **14** (1976) 345.
5. A. E. Stiegman and D. R. Tyler, Comments on Inorg. Chem. **5** (1986) 215.
6. J. K. Kochi, J. Organomet. Chem. **300** (1986) 139.
7. W. C. Trogler, Int. J. Chem. Kinet. **19** (1987) 1025.
8. N. G. Connelly and W. E. Geiger, Adv. Organomet. Chem. **23** (1984) 1; W. E. Geiger and N. G. Connelly, Adv. Organomet. Chem. **24** (1985) 87.
9. W. E. Geiger, Prog. Inorg. Chem. **33** (1985) 275.
10. M. C. Baird, Chem. Rev. **88** (1988) 1217.
11. D. Astruc, Chem. Rev. **88** (1988) 1189.
12. M. S. Wrighton and D. S. Ginley, J. Am. Chem. Soc. **97** (1975) 2065.
13. B. H. Byers and T. L. Brown, J. Am. Chem. Soc. **99** (1977) 2527.
14. L. A. Hanlan, H. Huber, E. P. Kündig, B. R. McGarvey, and G. A. Ozin, J. Am. Chem. Soc. **97** (1975) 7054.
15. G. L. Geoffroy and M. S. Wrighton, Organometallic Photochemistry, Academic Press, New York, 1979.
16. T. J. Meyer and J. V. Caspar, Chem. Rev. **85** (1985) 187.
17. D. C. Roe, Organometallics **6** (1987) 942.
18. N. P. Forbus and T. L. Brown, Inorg. Chem. **20** (1981) 4343.
19. M. Absi-Halabi, J. D. Atwood. N. P. Forbus, and T. L. Brown, J. Am. Chem. Soc. **102** (1980) 6248.
20. T. M. Bockman and J. K. Kochi, J. Am. Chem. Soc. **110** (1988) 1294.
21. T. L. Brown, Organometallic Radical Processes, W. C. Trogler, ed., Elsevier, Amsterdam, 1990.
22. J. P. Fawcett, R. A. Jackson, and A. Poë, J. Chem. Soc., Chem. Commun. (1975) 733; A. Poë, Transition Met. Chem. **7** (1982) 65.
23. J. D. Atwood, Inorg. Chem. **20** (1981) 4031; S. P. Schmidt, W. C. Trogler, and F. Basolo, Inorg. Chem. **21** (1982) 1698; E. L. Muetterties, R. R. Burch, and A. M. Stolzenberg, Ann. Rev. Phys. Chem. **33** (1982) 89; A. Pöe and C. V. Sekhar, J. Am. Chem. Soc. **107** (1985) 4874.
24. A. Fox, J. Malito, and A. Poë, J. Chem. Soc., Chem. Commun. (1981) 1052.
25. J. M. Hanckel, K.-W. Lee, P. Rushman, and T. L. Brown, Inorg. Chem. **25** (1986) 1852.
26. C. A. Tolman, Chem. Rev. **77** (1977) 313.
27. H. Yesaka, T. Kobayashi, K. Yasufuku, and S. Nagakura, J. Am. Chem. Soc. **105** (1983) 6249.
28. T. R. Herrinton and T. L. Brown, J. Am. Chem. Soc. **107** (1985) 5700.
29. J. K. Kochi, Organometallic Radical Processes, W. C. Trogler, ed., Elsevier, Amsterdam, 1990.
30. S. B. McCullen and T. L. Brown, J. Am. Chem. Soc. **104** (1982) 7496.
31. H. W. Walker, G. B. Rattinger, R. L. Belford, and T. L. Brown, Organometallics **2** (1983) 775.
32. S. B. McCullen, H. W. Walker, and T. L. Brown, J. Am. Chem. Soc. **104** (1982) 4007.
33. L. S. Crocker, D. M. Heinekey, and G. K. Schulte, J. Am. Chem. Soc. **111** (1989) 405.
34. P. K. Baker, N. G. Connelly, B. M. R. Jones, J. P. Maher, and K. R. Somers, J. Chem. Soc., Dalton Trans. (1980) 579.
35. N. G. Connelly and K. R. Somers, J. Organomet. Chem. **113** (1976) C39.
36. P. K. Baker, K. Broadley, and N. G. Connelly, J. Chem. Soc., Chem. Commun. (1980) 775.
37. P. K. Baker, K. Broadley, and N. G. Connelly, J. Chem. Soc., Dalton Trans. (1982) 471.
38. M. J. Therien and W. C. Trogler, J. Am. Chem. Soc. **108** (1986) 3697.
39. S. W. Blanch, A. M. Bond, and R. Colton, Inorg. Chem. **20** (1981) 755.

40. M. J. Therien, C.-L. Ni, F. C. Anson, J. G. Osteryoung, and W. C. Trogler, J. Am. Chem. Soc. **108** (1986) 4037.
41. R. N. Bagchi, A. M. Bond, C. L. Heggie, T. L. Henderson, E. Mocellin, and R. A. Seikel, Inorg. Chem. **22** (1983) 3007.
42. Q.-Z. Shi, T. G. Richmond, W. C. Trogler, and F. Basolo, J. Am. Chem. Soc. **104** (1982) 4032; **106** (1984) 71.
43. T. G. Richmond, Q.-Z. Shi, W. C. Trogler, and F. Basolo, J. Chem. Soc., Chem. Commun. (1983) 650.
44. T. G. Richmond, Q.-Z. Shi, W. C. Trogler, and F. Basolo, J. Am. Chem. Soc. **106** (1984) 76.
45. C. J. Pickett, and D. Pletcher, J. Chem. Soc., Dalton Trans. (1975) 879.
46. R. J. Klingler and J. K. Kochi, J. Am. Chem. Soc. **102** (1980) 4790.
47. R. N. Bagchi, A. M. Bond, R. Colton, D. L. Luscombe, and J. E. Moir, J. Am. Chem. Soc. **108** (1986) 3352.
48. R. N. Bagchi, A. M. Bond, G. Brain, R. Colton, T. L. E. Henderson, and J. E. Kevekordes, Organometallics **3** (1984) 4.
49. J. W. Hershberger, R. J. Klingler, and J. K. Kochi, J. Am. Chem. Soc. **104** (1982) 3034.
50. A. M. Bond, R. Colton, and K. McGregor, Inorg. Chem. **25** (1986) 2378.
51. F. L. Wimmer. M. R. Snow, and A. M. Bond, Inorg. Chem. **13** (1974) 1617.
52. A. M. Bond, R. Colton, and J. J. Jackowski, Inorg. Chem. **14** (1975) 274.
53. A. M. Bond, S. W. Carr, and R. Colton, Organometallics **3** (1984) 541.
54. A. M. Bond, R. Colton, and M. J. McCormick, Inorg. Chem. **16** (1977) 155.
55. A. M. Bond, R. Colton, and M. E. McDonald, Inorg. Chem. **17** (1978) 2843.
56. A. M. Bond, S. W. Carr, and R. Colton, Inorg. Chem. **23** (1984) 2343.
57. A. M. Bond, R. Colton, and J. E. Kevekordes, Inorg. Chem. **25** (1986) 749.
58. R. D. Rieke, H. Kojima, T. Saji, P. Rechberger, and K. Öfele, Organometallics **7** (1988) 749. K. A. Connor, R. A. Walton. Inorg. Chem. **25** (1986) 4422.
59. F. Basolo, Inorg. Chim. Acta **100** (1985) 33; J. M. O'Connor and C. P. Casey, Chem. Rev. **87** (1987) 307.
60. R. M. Kowaleski, A. L. Rheingold, W. C. Trogler, and F. Basolo, J. Am. Chem. Soc. **108** (1986) 2460.
61. E. U. van Raaij and H.-H. Brintzinger, J. Organomet. Chem. **356** (1988) 315.
62. (a) R. M. Kowaleski, W. C. Trogler, and F. Basolo, Gazz. Chim. Ital. **116** (1986) 105.
(b) R. M. Kowaleski, F. Basolo, W. C. Trogler, and R. D. Ernst, J. Am. Chem. Soc. **108** (1986) 6046.
63. R. M. Kowaleski, F. Basolo, W. C. Trogler, R. W. Gedridge, T. D. Newbound, and R. D. Ernst, J. Am. Chem. Soc. **109** (1987) 4860.
64. R. D. Ernst, Chem. Rev. **88** (1988) 1255.
65. N. C. Hallinan, G. Morelli, and F. Basolo, J. Am. Chem. Soc. **110** (1988) 6585.
66. J. W. Hershberger, R. J. Klingler, and J. K. Kochi, J. Am. Chem. Soc. **105** (1983) 61.
67. P. M. Zizelman, C. Amatore, and J. K. Kochi, J. Am. Chem. Soc. **106** (1984) 3771.
68. C. Amatore, Organometallic Radical Processes, W. C. Trogler, ed., Elsevier, Amsterdam, 1990.
69. N. N. Turaki and J. M. Huggins, Organometallics **5** (1986) 1703.
70. K. M. Doxsee, R. H. Grubbs, and F. C. Anson, J. Am. Chem. Soc. **106** (1984) 7819.
71. C. G. Zoski, D. A. Sweigart, N. J. Stone, P. H. Rieger, E. Mocellin, T. F. Mann, D. R. Mann, D. K. Gosser, M. M. Doeff, and A. M. Bond, J. Am. Chem. Soc. **110** (1988) 2109.
72. S. P. Gubin and V. S. Khandkarova, J. Organomet. Chem. **22** (1970) 449.
73. M. K. Lloyd, J. A. McCleverty, J. A. Connor, and E. M. Jones, J. Chem. Soc., Dalton Trans. (1973) 1768.
74. S. N. Milligan, I. Tucker, and R. D. Rieke, Inorg. Chem. **22** (1983) 987.
75. R. D. Rieke, S. N. Milligan, I. Tucker, K. A. Dowler, and B. R. Willeford, J. Organomet. Chem. **218** (1981) C25.
76. N. J. Stone, D. A. Sweigart, and A. M. Bond, Organometallics **5** (1986) 2553.
77. C. Degrand, A. Radecki-Sudre, and J. Besancon, Organometallics **1** (1982) 1311; C. Degrand and A. Radecki-Sudre, J. Organomet. Chem. **268** (1984) 63.

78. M. G. Peterleitner, M. V. Tolstaya, V. V. Krivykh, L. I. Denisovitch, and M. I. Rybinskaya, J. Organomet. Chem. **254** (1983) 313.
79. T. Ikeshoji and V. D. Parker, Acta. Chem. Scand. **B37** (1983) 715.
80. T. Ikeshoji and V. D. Parker, Acta. Chem. Scand. **A38** (1984) 175.
81. T. Ikeshoji and V. D. Parker, Bull. Chem. Soc. Jpn. **57** (1984) 1112.
82. N. A. Cooley, P. T. F. MacConnachie, and M. C. Baird, Polyhedron **7** (1988) 1965.
83. J. V. Caspar and T. J. Meyer, J. Am. Chem. Soc. **102** (1980) 7794; A. J. Dixon, M. A. Healy, M. Poliakoff, and J. J. Turner, J. Chem. Soc., Chem. Commun. (1986) 994; A. J. Dixon, S. J. Gravelle, L. J. van de Burgt, M. Poliakoff, J. J. Turner, and E. Weitz, J. Chem. Soc., Chem. Commun. (1987) 1023.
84. N. G. Connelly, M. D. Kitchen, R. F. D. Stansfield, S. M. Whiting, and P. Woodward, J. Organomet. Chem. **155** (1978) C34; N. G. Connelly, R. L. Kelly, M. D. Kitchen, R. M. Mills, R. F. D. Stansfield, M. W. Whiteley, S. M. Whiting, and P. Woodward, J. Chem. Soc., Dalton Trans. (1981) 1317; J. R. Hamon, D. Astruc, E. Roman, P. Batail, and J. J. Mayerle, J. Am. Chem. Soc. **103** (1981) 2431.
85. P. S. Waterman and W. P. Giering, J. Organomet. Chem. **155** (1978) C47.
86. P. M. Triechel and L. D. Rosenheim, J. Am. Chem. Soc. **103** (1981) 691.
87. L. Brammer, N. G. Connelly, J. Edwin, W. E. Geiger, A. G. Orpen, and J. B. Sheridan, Organometallics **7** (1988) 1259.
88. N. G. Connelly, A. R. Lucy, J. D. Payne, A. M. R. Galas, and W. E. Geiger, J. Chem. Soc., Dalton Trans. (1983) 1979; M. J. Freeman, A. G. Orpen, N. G. Connelly, I. Manners, and S. J. Raven, J. Chem. Soc., Dalton Trans. (1985) 2283.
89. N. G. Connelly, P. G. Graham, and J. B. Sheridan, J. Chem. Soc., Dalton Trans. (1986) 1619.
90. R. J. McKinney, J. Chem. Soc., Chem. Commun. (1980) 603.
91. R. S. Iyer and J. P. Selegue, J. Am. Chem. Soc. **109** (1987) 910.
92. E. Fonseca, W. E. Geiger, T. E. Bitterwolf, and A. L. Rheingold, Organometallics **7** (1988) 567.
93. N. J. Coville, Organometallic Radical Processes, W. C. Trogler, ed., Elsevier, Amsterdam, 1990.
94. R. T. Baker, J. C. Calabrese, P. J. Krusic, M. J. Therien, and W. C. Trogler, J. Am. Chem. Soc. **110** (1988) 8392.
95. M. G. Richmond and J. K. Kochi, Inorg. Chem. **25** (1986) 656.
96. D. R. Tyler, Organometallic Radical Processes, W. C. Trogler, ed., Elsevier, Amsterdam, 1990.
97. H. H. Ohst and J. K. Kochi, Inorg. Chem. **25** (1986) 2066.
98. T. M. Bockman and J. K. Kochi, J. Am. Chem. Soc. **109** (1987) 7725.
99. F. Basolo, and R. G. Pearson, Mechanisms of Inorganic Reactions, 2nd edn. Wiley, New York, 1967.
100. J. D. Atwood, Inorganic and Organometallic Reaction Mechanisms, Brooks/Cole Publishing Co., Monterey, CA, 1985.
101. R. R. Corderman and J. L. Beauchamp, Inorg. Chem. **16** (1977) 3135.
102. R. N. McDonald and P. L. Schell, Organometallics **7** (1988) 1806.
103. R. N. McDonald and P. L. Schell, Organometallics **7** (1988) 1820.
104. A. E. Stiegman and D. R. Tyler, Coord. Chem. Rev. **63** (1985) 217.
105. H. Taube, Electron Transfer Reactions of Complex Ions in Solution, Academic Press, New York, 1970.
106. M. Schneider and E. Weiss, J. Organomet. Chem. **121** (1976) 365.
107. J. H. Osborne, A. L. Rheingold, and W. C. Trogler, J. Am. Chem. Soc. **107** (1985) 7945; J. H. Osborne, A. L. Rheingold, and W. C. Trogler, J. Am. Chem. Soc. **107** (1985) 6292.
108. A. M. Bond and R. Colton, Inorg. Chem. **15** (1976) 2036.
109. D. S. Richeson, J. F. Mitchell, and K. H. Theopold, J. Am. Chem. Soc. **109** (1987) 5868.
110. B. J. Thomas and K. H. Theopold, J. Am. Chem. Soc. **110** (1988) 5902.
111. K. M. Nicholas and M. Rosenblum, J. Am. Chem. Soc. **95** (1973) 4449.

112. S. N. Anderson, C. W. Fong, and M. D. Johnson, J. Chem. Soc., Chem. Commun. (1973) 163.
113. P. L. Bock, D. J. Boschetto, J. R. Rasmussen, J. P. Demers, and G. M. Whitesides, J. Am. Chem. Soc. **96** (1974) 2814.
114. A. Cameron, V. H. Smith, and M. C. Baird, Organometallics **2** (1983) 465.
115. R. J. Klingler and J. K. Kochi, J. Organomet. Chem. **202** (1980) 49.
116. R. S. Bly, G. S. Silverman, and R. K. Bly, Organometallics **4** (1985) 374.
117. D. L. Reger and E. Mintz, Organometallics **3** (1984) 1759.
118. D. L. Reger, E. Mintz, and L. Lebioda, J. Am. Chem. Soc. **108** (1986) 1940.
119. J. B. Sheridan, S.-H. Han, and G. L. Geoffroy, J. Am. Chem. Chem. Soc. **109** (1987) 8097.
120. J. B. Sheridan, J. R. Johnson, B. M. Handwerker, G. L. Geoffroy, and A. L. Rheingold, Organometallics **7** (1988) 2404.
121. R. H. Magnuson, S. Zulu, W.-M. Tsai, and W. P. Giering, J. Am. Chem. Soc. **102** (1980) 6887.
122. R. H. Magnuson, R. Meirowitz, S. J. Zulu, and W. P. Giering, J. Am. Chem. Soc. **104** (1982) 5790.
123. R. H. Magnuson, R. Meirowitz, S. J. Zulu, and W. P. Giering, Organometallics **2** (1983) 460.
124. M. N. Golovin, R. Meirowitz, Md. M. Rahman, H. Y. Liu, A. Prock, and W. P. Giering, Organometallics **6** (1987) 2285.
125. M. J. Therien and W. C. Trogler, J. Am. Chem. Soc. **109** (1987) 5127.
126. D. Milholová and A. Vlček, J. Organomet. Chem. **240** (1982) 413.
127. T. C. Flood, Topics in Stereochem. **12** (1981) 37.
128. J. P. Collman, L. S. Hegedus, J. R. Norton, and R. G. Finke, Principles and Applications of Organotransition Metal Chemistry, University Science Books, Mill Valley, CA, pp. 358–360, 1987.
129. J. Morrow, D. Catheline, M.-H. Desbois, J.-M. Manriquez, J. Ruiz, and D. Astruc, Organometallics **6** (1987) 2605.
130. C. A. Tolman, Chem. Soc. Rev. **1** (1972) 337.
131. D. J. Kuchynka, C. Amatore, and J. K. Kochi, Inorg. Chem. **25** (1986) 4087.
132. B. T. Donovan and W. E. Geiger, J. Am. Chem. Soc. **110** (1988) 2335.
133. F. Mao, D. R. Tyler, and D. Keszler, J. Am. Chem. Soc. **111** (1989) 130.
134. S. Stinson, Chem. Eng. News, Dec. **7** (1987) 32.
135. F. Williams, Chem. Eng. News, May **9** (1988) 2.
136. R. D. Harcourt, Chem. Eng. News, Oct. **3** (1988) 5.
137. C. F. Aten, Chem. Eng. News, Oct. **17** (1988) 3.
138. F. Williams, Chem. Eng. News, Jan. **9** (1989) 2.
139. R. L. Harlow, P. J. Krusic, R. J. McKinney, and S. S. Wreford, Organometallics **1** (1982) 1506.
140. R. L. Harlow, R. J. McKinney, and J. F. Whitney, Organometallics **2** (1983) 1839.
141. M. J. Therien and W. C. Trogler, J. Am. Chem. Soc. **110** (1988) 4942; G. F. Holland, M. C. Manning, D. E. Ellis, and W. C. Trogler, J. Am. Chem. Soc. **105** (1983) 2308.
142. R. M. Kowaleski, F. Basolo, J. H. Osborne, and W. C. Trogler, Organometallics **7** (1988) 1425.
143. P. J. Krusic, R. T. Baker, J. C. Calabrese, J. R. Morton, K. F. Preston, and Y. LePage, J. Am. Chem. Soc. **111** (1989) 1262; R. T. Baker, P. J. Krusic, J. C. Calabrese, and D. C. Roe, Organometallics **5** (1986) 1506.
144. M. N. Golvin, Md. M. Rahman, J. E. Belmonte, and W. P. Giering, Organometallics **4** (1985) 1981.
145. K. Dahlinger, F. Falcone, and A. J. Poë, Inorg. Chem. **25** (1986) 2654.
146. A. J. Poë, Pure. Appl. Chem. **60** (1988) 1209.
147. Q.-Z. Shi, T. G. Richmond, W. C. Trogler, and F. Basolo, Inorg. Chem. **23** (1984) 957.

CHAPTER 10

Photochemically Generated Nineteen-Electron Organometallic Complexes

David R. Tyler

Department of Chemistry, University of Oregon, Eugene, Oregon 97403, U.S.A.

Table of Contents

1. Introduction

1.1 Outline and scope

This chapter discusses the reaction intermediates that form when 17-electron organometallic radicals react with 2-electron donor ligands [1]:

$$ML_n + L' \underset{}{\overset{K_{eq}}{\rightleftharpoons}} ML_nL' \quad (1)$$

17e⁻ 19e⁻

$ML_n = CpMo(CO)_3, CpW(CO)_3, CpFe(CO)_2, Mn(CO)_5, Co(CO)_4$

$L' = PR_3, P(OR)_3, NR_3, Cl^-, Br^-, I^-, CH_3CN$, THF, and other coordinating solvents

The adducts that form in these reactions are commonly called "19-electron complexes." Like much of organometallic radical chemistry, 19-electron complexes were essentially unknown ten years ago [2]. Their discovery and the subsequent development of their chemistry coincided with the emerging interest in organometallic radical chemistry that began in the mid to late 1970s [3]. It is an indication of how far the chemistry of 19-electron complexes has progressed that, today, these species are routinely proposed as intermediates in radical reactions, and they are even discussed in textbooks [4]. The history of 19-electron adducts – from discovery to inclusion in textbooks – is typical for that of many reaction intermediates: their existence was originally hypothesized to explain unusual reaction phenomena; next, the complexes were generated *in situ* and their reactivity was studied; and finally, stable derivatives and model complexes were synthesized and studied in order to gain insights into these species.

The outline of this chapter follows the historical development of 19-electron complexes. In the first section, several of the reaction systems that led to the hypothesis of 19-electron complexes are discussed, and evidence for these species is presented. This is followed by sections on the reactivity and the thermodynamics of these complexes. In the last section, the generation of stable 19-electron complexes is discussed and the electronic and geometric structures of these and the more reactive 19-electron complexes are discussed. The emphasis throughout is on photogenerated 19-electron complexes. Note that 19-electron metallocene complexes of the type that Astruc has studied and recently reviewed will not be discussed here [5].

1.2 Definition of the phrase "19-electron complex"

It is imperative at the outset to state exactly what is meant by the phrase "19-electron complex" as it is used in this review. When using this phrase, the potential exists for considerable confusion because the structures (both geometric and electronic) of the 19-electron species are, in general, not known. Thus, these complexes might not necessarily be "19-electron complexes" in the traditional sense of the phrase [6]. For example, consider the "19-electron" $CpMo(CO)_3PR_3$ complex. Several reasonable possibilities for the structure of this species are shown below:

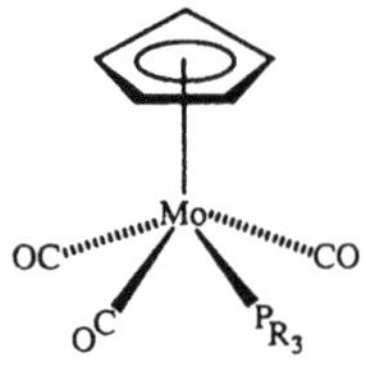

I

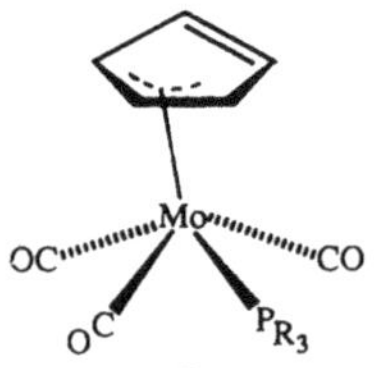

II

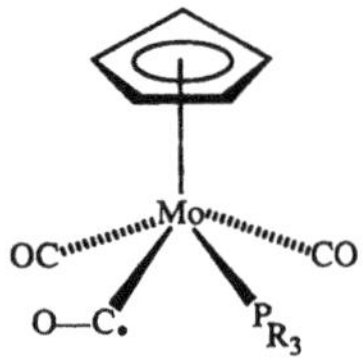

III

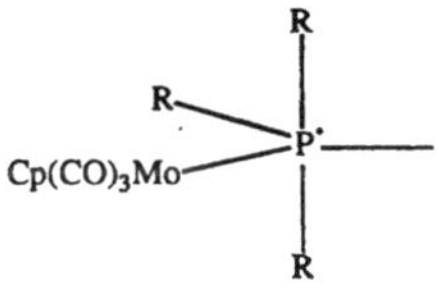

IV

Picture I shows a structure for what might be called the authentic 19-electron structure of this complex. If one counts electrons the traditional way [6] then the number of valence electrons associated with the metal in this structure is 19. (A discussion of which orbitals accommodate the unpaired electron is postponed until section 6.) Shown in drawing II is an alternative structure in which the 19-electron configuration at the metal is relieved by a slipped (η^3) Cp ring [7]; the complex would now traditionally be described as a 17-electron complex. In structures III and IV, the unpaired electron is also removed from the metal center by structural rearrangement of one of the ligands. Thus, in III, one of the CO ligands is bent, making the CO a 1-electron donor, analogous to NO [8]. The formal electron count for this species is 18. In structure IV, the structure of the PR_3 group has changed to that of a phosphoranyl radical [9], and the unpaired electron is now formally associated with the phosphorus atom. In this complex, the metal center formally also has 18 electrons. The point to this discussion is that, depending on the structure of the "19-electron complex," the metal center may or may not formally have 19 electrons. To avoid confusion, we state emphatically that the phrase "19-electron complex" is used to describe the *adducts* that form when 17-electron radicals react with 2-electron ligands. No implication of geometric or electronic structure is implied by this definition.

One final category of 19-electron complexes is the so-called "$18 + \delta$" class. The term "$18 + \delta$" was coined by Brown to describe those 19-electron complexes that are essentially 18-electron complexes with reduced ligands. Generally, 19-electron adducts will adapt this electronic structure if a ligand is present that has unfilled low energy orbitals. Examples of $18 + \delta$ complexes are increasingly numerous, and the following are just a few of the many that have recently been reported: $Co(CO)_3L_2$ (L_2 = 2,3-bis(diphenylphosphino)maleic anhydride) [10], $W(CO)_5$(4-cyanopyridine)$^-$ [11], $ReCO)_4$(1,2-diketone) [12], $Mn(CO)_4$(t-Bu-DAB) (DAB = 1,4-diaza-1,3-butadiene) [13], $Mo(CO)_4(bpy)^-$ [14]. If a 19-electron adduct is known to be an $18 + \delta$ complex then the term $18 + \delta$ is preferred because it provides more electronic structural information.

A more complete discussion of the electronic and geometric structures of 19-electron complexes is found in Section 6.

2. Photochemical Generation of 19-Electron Complexes

Much of the evidence for the formation of 19-electron complexes in photochemical systems comes from the study of the disproportionation reactions of metal-metal bonded carbonyl dimers. In this section, the disproportionation reactions are briefly reviewed, and the evidence they provide for the existence of 19-electron complexes is discussed.

2.1 Disproportionation reactions of metal carbonyl dimers

The photochemical disproportionation reactions of the metal-metal bonded carbonyl dimers pose a number of interesting questions for the mechanistic chemist, and for that reason these reactions have been studied since at least 1960 [15]. The disproportionation reactions, which involve a formal change in the metal atom oxidation states of ± 1, are exemplified by the reaction of $Cp_2Mo_2(CO)_6$ with phosphines [16]:

$$Cp_2Mo_2(CO)_6 + 2PR_3 \xrightarrow[\text{benzene}]{h\nu} CpMo(CO)_2(PR_3)_2^+ + CpMo(CO)_3^- + CO \quad (2)$$

Analogous reactions have been reported for essentially all the metal-metal bonded carbonyl dimers, including $Cp_2W_2(CO)_6$, $Mn_2(CO)_{10}$, $Co_2(CO)_8$, and $Cp_2Fe_2(CO)_4$ [17].

The disproportionation reactions and their ionic products are something of an anomaly in the field of metal-metal bond photochemistry. Generally, the photochemical reactions of the dimer complexes do not give ionic (i.e. electron transfer) products because irradiation results in homolytic cleavage of the metal-metal bond with the subsequent formation of metal radical intermediates (Eq. 3) [18, 19].

$$M_2L_{2n} \xrightarrow{h\nu} \underset{17e^-}{2ML_n} \quad (3)$$

ML_n = $Mn(CO)_5$, $Re(CO)_5$, $CpM(CO)_3$ (M = Cr, Mo, W), $CpFe(CO)_2$, $Co(CO)_4$

The photochemistry is thus usually typical of the chemistry of metal-centered radicals. (Note that M—CO bond dissociation also occurs in most cases [19].) Common photochemical reactions in nonpolar solvents include halogen atom abstraction from halocarbon molecules (Eq. 4), hydrogen atom abstraction from H-atom donors (Eq. 5), metal-metal cross coupling reactions (Eq. 6), and substitution reactions (Eq. 7 and Scheme I). The products of these reactions are all neutral species, and no intermetal electron transfer has occurred, as it has in the disproportionation reactions.

$$CpW(CO)_3 + C_6H_5I \rightarrow CpW(CO)_3I + C_6H_5 \quad [20] \quad (4)$$

$$Re(CO)_3(PR_3)_2 + HSnBu_3 \rightarrow Re(CO)_3(PR_3)_2H + \cdot SnBu_3 \quad [21\text{–}23] \quad (5)$$

$R = C_6H_{11}$

$$Mn_2(CO)_{10} + Re_2(CO)_{10} \xrightarrow{h\nu} (2)\ MnRe(CO)_{10} \quad [24] \quad (6)$$

$$Mn_2(CO)_{10} + 2PPh_3 \xrightarrow{h\nu} Mn_2(CO)_8(PPh_3)_2 \quad [25] \tag{7}$$

$$Mn_2(CO)_{10} \underset{k=3.9\times10^9 M^{-1}sec^{-1}}{\overset{h\nu}{\rightleftharpoons}} 2Mn(CO)_5$$

$$Mn(CO)_5 + PPh_3 \xrightarrow{k=1.7\times10^7 M^{-1}sec^{-1}} Mn(CO)_4PPh_3 + CO$$

$$2Mn(CO)_4PPh_3 \rightarrow Mn_2(CO)_8(PPh_3)_2$$

Overall: $Mn_2(CO)_{10} + 2PPh_3 \xrightarrow{h\nu} Mn_2(CO)_8(PPh_3)_2 + 2CO$

Scheme I

Early mechanistic studies on the disproportionation reactions were generally quite speculative because little definitive mechanistic evidence was available [15]. Hieber and coworkers, who initially observed the thermal disproportionation of $Mn_2(CO)_{10}$ in neat pyridine, postulated that the disproportionation reactions were solvent effects [15a, b]. They suggested that polar solvent molecules induced a dipole in the metal-metal bond, and that base attack on this dipolar intermediate produced ionic products. The idea of a solvent induced dipole was carried over to explain the photochemical disproportionation of these dimers; it was proposed that excitation of this dipolar species resulted in heterolysis of the metal-metal bond. Later work on these reactions showed that they occurred in nonpolar solvents with relatively small ligand concentrations, thus discounting the possibility that the disproportionation reactions were the result of a polar solvent effect [16].

Further investigation of the photochemical disproportionation reactions revealed that the quantum yields for the reactions were typically greater than one; under some conditions quantum yields as high as 100 were obtained [16, 26]. These results were indicative of a chain reaction, and the pathway in Scheme II was originally proposed (shown for the case of the $Cp_2Mo_2(CO)_6$ complex in benzene solution).

$$Cp_2Mo_2(CO)_6 \overset{h\nu}{\rightleftharpoons} 2CpMo(CO)_3 \quad \text{initiation} \tag{8}$$

$$CpMo(CO)_3 + L \rightarrow CpMo(CO)_2L + CO \tag{9}$$

$$CpMo(CO)_2L + Cp_2Mo_2(CO)_6 \rightarrow \underset{16e^-}{CpMo(CO)_2L^+} + Cp_2Mo_2(CO)_6^- \tag{10}$$

$$Cp_2Mo_2(CO)_6^- \rightarrow CpMo(CO)_3^- + CpMo(CO)_3 \tag{11}$$

$$\underset{16e^-}{CpMo(CO)_2L^+} + L \rightarrow CpMo(CO)_2L_2^+ \tag{12}$$

(Various radical coupling termination steps are not shown.)

$$\text{Overall: } Cp_2Mo_2(CO)_6 + 2L \rightarrow CpMo(CO)_3L_2^+ + CpMo(CO)_3^- + CO \tag{13}$$

Scheme II

Note that the primary photoprocess in this scheme is homolysis of the metal-metal bond with the subsequent formation of 17-electron metal radicals. Following substitution (Eq. 9), the next step is electron transfer from the substituted 17-electron radical to an unreacted dimer molecule (Eq. 10). Reduction of metal carbonyl dimers is known to cause fragmentation (Eq. 11) giving the anionic product of the reaction and the 17-electron radical [27]. The latter species is recycled and the chain continues.

Although orthodox in all its features, this mechanism was intuitively unsatisfying because the 17-electron complex is the species doing the electron transfer in the reaction. That a 17-electron complex would be the reductant seems unlikely because the dimers are difficult to reduce: their reduction potentials are usually on the order of -1.5 volts or lower vs. Ag/AgCl [28]. Chemical intuition suggests that the 17-electron radicals are not strong reductants because they are electron deficient.

An alternative mechanism for the disproportionation reaction was proposed by Tyler and Stiegman (Scheme III) [16]. This mechanism is similar to the pathway in Scheme II but it differs in one important aspect: the electron transfer occurs not from a 17-electron complex but from a 19-electron complex. Note that the 19-electron complex is formed by association of a 17-electron complex with a ligand (Eq. 16). The 19-electron complex is electronically supersaturated, and it should be an excellent reductant, capable of reducing the dimer.

$$Cp_2Mo_2(CO)_6 \overset{h\nu}{\rightleftharpoons} 2CpMo(CO)_3 \quad \text{initiation} \qquad (14)$$

$$\underset{17}{CpMo(CO)_3} + L \rightarrow \underset{17}{CpMo(CO)_2L} + CO \qquad (15)$$

$$\underset{17}{CpMo(CO)_2L} + L \rightleftharpoons \underset{19}{CpMo(CO)_2L_2} \qquad (16)$$

$$CpMo(CO)_2L_2 + Cp_2Mo_2(CO)_6 \rightarrow CpMo(CO)_2L_2^+ + Cp_2Mo_2(CO)_6^- \qquad (17)$$

$$Cp_2Mo_2(CO)_6^- \rightarrow CpMo(CO)_3^- + CpMo(CO)_3 \qquad (18)$$

(Various radical coupling and electron transfer chain termination steps are not shown.)

$$\text{Overall: } Cp_2Mo_2(CO)_6 + 2L \xrightarrow{h\nu} CpMo(CO)_2L_2^+ + CpMo(CO)_3^- + CO \qquad (19)$$

Scheme III

Several studies in the literature supported the proposal that 19-electron intermediates form in the disproportionation reactions. The first was a study by Poë showing that the ligand substitution reactions of the 17-electron $Re(CO)_5$ radical were associatively activated [29]. No 19-electron intermediates were detected in these reactions (the 19-electron complex was proposed to be a transition state), but the concept of associatively activated reactions of 17-electron complexes was born with this report. Subsequent work by Trogler and Basolo showed that the 6-coordinate $V(CO)_6$ complex also substituted via an associatively activated pathway [30], and once again a 19-electron transition state was invoked. In further support of the concept of 19-electron complexes, Bergman, Jones, and Huggins reported the following mechanism for the $CpV(CO)_3H^-$-catalyzed substitution of several metal carbonyl complexes (Eq. 24) [31]:

$$\dashrightarrow CpV(CO)_3H^- + ML_n \rightarrow CpV(CO)_3H + ML_n^- \quad (20)$$
18 18 17 19

$$ML_n^- \rightarrow ML_{n-1}^- + L \quad (21)$$
19 17

$$ML_{n-1}^- + L' \rightarrow ML_{n-1}L'^- \quad (22)$$
17 19

$$ML_{n-1}L'^- + CpV(CO)_3H \rightarrow ML_{n-1}L' + CpV(CO)_3H^- \quad (23)$$
19 17 18 18

$$\text{Overall: } ML_n + L' \rightarrow ML_{n-1}L' + L \quad (24)$$

$ML_n = CpV(CO)_4, CpFe(CO)(PPh_3)(COCH_3)$

Scheme IV

Note that two key steps in the scheme are the reaction of a 17-electron radical with a ligand to form a 19-electron adduct (Eq. 22) and the subsequent electron transfer (Eq. 23).

2.2 Evidence for 19-electron intermediates

Evidence that 17-electron radicals can react with ligands to form 19-electron complexes came from the analysis of disproportionation reaction products using monodentate, bidentate, and tridentate ligands [16, 32, 33]. Consider the reaction of $Cp_2Mo_2(CO)_6$ with various ligands in CH_2Cl_2 solution:

$$Cp_2Mo_2(CO)_6 + L \xrightarrow[CH_2Cl_2]{h\nu} CpMo(CO)_3L^+ + CpMo(CO)_3^- \quad (25)$$

(Note that in CH_2Cl_2 solution, the cationic product is $CpMo(CO)_3L^+$ rather than $CpMo(CO)_2L_2^+$ as in benzene [33].) As in benzene solution, two pathways are possible: one involving a 17-electron reductant (Scheme V) and the other a 19-electron reductant (Scheme VI). If the reaction follows the pathway in Scheme VI then the reducing agent is the 19-electron $CpMo(CO)_3L$ complex, and the cationic product of the reaction is $CpMo(CO)_3L^+$, as found. On the other hand, if the pathway in Scheme V is followed then the 17-electron $CpMo(CO)_2L$ radical is the reducing agent and the 16-electron $CpMo(CO)_2L^+$ species should form transiently. In the presence of a large excess of monodentate ligand, the reaction of $CpMo(CO)_2L^+$ with the ligand will be competitive with the reaction with CO (Eq. 30) and some $CpMo(CO)_2L_2^+$ should form:

$$CpMo(CO)_2L^+ + L \rightarrow CpMo(CO)_2L_2^+ \quad (36)$$
$16e^-$

Experimentally, it was found that the only cationic product formed in the reaction was $CpMo(CO)_3L^+$, a result consistent with the intermediate formation of the 19-electron species [16]. Using bidentate ligands and similar reasoning, it was possible to rule out the possibility that the reducing species is the $CpMo(CO)_3$ radical [16, 32]. A summary of the results using various ligands in CH_2Cl_2 and benzene is shown in Table 1. In every case a 19-electron species is the only intermediate consistent with all of the results.

$$Cp_2Mo_2(CO)_6 \underset{}{\overset{h\nu}{\rightleftharpoons}} 2CpMo(CO)_3 \quad \text{initiation} \tag{26}$$
$$\qquad\qquad\qquad\qquad 17$$

$$\rightarrow CpMo(CO)_3 + L \rightarrow CpMo(CO)_2L + CO \tag{27}$$
$$17 \qquad\qquad\qquad 17$$

$$CpMo(CO)_2L + Cp_2Mo_2(CO)_6 \rightarrow CpMo(CO)_2L^+ + Cp_2Mo_2(CO)_6^- \tag{28}$$
$$17 \qquad\qquad\qquad\qquad\qquad 16$$

$$CpMo_2(CO)_6^- \rightarrow CpMo(CO)_3^- + CpMo(CO)_3 \tag{29}$$

$$CpMo(CO)_2L^+ + CO \rightarrow CpMo(CO)_3L^+ \tag{30}$$
$$16$$

(Various radical coupling and electron transfer chain termination steps are not shown.)

$$\text{Overall: } Cp_2Mo_2(CO)_6 + L \xrightarrow{h\nu} CpMo(CO)_3L^+ + CpMo(CO)_3^- \tag{25}$$

Scheme V

$$Cp_2Mo_2(CO)_6 \underset{k_{-1}}{\overset{h\nu}{\rightleftharpoons}} 2CpMo(CO)_3 \quad \text{initiation} \tag{31}$$
$$\qquad\qquad\qquad\qquad 17$$

$$\rightarrow CpMo(CO)_3 + L \underset{k_{-2}}{\overset{k_2(K_{eq})}{\rightleftharpoons}} CpMo(CO)_3L \tag{32}$$
$$17 \qquad\qquad\qquad\qquad 19$$

$$CpMo(CO)_3L + Cp_2Mo_2(CO)_6 \xrightarrow{k_3} CpMo(CO)_3L^+ + Cp_2Mo_2(CO)_6^- \tag{33}$$
$$19$$

$$Cp_2Mo_2(CO)_6^- \xrightarrow{k_4} CpMo(CO)_3^- + CpMo(CO)_3 \tag{34}$$

One of several termination steps:

$$CpMo(CO)_3 + CpMo(CO)_3 \xrightarrow{k_5} CpMo(CO)_3L^+ + CpMo(CO)_3^- \tag{35}$$
$$19 \qquad\qquad\quad 17$$

$$\text{Overall: } Cp_2Mo_2(CO)_6 + L \xrightarrow{h\nu} CpMo(CO)_3L^+ + CpMo(CO)_3^- \tag{25}$$

Scheme VI

In addition to the original evidence presented above for 19-electron complexes, numerous other experiments now support the existence of these species. Perhaps the strongest evidence is the isolation of adducts formed by the reaction of 17-electron radicals with ligands. An example is shown in the equation below [10, 34].

$$Co_2(CO)_8 \rightleftharpoons 2Co(CO)_4 \xrightarrow{L_2} Co(CO)_3L_2 + CO \tag{37}$$
$$\qquad\qquad 17e^- \qquad\qquad 19e^-$$

$L_2 =$

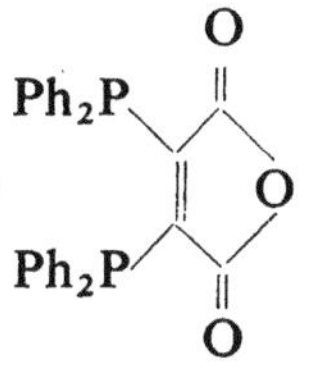

Adducts such as this are stable for special reasons, and this theme is developed in Section 6.

Table 1. Predicted and Actual Products from the Disproportionation of $Cp_2Mo_2(CO)_6$ with Monodentate, Bidentate, and Tridentate Ligands

Reducing Species	Ligand	Predicted Product	Actual Product
In CH_2Cl_2			
$CpMo(CO)_3$	L	$CpMo(CO)_3L^+$	$CpMo(CO)_3L^+$
	$\overset{\frown}{L\,L}$	$CpMo(CO)_3L^{+a}$)L	$CpMo(CO)_2(L_2)^+$
$CpMo(CO)_2L$	L	$CpMo(CO)_2L_2^+$	$CpMo$*b*$CO)_3L^+$
	$\overset{\frown}{L\,L}$	$CpMo(CO)_2(L_2)^+$	$CpMo(CO)_2(L_2)^+$
$CpMo(CO)_3L$	L	$CpMo(CO)_3L^+$	$CpMo(CO)_3L^+$
	$\overset{\frown}{L\,L}$	$CpMo(CO)_2(L_2)^{+b}$	$CpMo(CO)_2(L)_2^+$
In Benzene			
$CpMo(CO)_2L$	L	$CpMo(CO)_2L_2^+$	$CpMo(CO)_2L_2^+$
	$\overset{\frown}{L\,L}$	$CpMo(CO)_2(L_2)^+$	$CpMo(CO)_2(L_2)^+$
	$\overset{\frown}{L\,L}\overset{\frown}{\,L}$	$CpMo(CO)_2(L_2)^+$)L	$CpMo(CO)_2(L_2)^+$)L
$CpMo(CO)L_2$	L	$CpMo(CO)L_3^+$	$CpMo(CO)_2L_2^+$
	$\overset{\frown}{L\,L}$	$CpMo(CO)(L_2)L^+$)L	$CpMo(CO)_2(L_2)^+$
	$\overset{\frown}{L\,L}\overset{\frown}{\,L}$	$CpMo(CO)(L_3)^+$	$CpMo(CO)_2(L_2)^+$)L
$CpMo(CO)_2L_2$	L	$CpMo(CO)_2L_2^+$	$CpMo(CO)_2L_2^+$
	$\overset{\frown}{L\,L}$	$CpMo(CO)_2(L_2)^+$	$CpMo(CO)_2(L_2)^+$
	$\overset{\frown}{L\,L}\overset{\frown}{\,L}$	$CpMo(CO)_2(L_2)^+$)L	$CpMo(CO)_2(L_2)^+$)L

[a] The (L_2))L or L)L nomenclature indicates one atom in the chelating ligand is uncoordinated.

[b] The $CpMo(CO)_3L^+$)L product is not predicted because the 19-electron $CpMo(CO)_3L$)L complex will substitute rapidly.

3. The Reactivity of Photogenerated 19-Electron Complexes

3.1 Electron-transfer reactivity

3.1.1 Outer-sphere reductions. The dominant chemistry of 19-electron complexes is electron transfer [35]. This reactivity is expected because one of the steps in the disproportionation pathway is reduction of an unreacted dimer molecule by a 19-electron intermediate (Eq. 17, Scheme III, Eq. 33, Scheme VI). The reduction potentials of the dimers are quite negative, −1.5 volts or below (vs. Ag/AgCl) [28], and thus the 19-electron complexes are powerful reductants. Tyler and co-workers showed that generation of 19-electron complexes in the presence of suitable substrates led to the reduction of the substrates [35]. For example, irradiation of $Cp_2Fe_2(CO)_4$ and dppe in the presence of the substrates shown in Eqs. 38–44 gave the indicated reduction products [1, 35, 36]. In each case the indicated product is the known product of the electrochemical or chemical reduction of the substrate. The Fe-containing organometallic product in each reaction was $CpFe(CO)dppe^+$, i.e. the oxidized form of the presumed 19-electron reductant CpFe(CO)dppe. The reduction potentials of the substrates are included with each reaction. Clearly, the 19-electron CpFe(CO)dppe complex is a powerful reductant.

$$CpMo(CO)_3Cl \xrightarrow[-1.4V]{Cp_2Fe_2(CO)_4/dppe/h\nu} CpMo(CO)_3^- + CpFe(CO)dppe^+ \quad (38)$$

$$Mn_2(CO)_{10} \xrightarrow[-1.7V]{Cp_2Fe_2(CO)_4/dppe/h\nu} Mn(CO)_5^- + CpFe(CO)dppe^+ \quad (39)$$

$$Cp_2Mo_2(CO)_6 \xrightarrow[-1.4V]{Cp_2Fe_2(CO)_4/dppe/h\nu} CpMo(CO)_3^- + CpFe(CO)dppe^+ \quad (40)$$

$$Fe(CN)_6^{3-} \xrightarrow[-1.4V]{Cp_2Fe_2(CO)_4/dppe/h\nu} Fe(CN)_6^{4-} + CpFe(CO)dppe^+ \quad (41)$$

$$Cp_2Co^+ \xrightarrow[-0.88V]{Cp_2Fe_2(CO)_4/dppe/h\nu} Cp_2Co + CpFe(CO)dppe^+ \quad (42)$$

$$MV^{2+} \xrightarrow[-0.46V]{Cp_2Fe_2(CO)_4/dppe/h\nu} MV^+ + CpFe(CO)dppe^+ \quad (43)$$

(MV^{2+} = methylviologen^{2+})

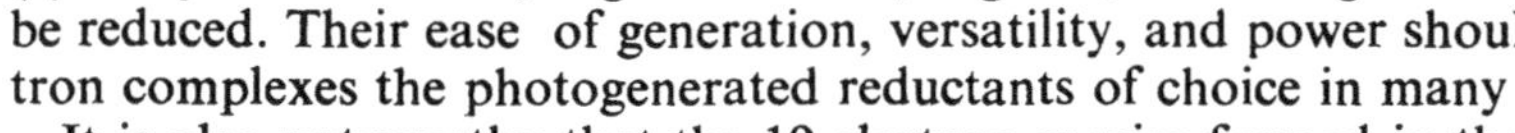

$$+ CpFe(CO)dppe^+ \quad (44)$$

In addition to their large oxidation potentials, two other features make the 19-electron complexes excellent reducing agents: (1) They are easy to generate; the dimer is simply irradiated at 520 nm in the presence of the substrate and the ligand. (2) They are versatile; organometallic, organic, and inorganic substrates can all be reduced. Their ease of generation, versatility, and power should make 19-electron complexes the photogenerated reductants of choice in many systems.

It is also noteworthy that the 19-electron species formed in these reactions are discrete complexes with sufficient lifetimes to undergo bimolecular reactions. Because many of the ligands that react with 17-electron species to form 19-electron species will also substitute the 17-electron complex, it is clear that substitution and electron transfer are competitive reaction pathways, e.g. [37]:

$$\underset{17}{CpFe(CO)_2} + L \rightleftharpoons \underset{19}{CpFe(CO)_2L} \begin{cases} \xrightarrow{-CO} \underset{17}{CpFe(CO)L} \\ \xrightarrow{S} CpFe(CO)_2L^+ + S^- \end{cases}$$

Scheme VII

A question of considerable importance concerning the 19-electron complexes is, what is their oxidation potential? These numbers have not been measured. However, the photochemical disproportionation reactions of the $Cp_2Mo_2(CO)_6$, $Cp_2Fe_2(CO)_4$, and $Mn_2(CO)_{10}$ dimers all proceed via the general mechanism in Scheme VI so it is possible to put a lower limit on the oxidation potentials for the 19-electron species that is equal to the negative of the reduction potential of the corresponding dimers (i.e. $CpFe(CO)L_2$: +2.2 V vs. Ag/Ag^+, $CpMo(CO)_3L$: +1.4 vs. Ag/Ag^+; $Mn(CO)_3(NR_3)_3$: =1.7 V vs. Ag/Ag^+) [28].

Rate constants for electron transfer have been measured in several systems. Thus, Hepp and Wrighton obtained the following rate constants [38]:

$$CpW(CO)_3 + CH_3CN + Cp_2Fe^+ \xrightarrow{k=1.4\times10^6 M^{-1}s^{-1}} Cp_2Fe + CpW(CO)_3CH_3CN^+ \quad (45)$$

$$Mn(CO)_5 + CH_3CN + Cp_2Fe^+ \xrightarrow{7.0\times10^6 M^{-1}s^{-1}} Cp_2Fe + Mn(CO)_5CH_3CN^+ \quad (46)$$

These equations were written the way they were because it is not clear if a 17- or a 19-electron complex is the reducing agent in these systems. Cp_2Fe^+ is a reasonable oxidizing agent; thus, there is no need to necessarily postulate that the 19-electron $ML_n(CH_3CN)$ species are the reductants in reactions 45 and 46. A different situation applies to the reduction of nonoxidizing agents, however. Meyer and Caspar reported rate constants for the reduction of a series of pyridinium ions [39]:

$$Mn(CO)_5(CH_3CN) + R\text{-}NC_5H_4X^+ \rightarrow Mn(CO)_5(CH_3CN)^+ + R\text{-}NC_5H_4X \quad (47)$$

Rate constants in the range 3.0×10^5 to 1.5×10^6 were obtained, varying monotonically with the redox potential of the pyridinium cation. In these reactions, coordination of a CH_3CN solvent molecule to $Mn(CO)_5$ is probably required for thermodynamic reasons before electron transfer will occur because the reduction potentials of many of the pyridinium cations are too negative to be reduced by a mild reductant like $Mn(CO)_5$; the 19-electron $Mn(CO)_5CH_3CN$ is expected to be a much better reductant than $Mn(CO)_5$. Prior coordination may also be necessary in Wrighton and Hepp's reductions of $C_7H_7^+$ and paraquat with $CpW(CO)_3$, $Mn(CO)_5$, or $Re(CO)_5$ radicals in CH_3CN [38].

All the electron transfer reactions discussed thus far are outer-sphere reactions; the 19-electron reductants are coordinatively saturated, so inner-sphere reactions of a 19-electron complex with a substrate are presumably impossible. Precedents for outer-sphere reductions using 19-electron complexes are found in the chemistry of the $Co(CN)_5^{3-}$ complex [40]. Thus, the reaction of CN^- which $Co(CN)_5^{3-}$ forms the electron-rich 19-electron $Co(CN)_6^{4-}$ species which is a good reductant:

$$Co(CN)_5^{3-} + CN^- \rightleftharpoons Co(CN)_6^{4-} \quad (48)$$

$$Co(CN)_6^{4-} + S \rightarrow Co(CN)_6^{3-} + S^- \quad (49)$$

$$S = Co^{III}(NH_3)_5X;\ X = PO_4^{3-}, CO_3^{2-}, SO_4^{2-}, NH_3, OAc^-$$

Nineteen-electron reductants need not necessarily always be generated by photolysis of metal-metal bonds. For example, the 19-electron reductant in Scheme IV is formed in a thermal reaction. This is an important point because in Section 5 evidence is presented which shows that many types of radicals react with nucleophiles to form analogues of the 19-electron complexes. These analogues also undergo electron transfer reactions.

3.1.2 Initiation of ETC reactions. The electron transfer reactions of 19-electron complexes find applications in the electron-transfer-chain (ETC) catalysis substitution reactions of metal carbonyl clusters [41]. In these reactions, a catalytic amount of reducing agent is used to labilize a metal carbonyl cluster toward substitution. An example is given in Eq. 50 [42].

$$Ru_3(CO)_{12} + PR_3 \xrightarrow{(e^-)} Ru_3(CO)_{11}PR_3 + CO \qquad (50)$$

The mechanism of this and similar reactions has been studied and the proposed pathway is shown in Scheme VIII.

$$Ru(CO)_{12} + e^- \rightarrow Ru_3(CO)_{12}^- \quad \text{initiation} \qquad (50)$$

$$Ru(CO)_{12}^- + PR_3 \rightarrow Ru_3(CO)_{11}PR_3^- + CO \qquad (51)$$

$$Ru_3(CO)_{11}PR_3^- + Ru_3(CO)_{12} \rightarrow Ru_3(CO)_{11}PR_3 + Ru_3(CO)_{12}^- \qquad (52)$$

Scheme VIII

The reaction is initiated by reduction of the cluster (Eq. 50). The reduced cluster is labile and substitution occurs (Eq. 51). Electron transfer from the substituted cluster to an unsubstituted cluster then follows (Eq. 52), and the chain is repeated. Typical reducing agents used in these reactions include Na and the benzophenone radical anion. However, Goldman and Tyler showed that 19-electron complexes can also be used as the reducing agent that initiates the reaction [43]. The 19-electron complexes were generated by irradiating catalytic amounts of $Cp_2Fe_2(CO)_4$ and dppe. Metal carbonyl complexes substituted by this method include $Ru_3(CO)_{12}$ $Os_3(CO)_{12}$, $Fe(CO)_5$, and $CpFe(CO)_2PBu_3^+$. The 19-electron complexes are such excellent reducing agents that virtually all ETC reactions initiated by reduction should be susceptible to initiation by this method.

Along similar lines, it has been reported that many of the metal-metal bonded carbonyl dimers are catalysts for the substitution of CO in metal carbonyl complexes [44]. For example, $Cp_2Fe_2(CO)_4$ is reported to catalyze the substitution of CNR for CO in the $CpFe(CO)_2I$ complex [44c]. Numerous schemes have been reported to account for the catalysis including the one shown in Scheme IX. (Note that species A is the unidentified catalyst in this reaction.)

$$Cp_2Fe_2(CO)_4 \xrightarrow{h\nu} A$$

$$A + CpFe(CO)_2I \rightarrow A\text{-}(\text{-}CpFe(CO)_2I)$$

$$A\text{-}(\text{-}CpFe(CO)_2I) + L \rightarrow A\text{-}(\text{-}CpFe(CO)(L)I) + CO$$

$$A\text{-}(\text{-}CpFe(CO)(L)I) \rightarrow A + CpFe(CO)(L)I$$

$$\text{net: } CpFe(CO)_2I + L \rightarrow CpFe(CO)(L)I + CO$$

Scheme IX

We suggest that, in many cases, the catalysis is by electron transfer catalysis, initiated by the formation of 19-electron complexes following homolysis of the metal-metal bond. The 19-electron complexes then act as reductants to initiate the substitution reactions according to Scheme VII. Further studies are needed to distinguish between the various pathways, and it is likely that, in many cases, multiple pathways are operating.

3.1.3 Reduction of CO_2. The most powerful photogenerated reductants are formed by irradiation of the $W_2(CO)_{10}^{2-}$ complex and a phosphine ligand [45]. Presumably the photogenerated reducing agent is the 19-electron $W(CO)_5PR_3^-$ complex, formed by the reaction of $W(CO)_5^-$ with PR_3; however, no mechanistic studies were done to establish this point with certainty. The $W_2(CO)_{10}^{2-}$ complex did disproportionate when irradiated in the presence of phosphines, and it was proposed that the mechanism of the disproportionation was analogous to the mechanism used by the neutral dimers; a 19-electron intermediate is therefore implied [46].

The reduction potential of the $W(CO)_5PR_3^-$ complex was not measured, but it was estimated at about −2.6 volts vs. SCE [47]. This 19-electron complex is thus a much more powerful reducing agent than the other 19-electron complexes, and experiments showed it was capable of reducing substrates that the others cannot. For example, irradiation of $W_2(CO)_{10}^{2-}$ and PPh_3 in the presence of CO_2 led to the reduction of the CO_2 (Eq. 53) [45].

$$CO_2 \xrightarrow[W_2(CO)_{10}^{2-}/PPh_3]{h\nu(\lambda > 420 nm)} HCO_2^- (20\%) + CO_3^{2-} (40\%)$$

$$HCO_3^- (20\%) + CO + W(CO)_5PPh_3 (70\text{–}90\%) \qquad (53)$$

None of the other photogenerated 19-electron complexes reduced CO_2. In addition, the $W_2(CO)_{10}^{2-}/PPh_3$ system was used to polymerize α-methylstyrene. An anionic polymerization was suggested.

3.1.4 Water-soluble 19-electron complexes as reducing agents. The 19-electron complexes discussed to this point have all been generated in nonaqueous solvents. Recent papers describe methods for water-solubilizing the 19-electron complexes. MacKenzie and Tyler reported a technique for reducing water-soluble substrates with 19-electron complexes in micellar solutions [48]. The procedure consisted of dissolving $Cp_2Fe_2(CO)_4$ and dppe in benzene solution. This solution was then added to an aqueous solution containing the substrate to be reduced (methylviologen or ferricyanide) and a surfactant. With the proper concentrations of reactants, a homogeneous emulsion formed on mixing. Irradiation of the solution in the usual manner resulted in the reduction of the substrate. No mechanistic work was done, but it was proposed that 19-electron complexes were forming in the benzene phase and transferring an electron to the substrate in the aqueous phase. From a practical point of view, the major problem with the system is that the concentrations of the species in solution must be kept low so as to avoid a separation of the phases.

A more direct method for reducing water-soluble substrates was reported by Avey and Tyler [49]. Their method was to use a water-soluble dimer as the precursor for the 19-electron complex. For this purpose they synthesized the $(CpCOO^-)_2W_2(CO)_6$ species. Irradiation of this complex and the water-soluble phosphine $PPh_2(C_6H_4SO_3)^-$ in aqueous solution containing substrates led to reduction of the substrates:

$$[Cp(COO^-)]_2W_2(CO)_6 + S \xrightarrow[H_2O]{h\nu, L} S^- \qquad (54)$$

$S = Fe(CN)_6^{3-}$, methylviologen, cytochrome $c(Fe^{III})$

Again, no mechanistic work was done in the aqueous solutions but it was noted that the photochemical reactivity of the $(CpCOO^-)_2W_2(CO)_6$ dimer in nonaqueous solution is similar to the reactivity of the "normal" dimers. It was speculated therefore that photogenerated 19-electron intermediates were involved in the reductions in Eq. 54.

3.2 Ligand dissociation reactions

Numerous studies have demonstrated that 19-electron complexes dissociate a ligand rapidly. (In fact, it is common to write the reaction of a 17-electron radical with a ligand to form a 19-electron complex as an equilibrium.) Electrochemical studies, in particular, provide abundant examples of ligand dissociation from 19-electron complexes:

$$Fe(CO)_5 \xrightarrow{e^-} [Fe(CO)_5^-] \rightarrow Fe(CO)_4^- + CO \quad [47, 50] \tag{55}$$

$$W(CO)_6 \xrightarrow{e^-} [W(CO)_6^-] \rightarrow W(CO)_5^- + CO \quad [51] \tag{56}$$

In many cases, ligand dissociation occurs in competition with electron transfer from the 19-electron complexes. Our own work on the photochemical disproportionation of the dimers illustrates this point, because both substitution and disproportionation products form in these reactions [16, 37]. Substitution occurs if the 19-electron intermediate dissociates a ligand and disproportionation results from electron transfer (Scheme VII).

It is now well established that, in the absence of unusual steric or electronic constraints, 17-electron radicals substitute by an associatively activated mechanism [52]. One of the fundamental questions concerning these substitution reactions is whether the 19-electron species formed are intermediates or transition states. Electron transfer reactivity can be used as a test for the formation of an intermediate because outer-sphere electron transfer cannot occur from a transition state. For example, $CpFe(CO)_2$ reacts with selected phosphines to form $CpFe(CO)PR_3$ [53]:

$$CpFe(CO)_2 + PR_3 \rightleftharpoons \underset{19e^-}{[CpFe(CO)_2PR_3]} \rightleftharpoons CpFe(CO)PR_3 + CO \tag{57}$$

For the case of PBu_3, Goldman and Tyler proposed that the 19-electron complex is an intermediate because PBu_3 will disproportionate the $Cp_2Fe_2(CO)_4$ dimer, and in addition, the $Cp_2Fe_2(CO)_4/PBu_3$ system can be used as a photogenerated reducing agent system [37]. Note, however, that this conclusion is inconsistent with the flash infrared spectroscopic results of Turner and Weitz, who found no evidence for a 19-electron intermediate in the bimolecular reaction of $CpFe(CO)_2$ with phosphines [53].

3.3 Substitution reactions

The few documented substitution reactions of 19-electron complexes are shown in the equations below.

$$Mn(CO)_4(DBOQ) + \text{t-Bu-DAB} \xrightarrow[\text{dark}]{RT} Mn(CO)_3(DBOQ)(\text{t-Bu-DAB-N}') + CO \quad [13, 54] \tag{58}$$

(DBOQ = 3,5-di-t-butyl-ortho-benzoquinone; the —N′ nomenclature for the DAB ligand indicates only one N atom is coordinated to the Mn)

$$Mo(CO)_4(bpy)^- + PPh_3 \xrightarrow{RT} Mo(CO)_3PPh_3(bpy)^- + CO \quad [14] \tag{59}$$

$$Mn(CO)_3(CH_3CN)_3 \xrightarrow[RT]{L} Mn(CO)_3(CH_3CN)_2L \xrightarrow{L} Mn(CO)_3(CH_3CN)L_2 \quad [55] \tag{60}$$

$L = PPh_3$ or PMe_2Ph

$$CpMo(CO)_2(\text{c-hex-DAB}) \xrightarrow[RT]{PPh_3} CpMo(CO)(PPh_3)(\text{c-hex-DAB}) + CO \quad [56] \tag{61}$$

$$Co(CO)_3L_2 + L' \rightarrow Co(CO)_2L_2L' + CO \quad [34] \tag{62}$$

L_2 = 2,3-bis(diphenylphosphino)maleic anhydride
$L' = PR_3, P(OR)_3, I^-$

The preceding section presented evidence that ligand dissociation is common for 19-electron complexes, and it is logical to propose that these substitution reactions follow dissociatively activated pathways. In fact, the mechanism of reaction 62 was studied, and the substitution was found to proceed by a pathway involving the initial dissociation of CO [34]:

$$Co(CO)_3L_2 \rightleftharpoons Co(CO)_2L_2 + CO$$

$$Co(CO)_2L_2 + L' \rightarrow Co(CO)_2L_2L'$$

Scheme X

The first-order rate constant for the reaction is 5.5×10^{-3} sec^{-1} (25°, CH_2Cl_2), and the activation parameters are $\Delta H^{\ddagger} = 23.8 \pm 0.6$ kcal mol^{-1} and $\Delta S^{\ddagger} = 11.1 \pm 2.2$ cal mol K^{-1} [34].

The $Co(CO)_3L_2$ complex was chosen for a study of the substitution reactions in 19-electron complexes because the molecule has a square pyramidal structure [10]; a vacant coordination site is thus available for attack by an incoming ligand. Therefore, if any 19-electron complex is going to react by an associatively activated pathway then this molecule is a prime candidate. However, there was no evidence for an associatively activated pathway, and it was concluded that if this molecule, especially chosen because it might undergo associatively activated reactions, reacts dissociatively then other 19-electron complexes are almost certain to react dissociatively.

3.4 Inner-sphere electron transfer reactions

An example of inner-sphere electron transfer was reported by Meyer and Caspar in the following reaction [39].

$$2\,Ru(bpy)_2Cl_2^+ + Mn_2(CO)_{10} + 2\,CH_3CN \xrightarrow{h\nu} 2\,Ru(bpy)_2(CH_3CN)Cl^+ + 2\,Mn(CO)_3Cl \tag{63}$$

Although no mechanistic details were reported for this reaction, one interpretation is that $Ru(bpy)_2Cl_2^+$ coordinates to a $Mn(CO)_5$ radical via a Cl bridge to give a 19-electron Mn center; inner-sphere electron transfer from Mn to Ru then follows:

$Mn_2(CO)_{10} \underset{}{\overset{h\nu}{\rightleftharpoons}} 2Mn(CO)_5$

$Mn(CO)_5 + Ru(bpy)_2Cl_2^+ \rightarrow [(CO)_5Mn{-}Cl{-}Ru(bpy)_2Cl]^+$

e⁻

$[(CO)_5Mn{-}Cl{-}Ru(bpy)_2Cl]^+ \rightarrow Mn(CO)_5Cl + Ru(bpy)_2Cl^+$

$Ru(bpy)_2Cl^+ + CH_3CN \rightarrow Ru(bpy)_2(CH_3CN)Cl^+$

Scheme XI

Other examples of inner-sphere reactions of this type can be cited wherever 17-electron radicals react with ligands to form $18 + \delta$ complexes. Recall that these species are those in which the unpaired electron is primarily ligand localized. The electron transfer in these reactions is formally from the 19-electron metal center to the ligand. Following are several of the numerous examples of this type of reaction. (See also reaction 37.)

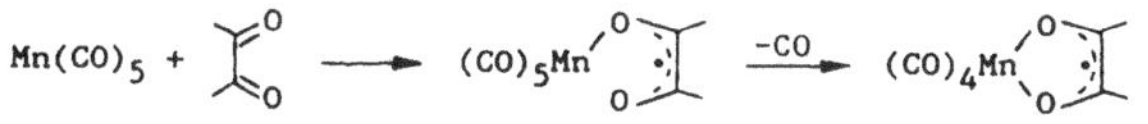

(64) [57]

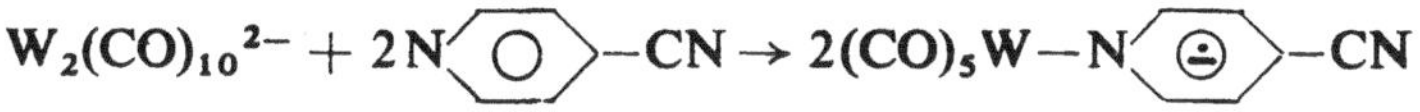

[11] (65)

$\overset{I}{Cp_2W_2}(CO)_6 + 2(R\text{-}DAB) \rightarrow 2Cp\overset{II}{W}(CO)_2(R\text{-}\overset{\ominus}{D}AB)$ [56] (66)

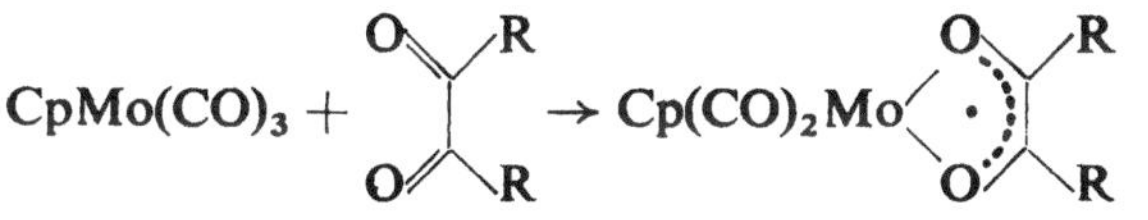

[58] (67)

Note in each case that the reaction is initiated by homolysis of the metal-metal bond to produce a 17-electron radical.

3.5 Ligand coupling reactions

The final reactivity of photogenerated 19-electron complexes to be discussed is the ligand coupling reaction. As discussed in more detail below, the unpaired electron in many 19-electron complexes is in an orbital that is primarily ligand in character. In some instances, the reactivity of the complex is thought to reflect the location of the extra electron. For example, Wrighton and Blaha reported the following photochemical reaction of $CpFe(CO)_2R$ ($R = CH_2C_6H_5$) with CO [59]:

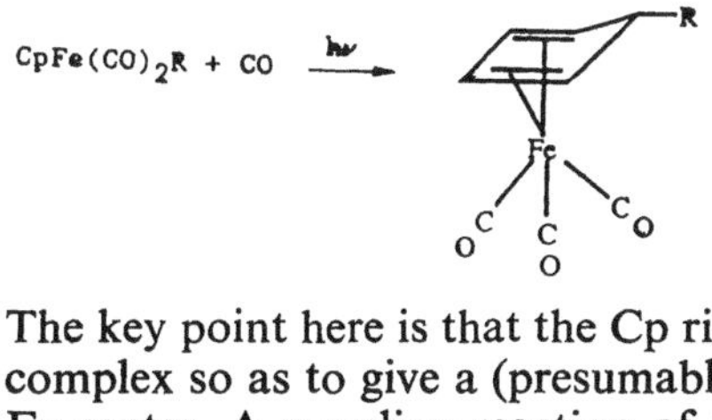

(68)

The key point here is that the Cp ring has slipped in the "19-electron" $CpFe(CO)_3$ complex so as to give a (presumably) more stable 18-electron configuration at the Fe center. A coupling reaction of the radical ring and the R radical is therefore possible.

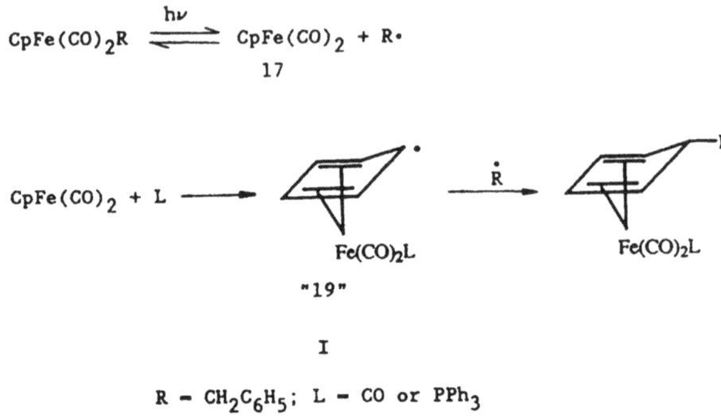

Scheme XII

Until recently, ligand coupling reactions between two photogenerated 19-electron complexes were unknown, probably because the concentration of 19-electron species is so low that bimolecular reactions between them are unlikely to occur before electron transfer or ligand dissociation (Scheme VII). However, Wrighton, Zou, and Ahmed recently reported [60] that irradiation of (η^5-cyclohexadienyl)$_2$ $Fe_2(CO)_4$ in the presence of CO or PPh_3 gave the ring-coupled product shown below (and its diastereomer):

$$(\eta^5\text{-cyclohexadienyl})_2Fe_2(CO)_4 + 2L \xrightarrow{h\nu}$$

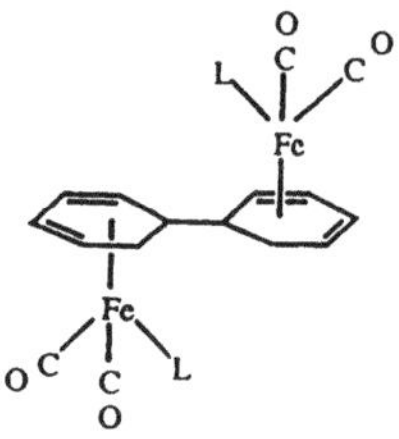

(69)

The proposed mechanism is shown in Scheme XIII and it involves the reaction of a photogenerated 17-electron (cyclohexadienyl)$Fe(CO)_2$ species with the ligand to form a 19-electron (cyclohexadienyl)$Fe(CO)_2L$ complex. Coupling of two of these species then gives the product.

$$(\eta^5\text{-}C_6H_7)_2Fe_2(CO)_4 \xrightleftharpoons{h\nu \text{ or } \Delta} 2(\eta\text{-}C_6H_7)Fe(CO)_2$$

$$(\eta\text{-}C_6H_7)Fe(CO)_2 + L \rightarrow$$

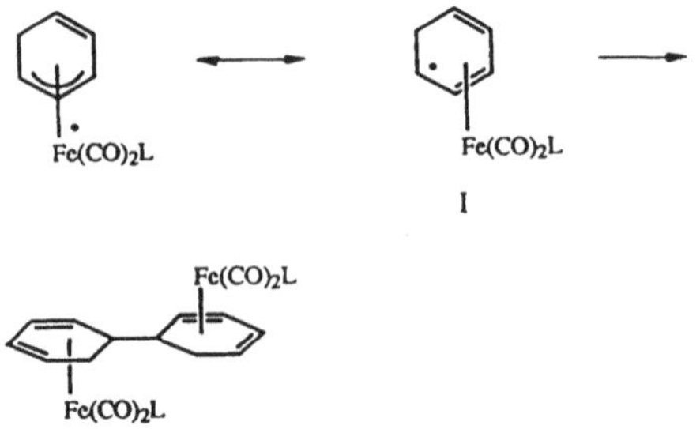

Scheme XIII

(Note that irradiation of the $(\eta^5\text{-cyclohexadienyl})_2Fe_2(CO)_4$ dimer is not necessary for the reaction in equation 69 to occur. The dimer and a small amount of the 17-electron radical are in equilibrium in solution. Irradiation of the solution speeds up the reaction, however. It is tempting to conclude that the 19-electron (cyclohexadienyl)$Fe(CO)_2L$ complex has the unpaired electron localized on the cyclohexadienyl ring, as shown in structure I in Scheme XIII, because the coupling occurs through the rings. However, Astruc recently concluded that it is unwise to draw conclusions about the location of the unpaired spin density on metal radicals based on the structure of the coupling products [5]. He showed that the location of the spin density in the transition state of the coupling reaction can be very different than in the ground state. If this is the case then reliable information about the location of the unpaired electron in 19-electron complexes would seem to be available only by direct spectroscopic measurement (using ESR, for example).

Electrochemical studies have shown that when 19-electron complexes couple, they usually couple through the ligands. This result is so common that Geiger made it one of his rules for predicting the coupling reactivity of organometallic radicals [61]. Typical examples of this reactivity are shown below [62]:

$$(\eta^5\text{-}C_6H_7)Fe(CO)_3^+ + e^- \rightarrow (\eta\text{-}C_7H_7)Fe(CO)_3 \rightarrow$$
$$(CO)_3Fe(\eta^4\text{-}C_6H_7\text{-}\eta^4\text{-}C_7H_6)Fe(CO)_3 \tag{70}$$

$$(\eta^7\text{-}C_7H_7)M(CO)_3^+ + e^- \rightarrow (\eta\text{-}C_7H_7)M(CO)_3 \rightarrow$$
$$(CO)_3M(\eta^6\text{-}C_7H_7\text{-}\eta^6\text{-}C_7H_7)M(CO)_3 \tag{71}$$

4. Thermodynamics of 19-Electron Complex Formation

The electrochemical reduction potentials of 18-electron organometallic complexes are generally quite negative [63]. This fact is consistent with the reactivity of 19-electron complexes discussed above showing that these species are powerful reducing agents. Clearly, the 19-electron complexes are high energy species, and it is reasonable to ask why 17-electron radicals should react with ligands to form such high energy complexes. To answer this question, Tyler, Philbin, and Granatir measured ΔG° for the reaction of $CpMo(CO)_3$ with X^- ($X = Cl, Br, I$) to form $CpMo(CO)_3X^-$ [64]. The procedure they used was to measure K_{eq} for the formation of the 19-electron complex from the 17-electron radical and the ligand:

$$\underset{17e^-}{CpMo(CO)_3} + X^- \overset{K_{eq}}{\rightleftharpoons} \underset{19e^-}{CpMo(CO)_3X^-} \tag{72}$$

$X = Cl, Br, I$

K_{eq} was obtained from a kinetic analysis of the disproportionation pathway. The quantum yield for the disproportionation reaction is related to the experimental parameters by the following expression [64, 65]:

$$\Phi = \phi\left(1 - \frac{k_{-1}}{k_{-1} + K[X^-]}\right) + \left(\frac{K'[X^-]}{(k_{-1} + K[X^-])^{1/2}}\right)(\phi/I_a)^{1/2} \tag{73}$$

where $K' = k_2k_3[Cp_2Mo_2(CO)_6]/k_{-2}$, $K = k_5k_2/k_{-2}$, ϕ = the quantum yield for homolysis, and I_a = the absorbed light intensity. (The rate constants refer to the pathway in Scheme VI with L = X^-.)
Note that the quantum yield for disappearance of the dimer is linearly proportional to $I_a^{-1/2}$. Numerous plots confirmed this relationship, and this was taken as evidence that the systems are well-behaved kinetically. The equation predicts a rather complicated dependence of the quantum yield on the concentration of the halide ligand. Data from the quantum yield dependence on the concentration of halide were fit to Eq. 73 by a non-linear least squares method. From the fit, a value for K was obtained. K is equal to k_5K_{eq}, where k_5 is the rate constant for the termination reaction between a 17-electron radical and a 19-electron species. The rate k_5 is not known but it was assumed to be diffusion controlled, and thus, lower limits on K_{eq} were obtained. The following values of $\Delta G°$ resulted: Cl^-, $\Delta G° = -2.1$ kcal/mol; $\Delta G° = -2.5$ kcal/mol; I^-, $\Delta G° = -3.1$ kcal/mol. The key point in these results is that the values for $\Delta G°$ are negative. (If k_5 is less than the diffusion controlled rate then the values for $\Delta G°$ are even more negative.) Thus, the reactions of $CpMo(CO)_3$ with X^- to form the 19-electron complexes are energetically downhill. This was perhaps a somewhat surprising result because one is accustomed to thinking of 17-electron species as acceptable intermediates but 19-electron complexes as high energy species. Note that the thermodynamics should be even more favorable with better ligands such as phosphines; however, $\Delta G°$ has not yet been measured using these ligands.

In other work related to the measurement of $\Delta G°$, Trogler and Therien [66] reported the equilibrium constant for the following reaction:

$$CpFe(CO)(PPh_3)CH_3^+ + py \underset{}{\overset{K_{eq} \approx 7}{\rightleftharpoons}} CpFe(CO)(PPh_3)(py)CH_3^+ \qquad (74)$$

The equilibrium constant was measured by electrochemical techniques, and once again the point is made that 19-electron complexes are thermodynamically reasonable species compared to 17-electron species and a ligand.

Numerous workers have proposed that the driving force for the formation of 19-electron complexes is the formation of the new metal-ligand bond [1]. To a first approximation, the new bond is of half order because the unpaired electron resides in the metal-ligand antibonding orbital (Fig. 1a). Even more favorable

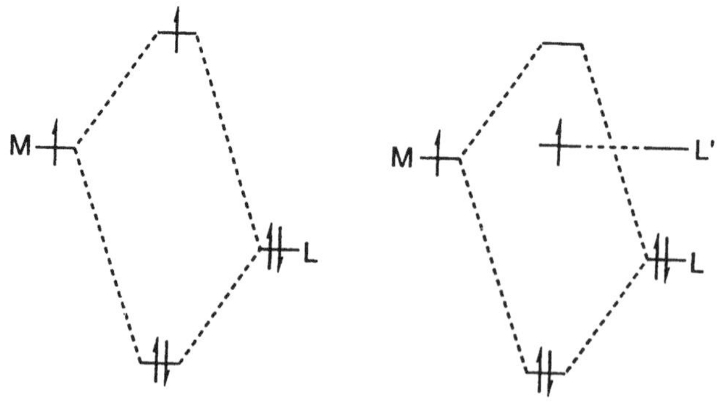

a b

thermodynamics are expected for those 19-electron complexes where the unpaired electron can be delocalized onto low energy ligands orbitals (e.g. $18 + \delta$ complexes, Fig. 1 b). These complexes are discussed further in Section 6 as part of the discussion of the structures of 19-electron complexes.

5. Generalization of the Principle of Adduct Formation Followed by Electron Transfer

One of the general principles to emerge from a study of 19-electron complexes is that metal radicals react with ligands (nucleophiles) to form adducts, which then undergo electron transfer. An interesting question is whether similar reactivity is found for other types of radicals. Tyler and Goldman studied this question using the $CpW(CO)_3CH_3$ complex [20]. This molecule is similar to the dimers studied in the disproportionation reactions, but with the obvious difference that one of the $CpM(CO)_3$ radicals in the dimer has been replaced by a CH_3 radical. Irradiation of this complex resulted in the following reaction (in addition to the expected CO substitution reaction):

$$Cp(CO)_3W{-}CH_3 \xrightarrow[PPh_3]{h\nu} CpW(CO)_3^- + CH_3{-}PPh_3^+ \quad (75)$$

$$Cp(CO)_3W{-}W(CO)_3Cp \xrightarrow[PPh_3]{h\nu} CpW(CO)_3^- + Cp(CO)_3W{-}PPh_3^+ \quad (76)$$

The products are analogous to the products that form in the disproportionation reactions (Eq. 76) but with the difference, once again, that CH_3 replaces one of the $CpM(CO)_3$ radicals.

The mechanism of reaction 75 was studied and found to be analogous to the dimer disproportionation mechanism; a key step is the reaction of the photogenerated CH_3 radical with PPh_3 to form the PPh_3CH_3 adduct, which then transfers an electron:

$$CpW(CO)_3CH_3 \overset{h\nu}{\rightleftharpoons} CpW(CO)_3 + \cdot CH_3$$

$$\cdot CH_3 + PPh_3 \rightleftharpoons \dot{P}Ph_3CH_3$$

(a phosphoranyl radical)

$$PPh_3CH_3 + CpW(CO)_3 \rightarrow PPh_3CH_3^+ + CpW(CO)_3^-$$

Scheme XIV

This reaction demonstrated that, at least in some instances, other types of radicals can react to form electronically supersaturated adducts that undergo electron transfer. Although not common, several other electron transfer reactions of phosphoranyl radicals have been reported. For example, phosphoranyl radicals were first proposed to explain the chain reaction of PPh_3 with bromoform (Eqs. 77 and 78) [67].

$$PPh_3 + \cdot CHBr_2 \rightarrow \dot{P}h_3PCHBr_2 \quad (77)$$

$$Ph_3\dot{P}CHBr_2 + CHBr_3 \rightarrow Ph_3PCHBr_2^+Br^- + \cdot CHBr_2 \quad (78)$$

Similar reactions of other phosphines and phosphites with halocarbons and with benzoyl peroxide, involving oxidation of phosphoranyl radicals, have also been reported [9].

Analogues of the 19-electron complexes are probably also forming in the reaction of $Co(CO)_4SnCl_3$ with phosphines [68]:

$$Co(CO)_4SnCl_3 + PR_3 \rightarrow Co(CO)_4^- + PR_3SnCl_3^+ \quad (79)$$

This reaction is also analogous to the disproportionation reactions of the dimer complexes; if an analogous mechanism is followed then the adduct has the formula $R_3P{-}SnCl_3$.

In one last example, note that the formation of 19-electron complexes via reaction of an organic radical with an organometallic nucleophile has been suggested. The reaction in Eq. 80 was proposed as a key step in the $NaCo(CO)_4$ catalyzed reaction in Eq. 81 [69].

$$C_6H_5 + Co(CO)_4^- \rightarrow \underset{19}{C_6H_5Co(CO)_4^-} \quad (80)$$

$$C_6H_5Br + CO + NaOC(CH_3)_2CH_2CH_3 \xrightarrow[NaCo(CO)_4 \; (2)H^+]{(1)h\nu} C_6H_5COOR \quad (81)$$

$R = C(CH_3)_2CH_2CH_3$ (70%); $R = H$ (17%)

6. Geometric and Electronic Structures of 19-Electron Complexes

6.1 Classes of 19-electron complexes

In the Introduction to this chapter, the point was made that several structures are possible for 19-electron complexes. It is conceptually useful to divide the structures into three limiting classes: Class I contains the σ^* complexes. These are defined [70] as 19-electron complexes in which the unpaired electron is occupying a M—L antibonding orbital. Examples include the $Mn(CO)_5Br^-$ [71] and $CpMo(CO)_3I^-$ [72] complexes [73]. Class II complexes are those that have a metal valence-electron count lower than 19 because of a ligand geometry change such as a slipped Cp ring [7], bent CO ligand [8], or a phosphoranyl radical structure. Examples of these types of molecules were given in the Introduction with reference to the $CpMo(CO)_3PR_3$ molecule. Finally, the class III 19-electron complexes are the $18 + \delta$ complexes. These are complexes in which the unpaired electron is located primarily in a low-energy ligand orbital (generally a π^* orbital). Examples include $Mo(CO)_4(bpy)^-$ [14] and $CpMo(CO)_2$(c-hex-DAB) (c-hex-DAB = c-$C_6H_{11}N{=}CHCH{=}N$-c-C_6H_{11}) [56, 74]. As discussed previously, molecules in the $18 + \delta$ class can be thought of as 18-electron complexes with reduced ligands.

Most work on the reactivity of 19-electron complexes has been on molecules in class III. Molecules in this category are generally the most stable because the unpaired electron is not in a M—L antibonding orbital and because no ligand has undergone a substantial structural change. In many cases, it is possible to isolate complexes in this category. However, it is important to emphasize that, although they are generally more unstable, molecules in the other classes have been detected. The point is that the categories above are not just hypothetical classifications devised to encompass all future findings.

6.2 Stable 19-electron complexes

Structural characterization of the 19-electron complexes is generally straightforward if the molecule is stable or can be stabilized at low temperature. For example,

Symons and Morton and Preston have characterized numerous class I (σ^*) 19-electron complexes at low temperature by esr spectroscopy [71–73]. In general, the molecules are generated by pulse radiolysis of the 18-electron analogues. The $18 + \delta$ complexes are frequently stable enough to characterize by standard spectroscopic techniques at room temperature. In many cases, these molecules have been characterized by X-ray crystallography [10].

6.3 Structural inferences from reactivity

When the 19-electron complexes are not stable, their structures must be inferred by their reactivity. Thus, we have already discussed the case of $CpFe(CO)_2L$ in Scheme XII. Addition of a radical to this complex yields the cyclopentadiene complex in reaction 68, from which it was inferred that the Cp ring has slipped to an η^4 geometry in the 19-electron complex (Structure I in Scheme XII). However, as discussed above, Astruc cautioned against using the structures of coupling products to determine the structures of metal radicals. In this case, therefore, it may be unwise to conclude that the $CpFe(CO)_2L$ radical has an η^4-Cp ring. Consistent with this conclusion, Wrighton, Wuu, and Zou later showed that the "19-electron" (indenyl)$Fe(CO)_3$ complex has the structure below [75]:

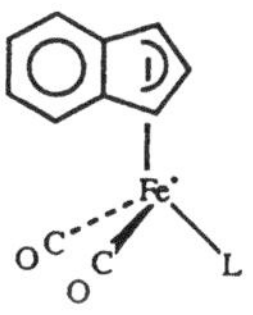

I

The alternative structure II below was ruled out because ESR spectra showed the unpaired electron in the (indenyl)$Fe(CO)_3$ complex is Fe localized. Similarly, structure III below was eliminated because the infrared spectrum of the (indenyl) $Fe(CO)_3$ complex is very similar to the spectrum for that of the 17-electron (η^3-C_3H_5)$Fe(CO)_3$ species.

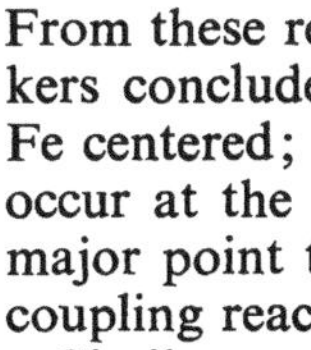

II

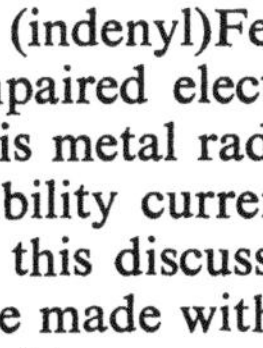

III

From these results with the (indenyl)$Fe(CO)_3$ complex, Wrighton and his coworkers concluded that the unpaired electron on the $CpFe(CO)_3$ complex may be Fe centered; coupling of this metal radical with the organic radical R could still occur at the ring – a possibility currently being investigated by Wrighton. The major point to be made in this discussion is that structural inferences based on coupling reactions should be made with caution.

Similar reactivity-based evidence suggests that the structure of the $CpMo(CO)_3$ $P(OR)_3$ complex may be that of a phosphoranyl radical. Haines and DuPreez reported that irradiation of the $Cp_2Mo_2(CO)_6$ dimer and phosphites leads to the usual ionic disproportionation products (Eq. 82) and to the phosphonite complex $CpMo(CO)_3P(O)(OR)_2$ (Eq. 83) [76].

$$Cp_2Mo_2(CO)_6 + P(OR)_3 \xrightarrow{h\nu} CpMo(CO)_3^- + CpMo(CO)_3P(OR)_3^+ \qquad (82)$$

$$Cp_2Mo_2(CO)_6 + P(OR)_3 \xrightarrow{h\nu} CpMo(CO)_3P(O)(OR)_2 + CpMo(CO)_3R \qquad (83)$$

The phosphonite product may be formed in an Arbuzov type reaction, as shown below, which implies that the 19-electron complex has a phosphoranyl radical type structure. (Such O—R cleavage reactions are common in phosphoranyl radicals [9].)

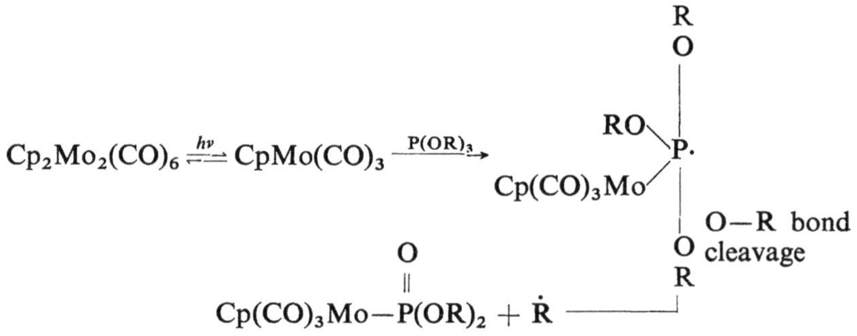

$\dot{R} + CpMo(CO)_3 \rightarrow CpMo(CO)_3R$

Scheme XV

Evidence for 19-electron complexes with bent CO structures comes from Kochi's electrochemical studies on the $Cr(CO)_6$ and $Mn(CO)_4(PPh_3)_2^+$ molecules [77–79]. Reduction of these molecules in the presence of $HSnBu_3$ gave the following formyl products:

$$Cr(CO)_6^- + HSnBu_3 \rightarrow Cr(CO)_5CHO^- + \cdot SnBu_3 \qquad (84)$$

$$Mn(CO)_4(PPh_3)_2 + HSnBu_3 \rightarrow Mn(CO)_3(PBu_3)_2CHO + \cdot SnBu_3 \qquad (85)$$

One interpretation of these reactions is that the 19-electron complexes have a bent CO ligand, making that CO a 1-electron donor, analogous to bent NO [80]:

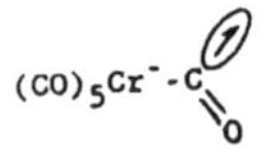

The bent CO gives the metal center an 18-electron configuration with the odd electron now situated on the carbon of the bent CO; H atom abstraction will then yield the formyl product directly.

It is important to stress that the structural conclusions above based on reactivity are only conjecture. In each of the cases discussed, other interpretations are possible. Astruc has also emphasized that the structures proposed above for the 19-electron complexes may only be transition states (or intermediates) along the reaction pathway; the location of the spin density in the ground state structures may be considerably different than that in the transition states. He cites as evidence that fact that it is not necessary to have a large spin density in the ground state in order to have a radical reaction at a given site. He proposes [5] that the

$CpFe(CO)_3$ species is a metal centered radical because it is isolobal [81] to the CpFe(arene) complexes. The benzyl radical in Eq. 68 or Scheme XII therefore combines preferably with the metal center. However, coupling at the metal center is reversible whereas coupling at the Cp ring is not reversible. Hence, the only product obtained in the net coupling reaction is the Cp-benzyl coupled species shown in Eq. 68. Astruc has clearly raised an interesting point regarding the appropriateness of using product studies to determine the structures of reactive 19-electron complexes. For the time being, however, the reactivity studies remain the only method for learning about the structures of these complexes. More definitive conclusions must wait until these 19-electron complexes can be directly observed by spectroscopy, either at low temperature or by using fast kinetics techniques.

7. Acknowledgments

The research from my laboratory described in this chapter was supported by the National Science Foundation, the Air Force Office of Scientific Research, and the Donors of the Petroleum Research Fund, administered by the American Chemical Society. The Sloan Foundation is also acknowledged for a Fellowship. Prof. Mark Wrighton is kindly thanked for sending me a preprint of reference 60. Marcie Wood is acknowledged for her assistance.

8. References

1. A. E. Stiegman and D. R. Tyler, Comments Inorg. Chem. **5** (1986) 215–245.
2. T. L. Brown, Ann. N. Y. Acad. Sci. **333** (1980) 80–89.
3. a. J. K. Kochi, Acc. Chem. Res. **7** (1974) 351–360.
   b. J. Halpern, Pure Appl. Chem. **58** (1986) 575–584.
   c. J. Halpern, Pure Appl. Chem. **51** (1979) 2171–2182.
   d. J. K. Kochi, Organometallic Mechanisms and Catalysis, Academic Press, New York, 1978.
4. J. P. Collman, L. S. Hegedus, J. R. Norton and R. G. Finke, Principles and Applications of Organotransition Metal Chemistry, University Science Books, Mill Valley, CA, 1987.
5. a. D. Astruc, Comments Inorg. Chem. **6** (1987) 61–84.
   b. D. Astruc, Acc. Chem. Res. **19** (1986) 377–383.
   c. D. Astruc, Chem. Rev. **88** (1988) 1189–1217.
6. All of the organometallic textbooks describe schemes for electron counting. See for example Ref. 4, Ch. 2. Recently a new scheme for oxidation states has been proposed. See E. A. Seddon and K. R. Seddon, The Chemistry of Ruthenium, Elsevier, Amsterdam, 1984, Ch. 2.
7. a. P. R. Schonberg, R. T. Paine, C. P. Campana and E. N. Duesler, Organometallics **1** (1982) 799–806.
   b. C. P. Casey and W. D. Jones, J. Am. Chem. Soc. **102** (1980) 6156–6157.
   c. M. E. Rerek and F. Basolo, J. Am. Chem. Soc. **106** (1984) 5908–5912.
   d. M. E. Rerek, L.-N. Ji and F. Basolo, J. Chem. Soc., Chem. Commun. (1983) 1208–1209.
   e. R. M. Kowaleski, A. L. Rheingold, W. C. Trogler and F. Basolo, J. Am. Chem. Soc. **108** (1986) 2460–2461.
8. Ligand rearrangements such as bent CO ligands may be more common than once thought in organometallic transformations. See Ref. 1. Also, see:
   a. F. Basolo, Coord. Chem. Rev. **43** (1982) 7–15.
   b. F. Basolo, Inorg. Chim. Acta **100** (1985) 33–39.
9. a. B. P. Roberts, Adv. Free Radical Chem. **6** (1980) 225–285.
   b. W. G. Bentrude, Phosphorus Sulfur **3** (1977) 109–130.
10. D. Fenske, Chem. Ber. **112** (1979) 363–375.

11. a. W. Kaim, Inorg. Chem. **23** (1984) 504–505.
b. W. Kaim, Z. Naturforsch. **396** (1984) 801–807.
12. a. K. A. M. Kreber and J. K. S. Wan, Trans. Met. Chem. **8** (1983) 253–254.
b. K. A. M. Kreber and J. K. S. Wan, J. Am. Chem. Soc. **103** (1981) 2101–2102.
c. K. A. M. Creber, K. S. Chen and J. K. S. Wan, Rev. Chem. Interm. **5** (1984) 37.
13. a. A. Alberti and A. Hudson, J. Organomet. Chem. **241** (1983) 313–319.
b. A. Alberti and A. Hudson, J. Organomet. Chem. **248** (1983) 197–204.
14. a. J. A. Conner, C. Overton and N. E. Murr, J. Organomet. Chem. **277** (1984) 227–284.
b. A. E. Alegria, O. Lozada and J. Sanchez, J. Organomet. Chem. **281** (1985) 229–236.
c. D. Miholova and A. A. Vlcek, J. Organomet. Chem. **279** (1985) 317–326.
15. a. W. Hieber and W. Schropp, Z. Naturforsch. Teil **B**, **15** (1960) 271–277.
b. W. Hieber, W. Beck and G. Zeitler, Angew. Chem. **73** (1961) 364–368.
c. H. Behrens, E. Ruyter and H. Wakamatsi, Z. Anorg. Allgem. Chem. **349** (1967) 241–250.
d. M. Herberbold, F. Wehrmann, D. Neugebauer and G. Huttner, J. Organomet. Chem. **152** (1978) 329–336.
e. D. M. Allen, A. Cox, T. J. Kemp, Q. Sultana and R. B. Pitts, J. Chem. Soc., Dalton Trans. (1976) 1189–1193.
16. A. E. Stiegman, M. Stieglitz and D. R. Tyler, J. Am. Chem. Soc. **105** (1983) 6032–6037.
17. A. E. Stiegman and D. R. Tyler, Coord. Chem. Rev. **63** (1985) 217–240.
18. G. L. Geoffroy and M. S. Wrighton, Organometallic Photochemistry, Academic Press, New York, 1979.
19. T. J. Meyer and J. V. Caspar, Chem. Rev. **85** (1985) 187–218.
20. A. S. Goldman and D. R. Tyler, J. Am. Chem. Soc. **108** (1986) 89–94.
21. B. H. Byers and T. L. Brown, J. Am. Chem. Soc. **97** (1975) 947–948.
22. B. H. Byers and T. L. Brown, J. Am. Chem. Soc. **99** (1977) 2527–2532.
23. N. W. Hoffman and T. L. Brown, Inorg. Chem. **17** (1978) 613–617.
24. M. S. Wrighton and D. S. Ginley, J. Am. Chem. Soc. **97** (1975) 2065–2072.
25. T. R. Herrinton and T. L. Brown, J. Am. Chem. Soc. **107** (1985) 5700–5703.
26. A. E. Stiegman and D. R. Tyler, J. Am. Chem. Soc. **104** (1982) 2944–2945.
27. W. L. Waltz, O. Hackelberg, L. M. Dorfman and A. Wojcicki, J. Am. Chem. Soc. **100** (1978) 7259–7264.
28. R. E. Dessy, P. M. Weissman and R. L. Pohl, J. Am. Chem. Soc. **88** (1966) 5117–5121. Note that these are irreversible potentials, and hence are only a rough guide to the thermodynamic potentials.
29. a. A. Fox, J. Malito and A. Poë, J. Chem. Soc., Chem. Commun. (1981) 1052–1053.
b. J. P. Fawcett, R. A. Jackson and A. Poë, J. Chem. Soc., Chem. Commun. (1975) 733–734.
30. a. Q.-Z. Shi, T. G. Richmond, W. C. Trogler and F. Basolo, J. Am. Chem. Soc. **104** (1982) 4032–4034.
b. Q.-Z. Shi, T. G. Richmond, W. C. Trogler and F. Basolo, J. Am. Chem. Soc. **106** (1984) 71–76.
31. W. D. Jones, J. M. Huggins and R. G. Bergman, J. Am. Chem. Soc. **103** (1981) 4415–4423.
32. A. S. Goldman and D. R. Tyler, J. Am. Chem. Soc. **106** (1984) 4066–4067.
33. C. E. Philbin, A. S. Goldman and D. R. Tyler, Inorg. Chem. **25** (1986) 4434–4436.
34. F. Mao, D. R. Tyler and D. Keszler, J. Am. Chem. Soc., **111** (1989) 130–134.
35. A. E. Stiegman, A. S. Goldman, D. B. Leslie and D. R. Tyler, J. Chem. Soc., Chem. Commun. (1984) 632–633.
36. D. R. Tyler, V. MacKenzie and A. S. Goldman in H. Yersin and A. Vogler (Eds.), Proc. of the Seventh Int. Symp. Photochemistry and Photophysics of Coordination Compounds, Schloss Elman, March 29–April 2, 1987, Springer-Verlag, Heidelberg, 1987, pp. 263–266.
37. A. S. Goldman and D. R. Tyler, Inorg. Chem. **26** (1987) 253–258.
38. A. F. Hepp and M. S. Wrighton, J. Am. Chem. Soc. **103** (1981) 1258–1261.
39. See Ref. 19, pp. 213–214.
40. a. J. Kwiatek, Catal. Rev. **1** (1967) 37–72.
b. J. Halpern, Chem. Eng. News **44** (1966) 68–75.
41. There is an enormous literature on ETC chemistry. For recent reviews see:

a. M. Julliard and M. Chanon, Chem. Rev. **83** (1983) 425–506.
b. M. Chanon, Bull. Soc. Chim. Fr. (1982) 197–240.
c. J. M. Saveant, Acc. Chem. Res. **13** (1980) 323–329.
d. M. Chanon and M. L. Tobe, Angew. Chem., Int. Ed. Engl. **21** (1982) 1–86.
e. M. Julliard and M. Chanon, Chem. Scr. **24** (1984) 11–21.
42. M. I. Bruce, J. G. Matisons and B. K. Nicholson, J. Organomet. Chem. **247** (1983) 321–343.
43. A. S. Goldman and D. R. Tyler, Inorg. Chim. Acta **98** (1985) L47–L48.
44. a. N. J. Coville, J. Organomet. Chem. **190** (1980) C84–C86.
b. N. J. Coville, J. Organomet. Chem. **218** (1981) 337–349.
c. N. J. Coville, M. O. Albers and E. Singleton, J. Chem. Soc., Dalton Trans. (1983) 947–953.
d. N. J. Coville, O. M. Albers, T. V. Ashworth and E. Singleton, J. Chem. Soc., Chem. Commun. (1981) 408–409.
e. N. J. Coville and E. A. Darling, J. Organomet. Chem. **277** (1984) 105–111.
f. M. O. Albers, N. J. Coville and E. Singleton, J. Organomet. Chem. **232** (1982) 261–266.
45. N. S. Silavwe and D. R. Tyler, Inorg. Chem., submitted for publication.
46. N. D. Silavwe, X. Pan and D. R. Tyler, Inorg. Chim. Acta **144** (1988) 123–128.
47. C. J. Pickett and D. Pletcher, J. Chem. Soc., Dalton Trans. (1975) 879–886.
48. V. MacKenzie and D. R. Tyler, J. Chem. Soc., Chem. Commun. (1987) 1783–1784.
49. A. Avey and D. R. Tyler, manuscript in preparation.
50. a. A. M. Bond, P. A. Dawson, B. M. Peake, B. H. Robinson and J. Simpson, Inorg. Chem. **16** (1977) 2199–2206.
b. P. J. Krusic, J. San Filippo, Jr., B. Hutchinson, R. L. Hance and L. M. Daniels, J. Am. Chem. Soc. **103** (1981) 2129–2131.
51. C. J. Pickett and D. Pletcher, J. Chem. Soc., Dalton Trans. (1976) 749–752.
52. W. C. Trogler, Int. J. Chem. Kinet. **19** (1987) 1025–1047.
53. A. J. Dixon, S. J. Gravelle, V. de Burgt, J. Lambertus, M. Poliakoff, J. J. Turner and E. Weitz, J. Chem. Soc., Chem. Commun. (1987) 1023–1025.
54. Note that an extensive chemistry of $18 + \delta$ DAB complexes also exists. See R. R. Andrea, W. G. J. deLange, T. vander Graaf, M. Rijkhoff, D. J. Stufkins and A. Oskam, Organometallics **7** (1988) 1100–1106 and references therein.
55. B. A. Narayanan, C. Amatore and J. K. Kochi, J. Chem. Soc., Chem. Commun. (1983) 397–399.
56. A. E. Bruce and D. R. Tyler, unpublished work.
57. a. A. Alberti and C. M. Camaggi, J. Organomet. Chem. **181** (1979) 355–363.
b. L. Posimeni, P. L. Zanonato and C. Corvaja, Inorg. Chim. Acta **37** (1979) 241–244.
58. M. Hanaya, S. Tero-Kubota and M. Iwaizumi, Organometallics **7** (1988) 1500–1504.
59. J. P. Blaha and M. S. Wrighton, J. Am. Chem. Soc. **107** (1985) 2694–2702.
60. C. Zou, K. J. Ahmed and M. S. Wrighton, J. Am. Chem. Soc., in press.
61. W. E. Geiger, T. Gennett, G. A. Lane, A. Salzer and A. L. Rheingold, Organometallics **5** (1986) 1352–1359.
62. J. A. Armstead, D. J. Cox and R. Davis, J. Organomet. Chem. **236** (1982) 213–219.
63. N. G. Connelly and W. E. Geiger, Adv. Organomet. Chem. **23** (1984) 1–93.
64. C. E. Philbin, C. A. Granatir and D. R. Tyler, Inorg. Chem. **25** (1986) 4806.
65. A. E. Stiegman and D. R. Tyler, J. Am. Chem. Soc. **107** (1985) 967–971.
66. W. C. Trogler and M. J. Therien, J. Am. Chem. Soc. **109** (1987) 5127–5133.
67. F. Ramairez and N. McKelvie, J. Am. Chem. Soc. **79** (1957) 5829–5830.
68. a. R. F. Heck, J. Am. Chem. Soc. **85** (1963) 657–661.
b. M. Absi-Halabi and T. L. Brown, J. Am. Chem. Soc. **99** (1977) 2982–2988.
69. J. Brunet, C. Sidot and P. Caubere, J. Organomet. Chem. **204** (1980) 229–241.
70. See Ref. 5 in Ref. 37.
71. O. P. Anderson and M. C. R. Symons, J. Chem. Soc., Chem. Commun. (1972) 1020–1021.
72. M. C. R. Symons, S. W. Bratt and J. L. Wyatt, J. Chem. Soc., Dalton Trans. (1983) 1377–1383.
73. For a discussion of other 19-electron complexes generated and stabilized at low temperature, see D. R. Tyler, Prog. Inorg. Chem., **36** (1988) 125–194.

74. For other 18 + δ complexes, see:
   a. M. J. Maroney and W. C. Trogler, J. Am. Chem. Soc. **106** (1984) 4144–4151.
   b. W. Kaim and S. Kohlmann, Inorg. Chem. **25** (1986) 3442–3448.
   c. W. Kaim, J. Organomet. Chem. **262** (1984) 171–178.
   d. W. Kaim, Inorg. Chim. Acta **53** (1981) L151–L153.
   e. W. Kaim, Inorg. Chem. **23** (1984) 3365–3368.
   f. A. E. Alegria, O. Lozada, H. Rivera and J. Sanchez, J. Organomet. Chem. **281** (1985) 229–236.
75. Y.-M. Wuu, C. Zou, and M. S. Wrighton, J. Am. Chem. Soc. **109** (1987) 5861–5862.
76. a. R. J. Haines and C. Nolte, J. Organomet. Chem. **24** (1970) 725–736.
   b. R. J. Haines, A. L. Du Preez and I. L. Marais, J. Organomet. Chem. **28** (1971) 97–104.
77. B. A. Narayanan and J. K. Kochi, J. Organomet. Chem. **272** (1984) C49–C53.
78. B. A. Narayanan, C. Amatore and J. K. Kochi, Organometallics **5** (1986) 926–935.
79. D. J. Kuchyuka, C. Amatore and J. K. Kochi, Inorg. Chem. **25** (1986) 4087–4097.
80. C. M. Lukehart, Fundamental Transition Metal Organometallic Chemistry, Brooks/Cole, Monterey, CA, 1985, p. 179.
81. R. Hoffmann, Angew. Chem. Int. Ed. Engl. **212** (1982) 711–724.

Abbreviations for Indicies

acac = acetylacetonate
AIBN = azoisobutyronitrile
Ar = aryl
bpy = 2,2′-bipyridine
Bu = n-C_4H_9
CIP = contact ion pair
COD = cyclooctadiene
Cp = cyclopentadienyl
Cp* = pentamethylcyclopentadienyl
DAB = diazabutadiene
dcype = 1,2-bis(dicyclohexylphosphino)-ethane
dppb = 1,4-bis(diphenylphosphino)butane
dppe = 1,2-bis(diphenylphosphino)ethane
dppm = 1,1-bis(diphenylphosphino)methane
dtc = dithiocarbamate
ESR = electron spin resonance
Et = C_2H_5
ETC = electron transfer catalysis
L = 2-electron donor ligand
Me = CH_3
nbd = norbornadiene
Ph = C_6H_5
phen = 1,10-phenanthroline
py = pyridine
R = alkyl
TCNE = tetracyanoethane
TCNQ = tetracyanoquinodimethane
tpb = tris(pyrazolyl)borate

Subject Index for Organometallic Radical Processes, W. C. Trogler, Ed.

A

B

C

D

E

J

K

L

M

N

O

S

Formula Index of Organometallic Radicals for Organometallic Radical Processes, W. C. Trogler, Ed.

Bei Fragen zur Produktsicherheit wenden Sie sich bitte an:
If you have any questions regarding product safety,
please contact:

Walter de Gruyter GmbH
Genthiner Straße 13
10785 Berlin
productsafety@degruytergbrill.com